Roselle

Roselle

Production, Processing, Products and Biocomposites

Edited by

S.M. SAPUAN
Professor of Composite Materials
Universiti Putra Malaysia, Serdang, Selangor, Malaysia

R. NADLENE
Fakulti Kejuruteraan Mekanikal
Universiti Teknikal Malaysia Melaka, Melaka, Malaysia

A.M. RADZI
Malaysia-Japan International Institute of Technology
Universiti Teknologi Malaysia, Kuala Lumpur, Malaysia

R.A. ILYAS
School of Chemical and Energy Engineering, Faculty of Engineering
Universiti Teknologi Malaysia, Johor Bahru, Johor, Malaysia

Centre for Advanced Composite Materials (CACM)
Universiti Teknologi Malaysia, Johor Bahru, Johor, Malaysia

Academic Press is an imprint of Elsevier
125 London Wall, London EC2Y 5AS, United Kingdom
525 B Street, Suite 1650, San Diego, CA 92101, United States
50 Hampshire Street, 5th Floor, Cambridge, MA 02139, United States
The Boulevard, Langford Lane, Kidlington, Oxford OX5 1GB, United Kingdom

Library of Congress Cataloging-in-Publication Data
A catalog record for this book is available from the Library of Congress

British Library Cataloguing-in-Publication Data
A catalogue record for this book is available from the British Library

ISBN: 978-0-323-85213-5

For information on all Academic Press publications visit our website at https://www.elsevier.com/books-and-journals

Publisher: Nikki Levy
Acquisitions Editor: Nancy J. Maragioglio
Editorial Project Manager: Lena Sparks
Production Project Manager: Niranjan Bhaskaran
Cover designer: Alan Studholme

Typeset by TNQ Technologies

List of Contributors

T.M.N. Afiq
Advanced Engineering Materials and Biocomposites Research Centre (AEMC)
Department of Mechanical and Manufacturing Engineering
Faculty of Engineering
Universiti Putra Malaysia
Serdang, Selangor, Malaysia

M. Ahiduzzaman
Department of Agro-Processing
Bangabandhu Sheikh Mujibur Rahman Agricultural University
Gazipur, Bangladesh

S.N. Ain
Fakulti Kejuruteraan Mekanikal
Universiti Teknikal Malaysia Melaka
Melaka, Malaysia

R.N.D. Aqilah
Fakulti Kejuruteraan Mekanikal
Universiti Teknikal Malaysia Melaka
Melaka, Malaysia

Mochamad Asrofi
Department of Mechanical Engineering
University of Jember
Jember, East Java, Indonesia

M.R.M. Asyraf
Department of Aerospace Engineering
Faculty of Engineering
Universiti Putra Malaysia
Serdang, Selangor, Malaysia

M.S.N. Atikah
Department of Chemical and Environmental Engineering
Faculty of Engineering
Universiti Putra Malaysia
Serdang, Selangor, Malaysia

A.M. Noor Azammi
Automotive Engineering Technology Section
UniKL-MFI
Bandar Baru Bangi, Selangor, Malaysia

Taofik Oladimeji Azeez
Department of Biomedical Technology
School of Health Technology
Federal University of Technology
Owerri, Nigeria

Innocent Ochiagha Eze
Department of Polymer and Textile Engineering
School of Engineering and Engineering Technology
Federal University of Technology
Owerri, Nigeria

Rabboni Mike Government
Department of Chemical Engineering
Faculty of Engineering
Nnamdi Azikiwe University
Awka, Anambra State, Nigeria

G. Guclu
Department of Food Engineering
Faculty of Agriculture
Cukurova University Adana
Adana, Turkey

M.M. Harussani
Advanced Engineering Materials and Composites (AEMC)
Department of Mechanical and Manufacturing Engineering
Faculty of Engineering
Universiti Putra Malaysia
Serdang, Selangor, Malaysia

Sulafa Hassan
Agricultural Product Processing and Storage Lab
School of Food and Biological Engineering
Jiangsu University
Zhenjiang, Jiangsu, China

M.D. Hazrol
Advanced Engineering Materials and Composites (AEMC)
Department of Mechanical and Manufacturing Engineering
Faculty of Engineering
Universiti Putra Malaysia
Serdang, Selangor, Malaysia

Xiaowei Huang
Agricultural Product Processing and Storage Lab
School of Food and Biological Engineering
School of Agricultural Equipment Engineering
Jiangsu University
Zhenjiang, Jiangsu, China

M.R.M. Huzaifah
Department of Crop Science
Faculty of Agricultural Science and Forestry
Universiti Putra Malaysia Bintulu Campus
Bintulu, Sarawak, Malaysia

R. Ibrahim
Innovation & Commercialization Division
Forest Research Institute Malaysia (FRIM)
Kepong, Selangor, Malaysia

Pulp and Paper Laboratory
Biomass Technology Programme
Forest Products Division
Forest Research Institute Malaysia
Kepong, Selangor, Malaysia

R.A. Ilyas
School of Chemical and Energy Engineering
Faculty of Engineering
Universiti Teknologi Malaysia
Johor Bahru, Johor, Malaysia

Centre for Advanced Composite Materials (CACM)
Universiti Teknologi Malaysia
Johor Bahru, Johor, Malaysia

A.K.M. Aminul Islam
Department of Genetics and Plant Breeding
Faculty of Agriculture
Bangabandhu Sheikh Mujibur Rahman Agricultural University
Gazipur, Bangladesh

Department of Agronomy
Faculty of Agriculture
Bangladesh Agricultural University
Mymensingh, Bangladesh

A.K.M. Mominul Islam
Department of Agronomy
Faculty of Agriculture
Bangladesh Agricultural University
Mymensingh, Bangladesh

Samuel Chidi Iwuji
Department of Biomedical Technology
School of Health Technology
Federal University of Technology
Owerri, Nigeria

Tahmina Sadia Jamini
Department of Genetics and Plant Breeding
Bangabandhu Sheikh Mujibur Rahman Agricultural University
Gazipur, Bangladesh

H. Kelebek
Department of Food Engineering
Faculty of Engineering
Adana Alparslan Turkes Science and Technology University Adana
Adana, Turkey

J.M. Khir
Department of Manufacturing Technology
Kolej Kumuniti Kuantan
Kuantan, Pahang, Malaysia

Lau Kia Kian
Laboratory of Biocomposite Technology
Institute of Tropical Forestry and Forest Products (INTROP)
Universiti Putra Malaysia
Serdang, Selangor, Malaysia

W. Kirubaanand
Advanced Engineering Materials and Composites (AEMC)
Department of Mechanical and Manufacturing Engineering
Faculty of Engineering
Universiti Putra Malaysia
Serdang, Selangor, Malaysia

Zhihua Li
Agricultural Product Processing and Storage Lab
School of Food and Biological Engineering
School of Agricultural Equipment Engineering
Jiangsu University
Zhenjiang, Jiangsu, China

Suzana Mali
Department of Biochemistry and Biotechnology
State University of Londrina
Londrina, Paraná, Brazil

Y. Martínez-Meza
Research and Graduate Studies in Food Science
Facultad de Química
Universidad Autónoma de Querétaro
Querétaro, Mexico

Mohsin Bin Mohamad
UKM-MTDC Symbiosis Programme
Universiti Kebangsaan Malaysia (UKM)
Bangi, Selangor, Malaysia

Bahaeldeen Babiker Mohamed
National Centre for Research (NCR)
Environment & Natural Resources Research Institute (ENRRI)
Khartoum, Sudan

Josphat Igadwa Mwasiagi
Department of Manufacturing
Industrial and Textile Engineering
School of Engineering
Moi University
Eldoret, Kenya

R. Nadlene
Fakulti Kejuruteraan Mekanikal
Universiti Teknikal Malaysia Melaka
Melaka, Malaysia

Centre for Advanced Research on Energy (CARe)
Universiti Teknikal Malaysia Melaka
Melaka, Malaysia

Prerana Nashine
Department of Mechanical Engineering
National Institute of Technology
Rourkela, Odisha, India

A. Nazrin
Advanced Engineering Materials and Composites (AEMC)
Department of Mechanical and Manufacturing Engineering
Faculty of Engineering
Universiti Putra Malaysia
Serdang, Selangor, Malaysia

Mohamad Bin Osman
Faculty of Plantation and Agrotechnology
Universiti Technology Mara (UiTM)
Shah Alam, Selangor, Malaysia

J. Pérez-Jiménez
Department of Metabolism and Nutrition
Institute of Food Science
Technology and Nutrition (ICTAN-CSIC)
Madrid, Spain

Thatayaone Phologolo
Department of Family and Consumer Science
University of Botswana
Gaborone, Botswana

A.M. Radzi
Malaysia-Japan International Institute of Technology
Universiti Teknologi Malaysia
Kuala Lumpur, Malaysia

Upendra Rajak
Department of Mechanical Engineering
Rajeev Gandhi Memorial College of Engineering and Technology
Nandyal, Andhra Pradesh, India

N.B. Razali
Section of Environmental Engineering Technology
Universiti Kuala Lumpur - Malaysian Institute of Chemical and Bioengineering Technology
78000 Alor Gajah, , Melaka,
Malaysia

Sanjay Mavikere Rangappa
Department of Materials and Production Engineering
The Sirindhorn International Thai-German Graduate School of Engineering
King Mongkut's University of Technology North Bangkok
Bangkok, Thailand

R. Reynoso-Camacho
Research and Graduate Studies in Food Science
Facultad de Química
Universidad Autónoma de Querétaro
Querétaro, Mexico

S.M. Sapuan
Advanced Engineering Materials and Biocomposites Research Centre (AEMC)
Department of Mechanical and Manufacturing Engineering
Faculty of Engineering
Universiti Putra Malaysia
Serdang, Selangor, Malaysia

Laboratory of Biocomposite Technology
Institute of Tropical Forestry and Forest Products (INTROP)
Universiti Putra Malaysia
Serdang, Selangor, Malaysia

S. Selli
Department of Food Engineering
Faculty of Agriculture
Cukurova University Adana
Adana, Turkey

O. Sevindik
Department of Food Engineering
Faculty of Engineering
Adana Alparslan Turkes Science and Technology University Adana
Adana, Turkey

Cukurova University Central Research Laboratory Adana
Adana, Turkey

S.F.K. Sherwani
Advanced Engineering Materials and Composites (AEMC)
Department of Mechanical and Manufacturing Engineering
Faculty of Engineering
Universiti Putra Malaysia
Serdang, Selangor, Malaysia

Jiyong Shi
Agricultural Product Processing and Storage Lab
School of Food and Biological Engineering
School of Agricultural Equipment Engineering
Jiangsu University
Zhenjiang, Jiangsu, China

Pankaj Shrivastava
Department of Mechanical Engineering
Prestige Institute of Engineering Management and Research
Indore, Madhya Pradesh, India

Suchart Siengchin
Department of Materials and Production Engineering
The Sirindhorn International Thai-German Graduate School of Engineering
King Mongkut's University of Technology North Bangkok
Bangkok, Thailand

D. Sivakumar
Fakulti Kejuruteraan Mekanikal
Universiti Teknikal Malaysia Melaka
Hang Tuah Jaya
76100 Durian Tunggal, Melaka, Malaysia

A. Suhrisman
Advanced Engineering Materials and Biocomposites Research Centre (AEMC)
Department of Mechanical and Manufacturing Engineering
Faculty of Engineering
Universiti Putra Malaysia
Serdang, Selangor, Malaysia

R. Syafiq
Advanced Engineering Materials and Composites (AEMC)
Department of Mechanical and Manufacturing Engineering
Faculty of Engineering
Universiti Putra Malaysia
Serdang, Selangor, Malaysia

J. Tarique
Advanced Engineering Materials and Composites (AEMC)
Department of Mechanical and Manufacturing Engineering
Faculty of Engineering
Universiti Putra Malaysia
Serdang, Selangor, Malaysia

Tikendra Nath Verma
Department of Mechanical Engineering
Maulana Azad National Institute of Technology
Bhopal, Madhya Pradesh, India

Z.M. Zahfiq
Advanced Engineering Materials and Composites (AEMC)
Department of Mechanical and Manufacturing Engineering
Faculty of Engineering
Universiti Putra Malaysia
Serdang, Selangor, Malaysia

Xiaodong Zhai
Agricultural Product Processing and Storage Lab
School of Food and Biological Engineering
School of Agricultural Equipment Engineering
Jiangsu University
Zhenjiang, Jiangsu, China

Junjun Zhang
Agricultural Product Processing and Storage Lab
School of Food and Biological Engineering
School of Agricultural Equipment Engineering
Jiangsu University
Zhenjiang, Jiangsu, China

Xiaobo Zou
Agricultural Product Processing and Storage Lab
School of Food and Biological Engineering
Jiangsu University
Zhenjiang, Jiangsu, China

Contents

CHAPTER 1

Roselle: Production, Product Development, and Composites

R.A. ILYAS • S.M. SAPUAN • W. KIRUBAANAND • Z.M. ZAHFIQ • M.S.N. ATIKAH • R. IBRAHIM • A.M. RADZI • R. NADLENE • M.R.M. ASYRAF • M.D. HAZROL • S.F.K. SHERWANI • M.M. HARUSSANI • J. TARIQUE • A. NAZRIN • R. SYAFIQ

1 INTRODUCTION

The total number of *Hibiscus* species, tropical and subtropical, exceeds 300 (Anderson, 2006). Jamaica sorrel (*Hibiscus sabdariffa*), or roselle, is a rare plant bred in many temperate climates for its seeds, stems, leaves, and calyces; the dried calyces are used to make drinks, syrups, jams, and jellies (Eslaminejad & Zakaria, 2011). Roselle is an annual plant that takes approximately 6 months to grow (Ansari, 2013). The morphologic features of this plant are shown in Fig. 1.1. The leaves of roselle are separated into three to five lobes on the stem and arranged alternately (Ansari, 2013). In each calyx lobe of the roselle flower, there is a notable center and two marginal ribs (Ansari, 2013). This trait puts the plant in the Furcaria group (America et al., 1993). The color of the flower ranges from white to pale yellow, with delicate and fleshy calyces, while the petals might differ from white to pink, red, yellow, orange, or purple (Ansari, 2013). The fruit's bright red color indicates it is a ripe fruit (Chin et al., 2016; Halimatul et al., 2007; Morton, 1987).

Roselle is recognized throughout the Indian subcontinent traditionally as "Mesta" and "Meshta" (Grubben & Denton, 2004; Halimatul et al., 2007; Udayasekhara Rao, 1996). In different nations, roselle is generally called by many names, as shown in Table 1.1 (Ansari, 2013). Owing to its market value as a natural food and staining component that could replace a variety of synthetic products, this plant has gained the attention of food, beverage, and pharmaceutical companies (Eslaminejad & Zakaria, 2011).

This chapter is a review of the production and applications of roselle plants and points out roselle as a promising crop for medicinal uses and polymer composites, which is an aspect that has not been widely studied to date.

2 ORIGIN AND DISTRIBUTION

Roselle was formerly grown in Sudan 6000 years ago, which emerged from Africa (Grubben & Denton, 2004). Today, the leading country in roselle production is Sudan, where it became a source for Sudanese tea (Mohammad et al., 2002). In the 17th century, it was brought as a vegetable to India and South America, and in Asia, it was bred for fiber processing use (Grubben & Denton, 2004). The evidence of its planting in India, Sri Lanka, Indonesia, Malaysia, and Thailand has started to emerge in the early 20th century (Ansari, 2013). A massive roselle farming was conducted in the Dutch East Indies (Indonesia today) in the 1920s, within a government-subsidized project to produce sugar sac fiber (Appell & Red, 2003). The plant is now common in all tropics, especially in the Western and Central African savanna (Grubben & Denton, 2004). Currently, the roselle plant can be found in Malaysia, India, Panama, Indonesia, Jamaica, Mexico, Guatemala, Australia, Philippines, Kenya, Madagascar, Mozambique, Malawi, Uganda, Somalia, Tanzania, Djibouti, Cambodia, Vietnam, Namibia, Gabon, Congo, Burundi, Rwanda, DR Congo, Myanmar, Thailand, Belize, China, Sudan, South Sudan, Egypt, Gambia, Senegal, Saudi Arabia, Bangladesh, Laos, Sri Lanka, Ghana, Nigeria, Brazil, and Cuba, as illustrated in Fig. 1.2.

3 ECOLOGY

The ideal region for roselle plants is a hot and humid tropical climate, as roselle is exceptionally vulnerable to frost and fog (Morton, 1987; Chin et al., 2016). This plant thrives in temperatures between 18°C and 35°C, where 25°C is the ideal temperature (Ansari,

[illegible]. https://doi.org/10.1016/B978-0-323-85213-5.00009-3

FIG. 1.1 Morphology of the roselle plant. **(A)** Stem, **(B)** leaf, **(C)** pink or yellow flower, **(D)** red fresh calyces, **(E)** fruit, and **(F)** dark brown seeds.

2013). Roselle development stops at 14°C (Grubben & Denton, 2004). Tropical and subtropical regions with a height of 3000 ft (900 m) above the sea level are the ideal location for roselle planting (Grubben & Denton, 2004). During the roselle growing phase, annual rainfall between 400 and 500 mm is essential (Islam, 2019). Owing to its sensitivity toward light, roselle, known for being a short-day plant, needs light phase regularly for 13 h a day in the first 4–5 months of its development (Grubben & Denton, 2004). Duke (1983) stated that when the sunlight exposure exceeds 13 h a day, flowers will cease to appear, while Grubben and Denton (2004) asserted that roselle plants flowered exceptionally with daylight that lasts less than 12 h. As claimed by Duke (1983), besides well-drained humus, this plant favors rich, fertile soils with a pH from 4.5 to 8.0. It can survive in a strong wind condition and floods (Ansari, 2013).

4 HARVESTING AND POSTHARVEST HANDLING

The roselle fruit must be harvested early enough before any woody matter appears in the pod or the calyx (Ansari, 2013; Rolfs, 1929). The harvested stems are soaked in water for 14 days, followed by a bark removal process (Ansari, 2013). The stems are, therefore, pounded to detach their fibers (Ansari, 2013). The pounded stems are washed, dried, and arranged to get the fiber, depending on size, color, and rigidity (Islam, 2019). Then sharp and round metal tubes are used to grab the seed capsules from the calyces (Ansari, 2013).

5 MORPHOLOGY AND TAXONOMY

Roselle (*H. sabdariffa*) is taxonomically classified in the Malvales order. It is also a member of Malvaceae family (Fig. 1.3). It is a known medicinal plant with a

TABLE 1.1
Names of Roselle in Different Regions (Ansari, 2013).

Regions	Vernacular Names
Thailand	Krachiap
Senegal	Bissap
Congo	Bissap
France	Bissap
Panama	Saril
Nigeria	Isapa
Namibia	Omutet
Myanmar	Chin baung
Mali	Dah, Dah bleni
Malaysia	Assam paya, Asam susur
Central America	Saríl or flor de Jamaica
Indonesia	Rosela
India	Mesta, Meshta
Bangladesh	Mesta, Meshta
The Gambia	Wonjo
Florida	Cranberry
Sudan	Karkade
Egypt	Karkade
Saudi Arabia	Karkade
Caribbean	Sorrel
Australia	Rosella or rosella fruit
Vietnam	cây quế mầu, cây bụp giấm, or cây bụt giấm
China	Luoshen hua

worldwide fame, and the plant can be found in almost all tropical countries such as Mexico, Egypt, Sudan, Vietnam, Philippines, Thailand, Indonesia, Malaysia, Saudi Arabia, and India.

The following is the morphologic description of a Roselle/Mesta: a hard and straight stem, unbranched, cylindric, often bristled, hardly smooth, and green, red, or regimented in different colors. It can grow up to 1 −5 m in height (Islam, 2019).

On new growth, the leaves are plain, becoming lobed alternate afterward, and they are also stipulated where the stipules are free lateral, having 0.5 −1.0 cm length and colored green or red (Islam, 2019). The stalk, or petiole, ranging from 4 to 14 cm in length with the color of green to red is pubescent on the abaxial surface where it bristles heavily or hairy slightly, colored green to deep red, and is either smooth or scabrous (Islam, 2019). Lamina are mostly 3 to 5, deeply palmately lobed, each lose ovate to oblong lanceolate, margin serrulose, apex-acute, pubescent and bristled along the veins on both the surface, scabrous or scaberulous, green to red, one green gland present in the mid vein on the undersurface (Islam, 2019). The roselle flowers are 8–10 cm in diameter, white to pale yellow with a dark red spot at the base of each petal, and have a stout fleshy calyx at the base, 1–2 cm wide, enlarging to 3–3.5 cm, fleshy and bright red as the fruit matures (Udayasekhara Rao, 1996).

The calyx has five sepals with lanceolate shaped. It connate below the middle in to a cup, with a lobe size of 1.5 to 2.0 cm (Islam, 2019).

Corolla (petal) is generally bell shaped and giant, spreading, and yellow as a whole or with a deep red center (Islam, 2019). Roselle has five petals, twisted, unrestrained, pubescent on the outside, and with glandular hairs that are 3–5 cm in length and located in the internal section (Islam, 2019).

Stamens are numerous, monadelphous, staminal column epipetros, truncate, 1.0 cm long, with glandular hairs, and yellow or red (Islam, 2019). The filaments extend from 0.1 to 0. 2 cm and are yellow to red, while the anthers are reinformed and the pollen is spiny (Islam, 2019). They are also characterized with having five carpels, oval ovary having a length of 0.3 −0.4 cm, commonly globular ovoid, with dense silky hairs, five chambered with many ovules in each chamber that are covered with hair and arranged in two to three rows, with five stigmas, capitate, and red or yellow exerted (Islam, 2019).

6 PROPAGATION OF ROSELLE

Roselle commonly reproduces by seeds, but it is also effectively developed by cutting (Ansari, 2013; Rolfs, 1929). Usually, seeding occurs at the beginning of the rainy season Ansari (2013) and can be done in two ways: direct seeding in the field and the seedbeds (Islam, 2019). Seeds are typically planted inside a warm greenhouse in early spring, and the germination takes a short time (Islam, 2019). When the seedlings have grown mature enough to manage, they are placed into independent containers (Islam, 2019). If they are set out to be annual crops, the seeds should be planted in fixed locations in early summer and covered with a frame or cloche until they grow further away (Islam, 2019). On the contrary, for perennials, it would be

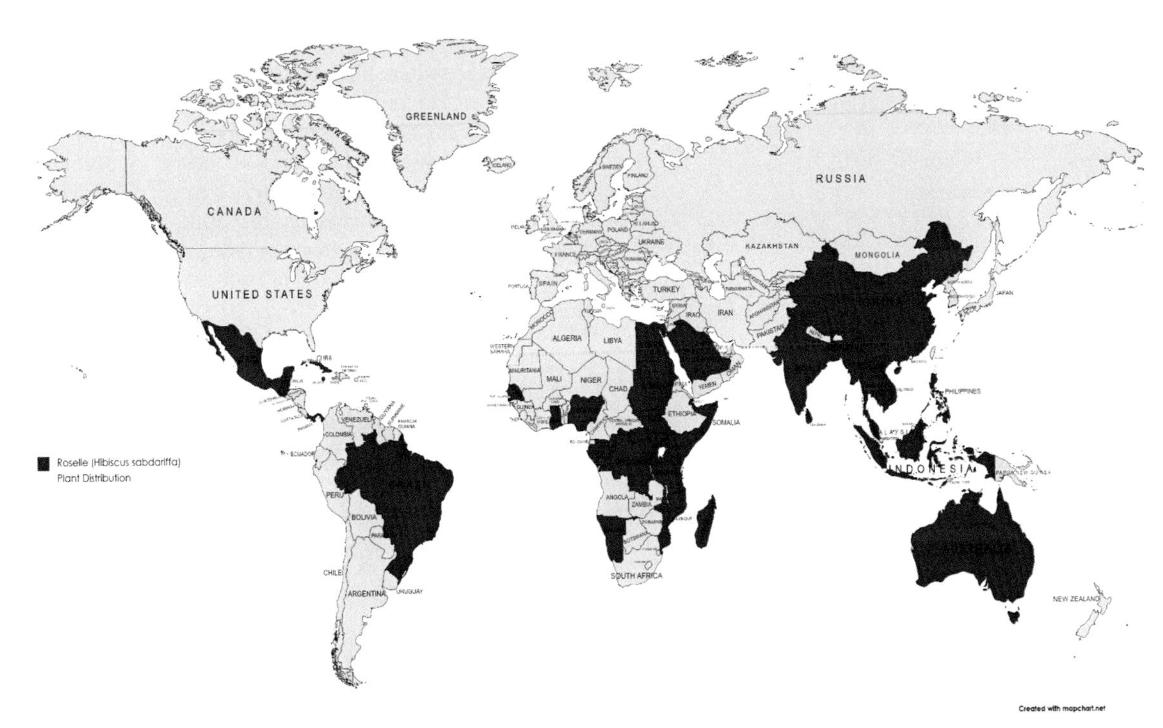

FIG. 1.2 Distribution of the roselle (*Hibiscus sabdariffa*) plant.

Scientific classification	
Kingdom	Plantae
Clade	Angiosperms
Clade	Eudicots
Clade	Rosids
Order	Malvales
Family	Malvaceae
Genus	*Hibiscus*
Species	*H. Sabdariffa*
Binomial name	*Hibiscus H. Sabdariffa*

FIG. 1.3 Taxonomy of the roselle plant.

better for them to be cultivated in the greenhouse during the first year and planted in early summer of the succeeding year (Islam, 2019). Fig. 1.4 shows the Mesta plant with leaves, flower, and fruits. The half-ripened woods are cut in July or August (Islam, 2019). Then they are overwintered in a warm greenhouse until the last anticipated frosts and are planted after that (Islam, 2019).

FIG. 1.4 Mesta plant with leaves, flower, and fruits (Islam, 2019).

7 BENEFITS OF ROSELLE IN THE MEDICAL FIELD

Since the beginning of humankind on Earth, the struggle with sickness begun and has continued into an eternal war. The discovery of antibiotics in the early 1960s turned this battle in favor of the humans; however, microbes returned a few years later with mutating strains that were resistant to nearly all innovative antibiotics. This obliged scientists to look for new alternatives to these adaptable microorganisms. The drastic increase in pathogen resistance to antibiotics currently contributes to the need for new antimicrobials (Abdallah, 2016; Falagas & Bliziotis, 2007; Viens & Littmann, 2015).

Plant species, particularly those prescribed for treating microbial infections, have been promising sources of new antimicrobials for a long time in traditional and common medicine in various societies (Abdallah, 2011). In Sudan, most people, like many African countries, still depend on traditional medications and medicinal herbs to cure diseases that are part of an informal health system, even though these common folk medicines are based on Islamic and West African medicines (WHO, 2001). Roselle is an annual African-born ubiquity that is known locally as Karkadeh in Southeast Asia and Central America (Mercedes et al., 2013; Voon et al., 2012).

H. sabdariffa is well known worldwide, and this plant's parts are widely used and recommended for use in traditional medicine in many countries such as the African countries, India, Mexico, Brazil, China, and Iran (Da-Costa-Rocha et al., 2014). The leaves are also consumed for their diuretic, antiseptic, digestive, purging, sedative, demulcent, and astringent properties (Obouayeba et al., 2014). The calyces are used in the treatment of high blood pressure and digestive disorders (Ewansiha, 2014; Voon et al., 2012; Wang et al., 2000). The seeds are seldom mentioned in traditional medicine system, compared with other plant parts, but they are roasted and consumed as food (Ismail et al., 2008). *H. sabdariffa*, however, is not satisfactory in scientific studies. It also has an abundance of polyphenolic compounds, such as flavonoids and phenolic acids (gallic and protocatechuic acid). In the red hibiscus calyces, a pigment called anthocyanin is present (Higginbotham et al., 2014).

A number of studies have begun to examine the antibacterial efficiency of the Sudanese roselle (*H. sabdariffa* L.) used in Sudanese folk medicine (Abdallah, 2016). In their experiment, the dried *H. sabdariffa* calyces were soaked in 80% v/v methanol to obtain methanol extract, which was evaluated using a disc diffusion system for five gram-negative and three gram-positive bacterial strains. The results of the test indicated that *the H. sabdariffa* calyx methanol extract contained powerful antibacterial agents, showed substantial inhibition zones for all tested gram-negative and gram-positive bacteria, competed for gentamicin, and showed substantially larger inhibition zones than penicillin, which showed mild to no implications. The results of the study conducted by (Abdallah, 2016) supported the widespread use of this popular plant in Sudanese folk medicine, particularly against some illnesses related to microbes.

Besides, researchers attempted to evaluate the impact of red rose (*H. sabdariffa* L.) antibacterial activity on *Staphylococcus* sp. to resolve staphylococcal infection (Agung et al., 2020). *Staphylococcus* sp. is one of the most prevalent skin-colonizing bacteria, mostly present in animals and humans. Humans have several distinct staphylococcal species (Kloos & Schleifer, 1983). The methodology for their experiment was using the decoction process, where the calyxes were extracted with fresh red roselle calyxes until the temperature reached 90° C for 30 min. The antibacterial effect was tested using *Staphylococcus epidermidis* ATCC 13228, *Staphylococcus warneri* ATCC 3340, and *Staphylococcus xylosus* ATCC 3342 in the agar diffusion method. The results of these tests showed that the roselle decoction possessed a wide range of antibacterial activity against all *Staphylococcus* sp. Together, these studies indicated that the invention of red roselle (*H. sabdariffa* L.) calyx decocted with a broad antistaphylococcal spectrum will strengthen the scientific evidence of this Indonesian traditional medicinal beverage and promote the production of new antibacterial drugs to resolve staphylococcal diseases, especially against resistant strains (Agung et al., 2020).

Several evidences indicated that postpartum hypertension complications included damage to the blood vessels, heart attacks, retinal injuries, renal failure, stroke, cerebral hemorrhage, pulmonary edema, brain disorders, liver necrosis, and renal disorders (Ikawati & Djumiani, 2012; Wright et al., 2002). The National Center for Complementary and Alternative Medicine of the National Institute of Health classified the various therapies and remedies into five classes, one of which was biological based therapies (BBT), a type of nutritional therapy with natural ingredients (Sherman, 2005). The dried roselle petals (*H. sabdariffa* L.) are one of the medicinal plants used to cure hypertension (Da-Costa-Rocha et al., 2014). The common mechanisms of action of this medicinal plant are controlling blood pressure through the effects of dilated blood vessels and reducing the ability of the kidneys to increase blood pressure (Hopkins et al., 2013).Therefore the objective of this study was to investigate the effect of dried roselle petals (*H. sabdariffa* L.) on reducing blood pressure in postpartum mothers who have used antihypertensive drugs (Ritonga, 2017). The results of this study suggested that both systolic and diastolic blood pressure variations were significant, which were consistent with previous findings, suggesting that there was a significant difference in the mean value of systolic and diastolic blood pressure before and after intervention in respondents who received the addition of roselle petals. Considering all these evidences, it seemed that for hypertension, doctors and nurses or health providers should apply the findings of this study in treating postpartum patients. This intervention is supposed to facilitate the healing process by combining antihypertensive medications with sedated roselle flower petals so that long-term pharmacologic drug consumption in postpartum women with hypertension can be avoided and complications from untreated puerperal hypertension can be easily prevented (Ritonga, 2017).

7.1 Traditional Medicines of Roselle

Roselle is being cultivated in tropical and subtropical countries and is known as a significant medicinal plant in many parts of the world (Eslaminejad Parizi et al., 2012). Tea from roselle can be used to regulate blood pressure; besides, its leaves are utilized as sources of mucilage in cosmetics and pharmaceutical products (Ansari, 2013). It also has been medicinally applied for the treatment of colds, hangovers, toothache, and urinary tract infections. People from Senegal have been using roselle leaf juice for treating conjunctivitis. In addition to being used as an anticorbic agent for the treatment of scurvy and in fever relief for its sedative, diuretic, and emollient characteristics, roselle leaves are used as a poultice for treating ulcers and soreness (Ansari, 2013). Gallaher et al. (2006) also reported that roselle root decoction is used for treating scurvy. Ethnobotany of the roselle plant is the study of the roselle plant parts and their practical uses through the traditional knowledge of local culture and people. Ethnobotanical information of the roselle plant revealed that it can be used for treating the after effects of drunkenness, decreasing blood viscosity, and treating gastrointestinal disorders, kidney stone, and liver damage. This is because the roselle plant possesses hypercholesterolemic, antitussive, antihypertensive, sedative, mild laxative, antifungal, antibacterial, uricosuric, diaphoretic, and diuretic properties (Alarcon-Aguilar et al., 2007; Alarcón-Alonso et al., 2012). Roselle is used in folk medicine as hot and cold beverages or drinks to treat hypertension, hypercholesterolemia, fever, liver diseases, and gastrointestinal disorders (Ojeda et al., 2010). Besides, the ripe calyces are used for making hot and cold drinks used medically for their antimicrobial, antispasmodic, and hypotensive effects and for relaxation of the uterine muscle (Khalid et al., 2012). For Malaysians, roselle is a popular health drink, consumed due to the high anthocyanin and vitamin C contents (Ansari, 2013). The bioactive compounds of roselle calyces, such as anthocyanins and proanthocyanidins, are responsible for reducing blood pressure. Also, quercetin in roselle was proven to affect the vascular endothelium, where nitric oxide increased kidney filtration and renal vasodilation (Alarcón-Alonso et al., 2012).

7.2 Nutritional Benefits of Roselle

Roselle, being a versatile plant, has various nutritional uses. First, the fresh calyx (the floral outer whorl), which is rich in citric acid and pectin, is eaten raw in salads, fried, used to flavor pastries, and used in other foods that are used for jelly preparation, soups, condiments, pickles, and so on. It is also often used to give red coloring to herbal teas and roasted to replace coffee. The roselle calyx can be cooked and sweetened with sugar and ginger can also be added to this pleasant and very famous beverage. The young leaves and tender stems of Roselle have a rhubarb-like acidic taste and can be used in making salads, as a potherb, and as an ingredient in curries. The seed is dried and powdered and can be used in oily soups and sauces. The oven-dried seeds were used as an aphrodisiac coffee substitute. Roselle root is edible and fibrous, and the seed contains 20% of oil (Islam, 2019; Puro et al., 2014).

The roselle plant parts are used in various food products, including leaves, seeds, roots, and fruits. The most commonly used part is the fleshy red calyces that are used to make fresh juice, wine, jam, jelly, syrup, gelatine, pudding, and ice cream. Furthermore, butter, pies, sauces, tarts, and other desserts are made with the dried and brewed tea and spices. The calyces have pectin that produces a solid jelly. Roselle's young leaves and tender stalks are consumable in salads or can be cooked alone as greens or mixed with other vegetables and/or meat. They are also added as seasoning to curries for having an acidic taste similar to rhubarb. The high-protein seeds are roasted and powdered and used in soups. The roasted seeds can be utilized in drinks; although the young roots are edible, they are extremely fibrous (refer to Table 1.2) (Islam, 2019; Puro et al., 2014).

The roselle plant's nutritional quality has been evaluated, where the carbohydrate content dominated with 68.7%, then came the crude fiber with 14.6%, and the ashes were the least with 12.2%, followed by others (Luvonga et al., 2010). The roselle plant is also plentiful in magnesium and potassium resources. There are also large levels of vitamins such as ascorbic acid, niacin, and pyridoxine present in roselle. Several researchers have documented different contents of minerals and ashes within the same roselle species. This might be due to the different types of soil that affects its properties (Adanlawo & Ajibade, 2006; Carvajal-Zarrabal et al., 2012; Falade et al., 2005; Nnam & Onyeke, 2003; Ojokoh, 2006). Herbal tea from roselle has been used for a long time to treat high blood pressure, liver damage, and fever but is poorly identified (Owoade et al., 2019). Choi and Mason (2000) stated that studies on balanced diet have shown that less fruit and vegetable intake is usually linked to the rising number of cancer (Islam, 2019).

TABLE 1.2
Nutritional Value per 100 g of Roselle (Islam, 2019).

ROSELLE/MESTA (RAW)	
NUTRITIONAL VALUE PER 100 G (3.5 OZ)	
Energy	205 kJ (49 kcal)
Carbohydrates	Carbohydrates
Fat	0.64
Protein	0.96 g
VITAMINS	
Vitamin A equiv.	14 μg (2%)
Thiamine (B_1)	0.011 mg (1%)
Riboflavin (B_2)	0.028 mg (2%)
Niacin (B_3)	0.31 mg (2%)
Vitamin C	12 mg (14%)
TRACE METALS	
Calcium	215 mg (22%)
Iron	1.48 mg (11%)
Magnesium	51 mg (14%)
Phosphorus	37 mg (5%)
Potassium	208 mg (4%)
Sodium	6 mg (0%)

Units: *μg*, micrograms; *mg*, milligrams; *IU*, International Unit.

7.3 Medical Uses of Roselle

Roselle is an astringent, aromatic, refreshing herb commonly used in the tropics. The leaves taste strongly mucilaginous and are used as a moisturizer and cough reliever. The flowers contain anthocyanin, the glycoside hibiscin, and gossypetin that have choleretic and diuretic effects, decrease the viscosity of the blood, induce intestinal peristalsis, and help lower blood pressure. Blossoms and flowers of roselle are used as a tonic for internal digestion and kidney processes. Roselle seed is a valuable food resource because it is rich in micronutrients and protein. It is also an excellent source of dietary fiber (Omobuwajo et al., 2000). According to Hainida et al. (2008), roselle seeds contain 18.3% of total dietary fiber. Roselle seed is effective in exerting the physiologic effects such as lowering the risk of cardiovascular disease, gastrointestinal disease, colon cancer, glycemic response, and obesity (Nyam et al., 2014). Roselle can also be used in the treatment of abscess, cancer, cough, fatigue, dyspepsia, dysuria, fever, heart attacks, scurvy, hypertension, hangover, neurosis, and strangury (Nnam & Onyeke, 2003). The reduced risk of cancer could be caused by polyphenol and anthocyanin (Briviba et al., 2001; Gao et al., 2002; Lin et al., 1999; Mei et al., 2005; Wang et al., 2003; Weisburger & Chung, 2002). Roselle plants are able to generate secondary metabolites in particular, namely, proteins, steroids, alkaloids, and so on, that increase their nutritional value (Islam, 2019; Umesha et al., 2013).

7.4 Medicinal Properties of Roselle Determined by the Biochemical Values

Roselle calyxes appear dark red, red, and green, and the most widely used are the red calyxes distinguished by the anthocyanin concentration. Cyanidin 3-sambubioside and delphinidin 3-sambubioside are the major anthocyanins (Islam, 2019). The roselle calyxes, seeds, and leaves are rich in minerals, amino acids, organic acids, carotene, vitamin C, and sugar at various proportions, depending on the variety and geographic region. Flavonoids, anthocyanidins, triterpenoids, alkaloids, and steroids were also found in roselle. In Table 1.3, the nutrient content of various parts of *H. sabdariffa* per 100 g is presented (Islam, 2019).

8 THE VARIOUS MEDICINAL PROPERTIES

8.1 Antimicrobial Properties

In disease diagnosis, Tolulope (2007) used roselle's aqueous methanol extract and stated the presence of flavonoids, alkaloids, saponins, and cardiac glycosides. Microbicidal activities were observed against *Bacillus stearothermophilus, Staphylococcus aureus, Micrococcus luteus, Serratia marcescens, Escherichia coli, Bacillus cereus, Clostridium sporogenes, Klebsiella,* and *Pseudomonas fluorescens*. The findings agreed on the usage of roselle plant for the treatment of abscesses, gallstones, cancer, and toxicity in conventional medicine. Isolation of *Listeria monocytogenes* and *Salmonella enterica* from food, clinical, and veterinary samples by Fullerton et al. (2011) showed that the roselle extract was successful as a possible antimicrobial for application in foods.

The antibacterial effects of roselle calyx aqueous extract (RW), roselle calyx ethanol extracts (RE), and protocatechuic acid (PA) against five food spoilage bacteria, namely, *B. cereus, S. aureus, L. monocytogenes, E. coli* O157:H7, and *Salmonella typhimurium* DT104, were investigated by Chao and Yin (2009). The result shows that after 3 days storage at a temperature of 25°C, the addition of RW, RE, and PA demonstrated dose-dependent inhibitory effects against all five test bacteria in apple juice and ground beef, in which RE showed greater antibacterial effects than RW. Chao and Yin (2009) concluded that RE and PA have a huge potential to be utilized as food additives to prevent contamination from these bacteria.

TABLE 1.3
Biochemical Values of Different Parts of the Roselle Plant (Chewonarin et al., 1999).

Nutrients	Calyxes	Seeds	Leaves
Protein (g)	2	28.9	3.5
Carbohydrates (g)	10.2	25.5	8.7
Fat (g)	0.1	21.4	0.3
Vitamin A (I.E.)	–	–	1000
Thiamine (mg)	0.05	0.1	0.2
Riboflavin (mg)	0.07	0.15	0.4
Niacin (mg)	0.06	1.5	1.4
Vitamin C (mg)	17	9	2.3
Calcium (mg)	150	350	240
Iron (mg)	3	9	5

8.2 Antioxidant Properties

The antioxidant activity and liver protection properties of a group of natural pigments known as roselle-hibiscus anthocyanins (HAs), present in the dried calyx, were studied. The HA antioxidant bioactivity and hepatotoxicity were tested in rat primary hepatocytes (Wang et al., 2000). The results revealed that HA significantly reduced lactate dehydrogenase leakage and malondialdehyde formation at concentrations of 0.10 mg/mL and 0.20 mg/mL and that hepatic enzyme markers' serum levels (aspartate aminotransferase and alanine) significantly decreased resulting in decreased oxidative liver harm. There have also been records of antioxidant involvement in cancerous cell lines (Akim et al., 2011). In their animal models, McKay et al. (2010) stated that extracts of roselle's calyx displayed hypocholesterolemic and antihypertensive properties. The antioxidant capacity of the three ratios of crude extract of ethanol (HS-E, soluble fraction of ethyl acetate; HS-R, residual fraction; HS-C, soluble fraction of chloroform) contained in dried flowers was assessed for their ability to quench free radicals and to inhibit the activity of xanthine oxidase (XO) (Tseng et al., 1997). The greatest free radical scavenging potential was shown by HS-E and the greatest impact of inhibition was shown by HS-C on the XO activity. Additionally, on rat primary hepatocytes, the bioactivities of antioxidants of these extracts were examined. Unscheduled DNA synthesis was found to significantly inhibit all fractions. These findings have shown that rat hepatocytes were protected from the genotoxicity and cytotoxicity caused by tert-butyl hydroperoxide (t-BHP) because of the dried flower extracts (HS-C and HS-E). The research on hepatoprotective and antioxidant effects on fish hepatocyte damage caused by carbon tetrachloride (CCl_4) verified the possible use of roselle extract as a medicinal product for hepatic disease treatment in aquaculture, as it showed high levels of glutamate oxalate transaminase, glutamate pyruvate transaminase, malondialdehyde, and lactate dehydrogenase and low levels of glutathione peroxidase and superoxide dismutase (Islam, 2019; Yin et al., 2011).

8.3 Anticancer Properties

Akim et al. (2011) tested roselle juice antiproliferative activity by using various cells, such as breast (MCF-7, MDA-MB-231), cervical (HeLa), and ovarian (Caov-3) cancer cell lines, and it was concluded that the highest antiproliferative capacity against cancer cells was in MCF-7. The effects to the human cancer cells (HL-60) were investigated with roselle anthocyanins (HA) and cell apoptosis was observed in a dose- and time-dependent manner (Chang et al., 2005). Hou et al. (2005) documented apoptosis of anthocyanin-induced leukemia cells by mediating mitochondrial pathways via reactive oxygen species. Protocatechuic acid (PCA), an insulated phenol compound from the dried bloom, was observed to suppress the survival of human promyelocytic leukemia (HL-60) in a concentration- and time-dependent manner (Tseng et al., 1997), and apoptosis was triggered by the reduction of phosphorylation and downregulation of the interpretation of Bcl-2 protein (Tseng et al., 2000). The study showed that cells experienced fragmentation of intranucleosomal DNA and a morphologic change in apoptosis during the operation of JNK/MAPK signaling pathways against gastric carcinomas (Lin et al., 2007). The action of methanol extract from roselle on seven cancer cell lines meant that the human gastric adenocarcinoma (AGS) cancer cells were most vulnerable to concentration, influencing the intrinsic and extrinsic apoptotic routes (Lin et al., 2005). The antioxidant capacity of roselle (*H. sabdariffa* L.) extracts was studied by Mohamed et al. (2007). Different roselle plant parts, such as seeds, stems, leaves, and sepals, were studied in terms of their antioxidant ability, which included water solubility, tocopherol material, and lipid-soluble antioxidant capacity. Roselle seed is a strong source of antioxidants, which is lipid soluble, μ-tocopherol in particular. Seed oil was extracted and characterized, and its physicochemical parameters were recorded as 8.63 meq/kg peroxide index, 2.24% acidity, 232 (k232) and 270 nm (k270), 1.46 and 3.19 extinction coefficients. The oxidative stability was 15.53 h; viscosity, 15.9 cP; refractive index, 1.477; and density, 0.92 kg/L. Roselle seed oil belonged to linoleic/oleic class, with C18:2 (40.1%), C18:1 (28%), C16:0 (20%), C18:0 (5.3%), and C19:1(1.7%) as the most abundant fatty acids. These sterols include campesterol (13.6%), β-sitosterol (71.9%), Δ-5-avenasterol (5.9%), clerosterol (0.6%), and cholesterol (1.35%). The average concentration of tocopherols included α-tocopherol (25%), γ-tocopherol (74.5%), and δ-tocopherol (0.5%). The worldwide characteristics of roselle seed oil suggested that vital industrial applications can occur, contributing to the traditional use of roselle sepals for karkade or roselle drink (Islam, 2019).

9 DIFFERENT PHYSIOLOGIC EFFECTS

9.1 Lipid Metabolism Effect

In a research conducted on hypertensive patients, the effect of roselle juice on the lipid profile, serum electrolytes, and creatinine suggested an upward trajectory of high-density lipid and total cholesterol. Consuming Roselle juice in different concentrations and durations could prevent renal stone disease (Islam, 2019; Kirdpon et al., 1994).

9.2 Antihypertensive Effect

Hypertension contributes to the development of cardiac ischemia, cardiac disorders, cerebrovascular disorders, and renal failure. (Haji Faraji and Haji Tarkhani, 1999) noted a substantial reduction in both systolic and diastolic due to the efficiency of aqueous extract in humans compared to the control group. From another study, the systemic pressure drop was noticeable and the diastolic pressure appeared unchanged (McKay & Blumberg, 2007). Studies on rats were also performed and the common assumption that roselle extract contains antihypertensive substances was endorsed (Odigie et al., 2003; Onyenekwe et al., 1999). The anthocyanin extract analyzed for its therapeutic efficiency, protection, and tolerability with antihypertensive medicines captopril (Herrera-Arellano et al., 2004) and lisinopril showed comparable results in humans and suggested that the synergistic mechanism of diuretic and ACE inhibition results in exerting hypotensive effects (Armando Herrera-Arellano et al., 2007; Islam, 2013, 2019).

9.3 Other Applications of Roselle

For various household purposes, fibers derived from the stem (called rosella hemp) are used for sacking, as twine, and as rope. Besides, the yellow color is also extracted from the floral petals. The seed of roselle itself contains 20% oil (Islam, 2019).

9.4 The Domestic Animal's Food Used for Medicinal Effects

Some animal tests were performed, in which roselle extract as acidifiers were found to have the ability to increase trypsin function, boost the feed conversion ratio, and boost digestion in postweaning pigs (Aphirakchatsakun et al., 2008). In order to search for the lipid to be oxidized by fat degradation, the effect of roselle calyx on egg development in the poultry, presence of thiobarbituric acid-reactive substances (TBARS), and ovarian consistency were examined in yolk and plasma. Store extract time was found to be a significant factor in reducing egg quality and increasing yolk TBARS (Islam, 2019; Sukkhavanit et al., 2011).

10 ROSELLE DRINK

Roselle is a caffeine-free herbal drink made from the dried fruit portion, known as calyx, in particular. It is red in color and has the flavor of berries (Dafallah & Al-Mustafa, 1996; Islam, 2019) (Fig. 1.5).

10.1 Steps in the Preparation of Calyx Drink

1. Pick and wash the fruits, then air-dry or oven-dry them at a temperature of 70° C for 3 days.
2. Then peel off the calyx and store in airtight containers.
3. Take 2 g of dried calyx for a drink and crush it into smaller pieces using a wooden roller.
4. Put them in a bag such as a tea bag or net, then add 8 ounces of boiling water, and soak them for 2–4 min; other aromas of choice such as a few drops of lemon juice can be added, if desired.
5. The dried calyx can also be cooled and served as roselle iced drink.

11 ROSELLE FIBER-REINFORCED POLYMER COMPOSITES

11.1 Natural Fiber

According to a reference module in the 2020 edition of Natural Fiber Composites written by Mohamed Zakriya and Ramakrishnan (2020), natural fibers from plant-

FIG. 1.5 Roselle drink.

based kingdom are renewable resources with several advantages, such as they impart the composite high specific stiffness and strength, are lightweight, are low cost, have a desirable fiber aspect ratio, have excellent mechanical properties, are biodegradable, and are abundantly available from natural sources. Artificial or synthetic fibers are petroleum-based products, originating from the nonrenewable sources, and exploration and drilling of petroleum has many negative effects on the environment. These advantages of natural plant fibers have allowed scientists and engineers to use them to reinforce polymer composites in order to reduce the utilization of forest sources and minimize the surplus of natural fibers (Abral, Atmajaya, et al., 2020; Abral, Basri, et al., 2019; Halimatul et al., 2019a; Nadlene et al., 2016). Natural plant fibers can be used as reinforcement in polymer composite materials, where the orientation of the fibers would improve the mechanical properties of the composite. Currently, natural fibers, such as oil palm fiber (Ayu et al., 2020), hemp (Arulmurugan et al., 2019), kenaf (Aisyah et al., 2019; Mazani et al., 2019), ginger (Abral, Ariksa, et al., 2019, 2020), sugarcane (Asrofi, Sapuan, et al., 2020; Asrofi, Sujito, et al., 2020; Jumaidin, Ilyas, et al., 2019), water hyacinth (Syafri, Sudirman, et al., 2019), corn (Ibrahim, Sapuan, et al., 2020; Sari et al., 2020), sugar palm (Atiqah, Ilyas, et al., 2019; Atiqah, Jawaid, et al., 2019; Halimatul et al., 2019b; Hazrol et al., 2020; Maisara et al., 2019; Norizan et al., 2020; Nurazzi et al., 2019; Nurazzi et al., 2020; Sapuan et al., 2020), jute (Islam et al., 2019), sisal (Yorseng et al., 2020), flax (Akonda et al., 2020), ramie (Syafri, Kasim, et al., 2019), cogon fiber (Jumaidin, Saidi, et al., 2019; Jumaidin et al., 2020), and cassava (Ibrahim, Edhirej, et al., 2020), are utilized as reinforcement agents in polymer composites. They are used for their various benefits, such as excellent mechanical properties, ease of processing, and low manufacturing cost compared to synthetic fibers (Ilyas & Sapuan, 2020a, 2020b; Nadlene et al., 2018; Nazrin et al., 2020). Natural plant fibers are mainly composed of cellulose, hemicellulose, lignin, pectin, and other waxy substances (Ilyas, Sapuan, Atikah, Asyraf, et al., 2020; Ilyas, Sapuan, Atiqah, Ibrahim, et al., 2020; Ilyas, Sapuan, Ibrahim, et al., 2020; Ilyas et al., 2017; Ilyas, Sapuan, Ibrahim, et al., 2019; Ilyas, Sapuan, & Ishak, 2018; Ilyas, Sapuan, Ishak, Zainudin, Atikah, 2018; Ilyas, Sapuan, Ishak, Zainudin, 2019). Reinforcement of natural fibers and polymers that have emerged in the field of polymer science has gained attention for use in a variety of advanced applications ranging from plastic packaging to the automotive industry (Atikah, Ilyas et al., 2019; Atikah, Jawaid, 2019; Ilyas, Sapuan, Atiqah, Asyraf, et al., 2020; Ilyas, Sapuan, Ibrahim, et al., 2019, 2020; Ilyas, Sapuan, Ishak, & Zainudin, 2018b, 2018c, 2018d; Ilyas, Sapuan, Sanyang, et al., 2018; Sanyang et al., 2018).

11.2 Roselle Fiber Production

Several researchers explored roselle bast fiber's potential as reinforcement materials because of its similar characteristics to other known natural fibers, such as jute. The fiber is located under roselle stalk's bark and extracted via a series of water retting steps, as shown in Fig. 1.6. High-quality fiber is obtainable by harvesting during the bud stage. The stalks were packed and water-treated for 3–4 days. The retted stalks were washed under flowing water and separated from the stalks, washed, and sun-dried (Nadlene et al., 2016; Radzi et al., 2018a, 2018b). The final product is shown in Fig. 1.6E.

11.3 Roselle Nanocellulose

Nanocellulose is a material produced from natural cellulose. Roselle had been recognized as the alternative source for the production of nanocellulose nanocomposites or polymer composites (Bandera et al., 2014; Mohamad Haafiz et al., 2013). Roselle-derived fibers mainly contain 59%–65% cellulose, 16%–20% hemicellulose, and 6%–10% lignin (Razali et al., 2015). Owing to their excellent thermal stability and tensile strength, biocompatibility, and environment-friendly biodegradability, many studies and research had been done vastly on the advancement of nanoparticles from renewable resources as reinforcement materials (Capron et al., 2017). Roselle fiber is a renewable and sustainable agricultural waste enriched with cellulose polysaccharides. Kian et al. (2018) showed that the isolation of nanocrystalline cellulose (NCC) from roselle-derived microcrystalline cellulose (MCC) is an alternative approach to recover the agricultural roselle plant residue. In the experiment, acid hydrolysis treatment with different reaction times was carried out to degrade the roselle-derived MCC to NCC. As evaluated from the morphologic investigations, the needle-shaped NCC nanostructures were studied using transmission electron microscopy and atomic force microscopy, while the irregular rod-shaped NCC structured were studied using field emission scanning electron microscopy. With 60 min hydrolysis time, X-ray diffraction analysis demonstrated the highest NCC crystallinity degree with 79.5% (Fig. 1.7). In thermal analysis by thermogravimetric analysis and differential scanning calorimetry, shorter hydrolysis time tended to produce NCC with higher thermal stability. Thus

FIG. 1.6 Extraction of roselle fiber: **(A)** roselle plant, **(B)** stalks in bundle form, **(C)** water retting process, **(D)** removal of the fibers from the stalks, and **(E)** final form of roselle fibers (Athijayamani et al., 2009a; Nadlene et al., 2016).

the NCC isolated from roselle-derived MCC has high potential to be used in pharmaceutical applications and biomedical field for nanocomposite fabrication.

11.4 Various Roselle Fiber-Reinforced Polymer Composites

Roselle had been recognized as the alternative source for the production of natural fiber-reinforced polymer composites. In the past few years, numerous researchers have conducted studies on surface modification of roselle fiber and the reinforcement behavior of roselle fiber within the polymer composites. Table 1.4 shows the various roselle fiber-reinforced polymer composites. In a study conducted by Singha and Thakura (2008) on the roselle fiber-incorporated phenol formaldehyde subjected to fiber loading, optimum mechanical properties were achieved at 30% of fiber loading. In another study carried out by Singha and Thakur (2009) on the effect of fiber size (particulate, <200 μm; short fiber, 3 mm; and long fiber, 6 mm) on the mechanical

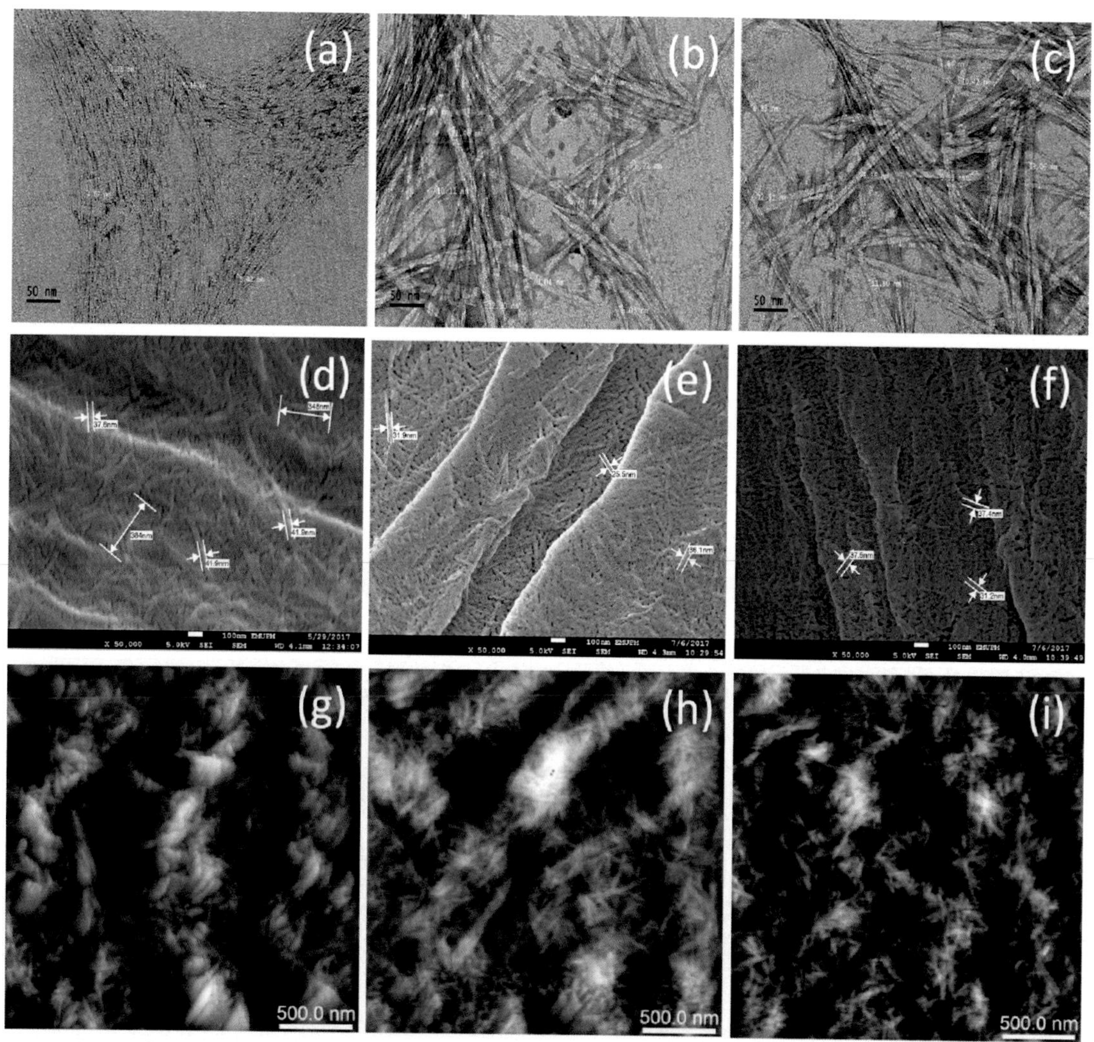

FIG. 1.7 **(A–C)** Transmission electron microscopic, **(D–F)** field emission scanning electron microscopic, and **(G–I)** atomic force microscopic images of nanocrystalline cellulose (NCC)-I **(A, D, G)**, NCC-II **(B, E, H)**, and NCC-III **(C, F, I)** (Kian et al., 2018). (Reproduced with copyright permission from Kian, L. K., Jawaid, M., Ariffin, H., & Karim, Z. (2018). Isolation and characterization of nanocrystalline cellulose from roselle-derived microcrystalline cellulose. *International Journal of Biological Macromolecules, 114*, 54–63. https://doi.org/10.1016/j.ijbiomac.2018.03.065, with permission from Elsevier.)

TABLE 1.4
Various Roselle Fiber-Reinforced Polymer Composites.

Polymer	Technique	Title	References
Roselle fiber-reinforced polyurethane	➢ Melt mixed mixer ➢ Hot press at 170°C	Influence of fiber contents on mechanical and thermal properties of roselle fiber-reinforced polyurethane composites	Radzi et al. (2017)
Roselle/sugar palm fiber-reinforced thermoplastic polyurethane	➢ Melt mixed mixer ➢ Hot press at 170°C	Effect of alkaline treatment on the mechanical, physical, and thermal properties of roselle/sugar palm fiber-reinforced thermoplastic polyurethane hybrid composites	Radzi et al. (2019a)
Roselle/sugar palm fiber-reinforced thermoplastic polyurethane hybrid	➢ Melt mixed mixer ➢ Hot press at 170°C	Water absorption, thickness, swelling, and thermal properties of roselle/sugar palm fiber-reinforced thermoplastic polyurethane hybrid composites	Radzi et al. (2019b)
Roselle/sugar palm fiber hybrid-reinforced polyurethane	➢ Melt mixed mixer ➢ Hot press at 170°C	Mechanical performance of roselle/sugar palm fiber hybrid-reinforced polyurethane composites	Radzi et al. (2018b)
Roselle fiber-reinforced polyurethane	➢ Melt mixed mixer ➢ Hot press at 170°C	Mechanical and thermal performances of roselle fiber-reinforced thermoplastic polyurethane composites	Radzi et al. (2018a)
Roselle fibers reinforced fillers for isostatic polypropylene (iPP)	➢ Co-rotating twin-screw extruder ➢ Injection molding	Mechanical properties of injection-molded isotactic polypropylene/roselle fiber composites	Junkasem et al. (2006)
Roselle fibers reinforced with sisal fibers for thermosetting matrix	➢ Fibers subjected to 10% sodium hydroxide solution treatment at different durations ➢ Scanning electron microscopic study	Influence of alkali-treated fibers on the mechanical properties and machinability of roselle and sisal fiber hybrid polyester composite	Athijayamani et al. (2009b)
Roselle fibers reinforced with vinyl ester composites	➢ Treated by alkalization and a silane coupling agent ➢ TGA analysis	The effects of chemical treatment on the structural, thermal, physical, mechanical, and morphologic properties of roselle fiber-reinforced vinyl ester composites	Nadlene et al. (2018)
Roselle fiber-reinforced vinyl ester	➢ Mechanical stirrer ➢ Hand lay-up processes	Mechanical and thermal properties of roselle fiber-reinforced vinyl ester composites	Razali et al. (2016)

Roselle/sugar palm fiber-reinforced vinyl ester	➢ Mechanical stirrer ➢ Hand lay-up processes	Mechanical properties and morphologic analysis of roselle/sugar palm fiber-reinforced vinyl ester hybrid composites	Razali et al. (2018)
Roselle fibers reinforced with urea formaldehyde	➢ Hot press at 165°C	Producing roselle (*Hibiscus sabdariffa*) particle board composites	Ghalehno and Nazerian (2011)
Roselle fiber-reinforced phenol-formaldehyde resin	➢ Hot press at 140°C	Fabrication and study of lignocellulosic *H. sabdariffa* fiber-reinforced polymer composites	Singha and Thakura (2008)
Roselle fiber-reinforced resorcinol-formaldehyde resin	➢ Hot press at 50°C for 12 h	Physical, chemical, and mechanical properties of *H. sabdariffa* fiber/polymer composite	Singha and Thakur (2009)
Roselle fiber-reinforced resorcinol-formaldehyde resin	➢ Hot press at 50°C for 12 h	Fabrication of *H. sabdariffa* fiber-reinforced polymer composites	Singha and Thakur (2008)
Roselle graft-copolymers (butyl acrylate (BA), methyl acrylate (MA), acrylonitrile (AN), and 4-vinyl pyridine (4-VP)) reinforced phenol-formaldehyde resin (PF)	➢ Stirring	Accreditation of novel roselle grafted fiber-reinforced biocomposites	Chauhan and Kaith (2012b)
Phenol-formaldehyde resin	➢ Hot press at 120°C for 10 min	Versatile roselle graft-copolymers: XRD studies and their mechanical evaluation after use as reinforcement in composites	Chauhan and Kaith (2012a)

TGA, thermogravimetric analysis; *XRD*, X-ray diffraction.

properties of roselle fiber-incorporated phenol formaldehyde, the tensile strength was highest in particulate fiber-reinforced phenol formaldehyde.

Chauhan and Kaith (2012a) studied an accreditation of novel roselle grafted fiber-reinforced biocomposites. The study was focused on the evaluation of the effect of versatile roselle graft-copolymer-reinforced phenol formaldehyde on hardness, stressed at the limit of proportion, modulus of rupture, and modulus elasticity properties. Besides, Nadlene et al. (2016) examined the fiber loading effect of roselle fiber-reinforced vinyl ester. From the results, they observed that the optimum fiber loading for mechanical properties of silane-treated roselle fiber was 20%. Moreover, several researchers have hybridized roselle fiber with other natural fibers to improve the mechanical and water barrier properties of polymer composites. Radzi et al. (2018b) conducted a study on the mechanical performance of roselle/sugar palm fiber-reinforced polyurethane hybrid composite polymers. From the investigation, they indicated that the hybridization of roselle fiber with sugar palm fiber in thermoplastic polyurethane enhanced the impact strength of the hybrid composite. The graft copolymers or the modification of the roselle fiber as well as reinforcement of roselle fiber within the polymer composites obtained were sensible solutions to solve the underutilization of renewable waste biomass. The utilization of roselle fiber within the polymer-matrix-based biocomposites was proved to enhance the mechanical strength and water barrier properties for various advanced applications including packaging, transportation, electronics, automotive, aerospace, and biomedical engineering and has served as a pioneer for the advancement of technology.

12 CONCLUSION

Roselle is probably native to Africa and now widely naturalized in the tropical and subtropical regions of the world, particularly in India and Southeast Asia. It is widely grown in Africa, Asia, Papua New Guinea, and the Pacific. China, Sudan, and Thailand are the leading producers and dominate the world supply. Mexico, Egypt, Senegal, Tanzania, Mali, and Jamaica are also important suppliers, but production is mostly used domestically. Roselle plants are rich in nutrients, vitamins, and minerals; have good health properties; and are of great benefit for human health. Many parts of roselle, including seeds, leaves, fruits, and roots, are used in various foods and medicines. Among them, the fleshy red calyces are the most popular. Roselle has been used as a therapeutic plant for centuries. Traditionally, the extract treats toothaches, urinary tract infections, colds, and even hangovers. In Senegal, the juice of leaves is used to treat conjunctivitis and, when pulverized, to soothe sores and ulcers. Root concoctions act as a potent laxative. Natives of various countries drink roselle tea to stabilize blood pressure and lower cholesterol levels. Besides using roselle for medicine, it is used to obtain its beneficial fiber. Recently, researchers and engineers have replaced synthetic fibers with natural fibers as the primary component of composites because of increasing environmental concerns. Roselle is one of the natural fibers that is suitable for this purpose. On the basis of the analysis of literature, detailed studies were carried out on the fiber extraction methods, morphology, properties, and potential surface treatments of roselle to improve its efficiency in the manufacturing of natural fiber-enhanced polymer composites. The improvements of roselle fiber-reinforced polymer composites using two-fiber hybrids together in a polymer composite have been comprehensively studied to be applied in automotive and structural applications. As roselle is hybridized with other fibers, the thermal properties, specifically thermal stability, could be altered and varied for each of the fiber combination in polymer composites. The roselle hybrid composites have significantly improved strength, great durability, and better thermal stability. Hence the roselle hybrid composites can be used for potential advanced applications in the packaging, construction and building, biomedical, automotive, electronic, and aerospace industries.

REFERENCES

Abdallah, E. M. (2011). Plants: An alternative source for antimicrobials. *Journal of Applied Pharmaceutical Science, 1*(6), 16–20.

Abdallah, E. M. (2016). Antibacterial efficiency of the Sudanese roselle (*Hibiscus sabdariffa* L.), a famous beverage from Sudanese folk medicine. *Journal of Intercultural Ethnopharmacology, 5*(2), 186–190. https://doi.org/10.5455/jice.20160320022623

Abral, H., Ariksa, J., Mahardika, M., Handayani, D., Aminah, I., Sandrawati, N., Pratama, A. B., Fajri, N., Sapuan, S. M., & Ilyas, R. A. (2020). Transparent and antimicrobial cellulose film from ginger nanofiber. *Food Hydrocolloids, 98*, 105266. https://doi.org/10.1016/j.foodhyd.2019.105266

Abral, H., Ariksa, J., Mahardika, M., Handayani, D., Aminah, I., Sandrawati, N., Sapuan, S. M., & Ilyas, R. A. (2019). Highly transparent and antimicrobial PVA based bionanocomposites reinforced by ginger nanofiber. *Polymer Testing*, 106186. https://doi.org/10.1016/j.polymertesting.2019.106186

Abral, H., Atmajaya, A., Mahardika, M., Hafizulhaq, F., Kadriadi, Handayani, D., Sapuan, S. M., & Ilyas, R. A. (2020). Effect of ultrasonication duration of polyvinyl alcohol (PVA) gel on characterizations of PVA film. *Journal of Materials Research and Technology, 9*(2), 2477–2486. https://doi.org/10.1016/j.jmrt.2019.12.078

Abral, H., Basri, A., Muhammad, F., Fernando, Y., Hafizulhaq, F., Mahardika, M., Sugiarti, E., Sapuan, S. M., Ilyas, R. A., & Stephane, I. (2019). A simple method for improving the properties of the sago starch films prepared by using ultrasonication treatment. *Food Hydrocolloids, 93*, 276–283. https://doi.org/10.1016/j.foodhyd.2019.02.012

Adanlawo, I. G., & Ajibade, V. A. (2006). Nutritive value of the two varieties of Roselle (*Hibiscus sabdariffa*) calyces soaked with wood ash. *Pakistan Journal of Nutrition, 5*(6), 555–557. https://doi.org/10.3923/pjn.2006.555.557

Agung, S., Kusuma, F., Zuhrotun, A., Hendriani, R., Indrayati, A., & Radianawati, A. (2020). Anti-staphylococcal effect of red roselle (*Hibiscus sabdariffa* L.) calyx decoction as an Indonesian folk medicine beverage. *Drug Invention Today, 13*(5), 3–8.

Aisyah, H. A., Paridah, M. T., Sapuan, S. M., Khalina, A., Berkalp, O. B., Lee, S. H., Lee, C. H., Nurazzi, N. M., Ramli, N., Wahab, M. S., & Ilyas, R. A. (2019). Thermal properties of woven kenaf/carbon fibre-reinforced epoxy hybrid composite panels. *International Journal of Polymer Science, 2019*, 1–8. https://doi.org/10.1155/2019/5258621

Akim, A., Ling, L. C., Rahmat, A., & Zakaria, Z. A. (2011). Antioxidant and anti-proliferative activities of Roselle juice on Caov-3, MCF-7, MDA-MB-231 and HeLa cancer cell lines. *African Journal of Pharmacy and Pharmacology, 5*(7), 957–965. https://doi.org/10.5897/AJPP11.207

Akonda, M. H., Shah, D. U., & Gong, R. H. (2020). Natural fibre thermoplastic tapes to enhance reinforcing effects in composite structures. *Composites Part A: Applied Science* and Manufacturing, 131(November 2019), 105822. https://doi.org/10.1016/j.compositesa.2020.105822

Alarcon-Aguilar, F. J., Zamilpa, A., Perez-Garcia, M. D., Almanza-Perez, J. C., Romero-Nuñez, E., Campos-Sepulveda, E. A., Vazquez-Carrillo, L. I., & Roman-Ramos, R. (2007). Effect of *Hibiscus sabdariffa* on obesity in MSG mice. *Journal of Ethnopharmacology, 114*(1), 66–71. https://doi.org/10.1016/j.jep.2007.07.020

Alarcón-Alonso, J., Zamilpa, A., Aguilar, F. A., Herrera-Ruiz, M., Tortoriello, J., & Jimenez-Ferrer, E. (2012). Pharmacological characterization of the diuretic effect of *Hibiscus sabdariffa* Linn (Malvaceae) extract. *Journal of Ethnopharmacology*. https://doi.org/10.1016/j.jep.2011.12.005

America, S., America, C., African, O., & Garden, G. E. (1993). *Islands of the Pacific Basin F 22N, 7*(4), 275–285.

Anderson, N. O. (2006). In *Flower breeding and genetics: Issues, challenges and opportunities for the 21st century*. Netherlands: Springer. https://doi.org/10.1007/978-1-4020-4428-1.

Ansari, M. (2013). An overview of the Roselle plant with particular reference to its cultivation, diseases and usages. *European Journal of Medicinal Plants, 3*(1), 135–145. https://doi.org/10.9734/EJMP/2013/1889

Aphirakchatsakun, W., Angkanaporn, K., & Kijparkorn, S. (2008). The effect of Roselle (*Hibiscus sabdariffa* Linn.) calyx as antioxidant and acidifier on growth performance in post-weaning pigs. *Asian-Australasian Journal of Animal Sciences*. https://doi.org/10.5713/ajas.2008.70242

Appell, S. D., & Red, S. (2003). *Hibiscus sabdariffa* the other "Cranberry". *Plants & Gardens News*.

Arulmurugan, M., Prabu, K., Rajamurugan, G., & Selvakumar, A. S. (2019). Impact of BaSO4 filler on woven Aloevera/Hemp hybrid composite: Dynamic mechanical analysis. *Materials Research Express, 6*(4), 045309. https://doi.org/10.1088/2053-1591/aafb88

Asrofi, M., Sapuan, S. M., Ilyas, R. A., & Ramesh, M. (2020). Characteristic of composite bioplastics from tapioca starch and sugarcane bagasse fiber: Effect of time duration of ultrasonication (Bath-Type). *Materials Today: Proceedings*. https://doi.org/10.1016/j.matpr.2020.07.254

Asrofi, M., Sujito, Syafri, E., Sapuan, S. M., & Ilyas, R. A. (2020). Improvement of biocomposite properties based tapioca starch and sugarcane bagasse cellulose nanofibers. *Key Engineering Materials, 849*, 96–101. https://doi.org/10.4028/www.scientific.net/KEM.849.96

Athijayamani, A., Thiruchitrambalam, M., Natarajan, U., & Pazhanivel, B. (2009). Effect of moisture absorption on the mechanical properties of randomly oriented natural fibers/polyester hybrid composite. *Materials Science and Engineering: A, 517*, 344–353. https://doi.org/10.1016/j.msea.2009.04.027

Athijayamani, A., Thiruchitrambalam, M., Natarajan, U., & Pazhanivel, B. (2009). Influence of alkali-treated fibers on the mechanical properties and machinability of Roselle and sisal fiber hybrid polyester composite. *Polymer Composites, 37*(9), 2832–2846. https://doi.org/10.1002/pc.20853

Atikah, M. S. N., Ilyas, R. A., Sapuan, S. M., Ishak, M. R., Zainudin, E. S., Ibrahim, R., Atiqah, A., Ansari, M. N. M., & Jumaidin, R. (2019). Degradation and physical properties of sugar palm starch/sugar palm nanofibrillated cellulose bionanocomposite. *Polimery, 64*(10), 27–36. https://doi.org/10.14314/polimery.2019.10.5

Atiqah, A., Jawaid, M., Sapuan, S. M., Ishak, M. R., Ansari, M. N. M., & Ilyas, R. A. (2019). Physical and thermal properties of treated sugar palm/glass fibre reinforced thermoplastic polyurethane hybrid composites. *Journal of Materials Research and Technology, 8*(5), 3726–3732. https://doi.org/10.1016/j.jmrt.2019.06.032

Ayu, R. S., Khalina, A., Harmaen, A. S., Zaman, K., Isma, T., Liu, Q., Ilyas, R. A., & Lee, C. H. (2020). Characterization study of empty fruit bunch (EFB) fibers reinforcement in poly(butylene) succinate (PBS)/starch/glycerol composite sheet. *Polymers, 12*(7), 1571. https://doi.org/10.3390/polym12071571

Bandera, D., Sapkota, J., Josset, S., Weder, C., Tingaut, P., Gao, X., Foster, E. J., & Zimmermann, T. (2014). Influence of mechanical treatments on the properties of cellulose nanofibers isolated from microcrystalline cellulose. *Reactive and Functional Polymers, 85*, 134–141.

Briviba, K., Abrahamse, S. L., Pool-Zobel, B. L., & Rechkemmer, G. (2001). Neurotensin- and EGF-induced metabolic activation of colon carcinoma cells is diminished by dietary flavonoid cyanidin but not by its glycosides. *Nutrition and Cancer, 41*(1–2), 172–179. https://doi.org/10.1207/s15327914nc41-1&2_24

Capron, I., Rojas, O. J., & Bordes, R. (2017). Behavior of nanocelluloses at interfaces. *Current Opinion in Colloid and Interface Science, 29*, 83–95. https://doi.org/10.1016/j.cocis.2017.04.001

Carvajal-Zarrabal, O., María, D., Barradas-Dermitz, Orta-Flores, Z., Margaret, P., Hayward-Jones, Nolasco-Hipólito, C., Aguilar-Uscanga, M. G., Miranda-Medina, A., & Bujang, K. Bin (2012). *Hibiscus sabdariffa* L., Roselle calyx, from ethnobotany to pharmacology. *Journal of Experimental Pharmacology, 4*(1), 25–39. https://doi.org/10.2147/JEP.S27974

Chang, Y. C., Huang, H. P., Hsu, J. D., Yang, S. F., & Wang, C. J. (2005). Hibiscus anthocyanins rich extract-induced apoptotic cell death in human promyelocytic leukemia cells. *Toxicology and Applied Pharmacology, 205*(3), 201–212. https://doi.org/10.1016/j.taap.2004.10.014

Chao, C. Y., & Yin, M. C. (2009). Antibacterial effects of Roselle calyx extracts and protocatechuic acid in ground beef and apple juice. *Foodborne Pathogens and Disease, 6*(2), 201–206. https://doi.org/10.1089/fpd.2008.0187

Chauhan, A., & Kaith, B. (2012a). Accreditation of novel Roselle grafted fiber reinforced bio-composites. *Journal of Engineered Fibers and Fabrics, 7*(2), 66–75. https://doi.org/10.1177/155892501200700210

Chauhan, A., & Kaith, B. (2012b). Versatile Roselle graft-copolymers: XRD studies and their mechanical evaluation after use as reinforcement in composites. *Journal of the Chilean Chemical Society, 57*(3), 1262–1266. https://doi.org/10.4067/s0717-97072012000300014

Chewonarin, T., Kinouchi, T., Kataoka, K., Arimochi, H., Kuwahara, T., Vinitketkumnuen, U., & Ohnishi, Y. (1999). Effects of Roselle (*Hibiscus sabdariffa* Linn.), a

Thai medicinal plant, on the mutagenicity of various known mutagens in *Salmonella typhimurium* and on formation of aberrant crypt foci induced by the colon carcinogens Azoxymethane and 2-Amino-1-methyl-6-phe. *Food and Chemical Toxicology*. https://doi.org/10.1016/S0278-6915(99)00041-1

Chin, K., Qi, Y., Chin, K. L., Malekian, F., Berhane, M., & Gager, J. (2016). Biological characteristics, nutritional and medicinal value of Roselle, *Hibiscus sabdariffa* biological characteristics, nutritional and medicinal value of Roselle, *Hibiscus sabdariffa*. *Agricultural Research and Extension Center, 70813*(March 2005), 603–604.

Choi, S. W., & Mason, J. B. (2000). Folate and carcinogenesis: An integrated scheme. *Journal of Nutrition, 130*(2), 129–132. https://doi.org/10.1093/jn/130.2.129

Da-Costa-Rocha, I., Bonnlaender, B., Sievers, H., Pischel, I., & Heinrich, M. (2014). *Hibiscus sabdariffa* L. - a phytochemical and pharmacological review. *Food Chemistry, 165*, 424–443. https://doi.org/10.1016/j.foodchem.2014.05.002

Dafallah, A. A., & Al-Mustafa, Z. (1996). Investigation of the anti-inflammatory activity of Acacia nilotica and *Hibiscus sabdariffa*. *The American Journal of Chinese Medicine, 24*(03n04), 263–269. https://doi.org/10.1142/S0192415X96000323

Duke, J. A. (1983). *Hibiscus sabdariffa* L. In *Handbook of energy crops*. Purdue University.

Eslaminejad Parizi, T., Ansaria, M., & Elaminejad, T. (2012). Evaluation of the potential of Trichoderma viride in the control of fungal pathogens of Roselle (*Hibiscus sabdariffa* L.) in vitro. *Microbial Pathogenesis*. https://doi.org/10.1016/j.micpath.2012.01.001

Eslaminejad, T., & Zakaria, M. (2011). Morphological characteristics and pathogenicity of fungi associated with Roselle (*Hibiscus sabdariffa*) diseases in Penang, Malaysia. *Microbial Pathogenesis, 51*(5), 325–337. https://doi.org/10.1016/j.micpath.2011.07.007

Ewansiha, J. (2014). Evaluation of the antimicrobial activity of roselle (*Hibiscus sabdariffa* L.) leaf extracts and its phytochemical properties. *Peak Journal of Medicinal Plants Research*.

Falade, O. S., Otemuyiwa, I. O., Oladipo, A., Oyedapo, O. O., Akinpelu, B. A., & Adewusi, S. R. A. (2005). The chemical composition and membrane stability activity of some herbs used in local therapy for anemia. *Journal of Ethnopharmacology, 102*(1), 15–22. https://doi.org/10.1016/j.jep.2005.04.034

Falagas, M. E., & Bliziotis, I. A. (2007). Pandrug-resistant gram-negative bacteria: The dawn of the post-antibiotic era? *International Journal of Antimicrobial Agents, 29*(6), 630–636. https://doi.org/10.1016/j.ijantimicag.2006.12.012

Fullerton, M., Khatiwada, J., Johnson, J. U., Davis, S., & Williams, L. L. (2011). Determination of antimicrobial activity of sorrel (*Hibiscus sabdariffa*) on *Escherichia coli* O157: H7 isolated from food, veterinary, and clinical samples. *Journal of Medicinal Food, 14*(9), 950–956. https://doi.org/10.1089/jmf.2010.0200

Gallaher, R. N., Gallaher, K., Marshall, A. J., & Marshall, A. C. (2006). Mineral analysis of ten types of commercially available tea. *Journal of Food Composition and Analysis, 19*, S53–S57. https://doi.org/10.1016/j.jfca.2006.02.006

Gao, X., Xu, Y. X., Divine, G., Janakiraman, N., Chapman, R. A., & Gautam, S. C. (2002). Disparate in vitro and in vivo anti-leukemic effects of resveratrol, a natural polyphenolic compound found in grapes. *Journal of Nutrition, 132*(7), 2076–2081. https://doi.org/10.1093/jn/132.7.2076

Ghalehno, M. D., & Nazerian, M. (2011). Producing roselle (*Hibiscus sabdariffa*) particleboard composites. *Ozean Journal of Applied Sciences, 4*(1), 1–5.

Grubben, G. J. H., & Denton, O. A. (2004). In G. J. H. Grubben, & O. A. Denton (Eds.), *Plant resources of tropical Africa 2: Vegetables. Prota foundation*. Netherlands/Backluys Publishers.

Hainida, E., Ismail, A., Hashim, N., Mohd-Esa, N., & Zakiah, A. (2008). Effects of defatted dried Roselle (*Hibiscus sabdariffa* L.) seed powder on lipid profiles of hypercholesterolemia rats. *Journal of the Science of Food and Agriculture, 88*(6), 1043–1050. https://doi.org/10.1002/jsfa.3186

Halimatul, S. M. N., Amin, I., Mohd-Esa, N., Nawalyah, A. G., & Siti Muskinah, M. (2007). *Protein quality of Roselle (*Hibiscus sabdariffa *L.) seeds*.

Haji-Faraji, M. H., & Haji-Tarkhani, A. H. (1999). The effect of sour tea (Hibiscus sabdariffa) on essential hypertension. *Journal of Ethnopharmacology, 65*(3), 231–236.

Halimatul, M. J., Sapuan, S. M., Jawaid, M., Ishak, M. R., & Ilyas, R. A. (2019a). Effect of sago starch and plasticizer content on the properties of thermoplastic films: Mechanical testing and cyclic soaking-drying. *Polimery, 64*(6), 32–41. https://doi.org/10.14314/polimery.2019.6.5

Halimatul, M. J., Sapuan, S. M., Jawaid, M., Ishak, M. R., & Ilyas, R. A. (2019b). Water absorption and water solubility properties of sago starch biopolymer composite films filled with sugar palm particles. *Polimery, 64*(9), 27–35. https://doi.org/10.14314/polimery.2019.9.4

Hazrol, M. D., Sapuan, S. M., Ilyas, R. A., Othman, M. L., & Sherwani, S. F. K. (2020). Electrical properties of sugar palm nanocrystalline cellulose reinforced sugar palm starch nanocomposites. *Polimery, 65*(05), 363–370. https://doi.org/10.14314/polimery.2020.5.4

Herrera-Arellano, A., Flores-Romero, S., Chávez-Soto, M. A., & Tortoriello, J. (2004). Effectiveness and tolerability of a standardized extract from *Hibiscus sabdariffa* in patients with mild to moderate hypertension: A controlled and randomized clinical trial. *Phytomedicine, 11*(5), 375–382. https://doi.org/10.1016/j.phymed.2004.04.001

Herrera-Arellano, A., Miranda-Sánchez, J., Ávila-Castro, P., Herrera-Álvarez, S., Jiménez-Ferrer, J. E., Zamilpa, A., Román-Ramos, R., Ponce-Monter, H., & Tortoriello, J. (2007). Clinical effects produced by a standardized herbal medicinal product of *Hibiscus sabdariffa* on patients with hypertension. A randomized, double-blind, lisinopril-controlled clinical trial. *Planta Medica, 73*(1), 6–12. https://doi.org/10.1055/s-2006-957065

Higginbotham, K. L., Burris, K. P., Zivanovic, S., Davidson, P. M., & Stewart, C. N. (2014). Antimicrobial activity of *Hibiscus sabdariffa* aqueous extracts against *Escherichia coli* O157:H7 and *Staphylococcus aureus* in a microbiological medium and milk of various fat concentrations. *Journal of Food Protection, 77*(2), 262–268. https://doi.org/10.4315/0362-028X.JFP-13-313

Hopkins, A. L., Lamm, M. G., Funk, J. L., & Ritenbaugh, C. (2013). *Hibiscus sabdariffa* L. in the treatment of hypertension and hyperlipidemia: A comprehensive review of animal and human studies. *Fitoterapia, 85*(1), 84–94. https://doi.org/10.1016/j.fitote.2013.01.003

Hou, D. X., Tong, X., Terahara, N., Luo, D., & Fujii, M. (2005). Delphinidin 3-sambubioside, a Hibiscus anthocyanin, induces apoptosis in human leukemia cells through reactive oxygen species-mediated mitochondrial pathway. *Archives of Biochemistry and Biophysics, 440*(1), 101–109. https://doi.org/10.1016/j.abb.2005.06.002

Ibrahim, M. I., Edhirej, A., Sapuan, S. M., Jawaid, M., Ismarrubie, N. Z., & Ilyas, R. A. (2020). Extraction and characterization of Malaysian cassava starch, peel, and bagasse, and selected properties of the composites. In R. Jumaidin, S. M. Sapuan, & H. Ismail (Eds.), *Biofiller-reinforced biodegradable polymer composites* (1st ed., pp. 267–283). CRC Press.

Ibrahim, M. I., Sapuan, S. M., Zainudin, E. S., Zuhri, M. Y., Edhirej, A., & Ilyas, R. A. (2020). Characterization of corn fiber-filled cornstarch biopolymer composites. In R. Jumaidin, S. M. Sapuan, & H. Ismail (Eds.), *Biofiller-reinforced biodegradable polymer composites* (1st ed., pp. 285–301). CRC Press.

Ikawati, Z., & Djumiani, S. (2012). Kajian keamanan pemakaian obat anti-hipertensi di Poliklinik usia lanjut Instalasi Rawat Jalan RS Dr Sardjito. *Pharmaceutical Sciences And Research (PSR), 5*(3), 150–169. https://doi.org/10.7454/psr.v5i3.3429

Ilyas, R. A., & Sapuan, S. M. (2020a). The preparation methods and processing of natural fibre bio-polymer composites. *Current Organic Synthesis, 16*(8), 1068–1070. https://doi.org/10.2174/157017941608200120105616

Ilyas, R. A., & Sapuan, S. M. (2020b). Biopolymers and biocomposites: Chemistry and technology. *Current Analytical Chemistry, 16*(5), 500–503. https://doi.org/10.2174/157341101605200603095311

Ilyas, R. A., Sapuan, S. M., Atikah, M. S. N., Asyraf, M. R. M., Rafiqah, S. A., Aisyah, H. A., Nurazzi, N. M., & Norrrahim, M. N. F. (2020). Effect of hydrolysis time on the morphological, physical, chemical, and thermal behavior of sugar palm nanocrystalline cellulose (*Arenga pinnata* (Wurmb.) Merr). *Textile Research Journal*. https://doi.org/10.1177/0040517520932393

Ilyas, R. A., Sapuan, S. M., Atiqah, A., Ibrahim, R., Abral, H., Ishak, M. R., Zainudin, E. S., Nurazzi, N. M., Atikah, M. S. N., Ansari, M. N. M., Asyraf, M. R. M., Supian, A. B. M., & Ya, H. (2020). Sugar palm (*Arenga pinnata* [Wurmb .] Merr) starch films containing sugar palm nanofibrillated cellulose as reinforcement: Water barrier properties. *Polymer Composites, 41*(2), 459–467. https://doi.org/10.1002/pc.25379

Ilyas, R. A., Sapuan, S. M., Ibrahim, R., Abral, H., Ishak, M. R., Zainudin, E. S., Atikah, M. S. N., Mohd Nurazzi, N., Atiqah, A., Ansari, M. N. M., Syafri, E., Asrofi, M., Sari, N. H., & Jumaidin, R. (2019). Effect of sugar palm nanofibrillated cellulose concentrations on morphological, mechanical and physical properties of biodegradable films based on agro-waste sugar palm (*Arenga pinnata* (Wurmb.) Merr) starch. *Journal of Materials Research and Technology, 8*(5), 4819–4830. https://doi.org/10.1016/j.jmrt.2019.08.028

Ilyas, R. A., Sapuan, S. M., Ibrahim, R., Abral, H., Ishak, M. R., Zainudin, E. S., Atiqah, A., Atikah, M. S. N., Syafri, E., Asrofi, M., & Jumaidin, R. (2020). Thermal, biodegradability and water barrier properties of bio-nanocomposites based on plasticised sugar palm starch and nanofibrillated celluloses from sugar palm fibres. *Journal of Biobased Materials and Bioenergy, 14*(2), 234–248. https://doi.org/10.1166/jbmb.2020.1951

Ilyas, R. A., Sapuan, S. M., & Ishak, M. R. (2018). Isolation and characterization of nanocrystalline cellulose from sugar palm fibres (*Arenga pinnata*). *Carbohydrate Polymers, 181*, 1038–1051. https://doi.org/10.1016/j.carbpol.2017.11.045

Ilyas, R. A., Sapuan, S. M., Ishak, M. R., & Zainudin, E. S. (2017). Effect of delignification on the physical, thermal, chemical, and structural properties of sugar palm fibre. *BioResources, 12*(4), 8734–8754. https://doi.org/10.15376/biores.12.4.8734-8754

Ilyas, R. A., Sapuan, S. M., Ishak, M. R., & Zainudin, E. S. (2018b). Water transport properties of bio-nanocomposites reinforced by sugar palm (*Arenga pinnata*) nanofibrillated cellulose. *Journal of Advanced Research in Fluid Mechanics and Thermal Sciences Journal, 51*(2), 234–246.

Ilyas, R. A., Sapuan, S. M., Ishak, M. R., & Zainudin, E. S. (2018c). Sugar palm nanocrystalline cellulose reinforced sugar palm starch composite: Degradation and water-barrier properties. *IOP Conference Series: Materials Science and Engineering, 368*, 012006. https://doi.org/10.1088/1757-899X/368/1/012006

Ilyas, R. A., Sapuan, S. M., Ishak, M. R., & Zainudin, E. S. (2018d). Development and characterization of sugar palm nanocrystalline cellulose reinforced sugar palm starch bionanocomposites. *Carbohydrate Polymers, 202*, 186–202. https://doi.org/10.1016/j.carbpol.2018.09.002

Ilyas, R. A., Sapuan, S. M., Ishak, M. R., & Zainudin, E. S. (2019). Sugar palm nanofibrillated cellulose (*Arenga pinnata* (Wurmb.) Merr): Effect of cycles on their yield, physic-chemical, morphological and thermal behavior. *International Journal of Biological Macromolecules, 123*. https://doi.org/10.1016/j.ijbiomac.2018.11.124

Ilyas, R. A., Sapuan, S. M., Ishak, M. R., Zainudin, E. S., & Atikah, M. S. N. (2018). Characterization of sugar palm nanocellulose and its potential for reinforcement with a starch-based composite. In *Sugar palm biofibers, biopolymers, and biocomposites* (pp. 189–220). CRC Press. https://doi.org/10.1201/9780429443923-10.

Ilyas, R. A., Sapuan, S. M., Sanyang, M. L., Ishak, M. R., & Zainudin, E. S. (2018). Nanocrystalline cellulose as reinforcement for polymeric matrix nanocomposites and its potential applications: A review. *Current Analytical Chemistry, 14*(3), 203–225. https://doi.org/10.2174/1573411013666171003155624

Islam, M. M. (2013). Biochemistry, medicinal and food values of jute (*Corchorus capsularis* L. and *C. olitorius* L.) leaf: A review. *International Journal of Enhanced Research in Science Technology & Engineering, 2*(11), 35–44

Islam, M. (2019). Food and medicinal values of Roselle (*Hibiscus sabdariffa* L. Linne Malvaceae) plant Parts: A review. *Open Journal of Nutrition and Food Sciences Review, 1*(1003), 14–20.

Islam, F., Islam, N., Shahida, S., Karmaker, N., Koly, F. A., Mahmud, J., Keya, K. N., & Khan, R. A. (2019). Mechanical and interfacial characterization of jute fabrics reinforced unsaturated polyester resin composites. *Nano Hybrids and Composites, 25,* 22–31. https://doi.org/10.4028/www.scientific.net/NHC.25.22

Ismail, A., Hainida, E., Ikram, K., Saadiah, H., & Nazri, M. (2008). Roselle (*Hibiscus sabdariffa* L.) seeds – nutritional composition, protein quality and health benefits. *Food, 2*(1), 1–16.

Jumaidin, R., Ilyas, R. A., Saiful, M., Hussin, F., & Mastura, M. T. (2019). Water transport and physical properties of sugarcane bagasse fibre reinforced thermoplastic potato starch biocomposite. *Journal of Advanced Research in Fluid Mechanics and Thermal Sciences, 61*(2), 273–281.

Jumaidin, R., Khiruddin, M. A. A., Asyul Sutan Saidi, Z., Salit, M. S., & Ilyas, R. A. (2020). Effect of cogon grass fibre on the thermal, mechanical and biodegradation properties of thermoplastic cassava starch biocomposite. *International Journal of Biological Macromolecules, 146,* 746–755. https://doi.org/10.1016/j.ijbiomac.2019.11.011

Jumaidin, R., Saidi, Z. A. S., Ilyas, R. A., Ahmad, M. N., Wahid, M. K., Yaakob, M. Y., Maidin, N. A., Rahman, M. H. A., & Osman, M. H. (2019). Characteristics of cogon grass fibre reinforced thermoplastic cassava starch biocomposite: Water absorption and physical properties. *Journal of Advanced Research in Fluid Mechanics and Thermal Sciences, 62*(1), 43–52.

Junkasem, J., Menges, J., & Supaphol, P. (2006). Mechanical properties of injection-molded isotactic polypropylene/Roselle fiber composites. *Journal of Applied Polymer Science, 101*(5), 3291–3300. https://doi.org/10.1002/app.23829

Khalid, H., Abdalla, W. E., Abdelgadir, H., Opatz, T., & Efferth, T. (2012). Gems from traditional north-African medicine: Medicinal and aromatic plants from Sudan. In *Natural products and bioprospecting*. https://doi.org/10.1007/s13659-012-0015-2

Kian, L. K., Jawaid, M., Ariffin, H., & Karim, Z. (2018). Isolation and characterization of nanocrystalline cellulose from Roselle-derived microcrystalline cellulose. *International Journal of Biological Macromolecules, 114,* 54–63. https://doi.org/10.1016/j.ijbiomac.2018.03.065

Kirdpon, S., Nakorn, S. N., & Kirdpon, W. (1994). Changes in urinary chemical composition in healthy volunteers after consuming roselle (*Hibiscus sabdariffa* Linn.) juice. *Journal of the Medical Association of Thailand = Chotmaihet Thangphaet*.

Kloos, W. E., & Schleifer, K. H. (1983). *Staphylococcus auricularis* sp. nov.: An inhabitant of the human external ear. *International Journal of Systematic Bacteriology, 33*(1), 9–14. https://doi.org/10.1099/00207713-33-1-9

Lin, H. H., Chen, J. H., Kuo, W. H., & Wang, C. J. (2007). Chemopreventive properties of *Hibiscus sabdariffa* L. on human gastric carcinoma cells through apoptosis induction and JNK/p38 MAPK signaling activation. *Chemico-Biological Interactions, 165*(1), 59–75. https://doi.org/10.1016/j.cbi.2006.10.011

Lin, H. H., Huang, H. P., Huang, C. C., Chen, J. H., & Wang, C. J. (2005). Hibiscus polyphenol-rich extract induces apoptosis in human gastric carcinoma cells via p53 phosphorylation and p38 MAPK/FasL cascade pathway. *Molecular Carcinogenesis, 43*(2), 86–99. https://doi.org/10.1002/mc.20103

Lin, J. K., Liang, Y. C., & Lin-Shiau, S. Y. (1999). Cancer chemoprevention by tea polyphenols through mitotic signal transduction blockade. *Biochemical Pharmacology, 58*(6), 911–915. https://doi.org/10.1016/S0006-2952(99)00112-4

Luvonga, W. A., Njoroge, M., Makokha, A., & Ngunjiri, P. (2010). Chemical characterisation of *Hibiscus sabdariffa* (Roselle) calyces and evaluation of its functional potential in the food industry. In *Proceedings of 2010 JKUAT scientific technological and industrialization conference* (pp. 631–638).

Maisara, A. M. N., Ilyas, R. A., Sapuan, S. M., Huzaifah, M. R. M., Nurazzi, N. M., & Saifulazry, S. O. A. (2019). Effect of fibre length and sea water treatment on mechanical properties of sugar palm fibre reinforced unsaturated polyester composites. *International Journal of Recent Technology and Engineering, 8*(2S4), 510–514. https://doi.org/10.35940/ijrte.b1100.0782s419

Mazani, N., Sapuan, S. M., Sanyang, M. L., Atiqah, A., & Ilyas, R. A. (2019). Design and fabrication of a shoe shelf from kenaf fiber reinforced unsaturated polyester composites. In *Lignocellulose for future bioeconomy* (pp. 315–332). Elsevier. https://doi.org/10.1016/B978-0-12-816354-2.00017-7 (Issue 2000).

McKay, D., & Blumberg, J. (2007). Hibiscus tea (*Hibiscus sabdariffa* L.) lowers blood pressure in pre-and mildly hypertensive adults. *The FASEB Journal,* 1–6. https://doi.org/10.3945/jn.109.115097.lowers

Mckay, D. L., Chen, C. Y., Saltzman, E., & Blumberg, J. B. (2010). Hibiscus sabdariffa L. tea (tisane) lowers blood pressure in prehypertensive and mildly hypertensive adults. *Journal of Nutrition, 140*(2), 298–303.

Mei, Y., Wei, D., & Liu, J. (2005). Modulation effect of tea polyphenol toward N-methyl-N′-nitro-N- nitrosoguanidine-induced precancerous gastric lesion in rats. *Journal of Nutritional Biochemistry, 16*(3), 172–177. https://doi.org/10.1016/j.jnutbio.2004.12.002

Mercedes, M. C., Javier, H. M., Gabriel, L. R. E., Yol, A.,S. M., S. R, L., & Javier, C. R. (2013). Influence of variety and extraction solvent on antibacterial activity of roselle (*Hibiscus sabdariffa* L.) calyxes. *Journal of Medicinal Plants Research, 7*(31), 2319–2322. https://doi.org/10.5897/JMPR12.1242

Mohamad Haafiz, M. K., Eichhorn, S. J., Hassan, A., & Jawaid, M. (2013). Isolation and characterization of microcrystalline cellulose from oil palm biomass residue. *Carbohydrate Polymers, 93*(2), 628–634. https://doi.org/10.1016/j.carbpol.2013.01.035

Mohamed Zakriya, G., & Ramakrishnan, G. (2020). Natural fibre composites. In *Natural fibre composites*. https://doi.org/10.1201/9780429326738

Mohamed, R., Fernández, J., Pineda, M., & Aguilar, M. (2007). Roselle (*Hibiscus sabdariffa*) seed oil is a rich source of γ-tocopherol. *Journal of Food Science*. https://doi.org/10.1111/j.1750-3841.2007.00285.x

Mohammad, O., Nazir, B. M., & Abdul Rahman, M.,H. S. (2002). *Roselle: A new crop in Malaysia. A grand international biotechnology event*.

Morton, J. F. (1987). Roselle. In *Fruits of warm climate*.

Nadlene, R., Sapuan, S. M., Jawaid, M., Ishak, M. R., & Yusriah, L. (2016). A review on Roselle fiber and its composites. *Journal of Natural Fibers, 13*(1), 10–41. https://doi.org/10.1080/15440478.2014.984052

Nadlene, R., Sapuan, S. M., Jawaid, M., Ishak, M. R., & Yusriah, L. (2018). The effects of chemical treatment on the structural and thermal, physical, and mechanical and morphological properties of Roselle fiber-reinforced vinyl ester composites. *Polymer Composites, 39*(1), 274–287. https://doi.org/10.1002/pc.23927

Nazrin, A., Sapuan, S. M., Zuhri, M. Y. M., Ilyas, R. A., Syafiq, R., & Sherwani, S. F. K. (2020). Nanocellulose reinforced thermoplastic starch (TPS), polylactic acid (PLA), and polybutylene succinate (PBS) for food packaging applications. *Frontiers in Chemistry, 8*(213), 1–12. https://doi.org/10.3389/fchem.2020.00213

Nnam, N. M., & Onyeke, N. G. (2003). Chemical composition of two varieties of sorrel (*Hibiscus sabdariffa* L.), calyces and the drinks made from them. *Plant Foods for Human Nutrition, 58*(3), 1–7. https://doi.org/10.1023/B:QUAL.0000040310.80938.53

Norizan, M. N., Abdan, K., Ilyas, R. A., & Biofibers, S. P. (2020). Effect of fiber orientation and fiber loading on the mechanical and thermal properties of sugar palm yarn fiber reinforced unsaturated polyester resin composites. *Polimery, 65*(2), 34–43. https://doi.org/10.14314/polimery.2020.2.5

Nurazzi, N. M., Khalina, A., Sapuan, S. M., & Ilyas, R. A. (2019). Mechanical properties of sugar palm yarn/woven glass fiber reinforced unsaturated polyester composites: Effect of fiber loadings and alkaline treatment. *Polimery, 64*(10), 12–22. https://doi.org/10.14314/polimery.2019.10.3

Nurazzi, N. M., Khalina, A., Sapuan, S. M., Ilyas, R. A., Rafiqah, S. A., & Hanafee, Z. M. (2020). Thermal properties of treated sugar palm yarn/glass fiber reinforced unsaturated polyester hybrid composites. *Journal of Materials Research and Technology, 9*(2), 1606–1618. https://doi.org/10.1016/j.jmrt.2019.11.086

Nyam, K.-L., Leao, S.-Y., Tan, C.-P., & Long, K. (2014). Functional properties of Roselle (*Hibiscus sabdariffa* L.) seed and its application as bakery product. *Journal of Food Science and Technology, 51*(12), 3830–3837. https://doi.org/10.1007/s13197-012-0902-x

Obouayeba, A. P., Djyh, N. B., Diabate, S., Djaman, A. J., N'Guessan, J. D., Kone, M., & Kouakou, T. H. (2014). Phytochemical and antioxidant activity of Roselle (*Hibiscus Sabdariffa* L.) petal extracts. *Research Journal of Pharmaceutical, Biological and Chemical Sciences, 5*(2), 1453–1465.

Odigie, I. P., Ettarh, R. R., & Adigun, S. A. (2003). Chronic administration of aqueous extract of *Hibiscus sabdariffa* attenuates hypertension and reverses cardiac hypertrophy in 2K-1C hypertensive rats. *Journal of Ethnopharmacology, 86*(2–3), 181–185. https://doi.org/10.1016/S0378-8741(03)00078-3

Ojeda, D., Jiménez-Ferrer, E., Zamilpa, A., Herrera-Arellano, A., Tortoriello, J., & Alvarez, L. (2010). Inhibition of angiotensin convertin enzyme (ACE) activity by the anthocyanins delphinidin- and cyanidin-3-O-sambubiosides from *Hibiscus sabdariffa*. *Journal of Ethnopharmacology*. https://doi.org/10.1016/j.jep.2009.09.059

Ojokoh, A. O. (2006). Roselle (*Hibiscus sabdariffa*) calyx diet and histopathological changes in liver of albino rats. *Pakistan Journal of Nutrition, 5*(2), 110–113. https://doi.org/10.3923/pjn.2006.110.113

Omobuwajo, T., Sanni, L., & Balami, Y. (2000). Physical properties of sorrel (*Hibiscus sabdariffa*) seeds. *Journal of Food Engineering, 45*(1), 37–41. https://doi.org/10.1016/S0260-8774(00)00039-X

Onyenekwe, P. C., Ajani, E. O., Ameh, D. A., & Gamaniel, K. S. (1999). Antihypertensive effect of Roselle (*Hibiscus sabdariffa*) calyx infusion in spontaneously hypertensive rats and a comparison of its toxicity with that in Wistar rats. *Cell Biochemistry and Function, 17*(3), 199–206. https://doi.org/10.1002/(SICI)1099-0844(199909)17:3<199::AID-CBF829>3.0.CO;2-2

Owoade, A. O., Adetutu, A., & Olorunnisola, O. S. (2019). A review of chemical constituents and pharmacological properties of *Hibiscus sabdariffa* L. *International Journal of Current Research in Biosciences and Plant Biology, 6*(04), 42–51. https://doi.org/10.20546/ijcrbp.2019.604.006

Puro, K., Sunjukta, R., Samir, S., Ghatak, S., Shakuntala, I., & Sen, A. (2014). Medicinal uses of Roselle plant (*Hibiscus sabdariffa* L.): A mini review. *Indian Journal of Hill Farming, 27*(1), 81–90.

Radzi, A. M., Sapuan, S. M., Jawaid, M., & Mansor, M. R. (2017). Influence of fibre contents on mechanical and thermal properties of Roselle fibre reinforced polyurethane composites. *Fibers and Polymers, 18*(7), 1353–1358. https://doi.org/10.1007/s12221-017-7311-8

Radzi, A. M., Sapuan, S. M., Jawaid, M., & Mansor, M. R. (2018a). Mechanical and thermal performances of Roselle fiber-reinforced thermoplastic polyurethane composites. *Polymer - Plastics Technology and Engineering, 57*(7), 601–608. https://doi.org/10.1080/03602559.2017.1332206

Radzi, A. M., Sapuan, S. M., Jawaid, M., & Mansor, M. R. (2018b). Mechanical performance of Roselle/sugar palm fiber hybrid reinforced polyurethane composites. *Bio-Resources, 13*(3), 6238–6249.

Radzi, A. M., Sapuan, S. M., Jawaid, M., & Mansor, M. R. (2019a). Effect of alkaline treatment on mechanical, physical and thermal properties of Roselle/sugar palm fiber reinforced thermoplastic polyurethane hybrid composites. *Fibers and Polymers, 20*(4), 847–855. https://doi.org/10.1007/s12221-019-1061-8

Radzi, A. M., Sapuan, S. M., Jawaid, M., & Mansor, M. R. (2019b). Water absorption, thickness swelling and thermal properties of Roselle/sugar palm fibre reinforced thermoplastic polyurethane hybrid composites. *Journal of Materials Research and Technology*. https://doi.org/10.1016/j.jmrt.2019.07.007

Razali, N., Salit, M. S., Jawaid, M., Ishak, M. R., & Lazim, Y. (2015). A study on chemical composition, physical, tensile, morphological, and thermal properties of Roselle fibre: Effect of fibre maturity. *BioResources, 10*(1), 1803–1823. https://doi.org/10.15376/biores.10.1.1803-1824

Razali, N., Sapuan, S. M., Jawaid, M., Ishak, M. R., & Lazim, Y. (2016). Mechanical and thermal properties of roselle fibre reinforced vinyl ester composites. *BioResources, 11*(4), 9325–9339. https://doi.org/10.15376/biores.11.4.9325-9339

Razali, N., Sapuan, S. M., & Razali, N. (2018). Mechanical properties and morphological analysis of Roselle/sugar palm fiber reinforced vinyl ester hybrid composites. In *Natural fibre reinforced vinyl ester and vinyl polymer composites* (pp. 169–180). Elsevier. https://doi.org/10.1016/B978-0-08-102160-6.00008-1.

Ritonga, N. (2017). Roselle flower (*Hibiscus sabdariffa*). *Belitung Nursing Journal, 3*(3), 229–237.

Rolfs, P. H. (1929). In P. H. Rolfs (Ed.), *Subtropical vegetable-gardening*. The Macmillan Company.

Sanyang, M. L., Ilyas, R. A., Sapuan, S. M., & Jumaidin, R. (2018). Sugar palm starch-based composites for packaging applications. In *Bionanocomposites for packaging applications* (pp. 125–147). Springer International Publishing. https://doi.org/10.1007/978-3-319-67319-6_7.

Sapuan, S. M., Aulia, H. S., Ilyas, R. A., Atiqah, A., Dele-Afolabi, T. T., Nurazzi, M. N., Supian, A. B. M., & Atikah, M. S. N. (2020). Mechanical properties of longitudinal basalt/woven-glass-fiber-reinforced unsaturated polyester-resin hybrid composites. *Polymers, 12*(10). https://doi.org/10.3390/polym12102211

Sari, N. H., Pruncu, C. I., Sapuan, S. M., Ilyas, R. A., Catur, A. D., Suteja, S., Sutaryono, Y. A., & Pullen, G. (2020). The effect of water immersion and fibre content on properties of corn husk fibres reinforced thermoset polyester composite. *Polymer Testing, 91*, 106751. https://doi.org/10.1016/j.polymertesting.2020.106751

Sherman, K. J. (2005). Complementary and alternative medicine in the United States. *Annals of Internal Medicine, 143*(9). https://doi.org/10.7326/0003-4819-143-9-200511010-00026

Singha, A. S., & Thakur, V. K. (2008). Fabrication of *Hibiscus sabdariffa* fibre reinforced polymer composites. *Iranian Polymer Journal (English Edition), 17*(7), 541–553.

Singha, A. S., & Thakur, V. K. (2009). Physical, chemical and mechanical properties of *Hibiscus sabdariffa* fiber/polymer composite. *International Journal of Polymeric Materials and Polymeric Biomaterials, 58*(4), 217–228. https://doi.org/10.1080/00914030802639999

Singha, A. S., & Thakura, V. K. (2008). Fabrication and study of lignocellulosic *Hibiscus sabdariffa* fiber reinforced polymer composites. *BioResources, 3*(4), 1173–1186. https://doi.org/10.15376/biores.3.4.1173-1186

Sukkhavanit, P., Angkanaporn, K., & Kijparkorn, S. (2011). Effect of Roselle (*Hibiscus sabdariffa* Linn.) calyx in laying hen diet on egg production performance, egg quality and TBARS value in plasma and yolk. *Thai Journal of Veterinary Medicine*.

Syafri, E., Kasim, A., Abral, H., & Asben, A. (2019). Cellulose nanofibers isolation and characterization from ramie using a chemical-ultrasonic treatment. *Journal of Natural Fibers, 16*(8), 1145–1155. https://doi.org/10.1080/15440478.2018.1455073

Syafri, E., Sudirman, M., Yulianti, E., Deswita, Asrofi, M., Abral, H., Sapuan, S. M., Ilyas, R. A., & Fudholi, A. (2019). Effect of sonication time on the thermal stability, moisture absorption, and biodegradation of water hyacinth (*Eichhornia crassipes*) nanocellulose-filled bengkuang (*Pachyrhizus erosus*) starch biocomposites. *Journal of Materials Research and Technology, 8*(6), 6223–6231. https://doi.org/10.1016/j.jmrt.2019.10.016

Tolulope, M. (2007). Cytotoxicity and antibacterial activity of methanolic extract of *Hibiscus sabdariffa*. *Journal of Medicinal Plants Research, 1*(1), 009–013.

Tseng, T. H., Kao, E. S., Chu, C. Y., Chou, F. P., Lin Wu, H. W., & Wang, C. J. (1997). Protective effects of dried flower extracts of *Hibiscus sabdariffa* L. against oxidative stress in rat primary hepatocytes. *Food and Chemical Toxicology, 35*(12), 1159–1164. https://doi.org/10.1016/S0278-6915(97)85468-3

Tseng, Tsui, H., Kao, T. W., Chu, C. Y., Chou, F. P., Lin, W. L., & Wang, C. J. (2000). Induction of apoptosis by *Hibiscus* protocatechuic acid in human leukemia cells via reduction of retinoblastoma (RB) phosphorylation and Bcl-2 expression. *Biochemical Pharmacology, 60*(3), 307–315. https://doi.org/10.1016/S0006-2952(00)00322-1

Udayasekhara Rao, P. (1996). Nutrient composition and biological evaluation of mesta (*Hibiscus sabdariffa*) seeds. *Plant Foods for Human Nutrition, 49*(1), 27–34. https://doi.org/10.1007/BF01092519

Umesha, S., Marahel, S., & Aberomand, M. (2013). Antioxidant and antidiabetic activities of medicinal plants: A short review. *International Journal of Research in Pharmacology & Pharmacotherapeutics, 3*(1), 40–53.

Viens, A. M., & Littmann, J. (2015). Is antimicrobial resistance a slowly emerging disaster? *Public Health Ethics, 8*(3), 255–265. https://doi.org/10.1093/phe/phv015

Voon, H. C., Bhat, R., & Rusul, G. (2012). Flower extracts and their essential oils as potential antimicrobial agents for food uses and pharmaceutical applications. *Comprehensive Reviews in Food Science and Food Safety, 11*(1), 34–55. https://doi.org/10.1111/j.1541-4337.2011.00169.x

Wang, S., DeGroff, V. L., & Clinton, S. K. (2003). Tomato and soy polyphenols reduce insulin-like growth factor-I-stimulated rat prostate cancer cell proliferation and apoptotic resistance in vitro via inhibition of intracellular signaling pathways involving tyrosine kinase. *Journal of Nutrition, 133*(7), 2367–2376. https://doi.org/10.1093/jn/133.7.2367

Wang, C. J., Wang, J. M., Lin, W. L., Chu, C. Y., Chou, F. P., & Tseng, T. H. (2000). Protective effect of *Hibiscus* anthocyanins against tert-butyl hydroperoxide-induced hepatic toxicity in rats. *Food and Chemical Toxicology, 38*(5), 411–416. https://doi.org/10.1016/S0278-6915(00)00011-9

Weisburger, J. H., & Chung, F. L. (2002). Mechanisms of chronic disease causation by nutritional factors and tobacco products and their prevention by tea polyphenols. *Food and Chemical Toxicology, 40*(8), 1145–1154. https://doi.org/10.1016/S0278-6915(02)00044-3

WHO. (2001). *Legal status of traditional medicine and complementary alternative medicine: A worldwide review*. Geneva: World Health Organization.

Wright, J. T., Bakris, G., Greene, T., Agodoa, L. Y., Appel, L. J., Charleston, J., Cheek, D. A., Douglas-Baltimore, J. G., Gassman, J., Glassock, R., Hebert, L., Jamerson, K., Lewis, J., Phillips, R. A., Toto, R. D., Middleton, J. P., & Rostand, S. G. (2002). Effect of blood pressure lowering and antihypertensive drug class on progression of hypertensive kidney disease: Results from the AASK trial. *Journal of the American Medical Association, 288*(19), 2421–2431. https://doi.org/10.1001/jama.288.19.2421

Yin, G., Cao, L., Xu, P., Jeney, G., & Nakao, M. (2011). Hepatoprotective and antioxidant effects of *Hibiscus sabdariffa* extract against carbon tetrachloride-induced hepatocyte damage in *Cyprinus carpio*. *In Vitro Cellular and Developmental Biology Animal, 47*(1), 10–15. https://doi.org/10.1007/s11626-010-9359-2

Yorseng, K., Rangappa, S. M., Pulikkalparambil, H., Siengchin, S., & Parameswaranpillai, J. (2020). Accelerated weathering studies of kenaf/sisal fiber fabric reinforced fully biobased hybrid bioepoxy composites for semi-structural applications: Morphology, thermo-mechanical, water absorption behavior and surface hydrophobicity. *Construction and Building Materials, 235*, 117464. https://doi.org/10.1016/j.conbuildmat.2019.117464

CHAPTER 2

Vegetable Mesta (*Hibiscus sabdariffa* L. var *sabdariffa*): A Potential Industrial Crop for Southeast Asia

A.K.M. AMINUL ISLAM • MOHAMAD BIN OSMAN • MOHSIN BIN MOHAMAD • A.K.M. MOMINUL ISLAM

1 INTRODUCTION

1.1 Background

Roselle (*Hibiscus sabdariffa* L.) belongs to the genus *Hibiscus* under the tribe Hibisceae of the family Malvaceae (Borssum-Waalkes, 1966). The genus *Hibiscus* contains more than 300 species that are grown throughout the tropical and subtropical regions of the world (Anderson & Pharis, 2003). Several species of *Hibiscus* are economically important, as they provide food, fiber, and medicine. Other species are valuable, as they have esthetic value (Wilson & Menzel, 1964). Indian sorrel or roselle (*H. sabdariffa* L.) is cultivated for use as a vegetable and/or to make drinks and jams. On the other hand, kenaf (*Hibiscus cannabinus*) is cultivated and extensively used for making fiber. Over the past few decades, two *Hibiscus* species (namely, roselle and kenaf) have been researched to evaluate them as new crops in different countries (Omalsaad et al., 2012). Indian sorrel is an annual or perennial plant. It is cultivated for many useful purposes such as edible calyx, leaves, and seeds or fibers and paper pulp (Adamson & O'Bryan, 1981; Wilson & Menzel, 1964).

1.2 Origin and Distribution

The plant is thought to be native to the tropical region of Africa (Andrews, 1952; Morton, 1987). According to other records, roselle has been cultivated in the Indian subcontinent and is believed to be native to Asia (Anonymous, 1959; Brown, 1954; Drury, 1873, pp. 244–245), but it probably originated in Africa. Wild forms of *H. sabdariffa* are available in Uganda and they very much resemble *Hibiscus machowii*. These wild forms are believed to be the wild progenitor of *H. sabdariffa*. It is not native to some areas such as Central America and West Indies, but it has become adapted to these environments (Britton & Wilson, 1924; Harris, 1913, p. 20; Morton, 1987; Robyns, 1965; Standley, 1923). Probably it was introduced in Central America (Mexico) in the 17th century by the Spaniards (Torres-Morán et al., 2011). Besides India and Bangladesh, vegetable-type roselle is grown mainly in North Africa (Egypt, Sudan, Senegal, Mali, Jamaica), East Africa (Mozambique, Zimbabwe), South Africa, Southeast Asia (Indonesia, Malaysia, Thailand), West Asia (Iran), South America (Argentina, Brazil, Guatemala, Peru, Cuba), North America (United States), North Asia (Russia, China), Western Europe (Spain, Italy), Caribbean (Haiti, West Indies) Australia, and New Guinea (Leung & Foster, 1996). It was brought to Malaysia from India, where roselle is usually cultivated (Anonymous, 1959; Brown, 1954; Drury, 1873, pp. 244–245). Around 100 years ago, roselle seeds may have been carried from Jamaica to the West Indies and Central America probably by African slaves. It was perhaps transported to Australia by colonial rulers India and Malaysia. The United States Department of Agriculture (USDA) introduced roselle to the Philippines where it has popularized as roselle drinks. Roselle cultivation is limited to latitudes between 25°N and 25°S at elevations up to 700 m in the tropical regions of the world (Fig. 2.1).

1.3 Vernacular Names of Roselle

Roselle (*H. sabdariffa*) has different vernacular names in the world. It is popularly known as roselle, Jamaica sorrel, Indian sorrel, Guinea sorrel, red sorrel, sorrel, sour-sour, jelly plant, lemon bush, or cranberry in English-speaking countries. Roselle is commonly known as Chukair or mesta (Ghani, 1998) in Bangladesh with several vernacular names such as lalmesta, patwa,

Roselle. https://doi.org/10.1016/B978-0-323-85213-5.00016-0

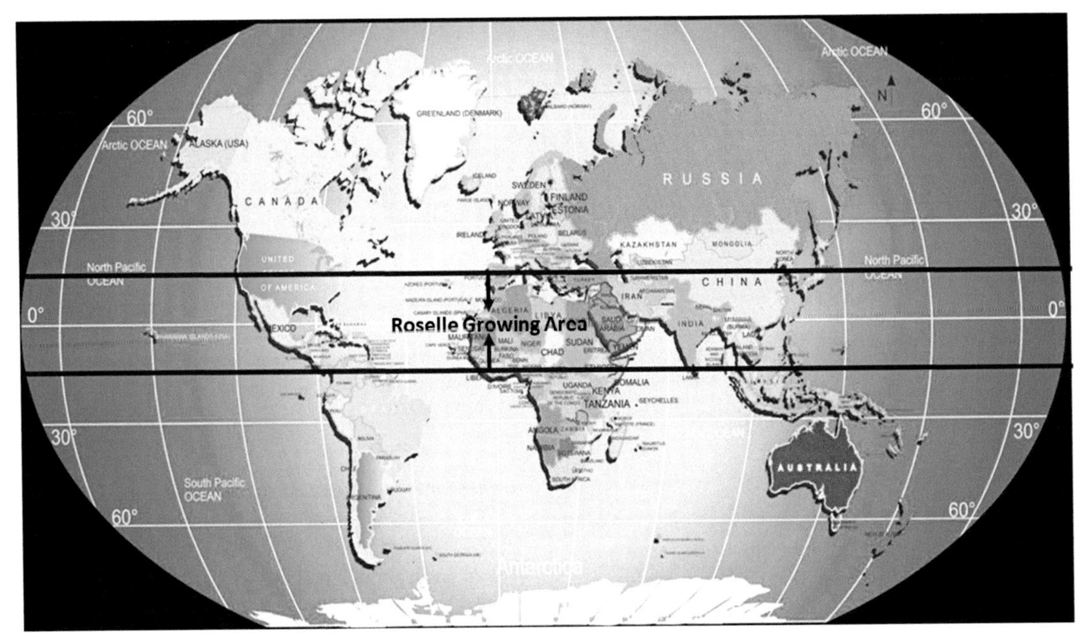

FIG. 2.1 An overview of roselle planting belts.

kharapata, etc. Roselle is called karkade or carcade in Arabic-speaking countries of north and east Africa. It is known by these names in Europe in the pharmaceutical and food industries (Morton, 1987). The common name of roselle is bissap in Senegal, Guinea Bissau, Mali, Burkina Faso, Ghana, Benin, Niger, Congo, and France; gamet walanda or rosella in Indonesia; Tagalog or Subanon in Philippines; slok chuu in Cambodia; and krachiap-daeng in Thailand. Locally it is known as 'asam susur', 'asam paya,' and 'asam kumbang' in Malaysia (Mohamad et al., 2006). In Australia, it is known as *rosella*. It is known as tengamora in Assam, gongura in Telugu, mathipuli in Kerala, chin baung in Burma, zobo in western Nigeria, chaye-torosh in Iran, flor de Jamaica in Mexico, saril in Panama, luoshen hua in China, cururu azedo in Portuguese, rosa de Jamaica in Spain, wonjo in Gambia, and zuring in Dutch (Chau et al., 2000).

1.4 Roselle Around the Globe: Malaysia and Bangladesh

Probably roselle (*H. sabdariffa*) originated in Africa and was later on domesticated in Sudan first for its seed and then for leaf and calyx production about 6000 years ago. It is an important leafy vegetable in the drier parts of West and Central Africa. It is common in the savanna region of tropical Africa, especially West and Central Africa. Large quantities of calyces of roselle are produced in Jamaica, Sudan, Senegal, Mali, and Chad for the preparation of beverages. Traditionally, roselle is consumed as a popular drink 'karkadeh' during Ramadan and is an economic crop in arid regions of Sudan. During Christmas, a refreshing beverage flavored with rum and ginger is prepared from the calyces of fresh roselle. In recent times, the use of roselle has been extended for the production of squash and chutney in Jamaica.

Roselle has been introduced in the 1990s and is a relatively new crop in Malaysia (Mohamad, Nazir, Abdul Rahman, & Herman, 2002a, 2002b). It is presently an important cash crop grown in the East Coast (Terengganu and Kelantan) of Malaysia and has spread to other states (Mohamad et al., 2006). Two varieties "Terengganu" and "Arab" are available in Malaysia for cultivation since 2008. Universiti Kebangsaan Malaysia (UKM) has released three new varieties named UKMR-1, UKMR-2, and UKMR-3 in April 2009. Mutation breeding was used to develop these three varieties using the 'Arab' variety as the parent.

The Central Research Institute for Jute and Allied Fibres (CRIJAF) has yielded a number of varieties of mesta in India, which are popular among the farmers for various situations and purposes. A mutant variety of mesta with a long and fleshy calyx was developed by CRIJAF and cultivated for the production of jam, jelly, and sauce.

In Bangladesh, mesta (*H. sabdariffa*) is one of the important fiber crops next to jute and kenaf. Mesta cultivation is widely scattered in homestead level throughout Bangladesh. It is traditionally cultivated for its leaves, fruits, seeds, and stem, but now it is being grown commercially for its fiber. The Bangladesh Jute Research Institute (BJRI) has been conducting researches on the agricultural aspects of roselle.

2 BOTANY OF ROSELLE

Roselle is grown worldwide as a common flowering plant consisting of ornamental plants, vegetables, medicinal plants, and forest trees. Therefore there is great diversity in the shape and size of the plants, ranging from an herb to a tall plant up to 30 m in height. One of these species is *H. sabdariffa* and its relative species in the genus *Hibiscus*, including kenaf, is grown widely in many countries around the world (Dempsey, 1975; Wilson & Menzel, 1964).

2.1 Taxonomic Classification

Roselle is probably a tetraploid species and a member of the family Malvaceae (Table 2.1) (Halimaton, 2005; Wilson, 1999). It is widely cultivated in humid regions of tropical and subtropical climate.

2.2 Botanic Aspects

Roselle is an attractive annual or perennial herb (Purseglove, 1974). It is a dicotyledonous herbaceous shrub, which is an indigenous edible medicinal plant (Musthafa et al., 2005). The morphology and cytology of some of the species have been much investigated, especially those from the section Furcaria (Skovsted 1935, 1941) (Fig. 2.2).

TABLE 2.1
Taxonomy of Roselle (*Hibiscus sabdariffa* L.).

Kingdom	Plantae
Subkingdom	Tracheobionta
Kingdom	Magnoliophyta
Division	Spermatophyta
Subdivision	Angiospermae
Class	Dicotyledonae
Subclass	Dilleniidae
Order	Malvales
Family	Malvaceae
Genus	Hibiscus
Species	*Hibiscus sabdariffa* L.

Heywood, V. H. (1978). *Flowering plants of the world* (pp. 94–95). Oxford University Press; Olubukola, A., & Illoh, H. C. (1996). Pollen grain morphology of some species of hibiscus. *Nigerian Journal of Botany*, 5, 5–14; USDA. (2000). *Plant database*.

2.2.1 Stem

The roselle plant grows up to 2–3 m in height, and it is an annual herb. The stem of roselle is erect and habitually branched. Roselle has glabrous, more or less even, and cylindric stem that is red or deep green (Morton, 1987) (Fig. 2.3).

2.2.2 Root

Roselle has a deep penetrating taproot (Duke, 1983). Plants produced by stem cuttings bear a fibrous root system instead of taproot (Fig. 2.4).

2.2.3 Leaf

The leaves are simple, ovate to lanceolate, arranged alternatively, often with serrated margins, and divided into three to seven lobes (Fig. 2.5). Leaves are dark green, some with red pigments. The shape of young leaves is oval but changes into finger shape when mature. Lower leaves are ovate and upper leaves are palmately lobed (three to five lobes). Leaves are also glabrous and palmate with long petioles (Mohamad et al., 2005).

2.2.4 Flower

Roselle produces hermaphrodite flowers with five petals and a short peduncle. Calyx diameter is 2.5–3.5 cm and length is 4.2–7.2 cm. The flowers are light yellow to pink, red, orange, and purple red with a reddish center at the base of the staminal column (Fig. 2.6). Roselle flowers grow in leaf axil or in the terminal raceme. The flowers are trumpet shaped, large ranging from 4 to 18 cm in diameter, and conspicuous. Joined sepals produce fleshy calyx, which is known as the roselle fruit. The calyx is varying in color, size, and shape. Calyx group is of red, dark red, green, and white types (Schippers 2000, pp. 1–214). Self-pollination occurs in roselle flower before opening. Owing to its cleistogamous nature, conventional hybridization is quite difficult in roselle (Mohamad et al., 2005).

2.2.5 Fruit

The fruit is a velvety five-lobed capsule, is 1–2 cm long, contains different numbers of seeds, and is enclosed in red, fleshy calyces. Capsules are green to purplish at the young stage and change to a different color, such as, brown when mature (Morton, 1987). Dry capsules contain several seeds in each lobe that are released when the capsule splits open at maturity. Fruit is fleshy and bright red (Fig. 2.7)

FIG. 2.2 Phenotypic differences in plant architecture for different types of roselle plants.

2.2.6 Seed

The seeds of roselle are kidney shaped and light brown in color with dark brown veins. Seeds are 3–5 mm long and covered with thick radial hairs (Morton, 1987) (Fig. 2.8). The weight per 1000 seeds is 25–29 g.

2.3 Life Cycle of Roselle

Roselle requires 5–6 months from planting to harvesting. Flowering in this herb occurs 2–3 months from planting (Fig. 2.9).

FIG. 2.3 Stem of roselle.

2.4 Reproductive Biology

Roselle (*H. sabdariffa*) is a self-pollinated hermaphrodite shrub. Roselle plants respond to photoperiodism. They do not flower at a short day length of 13.5 h but flower at a day length of 11 h. Very little amount (0.02%) of outcrossing is found in roselle. This outcrossing rate is much lower in comparison to estimates of natural cross-pollination (0.20% and 0.68%) (Mohamad et al., 2002a, 2002b).

2.5 Physicochemical Characteristics and Food Value of Roselle

Roselle is found to contain high amounts of ascorbic acid (vitamin C) and is characterized as a highly acidic fruit with low sugar content. It is also rich in riboflavin (B_2), niacin, calcium, and iron (Qi et al., 2005). Roselle calyx is potentially a good source of anthocyanin and antioxidants including flavonoids, gossypetine, and hibiscetine (Hong & Wrostlad, 1990; Vilasinee et al., 2005; Chau et al., 2000). Anthocyanin is the major source of antioxidants in roselle (Tsai et al., 2002). Hydroxycitric acid, anthocyanin, and ascorbic acid extracted from calyx have been believed to be valuable among other chemical components of roselle parts. Anthocyanins are a group of colorful pigments that are responsible for the attractive pigments found in any fruits, flowers, leaves, and roots (Jennifer, 2009). Anthocyanins have high thermostability (Odake et al., 1992) and are responsible for antioxidant activity (Furuta et al., 1998). These pigments that produce color are widely used as food colors (Tsai & Ou, 1996). Roselle leaf extract contains fiber, ash, calcium, phosphorus, iron, carbohydrate, protein, fat, thiamin, β-carotene, riboflavin (B_2), and niacin (Ali-Bradeldin et al., 2005; Morton, 1987; Watt & Breyer-Brandwijk, 1962). Reducing sugars, glycosides, acids, alkaloid hibiscic acid, flavonoid, and resins have been isolated from flower extract (Asolkar et al., 1992). The seeds of roselle are rich in oil, proteins, dietary fiber, carbohydrates, and

FIG. 2.4 Root system of roselle.

FIG. 2.5 Variation in the leaf shape of different types of roselle: **(A)** single lobe, **(B)** two lobes, **(C)** three lobes, **(D)** four lobes, and **(E)** five lobes.

FIG. 2.6 Variation in the flower color of roselle.

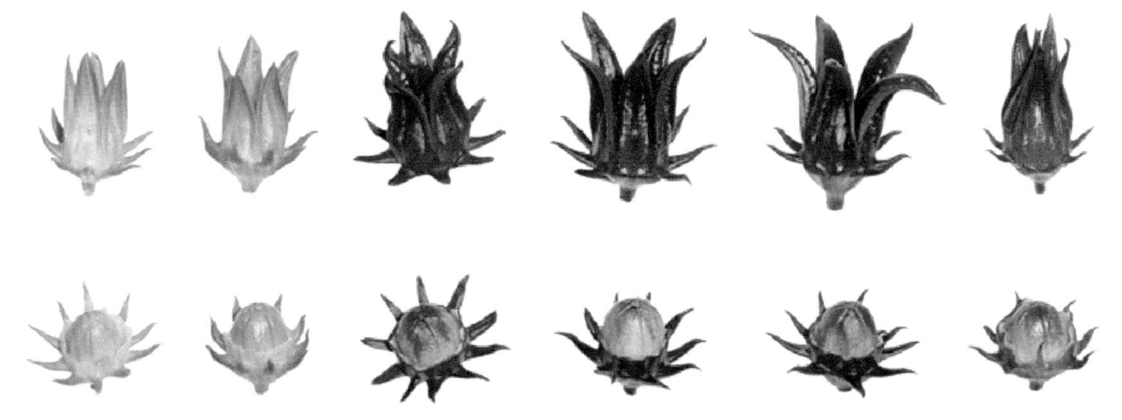

FIG. 2.7 Variation in the fruit color and shape of roselle.

FIG. 2.8 Structure, color, and shape of roselle seeds.

FIG. 2.9 Life cycle of roselle plant.

fats (Abu-Tarboush et al., 1997). It is also a good source of minerals such as Mg, Ca, lysine, and unsaturated fatty acids (Samy, 1980). Food value per 100 g of edible roselle is presented in Table 2.2.

3 GENETICS AND BREEDING OF ROSELLE

Genetic variability is the key factor for any crop improvement through plant breeding. Roselle (*H. sabdariffa*) is a tetraploid ($2n = 4x = 72$) and its chromosome number is allied with diploid ($2n = 2x = 36$) kenaf (*H. cannabinus*) (Wilson & Menzel, 1964). *H. sabdariffa* var *sabdariffa* and *H. sabdariffa* var *altissima* are the two botanic types of roselle that are cultivated for their fleshy calyx and phloem fiber (Purseglove, 1974).

3.1 Genetic Resources

Roselle cultivars are common throughout the tropical and subtropical countries. There are three different types of roselle found in Senegal: are 'bissap', green, and red types. Fifty accessions of roselle have been preserved in the Plant Genetic Resources Unit of the Agricultural Research Corporation, Wad Medani, Sudan, from the local collections. Characterization of these genotypes showed significant variation in leaf shape, stem color, fruit shape, calyx color, and size. The 'bissap' type is characterized by its unique color, size, and shape of leaves and fruits. The germplasm of the 'bissap' type is stored and maintained at the Seed Production Unit, Horticultural Development Centre (CDH), Dakar. Other germplasm collections of *H. sabdariffa* are stored and maintained in Bangladesh, United States, India, and Nigeria (Table 2.3).

Roselle (*H. sabdariffa* var *altissima*) is cultivated for its fiber in India, the East Indies, and other tropical countries. *H. sabdariffa* var *sabdariffa* is normally grown for its edible calyces or as an ornamental plant. *Hibiscus acetosella* and *Hibiscus surattensis* L. are two wild ancestors of roselle (Fig. 2.10). *H. acetosella* is frequently confound with roselle, an ornamental plant from tropical Africa, and is described as fake roselle. Leaves of new varieties look like the leaves of Japanese maples and are acutely serrated. *H. surattensis* L. is a shrubby annual that is weak stemmed, tenderly velvety, and thinly armed with spine and is a wild relative of roselle.

TABLE 2.2
Food Value of Roselle (*Hibiscus sabdariffa* L.) per 100 g of Calyx.

Constituents	Fresh Calyces	Fresh Leaves	Seeds
Moisture	9.20 g	85.60 g	8.2 g
Protein	1.15 g	3.30 g	19.6 g
Fat	2.61 g	0.30 g	16.0 g
Fiber	12.00 g	10.00%	11.0 g
Energy	44 kcal	43 kcal	411 kcal
Ash	6.90 g	1.00 g	-
Calcium	12.63 mg	213.00 mg	356 mg
Phosphorus	273.20 mg	93.00 mg	462 mg
Iron	8.98 mg	4.80 mg	4.2 mg
Carotene	0.03 mg	4135 μg	–
Thiamine	0.12 mg	0.2 mg	0.1 mg
Riboflavin	0.28 mg	0.45 mg	0.15 mg
Niacin	3.77 mg	1.2 mg	1.4 mg
Ascorbic acid	6.70 mg	54 mg	Trace
Carbohydrates	10.00 g	9.20 g	51.3 g

Morton, J. (1987). Roselle. In: *Fruits of warm climates* (pp. 281–286). Web Publications Purdue University. ISBN: 0-9610184-1-0. www.hort.purdue.edu/newcrop/morton/roselle.html; Morton, J. F., & Dowling, C. F. (1987). *Roselle: Fruits of warm climates* (505 pp). (Internet) Julia F. Morton, Miami, United States. http://www.hort.purdue.edu/newcrop/morton/roselle.html; Leung, W. T. W., Busson, F., & Jardin, C. (1968). *Food composition table for use in Africa* (306 pp). FAO.

TABLE 2.3
Roselle Accessions Stored and Maintained in Different Institutes of the World.

Sl. No.	Institute	Number of Accessions Stored and Maintained
1	Plant Genetic Resources unit of the Agricultural Research Corporation, Wad Medani, Sudan	50
2	Bangladesh Jute Research Institute, Dhaka, Bangladesh	320
3	USDA Southern Regional Plant Introduction Station, Griffin, GA, United States	95
4	Central Research Institute for Jute and Allied Fibres, Barrackpore, India	75
5	National Horticultural Research Institute, Ibadan, Nigeria	11

A large number of new varieties of roselle have been developed throughout the world, including Malaysia, Indonesia, Thailand, Bangladesh, China, India, Philippines, Africa, Sudan, Jamaica, and United States. The leading germplasm collections are stored in Maryland, United States, and Australia. From Indochina-Indonesia to African Centers of Diversity, the best known varieties are reported to be Victor, Rico, and Archer (Crane, 1949; Duke, 1993).

3.2 Breeding of Roselle

Plant hybridization followed by selection is the most important process in any genetic change to develop better genotypes. Roselle breeding through hybridization has received little attention. Cultivation of roselle in Asia, e.g., India, Sri Lanka, Thailand, Malaysia, and Java, has been reported from the beginning of 20th century for fiber production. It is now available all over the tropics. The fiber-producing genotypes have developed from the vegetable variety in the course of a long process and may be interogressed by *Hibiscus asper* and *H. machowii*. The tall fiber-yielding type was described by Webster (1914) in Thailand as *H. sabdariffa* var *altissima* and isolated from African seeds. This variety was brought to India by chance as a single seed in admixture

FIG. 2.10 Wild relatives of roselle: **(A)** *Hibiscus surattensis,* **(B)** *Hibiscus acetosella,* **(C)** *Hibiscus radiates*, and **(D)** *Hibiscus hispidissimus*.

with the shipment of other seeds from Indonesia and was first reported by Khan (1930) as a new type of roselle. Sudan carried out breeding of roselle with the objectives of improvement of new cultivars with better yield and calyx quality from local germplasm. In Senegal, breeding programs have been carried out with the objectives to improve leaf yield of green cultivars and taste and yield of calyx for commercial production.

There is a very limited number of germplasm collections available for breeding in Malaysia because roselle was an introduced species. Systematic researches and studies on the roselle plant have been initiated in Malaysia in 1993 by a group of investigators from the Institute of Higher Education, University of Malaya (Rusmawati, 2004). After a while, the Malaysian Agricultural Research and Development Institute (MARDI) also participated with Universiti Malaya research group. Investigations started while Arab, Terengganu, and MARDI were the only germplasm accession available for breeding. But to improve roselle productivity and industry in Malaysia, genetic variation was essential. A breeding program was initiated at UKM in 1999 to increase genetic variability in the germplasm.

Roselle requires longer time for fixation of segregating populations compared with diploid species in order to achieve homozygosity, as it is a tetraploid. In addition, conventional breeding is very difficult in roselle, as it has a cleistogamous flower. To overcome this problem, induced mutation breeding was used to develop a new variety of roselle. Because of these reasons, the use of induced mutations started to generate new genetic variability. A mutation breeding program was initiated at UKM in collaboration with the Malaysian Nuclear Agency in 1999 to obtain superior genotypes (Mohamad et al., 2002a, 2002b). Induced mutation technique is a branch of biotechnology that has been widely used in plant breeding (Mohamad et al., 2005) to create novel characters. Mutation breeding creates genetic variability and increases the scope of selection in order to develop high-yielding varieties of roselle. Mutation breeding at UKM resulted in three breeding lines, namely, UKMR-1, UKMR-2, and UKMR-3. These promising lines were developed through induced mutation from the Arab variety as the parent. Despite the obtained promising results, in many respects, the amount of researches is still considered insufficient to support the growing roselle industry in Malaysia (Mohamad et al. 2005, 2008). At present, UKM maintains a working germplasm collection and also conducts agronomic research and crop improvement.

3.3 Genetic Identification Based on Molecular Markers

Several DNA-based markers (restriction fragment length polymorphism [RAPD], random-amplified polymorphic DNA [RFLP], amplified fragment length polymorphism [AFLP], and simple sequence repeat [SSR]) are commonly used for ecologic, evolutionary, taxonomic, phylogenic, and genetic studies in plant sciences (Ayad et al., 1995). Genetic variation between two species was identified by using RAPD markers with three selected primers. A total of 62 DNA fragments from different size ranges were amplified by using universal M13 primer and two random chloroplast primers Chl1 and Chl4. The polymorphism generated by three primers ranged from 82% to 90%. Dendrogram generated from data showed two major groups, where the first group (A) contains the accession of kenaf while the second group (B) contains all accessions of roselle and two accessions of kenaf (Omalsaad et al., 2012). RAPD markers have the potential to identify the chloroplast and mitochondrial genome of two species and to characterize the genetic variation within the varieties. In another study, genetic relationships were assessed among 17 accessions (7 samples of kenaf and 10 samples of roselle) by molecular markers inter-retrotransposon amplified polymorphism (IRAP) and retrotransposon-microsatellite amplified polymorphism (REMAP) using four primers. A total of 168 fragments were obtained from the combination of primer IRAP and REMAP. Fragment size ranged from 150 to 2267 bp. The polymorphism generated by three primers varies from 96.65% to 100%. The dendrogram generated three major clusters, where cluster A comprises all roselle accessions and cluster C comprises all kenaf accessions. But cluster B comprises a roselle accession Bengkalis and kenaf accession Kho Khen (Omalsaad et al., 2014). Information about the study of genetic relationships among roselle and kenaf is very useful in planning breeding programs in the future to improve the existing cultivars. In addition, this genetic knowledge can assist in the conservation of germplasm accessions of roselle and kenaf.

4 CULTIVATION TECHNOLOGY OF ROSELLE

4.1 Soil Requirements

Roselle can easily grow in a wide range of soil and adapt to even poor soil or sandy soil. But it prefers fairly fertile sandy loam neutral soils and requires well-drained soil. Roselle is reported to tolerate soil pH from 4.4 to 8.0.

4.2 Climatic Requirements

Roselle can grow well to a wide range of climatic conditions with no trouble. It is normally grown in field environments under inclusive daylight (Mohamad et al., 2005). Roselle has the best produce in tropical and subtropical regions and is sensitive to frost and fog. It is sensitive to photoperiod that produces flowers when the day length is shorter than 12 h. Roselle requires 13 h of day length for the period of vegetative development to prevent premature flowering. Roselle can stand against submergence, heavy storm, or wet soil, but it cannot grow up under shade (Duke, 1983; Morton, 1987). The temperature requirement for roselle ranges from 18°C to 35°C, and roselle stops growing at 14°C, with death occurring after 15 days. Roselle requires night temperatures not less than 21°C for 4–8 months. Roselle best grows in the areas with 800–1600 mm of annual rainfall. It requires a minimum of 100–150 mm rainfall per month for the period of vegetative growth. Roselle can withstand dry periods in the final months of growth and development. High humidity and rain in the harvest time can cause drying and reduce the yield and quality of the calyces. However, yield of some cultivars partially resistant to drought has been satisfactory in wet humid regions.

4.3 Propagation of Roselle

Roselle is generally reproduced by seed, but it can also be propagated readily by stem cuttings or microcutting. The second method produces shorter plants ideal for intercropping with trees but comparatively lowers the yield of calyces. Germination of seeds is usually fairly rapid.

4.4 Land Preparation and Planting

Deep ploughing is recommended in heavy soil during seed bed or land preparation for cultivating roselle, as it is a deep-rooted crop. In light or sandy soils, shallow ploughing is recommended. In rainfed conditions, roselle crops are sown at the commencement of the rainy season. Roselle plants are often permitted to cultivate with other crops and are intercropping with sorghum, millet, cowpea, groundnut, sweet potato, or yam. Lots of farmers plant roselle by the side of field borders or outline subplots inside the field.

Seeds of roselle can be directly sown, whereby three to five seeds are sown per hole. Germination starts after 2–3 days of sowing. The spacing among plants depends on the total growth of the roselle variety, time of planting, type of soil, and environment. Depending on the soil type, the seed rate of roselle is 11–22 kg per ha, spacing is 15 cm × 15 cm, depth of planting is about 0.5 cm, and three to five seeds per drilled hole. Line sowing of roselle seeds is recommended rather broadcasting is not for the reason that of asymmetric stand. For leafy vegetable, roselle seeds are either broadcast or drilled, with a spacing of 60–90 cm between rows and 40–60 cm between plants. For calyx production, the spacing should be wider, up to 100 cm apart. Seeds can be sown in a nursery under the shade and transplant into the field at 25–30 days after germination. But transplantation is supposed to be carried out before the plants are more than 20 cm tall, even if the plants continue to grow. Reduced planting rate produces a larger calyx. Planting can be done with a modern grain drill and then thinned by hand, or the seeds can be hand-planted. For fiber production, seeds are drilled more closely, at 15 cm × 20 cm or 10 cm × 30 cm (Fig. 2.11).

FIG. 2.11 Roselle plantations in the field: **(A)** line planting, **(B)** Universiti Kebangsaan Malaysia, and **(C)** Terengganu.

4.5 Varieties of Roselle

There are over 100 cultivars or seed varieties of *H. sabdariffa*. The major commercial varieties are those grown in China, Thailand, Mexico, and Africa, principally Sudan, Senegal, and Mali. Morton, in 1987, reported that Wester, in 1920, has described three named, edible varieties that were grown at that time in the Philippines: Rico (named in 1912), Victor (a superior selection from seedlings grown at the Subtropical Garden in Miami in 1906), and Archer (sometimes called "white sorrel"), which resulted from the seeds sent to Wester by A.S. Archer of the island of Antigua. Another roselle selection that originated in 1914 at the Lamao Experiment Station was named 'Temprano' because of its early flowering. A strain with dark-red, plump but stubby calyces (the sepals are scarcely longer than the seed capsule) is grown in the Bahamas (Abu-Tarboush et al., 1997). Other reported roselle varieties are Terengganu, UKMR-1, UKMR-2, and UKMR-3 from Malaysia; BJRI Mesta 1 from Bangladesh; and H.S. 4288 and H.S. 7910 (Ujjal) from India.

4.6 Nutrient Management

Roselle responds well to fertilizers, although it is a low-input, low-labor crop. The recommended fertilizer dose is 15 kg N, 15 kg P, and 15 kg K/ha. The following fertilizer rates are also recommended for roselle: 80 kg N/ha, 36–54 kg P_2O_5/ha and 75–100 kg K_2O/ha. To produce a large calyx, 1200–2500 kg of manure is added per hectare. Calyx production is greater when plants are fertilized at thinning (20–30 days after planting) than when the applications are split and performed during the vegetative stage and at flowering. However, chemical fertilizers are rarely applied, as they are too costly under the uncertain climatic conditions where roselle is grown.

4.7 Weeding and Irrigation

Roselle can be grown both as a rainfed field crop and as an irrigated vegetable crop. Roselle grown in home gardens as a leafy vegetable or for leaves and calyces is cultivated under irrigation, mostly done manually with watering cans. Removing tall-growing weeds and grasses may be required once or twice, yet as soon as the crop is fully grown, no further weeding is necessary. Weeding is rarely practiced, but if done, it results in higher calyx yields. A single round of weeding and thinning is usually performed 20–30 days after planting.

4.8 Pest and Disease Management

Roselle is susceptible to most diseases affecting roots, and the stem rot caused by several *Phytophthora* spp. leads to plant losses. *Phytophthora parasitica* var *sabdariffae* causes stem burn (also called collar rot or stem canker), resulting in purplish black discoloration around the stem 30 cm above the ground and sudden wilting of the plant. Plant and calyx senescence due to this pathogen has been observed in Central Africa, Nigeria, the Caribbean region, and India. Leaf spot caused by *Cercospora hibisci* and *Phoma sabdariffae* is also common. Roselle types with green leaves are susceptible to powdery mildew (*Oidium abelmoschi*), whereas types with red leaves are partially resistant. A viral disease is reported from Nigeria, causing hard cracking leaves. Roselle is rather resistant to root-knot nematodes (*Meloidogyne* spp.) but not to free-living nematodes (*Heterodera* spp.).

Damage done to hibiscus by insects is minor but it does exist. A coleopteran insect, flea beetle (*Nisotra orbiculate*) feeds on the leaf lamina of roselle. Although it has not proved so serious, it needs proper control. It can be controlled by dusting Sahajanand Health Care (SHC) (10%) powder on the leaves. The insects of the order Hemiptera are a minor problem, including the mealy bugs and the leafhopper. Because of the attack of mealy bugs (*Maconellicoccus hirsutus*), the tip of the plant is crowded with leaves and further growth of the plant is completely checked. This may be controlled by a spray of parathion (0.04%) or fenitrothion (0.04%).

All types of infestation can be avoided by prevention techniques and possibly by using a suitable crop rotation system.

4.9 Harvesting

Harvest is timed according to the ripeness of the seed. Calyces are harvested after the flower has dropped (two to three weeks after flowering) but before the seed pod has dried and opened. The more time the capsule remains on the plant after the seeds begin to ripen, the more susceptible the calyx is to sores, sun cracking, and general deterioration in quality. All harvesting is done by hand. Different harvesting methods are in use today. In Mexico the entire plant is cut down and taken to a nearby location to be stripped of the calyces. In Malaysia and China, only ripe calyces are harvested with clippers, leaving the stalks and immature calyces to ripen in the field. The field is harvested approximately every 10 days until the end of the growing season. The calyx is separated from the seed pod by hand, or by pushing a sharp-edged metal tool through the fleshy tissue of the calyx separating it from the seed pod. Special care must be taken during the harvesting operation to avoid contamination by extraneous material.

4.10 Yield

For leafy branches, yields up to 20 t/ha from three cuttings have been reported. Fresh calyx yields range from 4 to 6.5 t/ha, or about 800–1200 kg/ha when dried to 12% moisture content. In Asia, fresh calyx yields up to 15 t/ha have been reported. A single roselle plant may yield as many as 250 calyces, or 1–1.5 kg fresh weight. With good care, roselle has been reported to yield more than 4 kg of fruits per plant. In Africa, average yields are much lower and variable because of the environmental conditions and extensive management. Sudan reports an average yield of dry calyces of 93 kg/ha. In Senegal the maximum production of calyx on a dry weight basis is 500 kg/ha. Average fiber yields from roselle are 1.5–2.5 t/ha, depending on the cultivar and management, and the seed yield ranges from 200 to 1500 kg/ha.

4.11 Handling After Harvest

Drying can be accomplished by different methods. Calyces collected after 15 days of flowering can be dried by forced air, sun, or artificial heat before marketing. Drying in the sun can lead to reduced quality. Plastic sheets are placed on the ground to avoid contamination with soil during sun drying, which also strongly reduces the value. Adequate ventilation is important during drying, and temperatures must remain below 43°C to be dried to a maximum water content of 12%. The drying ratio is 10:1.1; that is, for every 100 kg of fresh calyx, 11 kg of dry calyx is produced. After the drying process, dried calyces can be packed and stored for a long time.

5 UTILIZATION OF ROSELLE

All parts of roselle are in value, but leaves and calyces have been highly valuable to humans (Bolade et al., 2009). Roselle is a vegetable plant that has economic importance and has gained popularity as an ornamental, medicinal, industrial, and food plant. In general, the calyx, leaf, seed, and stem of roselle are used to produce beverages, tea, oil, and fiber, respectively. Many parts of the plant are claimed to have various medicinal values and have been used for such purposes in several countries ranging from Mexico through Africa and India to Thailand (Wikipedia, 2010). More details on different roselle part utilization and benefits are provided in the following sections.

5.1 Calyx

Fruits (calyces: the outer whorl of the flower) are eaten (Watt & Breyer-Brandwijk, 1962) raw in salads, cooked, and used as a flavoring in cakes. The calyx is rich in citric acid and pectin, so it is useful in making jams, jellies, soups, sauces, pickles, puddings, etc. Calyces are used in both fresh and dry forms. The dried calyx is used to prepare an infusion or an aqueous macerate, which has a high content of calcium, niacin, riboflavin, and iron (Ali-Bradeldin et al., 2005; Morton, 1987). A refreshing and very popular beverage can be made by boiling the calyx, sweetening it with sugar, and adding ginger. The calyces are used to make cold and hot beverages in many of the world's tropical and subtropical countries.

Roselle is the source of a red beverage known as jamaica in Mexico. In Switzerland the calyx is called karkade and used in jams, jellies, sauces, and wines. Elsewhere in the tropics the fleshy calyces are used fresh for making roselle wine, jelly, syrup, gelatin, refreshing beverages, pudding, and cakes, and dried roselle is used in making tea, jelly, marmalade, ices, ice cream, sherbets, butter, pies, sauces, tarts, and other desserts. Calyces are used in the West Indies to color and flavor rum (Duke, 1983), and in Nigeria, a nonalcoholic beverage called 'soborodo' or 'zobo' is made from dried roselle calyx. The drink serves as a cheaper alternative to the industrially produced carbonated soft drinks that are also available in every nook and cranny of the country (Bolade et al., 2009). In countries like India, roselle calyces are utilized in producing refreshing beverages, jellies, jam, sauces, and food preserves (Clydescale et al., 1979). The drink is becoming popular because it is easily processed at home. It serves as an income-generating source for many women (Fasoyiro et al., 2005).

Roselle calyx is used in traditional medicine as a digestive agent, purgative, and diuretic (Osuntogun & Aboaba, 2004). The roselle plant parts have also been reported to be folk remedy for cancer, obesity, diabetes, and hypertension (Duke, 1983; Hamdan & Afifi, 2004; Odigie et al., 2003; Perry, 1980; Chau et al., 2000). Tea of *Hibiscus* calyces is effective in lowering blood pressure (11.2% reduction) in people suffering from hypertension (Herrera-Arellano et al., 2007; McKay et al., 2010). More recently, human studies have shown that *H. sabdariffa* can lower cholesterol levels (Tzu-Li Lin et al., 2007).

5.2 Leaf

The young leaves are eaten as cooked vegetables, especially with soup (Aliyu, 2000). Tender leaves are eaten as salad and as a potherb and are used for seasoning curries (Duke, 1983; Watt & Breyer-Brandwijk, 1962). Medicinally, leaves are emollient and are used in Burma for this purpose, and they are also much used in Guinea as a diuretic, refrigerant, sedative, and antiscorbutic. Angolans use the mucilaginous leaves as an emollient and as a soothing cough remedy (Duke, 1983; Perry, 1980). A lotion made from roselle leaves is used on sores and wounds (Morton, 1987). The heated leaves are applied to cracks in the feet and on boils and ulcers to speed maturation. The leaves and flowers are used as a tonic tea for digestive and kidney functions (Qi et al., 2005).

5.3 Stem

Roselle is cultivated primarily for the bast fiber obtained from the stems. The fiber strands, up to 1.5 m long, are used for cordage and as a substitute for jute in the manufacture of burlap. As in Asia, roselle fiber is locally used in West Africa, but on a very small scale. The bast fiber is a good substitute for jute and it is used for making twine, cordage, rope, netting, and sacks. The bast fiber and sometimes the whole stem are used in the paper industry in the United States and Asia. Stalks are eaten as salad and potherb. Stems are also used for seasoning curries (Duke, 1983; Watt & Breyer-Brandwijk, 1962). The plant has cathartic properties (Yusuf et al. 2009). In West Africa, roselle plants are often used as fodder for livestock.

5.4 Flower

In some countries, roselle flowers are used for decorative purposes (Sanchez-Mendoza et al., 2007). Flowers contain gossypetin, anthocyanin, and the glucoside hibiscin, which may have diuretic and choleretic effects, decreasing the viscosity of the blood, reducing blood pressure, and stimulating intestinal peristalsis (Duke, 1983; Perry, 1980). It is also used as a tonic tea for digestive and kidney functions (Qi et al., 2005).

5.5 Root

The young root is edible but is very fibrous, mucilaginous, and without much flavor. In Philippines the bitter root is used as an aperitifs and tonic (Perry, 1980). No more information has been reported about the utilization and benefits of the roots in roselle plant.

5.6 Seed

The seeds are high in protein and can be pounded after roasting to be used in oily soups or sauces. The oil of roselle seed is extracted and used for cooking in Chad, Tanzania, and China. The seed yields 20% oil (Aliyu, 2000). Oil extracted from the seed is a substitute for castor oil. In China the seeds are used for their oil and the plant is used for medicinal properties. In Burma the seeds are used for debility. Taiwanese regard the seed as a diuretic, laxative, and tonic. In Nigeria, people ferment roselle seeds to make a cake used as 'sorrel meat'. The oil is also used as an ingredient in paints. Seed oil possesses antimicrobial activity (Asolkar et al., 1992). The fatty acid extracted of roselle seeds is also used for the treatment of male sterility (Comhaire & Mahmoud, 2003). Seeds are roasted and ground into a powder then used in oily soups and sauces. It has been reported that the seed oil is rich in phytosterols and tocopherols (Hussein et al., 2010).

6 PRODUCTS AND SERVICES OF ROSELLE

6.1 Food

Roselle is the source of a red beverage known as 'jamaica' in Mexico. Calyx, called karkade in Switzerland, is used in jams, jellies, sauces, and wines. In the West Indies and elsewhere in the tropics, the fresh fleshy calyces are used for making roselle wine, jelly, syrup, gelatin, refreshing beverages, pudding, and cakes (Nyam et al., 2014) and dried roselle is used in making tea, jelly, marmalade, ices, ice cream, sherbets, butter, pies, sauces, tarts, and other desserts (Fig. 2.12). Tender leaves and stalks are eaten as salad and potherb. The leaves are used for seasoning curries and are also consumed as a vegetable.

FIG. 2.12 Products of roselle: **(A)** juice, **(B)** jam, **(C)** roselle two-in-one instant drink, and **(D)** dried roselle calyces labeled "Flor de Jamaica".

Hibiscus tea is an herbal tea made from the sepals (calyces), which are the small structures at the base of the flower. Hibiscus tea is widely consumed as a drink on its own from Southeast Asia to Africa and the Caribbean; each of these regions has its own unique twist on how the drink is prepared. In Latin America, the drink is known as agua de flor de Jamaica, sometimes shortened to *agua de Jamaica, rosa de Jamaica*, or just *Jamaica*. In Panama, hibiscus tea is called saril. In North Africa, especially Egypt, the drink is called karkadé (كركديه), which just means *hibiscus* in Arabic. In some regions, hibiscus is also blended with black tea. Hibiscus tea has a strong sour taste and a distinctive flavor. It is popular as a beverage and is generally nutritious, but it also has powerful health benefits, including the potential to significantly lower blood pressure.

Thai people have grown roselle and used it to make tea because it lowers cholesterol, helps treat heart disease, is a diuretic, is loaded with antioxidants, and, in some instances, is even used to help treat cancer. In other Asian countries, roselle is used to help promote weight loss and to help with circulation problems.

6.2 Apiculture

Flowers are a source of pollen and nectar for bees.

6.3 Fuel

H. sabdariffa stem and wood are potential raw materials for charcoal making. Roselle seed oil is rich in linoleic acid and is a candidate source for vegetable oils. Oil of roselle seeds can be used for biodiesel production (Ahmed et al., 2016; Nakpong & Wootthikanokkhan, 2010).

6.4 Fiber

Roselle is cultivated primarily for the bast fiber obtained from the stems. The fiber strands, up to 1.5 m long, are used for cordage and as a substitute for jute in the manufacture of burlap.

6.5 Poison

H. sabdariffa is toxic to *Schistosoma mansoni* at 50–100 ppm, showing both miracidicidal and cercaricidal activity.

6.6 Ornamental

The roselle is a multiple-use species with beautiful flowers. The ornamental value of roselle is of recent interest as a garden plant or cut flower. The decorative red stalks with ripe red fruits are exported to Europe where they are used in flower arrangements.

7 ECONOMICS AND MARKET OF ROSELLE

7.1 Production and International Trade

International trade of roselle calyces has increased steadily over the past decades, with 15,000 t per year now entering the world market. Germany and the United States are large importers. In 1998, the US and German importers paid US $1200–$1700 per ton for Egyptian and Sudanese roselle; prices of Chinese roselle were lower. Prices fluctuate because of high variability in supply. A decrease in product quality in Thailand and China due to excessive precipitation caused world prices to soar to US $4000 per ton in 2003. Sudan is the most important roselle producer in Africa, the annual area fluctuating between 11,000 and 57,000 ha depending on the amount of rainfall and prices. In 1995, Sudan reported exports of 32,000 t. In Sudan, smallholder farmers traditionally grow roselle in plots ranging from under 0.25 to 2 ha, but some growers have up to 20 ha. Sudanese roselle is viewed as the superior quality, but the United States trade embargo and large-scale production in Mexico, Thailand, and China have led to shifts in the market. Jamaica and Egypt also export roselle. Senegal and Mali are major producers, but the vast majority of their production is for domestic consumption or sold on the local markets. Fluctuations in export prices for cash crops such as cotton have led many West African farmers to diversify production, e.g., by growing roselle for the domestic market. In Asia, roselle is primarily a fiber crop, accounting for 20% of jutelike fiber production or 700,000 t per year. There are currently 18 companies involved in the production, processing, and marketing of roselle products for the local market in Malaysia. The total market value of the roselle industry in Malaysia is RM10.0–15.0 million, with about 65%–80% of the value staying with the processors. The domestic market consumes 500 t of roselle calyces, of which over 80% is processed for juice and drink, and the current annual export value of fresh calyces is RM2.5 million. The UKM is establishing a startup company (STU) under a framework called the UKM-MTDC Symbiosis Programme. The technology provider or innovator will be given incentive in the form of equity of up to 20%. The STU will be funded up to RM2.5 million over 2 years, with the expectation that it will function as an independent company thereafter (http://www.raidah.com.my/symbiosis/ukm/index2.html).

7.2 Importers

The primary importers of the dried calyx of *H. sabdariffa* are the United States and Germany. England is not a regular importer of the product, as imported herbal tea from Germany satisfies most market demand. There are no statistics for the volume and value of dried hibiscus imported into these markets; information in this survey has been gathered mainly through primary interviews with importers. The major clients for roselle importers are herbal tea manufacturers, as the dried calyx is used as part of the base for most herbal teas, along with apple peel, orange peel, and lemon twist. The United States imported over 5000 million tons (US $22 million) of plants and plant parts for use in herbal teas in 1998, an increase of 78% in volume and 156% in value from the 1994 levels (*Source: USA Trade*). Germany is an even larger importer of herbal tea ingredients. In 1997, imports totaled nearly 43,000 million tons or ECU 90 million (US $97 million) for plants and plant parts for use in herbal teas, medicines, and perfumes, an increase of 41% in volume and 72% in value from the 1993 levels.

7.3 Supply

Roselle is an important leafy vegetable in the drier parts of West and Central Africa. In Senegal, Mali, Chad, and Sudan, large quantities of calyces are produced for the preparation of beverages. Sudan is the major country in tropical Africa producing dried roselle calyces for local consumption and export, mainly in the Kordofan and Darfur regions in the west of the country. Roselle is available in the international market from Thailand, Sudan, China, Mexico, and various other smaller producing nations including Egypt, Senegal, Tanzania, Mali, and Jamaica. China is the dominant supplier to the United States according to the importers surveyed. Thailand, Mexico, and Egypt supply smaller amounts. The preference is for Sudanese product at a considerable markup. Sudan dominates the German import market. German herbal tea manufacturers consider Sudanese roselle to have the perfect color blend and taste for herbal tea bases, and also roselle from China and Thailand and much smaller quantities from Egypt and Mexico are used.

7.4 Prices

Hibiscus has very different qualities depending on where it is grown. The most desirable product is from Sudan and Thailand, but hibiscus from these two countries is also very different. The market is very volatile from one crop year to the next, based on the thinking patterns of farmers around the world. One crop year may yield high prices and encourage many farmers to plant hibiscus the following year. Supply then outpaces demand and prices fall. When the prices fall, farmers may switch to other crops, causing a supply shortage and thereby increasing prices again. Roselle prices vary markedly depending on the quantity purchased from $4.00 to $4.90 per pound.

8 PROSPECTS OF ROSELLE AS AN INDUSTRIAL CROP FOR SOUTHEAST ASIA

Roselle is an annual shrub that originated in India and was transferred to Malaysia (Morton, 1987) and grown all over the tropics and subtropics. It is one of the major cash crops in China, Sudan, and Thailand. There are two main subspecies *H. sabdariffa* var *sabdariffa* and *H. sabdariffa* var *altissima* cultivated for food and fiber. Roselle has wide uses in the food and beverage industry.

Roselle calyx and flower are used to prepare juice, tea, jam, jelly, wine, cake, biscuit, pudding, ice cream, gelatin, syrup, etc. Seed oil of roselle can be used as feedstock for biodiesel. Leaves of roselle are also consumed as vegetables. As a tropical crop, roselle has a huge potential for good yield and quality production. Roselle flower, calyx, and seed have good market opportunity as industrial crops in Southeast Asia and can significantly contribute to the agricultural economy based on their diverse uses.

ACKNOWLEDGMENT

Authors are thankful to the authority of Bangabandhu Sheikh Mujibur Rahman Agricultural University, Gazipur 1706, Bangladesh, for their support.

REFERENCES

Abu-Tarboush, H. M., Ahmed, S. A. B., & Al Kahtani, H. A. (1997). Some nutritional properties of karkade (*Hibiscus sabdariffa*) seed products. *Cereal Chemistry, 74*, 352–355.

Adamson, W. C., & O'Bryan, J. E. (1981). Inheritance of photosensitivity in roselle, *Hibiscus sabdariffa*. *Journal of Heredity, 72*, 443–444.

Ahmed, N. B., Abdalla, B. K., Elamin, I. H. M., & Elawad, Y. I. (2016). Biodiesel production from roselle oil seeds and determination the optimum reaction conditions for the transesterification process. *International Journal of Engineering Trends and Technology, 39*(2), 105–111.

Ali-Bradeldin, H., Al-Wabel, N., & Gerald, B. (2005). Phytochemical, pharmacological and toxicological aspects of *Hibiscus sabdariffa* L: A review. *Phytotherapy Research, 19*, 369–375.

Aliyu, L. (2000). Roselle (*Hibiscus sabdariffa* L.). Production as affected by pruning and sowing date. *Journal of Applied Agricultural Technology, 6*, 16–20.

Anderson, N., & Pharis, J. (2003). Kenaf fiber: A new basket liner. *Minnesota Commercial Flower Growers Bulletin, 52*(3), 7–9.

Andrews, F. W. (1952). *The flowering plants of the Anglo-Egyptian Sudan*. Pub'd for the Sudan Gov't by T. Buncle & Co., Ltd., Arbroath, Scotland. 2: 28 & 30.

Anonymous. (1959). *The wealth of India: Raw materials* (Vol. 5, pp. 92–96). New Delhi: Council of Sci. & Indus. Res..

Asolkar, L. V., Kakkar, K. K., & Chakre, O. J. (1992). *"Second supplement to glossary of Indian medicinal plants with active principles." part-1 (A-K)*. New Delhi: CSIR.

Ayad, W. G., Hodking, A., Jaradat, A., & Rao, V. R. (1995). Molecular genetic techniques for plant genetic resources. In *IPGRI workshop: Rome, Italy, October 9–11, 1995*.

Barssum – Waalkes, J. Van (1966). Malesian malvacea revised. *Blumea, 14*, 1–251.

Bolade, M. K., Oluwalana, I. B., & Ojo, O. (2009). Commercial practice of roselle (*Hibiscus sabdariffa* L.) beverage production: Optimization of hot water extraction and sweetness level. *Department of Food Science and Technology. Nigeria, 5*(1), 126–131.

Britton, N. Y., & Wilson, P. (1924). Botany of Porto Rico and the Virgin Islands. Pts. 1-4. Sci. Surv. Of Puerto Rico and the Virgin Isls. *New York Academy of Sciences. New York, 5*, 563.

Brown, W. H. (1954). Useful plants of the Philippines. In *Tech. Bull. Philippine Dept. Agr. & Nat. Manila, 2* pp. 416–418).

Chau, J. W., Jin, M. W., Wea, L. L., Chia, Y. C., Fen, P. C., & Tsui, H. T. (2000). Protective effect of Hibiscus anthocyanins against tert butyl hydroperoxide-induced hepatic toxicity in rats. *Food and Chemical Toxicology, 38*(5), 411–416.

Clydescale, F. M., Main, J. H., & Francis, F. J. (1979). Roselle (*Hibiscus sabdariffa*) anthocyanins as colourants for beverages and gelatin deserts. *Journal of Food Protection, 42*, 204–267.

Comhaire, F. H., & Mahmoud, A. (2003). The role of food supplements in the treatment of the infertile man. *Reproductive BioMedicine Online, 7*(4), 385–391.

Crane, J. C. (1949). Roselle—a potentially important plant fiber. *Economic Botany, 3*, 89–103.

Dempsey, J. M. (1975). *Fiber crops*. Gainesville: The University Presses of Florida.

Drury, H. (1873). *The useful plants of India*. London: William, H. Allen & Co.

Duke, J. A. (1983). *Handbook of energy crops*. http://www.hort.purdue.edu/newcrop/duke_energy. (Accessed 7 January 1998).

Duke, J. A. (1993). *Alternative cash crops*. CRC Handbook CRC Press, 544pp.

Fasoyiro, S. B., Ashaye, O. A., Adeola, A., & Samuel, F. O. (2005). Chemical and storability of fruit-flavoured (*Hibiscus sabdariffa*) drinks. *Institute of Agricultural Research and Training. Nigeria, 1*(2), 165–168.

Furuta, S., Suda, I., Nishiba, Y., & Yamakawa, O. (1998). High tert-butylperoxyl radical scavenging activities of sweetpotato cultivars with purple flesh. *Food Science and Technology International. Tokyo, 29*, 33–35.

Ghani, A. (1998). Medicinal plants of Bangladesh: Chemical constituents and uses. *Asiatic Society of Bangladesh*, 84–85.

Halimaton, S. O. (2005). *Kajian kromosom dan pembinaan deskriptor bagi tanaman roselle (Hibiscus sabdariffa L.)*. Bachelor Thesis of Science. UKM.

Hamdan, I. I., & Afifi, F. U. (2004). Studies on the *in vitro* and *in vivo* hypoglycemic activities of some medicinal plants used in treatment of diabetes in Jordanian traditional medicine. *Journal of Ethnopharmacology, 93*, 117–121.

Harris, W. (1913). *Notes on fruits and vegetables in Jamaica*. Kingston: Gov't Printing Office.

Herrera-Arellano, A., Miranda-Sanchez, J., Avila-Castro, P., Herrera-Alvarez, S., Jiménez-Ferrer, J. E., Zamilpa, A., Román-Ramos, R., Ponce-Monter, H., & Tortoriello, Jaime (2007). Clinical effects produced by a standardized herbal medicinal product of *Hibiscus sabdariffa* on patients with hypertension. A randomized, double-blind, lisinopril-controlled clinical trial. *Planta Medica, 73*(1), 6–12.

Heywood, V. H. (1978). *Flowering plants of the world* (pp. 94–95). Oxford, London: Oxford University Press.

Hong, V., & Wrostlad, O. (1990). Use of HPLC separation/photodiode array detection for characterization of anthocyanin. *Journal of Agricultural and Food Chemistry, 38*, 708–715.

Hussein, R. M., Yasser, E. S., Amr, E., & Awad, M. N. (2010). Biochemical and molecular characterization of three colored types of roselle (*Hibiscus sabdariffa* L.). *Journal of American Science, 6*(11), 726–733.

Jennifer, A. W. (2009). *The effect of anthocyanin acylation on the inhibition of HT-29 colon cancer cell proliferation*. Thesis of Master of Science. Ohio State University.

Khan, A. R. (1930). Roselle fiber studied in Netherland East Indies. *Cord Age, 31*(2), 28.

Leung, W. T. W., Busson, F., & Jardin, C. (1968). *Food composition table for use in Africa*. Rome, Italy: FAO, 306 pp.

Leung, A. Y., & Foster, S. (1996). *Encyclopedia of common natural ingredients used in food, drugs and cosmetics* (p. 2). New York: John Wiley and Sons.

McKay, D. L., Chen, C. Y., Saltzman, E., & Blumberg, J. B. (2010). *Hibiscus sabdariffa* L. tea (tisane) lowers blood pressure in prehypertensive and mildly hypertensive adults. *Journal of Nutrition, 140*(2), 298–303.

Mohamad, O., Halimaton Saadiah, O., & Noor Baiti, A. A. (2008). HCA from Gracinia species an anti-obesity agent. *Poster ITEX*.

Mohamad, O., Herman, S., Halimaton Saadiah, O., Ramadan, G., Aulia Rani, A., Elfi, K., Bakhendri, S., Chang, S. V., Wong, W. H., Noor Baiti, A. A., Lee, S. H., Mamot, S., Aminah, A., Ahmad Mahir, M., Nazir, B. M., & Abul Rahman, M. (2006). *Mutation breeding of roselle. Poster. Malaysia agriculture, horticulture and agrotourism exhibition*.

Mohamad, O., Herman, S., Nazir, B. M., Aminah, A., Mamot, S., Bakhendri, S., & Abdul, R. M. (2005). Mutation breeding of roselle in Malaysia. In *Paper presented at FNCA 2005 workshop on mutation breeding, 5-9 Dec, Kula Lumpur* (pp. 1–7).

Mohamad, O., Nazir, B. M., Abdul Rahman, M., & Herman, S. (2002a). Roselle: A new crop in Malaysia. *Buletin PGM*, 12–13.

Mohamad, O., Nazir, B. M., Abdul Rahman, M., & Herman, S. (2002b). Roselle: A new crop in Malaysia. *Bulletin Genetics Society. Malaysia, 7*(1–2), 12–13.

Morton, J. (1987). Roselle. In *Fruits of warm climates* (pp. 281–286). Miami, FL: Web Publications Purdue University, ISBN 0-9610184-1-0. www.hort.purdue.edu/newcrop/morton/roselle.html.

Morton, J. F., & Dowling, C. F. (1987). *Roselle: Fruits of warm climates* (Internet) Julia F. Morton, Miami, United States. 505 pp http://www.hort.purdue.edu/newcrop/morton/roselle.html.

Musthafa, M. E., Perumal, S., Ganapathy, S., Tamilarasan, M., Kadiyala, B. D., Ramar, S., Selvaraju, S., & Govindarajaha, V. (2005). Influence of *Hibiscus sabdariffa* (Gongura) on the levels of circulatory lipid peroxidation products and liver marker enzymes in experimental hyperammonemia. *Journal of Applied Biomedicine, 4*(1), 53–58.

Nakpong, P., & Wootthikanokkhan, S. (2010). Roselle (*Hibiscus sabdariffa* L.) oil as an alternative feedstock for biodiesel production in Thailand. *Fuel, 89*(8), 1806–1811.

Nyam, K. L., Leao, S. Y., Tan, C. P., & Long, K. (2014). Functional properties of roselle (*Hibiscus sabdariffa* L.) seed and its application as bakery product. *Journal of Food Science & Technology, 51*(12), 3830–3837.

Odake, K., Terahara, N., Saito, N., Toki, K., & Honda, T. (1992). Chemical structures of two anthocyanins from purple sweetpotato, Ipomoea batatas. *Phytochemistry, 31*(6), 2127–2130.

Odigie, I. P., Ettarh, R. R., & Adigun, S. A. (2003). Chronic administration of aqueous extract of *Hibiscus sabdariffa* attenuates hypertension and reverses cardiac hypertrophy in 2K-1C hypertensive rats. *Journal of Ethnopharmacology, 86*(2–3), 181–185.

Olubukola, A., & Illoh, H. C. (1996). Pollen grain morphology of some species of Hibiscus. *Nigerian Journal of Botany, 9*, 9–14.

Omalsaad, M., Osman, M., & Aminul Islam, A. K. M. (2012). Characterization of roselle (*Hibiscus sabdariffa* L.) and kenaf (*Hibiscus canabinus* L.) accessions from different origin based on morpho-agronomic traits. *International Journal of Plant Breeding, 6*(1), 1–6.

Omalsaad, M., Osman, M., Jahan, M. A., & Islam, A. K. M. A. (2014). Genetic relationship between roselle (*Hibiscus sabdariffa* L.) and kenaf (*Hibiscus cannabinus* L.) accessions through optimization of PCR based RAPD method. *Emirates Journal of Food and Agriculture, 26*(3), 247–258. https://doi.org/10.9755/ejfa.v26i3.16498

Osuntogun, B., & Aboaba, O. O. (2004). Microbiological and physico-chemical evaluation of some non-alcoholic beverages. *Pakistan Journal of Nutrition, 3*(3), 188–192.

Perry, L. M. (1980). *Medicinal plants of east and Southeast Asia*. Cambridge: MIT Press.

Purseglove, J. W. (1974). *Tropical crop: Dicotyledon*. London: Longman.

Qi, Y., Kit, L. C., Malekian, F., Berhane, M., & Gager, J. (2005). Biological characteristics, nutritional and medicinal value of roselle, *Hibiscus sabdariffa*. *Circular – Urban Forestry Natural Resources and Environment, 604*.

Robyns, A. (1965). *Hibiscus sabdariffa* L. (n: Malvaceae, Fam. 115, Pt. VI of Flora of Panama, by R.E. Woodson, Jr., R. W. Schery and collaborators) *Annals of the Missouri Botanical Garden, 52*(4), 506–508.

Rusmawati, C. M. (2004). Tanaman Roselle. Lumayan dan berkhasiat tinggi. Rencana Patriot Pertanian. *Bank Pertanian Malaysia, 3*(11), 16–22.

Samy, M. S. (1980). Chemical and nutritional studies on roselle seeds (*Hibiscus sabdariffa* L.). *Zeitschrift fur Ernahrungswissenschaft, 19*(1), 47–49.

Sanchez-Mendoza, J., Dominguez-Lopez, A., Navarro-Galindo, S., & Lopez-Sandoval, J. A. (2007). Some physical properties of Roselle (*Hibiscus sabdariffa* L.) seeds as a function of moisture content. *Journal of Food Engineering, 87*, 391–397.

Schippers, R. R. (2000). *African indigenous vegetable: An overview of the cultivated species* (pp. 1–214). Chatham, UK: National Resources Institute/ACP-EU Technical Center of Agricultural and Rural Cooperation.

Skovsted, A. (1935). Chromosome numbers in the Malvaceae I. *Journal of Genetics, 31*, 263–293.

Skovsted, A. (1941). Chromosome numbers in the Malvaceae II. *Comptes Rendus des 46 Travaux Laboratoire Carlsberg. Serie Physiologie, 23*, 1995 – 242.

Standley, P. C. (1923). Trees and shrubs of Mexico (3). In *Contrib. U. S. National Herbarium. Smithsonian Inst., Washington D. C. P.* (Vol. 23, p. 779).

Torres-Morán, M. I., Escoto-Delgadillo, M., Ron-Parra, J., Parra-Tovar, G., Mena-Munguía, S., Rodríguez-García, A., Rodríguez-Sahagún, A., & Castellanos-Hernández, Y. O. (2011). Relationships among twelve genotypes of roselle (*Hibiscus sabdariffa* L.) cultivated in western Mexico. *Industrial Crops and Products, 34*, 1079–1083.

Tsai, P. J., McIntosh, J., Pearce, P., Camden, B., & Jordan, B. R. (2002). Anthocyanin and antioxidant capacity in Roselle (*Hibiscus sabdariffa* L.) extract. *Food Research International, 35*, 351–356.

Tsai, P. J., & Ou, A. S. M. (1996). Colour degradation of dried roselle during storage. *Food Science, 23*, 629–640.

Tzu-Lilin, Lin, H. H., Chen, C. C., Lin, M. C., Chou, M. C., & Wang, C. J. (2007). *Hibiscus sabdariffa* extract reduces serum cholesterol in men and women. *Nutrition Research, 27*(3), 140–145.

USDA. (2009). *Plant Database*. United States Department of Agriculture (USDA).

Vilasinee, H., Anocha, U., Noppaawan, P. M., Nuntavan, B., Hitoshi, S., Angkana, H., & Chuthamanee, S. (2005). Antioxidant Effects of aqueous extracts from dried calyx of *Hibiscus sabdariffa L.* (Roselle) *in vitro* using rat low-density lipoprotein (LDL). *Biological and Pharmaceutical Bulletin, 28*(3), 481–484.

Watt, J. M., & Breyer-Brandwijk, M. G. (1962). *The medicinal and poisonous plants of southern and eastern Africa* (2nd ed.). Edinburgh and London: E.&S. Livingstone, Ltd.

Webster, P. J. (1914). New varieties of roselle. *Philippine Agricultural Review, 7*, 266–269.

Wikipedia. (2010). *Roselle*. http://en.wikipedia.org/wiki/Roselle_ (plant. (Accessed 20 June 2010).

Wilson, F. D. (1999). Revision of Hibiscus section Furcaria (Malvaceae) in Africa and Asia. *Bulletin of the Natural History Museum: Botany Series, 29*, 47–79.

Wilson, F. D., & Menzel, M. Y. (1964). Kenaf (*Hibiscus cannabinus*),Roselle (*Hibiscus sabdariffa*). *Economic Botany, 18*, 80–91.

Yusuf, M., Chowdhury, J. U., Hoque, M. N., & Begum, J. (2009). *Medicinal plants of Bangladesh.* BCSRI, Chittagong-4220. Bangladesh. 1–692 pp.

FURTHER READING

Mohamad, O., Herman, S., Nazir, B. M., Shamsudin, S., & Takim, M. (2003). A dosimetry study using gamma irradiation on two accessions, PHR and PHI, in mutation breeding of roselle. (*Hibisc sabdariffa* L.). In *Paper presented at 7th MSAB symposium on applied biology, 3–4 June, Sri Kembangan* (pp. 1–10).

CHAPTER 3

Growing and Uses of *Hibiscus sabdariffa* L. (Roselle): A Literature Survey

JOSPHAT IGADWA MWASIAGI • THATAYAONE PHOLOGOLO

1 INTRODUCTION

Hibiscus sabdariffa L. (roselle) belongs to the family Malvaceae. Roselle plant is common in the tropical regions of Asia, America, and Africa. It is an erect annual herb, which can grow up to 2.5 m in height, and it is known by different names from region to region, for example; *mesta* in India, *asam susur* in Malaysia, *krachiap* in Thailand, and *karkade* in Arabic (Plotto, 2004; Mati and Hugo, 2011). As shown in Fig. 3.1, apart from the stem and leaves, roselle has a calyx that has been used for many of the infusion products made from it. Roselle can be grown for its seed, flower, calyx, root, leaves, core, and bast (bark) fibers. In Asia (China, Thailand, and India), the roselle plant is

FIG. 3.1 Roselle plant. (Mwasiagi, J. I., Yu, C. W., Phologolo, T., Waithaka, A., Kamalha, E., & Ochola J. R. (2014). Characterization of the Kenyan *Hibiscus sabdariffa* L. (Roselle), bast fiber. *Fibres and Textiles in Eastern Europe, 22*(3(105)), 31–34. http://www.fibtex.lodz.pl/2014/3/31.pdf, used with permission.)

grown for its bast fibers. The use of roselle core fiber for the manufacture of paper has also been reported in some Asian countries (Dutt et al., 2009). In other parts of the world, especially Africa, the roselle variety grown is mainly for the roots, leaves, calyces, and flowers.

The leaves and roots have also been used for a variety of applications, which includes food (vegetables), beverages, and medicinal products. The use of roselle is widespread, and it has the hallmark of an indigenous product that exhibits application from region to region. Roselle plant has drawn the attention of many scientists, who have taken efforts to try and study the usage of the plant. Reports indicated that the roselle plant has been used in human and animal food products, as a medical and pharmacologic product, and in other industrial applications, which include paper making and textile products. This chapter gives a literature review of the growing and selected uses of roselle plant.

2 GROWING THE ROSELLE PLANT

Roselle is an annual plant that grows both in the tropical and subtropical regions. Roselle grows wildly in many parts of Africa. There are commercial types of roselle grown in several parts of Asia, Africa, America (North, Central, and South), and Australia (Plotto, 2004). Looking at commercial production, Sudan produces high-quality roselle but in small quantities, whereas China and Thailand are the leading producers in terms of quantity. Apart from commercial farming, there are many African peasant farmers, especially in the western and eastern parts of Africa, who grow roselle as a vegetable or for making juice. In some areas, roselle grows wildly in the natural forests and it is used as a medicinal plant. The roselle plant needs well-drained soils (Changdee et al., 2009) with low salinity (Azooz, 2009) for optimum growth and yield. Growing roselle enriches the soil fertility with an increase in nitrogen, phosphorus, and potassium levels (Maqsudul et al., 2001).

Roselle plant can be grown by planting seeds. Apart from the traditional method of keeping the seeds fully dried and then storing them for the next planting season, the seeds can be flushed with nitrogen and then stored to mitigate deterioration and hence ensure better germination (Nurul and Mikael, 2018). Studies of the efficiency of regeneration of the plant using the shoot apex as explants have been conducted and proven to be a viable way of roselle plant propagation (Gomez et al., 2008). The growing of roselle can be optimized by following proper planting periods (dates) and intercropping (Ayipio et al., 2018; Nnebue et al., 2014; Talukder et al., 2001), using irrigation (Babatunde & Mofoke, 2006), and proper selection and application of fertilizers (Abo-Baker and Mostafa, 2011). The use of composts has been reported by Syahnaz et al. (2019), who used food waste and nanoparticles made from magnetite (Fe_3O_4), during farming of roselle. The plants reported better growth than the control sample (which did not get fertilizer treatment). Use of fertilizers (either organic or inorganic) is, however, important because it leads to larger calyces (Plotto, 2004). Roselle, like any other plant, is prone to attacks by pests, fungi, and diseases. The attack of roselle plant by vascular wilt was reported in the humid forest regions of southwestern Nigeria (Amusa et al., 2005). This is a serious threat to the roselle plants, which have been a source of food and medicine for the people living around the forest, and needs urgent attention to avert loss of the plants. The control of the vascular wilt can be achieved by proper plant management, which may include elimination of affected plants. Breeding of resistant plant varieties may, however, offer long-term solution. While interventions may be easy to implement for the commercially farmed plants, there is need for consideration of the plants that grow wildly in the forests. Governments and local authorities may have to consider an intervention to avoid the loss of the roselle plants growing wildly.

3 USES OF ROSELLE PLANT

3.1 Use of Roselle in Human and Animal Diets

The use of roselle in human and animal food and drinks shows its nutritional values in leaves, calyx, roots, stem, and barks, which points to the fact that the whole of the roselle plant has a commercial value. Roselle plant has found various applications in different parts of the world. Extracts from the calyx and petals have a sweet and tart flavor and contain minerals, vitamins, and bioactive compounds. In Kano, Nigeria, zobo (also known as sobo), a drink rich in vitamin C, is normally prepared by collecting roselle calyces, grinding them to a fine powder and then boiling in water, and sieving the mixture; the obtained filtrate is used as juice (zobo) that is commonly drank and believed to be a healthy beverage. People prefer the drink to be prepared to have a dark red color, and it is believed that zobo can help manage obesity and high blood pressure. Investigations to establish the contents of the zobo drink were carried out by Olayemi et al. (2011), who tested zobo for protein, pH, minerals, and ash contents. The pH of the drink made from zobo was found to be around 2.5, which is on the acidic side. The vitamin C contents were reported to be 6.0 mg/g. Minerals that included calcium (2.7 ppm), magnesium (9.2 ppm), potassium (220 ppm), sodium (35 ppm), and iron (0.82 ppm) were reported as the contents of zobo. The drink (zobo) was also found to contain protein (0.06 ppm) and 14.1% of ash. Therefore the local belief that zobo contains vitamin C is actually true. The presence of minerals such as magnesium and potassium may be the reason why zobo is believed to have curative properties. This finding may vindicate the traditional belief of the people of Kano in northern Nigeria who have always attributed curative powers to zobo beverage and designated it as a social drink as part of preventive medicine. In another study, Aina and Shodipe (2006) established that the zobo drink should be prepared at a controlled temperature and stored in glass or plastic containers rather than flexible polythene packets to preserve the quality of the drink. Further studies carried out by Preciado-Saldana et al. (2019) on the preparation of beverages from hibiscus leaves and calyces elucidated the important factors that affect hibiscus beverages, which include temperature, leaf-to-water ratio, and infusion time, and need to be balanced to ensure that high levels of phenolic compounds are extracted while at the same time keeping the taste of the beverage within acceptable levels. While the taste of the beverages seem to be more subjective and even have a gender preference (Monteiro et al., 2017), more research need to be carried out to ensure that the full benefits of hibiscus beverages are optimized. Some of the research and innovation has explored the use of blending with other fruits. Although from a theoretic point of view any fruit may be blended with roselle juice, caution should be taken to ensure that any fruit that will be adversely affected by the acidic nature of roselle juice should be avoided. Blending of roselle beverages and mango has been reported by Mgaya-Kilima et al. (2015). Investigations of the shelf life of the mango-roselle beverage for periods ranging from 2 to 6 months revealed that the acidity, color, phenol levels, and vitamin C content decreased with time. The deterioration of the blended juice could however be reduced by refrigeration. Roselle calyx has also been used in traditional food preparations where the ground powder prepared from dried calyx is soaked overnight in wood ash solution (Adanlawo & Ajibade, 2006). The soaked roselle calyx showed increased protein contents and reduced antinutritional contents.

The use of roselle calyx as a food coloring has also been reported. The extraction of food colors in roselle can be done using ethanol extraction or aqueous extraction. Roselle food coloring contains less ash and has high shelf life, and it also exhibits low pH values (high acidity) (Oluwaniyi et al., 2009). In India, extracts of roselle calyx have been incorporated in yogurt. This is done to improve the color and taste of the yogurt and has been marketed as a healthy product. The addition of roselle extracts also provides vitamins and other bioactive compounds to the yogurt. Roselle calyx extracts improved the organoleptic qualities and antioxidant property of the yogurt, while at the same time decreased the exudation of protein (Rasdhari et al., 2008). Apart from the calyx and leaves, the seeds from roselle plant are normally cooked and eaten as an indigenous nonanimal protein (meat substitute); for example, in Sudan the seed is cooked and then fermented. The resultant product was reported to be rich in proteins. The seed can also be eaten just after boiling. Studies undertaken by Gasim et al. (2008a,b) indicated that boiling of roselle seeds significantly increases the protein content, while reducing starch and soluble carbohydrate levels. In Ghana, Niger, and several other countries in West Africa, roselle leaves are eaten as vegetables and they are reported to contain minerals, which include calcium and magnesium (Glew et al., 2010).

The use of roselle is not limited to human diets. It has also found usage in animal feeds. A study by Ibrahim et al. (2007) on the impacts of roselle calyx extracts on the growth, hematologic, and serobiochemical parameters of Bovans chicks reported that 10% of roselle calyx exhibited some toxicity, which was however not fatal to the birds. The birds (Bovans chicks) reported reduced weight gain, therefore leading to poor utilization of feeds. Reports from Sokoto, Nigeria, about the use of roselle as a chicken feed, however, gave some favorable results (Habibullah et al., 2007). Lower levels of roselle calyx extract showed an increased cholesterol level in egg serum and yolk, when added to the chicken feed of egg-laying hens. Therefore it can be concluded that the use of roselle calyx extract as part of chicken feed is dependent on the concentration of the extract, which needs to be optimized. Lower concentrations may have some beneficial results, whereas higher levels could produce unfavorable results. The study of the effect of roselle products on other animals has attracted several researchers. A study of the use of roselle seed as a protein supplement in sheep diet was carried out in Sudan by Beshir and Babiker (2009). Addition of roselle to the sheep diet showed better feed intake leading to an increase in feed conversion efficiency and final body weight. Other investigations on the reproductive health of animals are also going on. The impact of roselle on the fetal development was investigated by Iyare and Adegoke (2008) in Lagos, Nigeria. A total of 18 in-bred virgin female rats aged between 10 and 12 weeks were used for the study. The investigation included adding the aqueous extracts of roselle to the drinking water consumed by the rats. The results of the abovementioned study indicated that consumption of roselle aqueous extract during pregnancy significantly increases postnatal weight gain and delayed the onset of puberty in the female offspring. The addition of roselle extracts to the daily feeds of rats recorded a beneficial health effect related to kidney treatment for rats (Ademiluyi et al., 2013). This could be an indication that the roselle extract could be beneficial to other mammals, and more studies need to be done in this direction to ascertain if extracts from roselle could be used for kidney treatments in human beings. While the mechanism of the working of hibiscus extracts in animals has not been fully recorded, it could be assumed that extracts from hibiscus have antioxidant property and when consumed as food or drink, they work on different parts of the body. This could be inferred by the research reported by Orji and Obi (2017), who reported that extracts from hibiscus reduced drug-induced alterations of liver functions in rats when tested with the use of paracetamol, a commonly available over-the-counter pain killer. These findings should be further investigated to establish whether the same results can be replicated in human beings, as it would present a low-cost approach for the prevention and/or cure of liver diseases.

3.2 Medical and Pharmacologic Applications

The use of roselle for medicinal and pharmacologic purposes has been reported by several researchers from different regions of the world. In the Baskoure region of Burkina Faso, the use of plants for medicinal purposes for daily healthcare is prevalent. A study undertaken to document the use of plants for medicinal purposes revealed that decoctions from roselle plant parts, which include roots, leaves, and calyces, are used for a variety of ailments, which include malaria, tonic fever, snake bites, stomach aches, and epilepsies (Nadembegaa et al., 2011). This information was gathered from traditional healers who normally pass their information from generation to generation using verbal methods. In another study undertaken in Kurdish Iraq, roselle, which is among the many medicinal plants sold in the Kurdish markets, is reportedly used for a variety of illnesses (obesity and hypertension) (Mati and Hugo, 2011). While the aforementioned reports could serve as examples of reports on the traditional usage of roselle in medicinal and pharmacologic products, some researchers have carried out laboratory experiments to try and verify the authenticity of the claims.

Research into the potency of roselle as an antioxidant can be broadly divided into two: methods of extracting phenolic compounds from roselle and the potency of roselle extracts as antioxidants. Investigations into the extraction of polyphenolic contents from roselle calyx using methanol, ethanol, acetone, and water as solvents showed ethanol and methanol to be better solvents than acetone and water. The study also indicated that phenols contributed more to the antioxidant activity of roselle than flavonoids (Chinedu et al., 2011). The quantification of phenolic compounds in roselle calyx was undertaken by Salvador et al. (2011), who were able to characterize the aqueous extracts of roselle and obtained 17 phenolic compounds in roselle extracts. These researchers used two methods of evaluating the antioxidant mechanisms, namely, the electron transfer mechanisms (i.e., reducing capacity) and the hydrogen atom transfer reactions. To further validate their research, Salvador et al. (2011) also compared the performance of roselle as an antioxidant with extracts from other compounds with proven antioxidant ability (olives and pomegranate), where roselle performance was rated as excellent.

The polyphenolic compounds found in roselle leaves and flowers have demonstrated an ability to act as antioxidants, a fact that has attracted the attention of many scientists. A study into the ability of extracts from sour tea made from roselle calyx to offer protection from lipid peroxidation induced by prooxidants (Fe(11), sodium nitroprusside, quinolinic acid) in rat's brain in vitro indicated that roselle tea extracts have high total phenol contents and chelating ability (Ganiyu, 2009). Therefore it can be concluded that roselle tea extracts exhibited high antioxidant and neuroprotective potentials. Extracts of roselle calyx have also shown good potential in being able to reduce cadmium-induced oxidative damage in the serum and tissues of rats (Omonkhua et al., 2009) and sodium arsenite-induced oxidative stress in rats. The study reported that roselle extracts have good chemopreventive and antioxidant properties (Usoh et al., 2005). The use of hibiscus extracts for effective antipoisoning intervention was further proven by Liu et al. (2006) who reported that the damage caused by carbon tetrachloride to the liver function of male Wistar rats was significantly reduced by using hibiscus extracts ranging in dosages from 1% to 5%. Further research on the antioxidant action of roselle extract on the functions of the liver and kidney was performed by Alkusi (2017). The ability of roselle extract to reduce oxidative damage in rats was however subject to a limiting dosage. Some level of dosage of roselle extract appears not to affect the functions of the liver and kidney (Nwachukwu et al., 2009), but other reports indicated that higher dosages of roselle extract could be toxic (Ukoha et al., 2015). It can be concluded that the roselle extract has an antioxidant effect on animals; therefore it is clear that the effect of the method of extraction and dosage taken is among the important factors that should be considered when studying the effect of roselle extracts on the kidney and liver functions of rats.

Apart from research on rats, other experiments have been reported on the effect of roselle extracts on other mammals. The potency of roselle calyx extracts from dried roselle flowers has been further tested for the prevention of 2,4-dinitrophenylhydrazine-induced tissue damage in rabbits (Ologundundu and Obi, 2005). The decoction of dried roselle flowers orally administered to the rabbits was able to not only prevent 2,4-dinitrophenylhydrazine-induced tissue damage but also reduce lipid peroxidative effect in the blood, brain, and liver in rabbits. The presence of antioxidants in roselle was also investigated by Umamaheswari and Givindan (2007), who used extracts from roselle leaves. The extracts from roselle leaves were obtained using methanol, ethanol, ethyl acetate, and chloroform. The extracts were used in the screening of hepatocellular carcinoma Hep 3B cell culture, using the cytotoxicity method. The results indicated that roselle has potent anticancer activity against Hep 3B cell lines. Therefore using roselle leaves as herbal therapy could have the potential of reducing chemotherapeutic side effects in cancer patients. A comparative research undertaken by Mazzio and Soliman (2009) showed that the efficiency of roselle extracts as a chemotherapeutic drug in anticancer treatment was medium to weak. The research of roselle in anticancer treatment is likely to attract more attention because the treatment is reported to exhibit limited or no side effects. Reports by Takuji et al. (2011) further validate the potency of roselle as an antioxidant. Protocatechuic acid, extracted from roselle, was tested using an animal (rats) model and it was proved that it can be used in conventional treatment programs to prevent cardiovascular diseases and cancer. The mechanism of prevention is based on its antioxidant properties and its ability to scavenge and increase the catalytic activity of endogenous enzymes, which are involved in the neutralization of free radicals. Further studies carried out on rabbits indicated that extracts from roselle calyx inhibit the angiotensin-converting enzyme (ACE) activity, hence validating roselle antioxidant properties (Deyanira et al., 2009). Owing to the inhibition of ACE activity, roselle has also been used to treat and manage blood pressure in humans (Muhammed et al., 2011).

Apart from its ability to reduce oxidative damage, the roselle calyx extract can be used for the management of obesity. As the economy in some of the developing countries improves, hitherto uncommon sicknesses, such as obesity, are on the increase. In western countries, obesity and overweight is a serious public health problem that has to be urgently addressed. The ability of roselle extract to manage obesity was demonstrated by using an animal (mice) model by Francisco et al. (2007). Obese and normal mice were orally fed extracts from roselle calyx for 60 days. During the study period, all the mice showed decreased body weight gain and increased liquid intake. In another study on the reduction of body weight by varying the amount of hibiscus extracts included in rat diets, Carvajal-Zarrabal et al. (2009) established that for percentages of hibiscus extracts ranging from 0% to 15% added to the daily diets of 40 Sprague-Dawley rats 6-week-old male rats), the higher the percentage of hibiscus extracts, the higher the reduction of body weight and the amount of food consumed, thus supporting the hypothesis that roselle can be used to manage obesity and body weight.

The diversity of the use of roselle plant can best be demonstrated by its use as an allelopathic product. Apart from its use in human and animal medicine, roselle can also be used as a weed control product. The aqueous methanolic extracts of the whole plant (roots, stem, and roots) obtained from roselle showed good allelopathic effects by its inhibitory effect on cress and alfalfa seedlings. These were reported as initial research findings, which are exciting, and need to be evaluated under field conditions (Suwitchayanona et al., 2015).

3.3 Roselle Extract Used for Coloring

Extracts from roselle calyx and petals contain anthocyanin, which is a pigment (coloring material) and also exhibits antioxidant activity. Extracts from the petals and the calyx of the roselle plant can be used to color various items, which include textiles, food, and other industrial products. Roselle extracts can be extracted using water at temperatures of 50–60°C, and the use of acidic conditions is advisable to increase the efficiency of the extraction (Inggrid et al., 2017). Extraction of the pigments can also be done by boiling or ethanol extraction. For food pigments, water extraction acidified with 2% citric acidic is commonly used. The pigments extracted from roselle can be stable especially when stored under refrigerated conditions (Manjula et al., 2018), which is common with most foods or beverages. Storage under higher temperatures may lead to the deterioration of color and other properties of the roselle extracts.

The pigments extracted from roselle calyx can also be applied to different materials including textiles, PVA coating materials, and poly(acrylamide-*co*-acrylic acid)-based and water-based coating materials (Lee et al., 2013a, 2013b). During the dyeing of textiles, different mordants (organic [e.g., aluminum sulfate, ferrous sulfate, potassium dichromate] and natural mordants) can be used. There are also different mordanting methods (pre, meta, and post) that can be used to dye textile substrates. Extraction of the coloring matter in roselle for use in textile dyeing can be implemented using organic solvents (methanol and ethanol) or aqueous methods. The acceptability of cotton fabric dyed using extracts from roselle calyx was reported by Ozougwu and Anyakoha (2016), who used aqueous extraction and aluminum sulfate to dye cotton fabric, where organoleptic attributes that included hue, texture (sight and feel), smell/odor, and evenness of shade were reported to be within acceptable standards. The use of extracts from *H. sabdariffa* as dyes needs further research so as to optimize the extraction temperature, material-to-liquor ratio, the section of the plant to be used, the solvents, the dyeing conditions (acidic or basic), and the type of mordants and mordanting methods. Dyestuff extracted from roselle have also been used in the study of dye-sensitized solar cells (Hammadi and Naji, 2014), which is an area that should be exploited because it will contribute not only to the increased use of the plant but also to the eco-friendly manufacturing of solar cells. Apart from cotton, the dyeing of wool by using extracts from roselle calyx and petals yielded different colors based on the type of mordant used (Kayabasi et al., 2001). The mordants used included copper sulfate, iron sulfate, alum of chrome, potassium bichromate, tannins, and zinc chloride.

3.4 Use of Roselle in Industrial Applications

The roselle stem may vary from 15 to 25 cm in diameter. The outer part of the stem is covered by a bark. The woody part of the stem can be used to make paper, while the bark contains textile fiber. The use of roselle for the manufacture of paper has been investigated by Olotuah (2006). Fibers extracted from roselle stem had a mean fiber length of 2.2 mm and slenderness and Runken ratio of 102 and 2.56, respectively. Although the Runken ratio was more than unity, the study reported that roselle can still be used to make paper of acceptable quality. The use of the Kraft method has been reported to give better quality paper from roselle when compared with non-Kraft methods (Dutt et al., 2009). The reports of the potential of using roselle as a paper-making raw material should attract many countries that are still using trees as the primary source of paper-making material. The trees used for making paper may need over 5 years before they can be harvested. The fact that roselle grows in less than 6 months should attract more research about its use as a paper-making material. With proper crop husbandry, roselle plants can be grown and the stem can be a source of paper-making material. The bast fiber obtained from roselle stem has been investigated by Eromosele et al. (1999), who reported that roselle fiber has an initial modulus of 2.92 N/tex and tenacity of 0.13 N/tex, which are lower than the equivalent tensile properties of other conventional bast fibers such as jute and hemp. Its specific work of rupture of 3.34 mN/tex was however higher than that of jute. The density of roselle fiber of 1.28 g/cm was within the range given by Modibbo et al. (2007), who also investigated the chemical characteristics of roselle fiber. The moisture absorption behavior of roselle fiber can be modified using chemical treatment. Raw roselle fibers showed the least moisture absorption because of the presence of gums and other nonfibrous contents covering the fiber. Treatment of the fiber with sodium hydroxide alters the absorption of the fiber. Like other bast fibers, the properties of the roselle bast fibers are affected by the growth environment, the age of the plant, and other plant husbandry conditions. The method used to extract the fiber will also affect the chemical and/or mechanical properties of the fibers. Therefore there is need to optimize the roselle-growing conditions and bast fiber extraction methods. The fiber obtained from roselle bark can be used in several applications, which include the manufacture of packaging material, baskets (see Fig. 3.2), and mats and as reinforcements in composite structures.

Roselle fiber can be used in the manufacture of grafted copolymers (Chauhan & Kaith, 2011). This can also serve as a utilization of waste of the short fibers, which may not be used for the manufacture of yarn. A study of the performance of roselle bark fiber in composites was undertaken by Singha and Thakur (2009), who reported that the mechanical properties of roselle fiber-reinforced phenolic (resorcinol-formaldehyde) resin matrix-based composites showed a decrease in mechanical properties when the fiber particle size

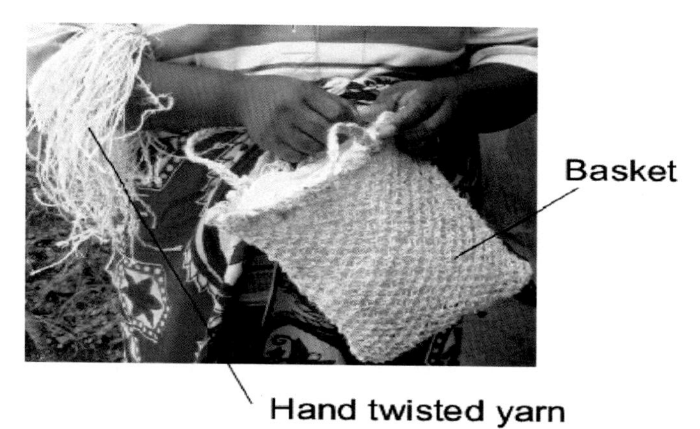

FIG. 3.2 Use of roselle fiber in the manufacture of baskets. (Mwasiagi, J. I., Yu, C. W., Phologolo, T., Waithaka, A., Kamalha, E., & Ochola J. R. (2014). Characterization of the Kenyan *Hibiscus sabdariffa* L. (Roselle), bast fiber. *Fibres and Textiles in Eastern Europe, 22*(3(105)), 31–34. http://www.fibtex.lodz.pl/2014/3/31.pdf, used with permission.)

increased. Good mechanical properties were obtained when roselle fibers were used in the particle form. Roselle composites however exhibited reduced low chemical resistance owing to the hydrophilic behavior of the fiber. The hydrophilic behavior of roselle fiber can be reduced by using chemical treatment as reported by Singha et al. (2009), who showed that silane treatment of roselle fiber reduced water absorption behavior. Other physiochemical properties were also affected in a manner that makes the fiber to be more suitable for use in composites.

4 CONCLUSIONS

This chapter is a review of the growing and use of the roselle plant. Roselle plant grows in the tropical and subtropical regions as both a cultivated and a wild plant, with its roots, stem, bark, leaves, flowers, and calyces having applications in human and animal diets and coloring foods, drinks, and textiles. It can be concluded that the roselle plant grows in many countries in all the continents and the whole plant (roots, stem, bark, leaves, seeds, and flowers [calyces]) can be used. Extracts from the leaves, petals, and calyces contain vitamins, dietary fibers, bioactive compounds, and minerals. The extract can be obtained by different methods but simple boiling can yield a good extract, which has a sweet and tart flavor and can be drank as a refreshing drink. The extracts from roselle plant has medicinal value ranging from weight control and management of cholesterol levels to use as antioxidants. Use of roselle extracts in treating liver and kidney problems in rats has provided encouraging results and could provide some pharmacologic reprieve for kidney and liver treatment. Extracts from roselle flowers can also be used to color food, drinks, and textile material. The stem can be used in several ways, which include using the bark as a bast fiber, while the core can be used to make paper. The bast fiber can be used for textile products, which include yarns, twines, mats, and textile composite materials.

REFERENCES

Abo-Baker, A. A., & Mostafa, G. G. (2011). Effect of bio-and chemical fertilizers on growth, sepals yield and chemical composition of Hibiscus sabdariffa at new reclaimed soil of South Valley area. *Asian Journal of Crop Science, 3*(1), 16–25. https://doi.org/10.3923/ajcs.2011.16.25

Adanlawo, I. G., & Ajibade, V. A. (2006). Nutritive value of the two varieties of roselle (*Hibiscus sabdariffa*) calyces soaked in wood ash. *Pakistan Journal of Nutrition, 5*(6), 555–557. https://doi.org/10.3923/pjn.2006.555.557

Ademiluyi, A. O., Oboh, G., Agbebi, O. J., & Akinyemi, A. J. (2013). Anthocyanin - rich red dye of *Hibiscus sabdariffa* calyx modulates cisplatin-induced nephrotoxicity and oxidative stress in rats. *International Journal of Biomedical Science, 9*(4), 243–248. https://www.ncbi.nlm.nih.gov/pmc/articles/PMC3884795/.

Aina, J. O., & Shodipe, A. A. (2006). Colour stability and vitamin C retention of roselle juice (*Hibiscus sabdariffa* L) in different packaging materials. *Nutrition & Food Science, 36*(2), 90–95. https://doi.org/10.1108/00346650610652295

Alkushi, A. G. (2017). Protective effect of sorrel extract on adult rats treated by carbon tetrachloride. *Pharmacognosy Research, 9*(2), 200–207. http://www.phcogres.com/text.asp?2017/9/2/200/204653.

Amusa, N. A., Adegbite, A. A., & Oladapo, M. O. (2005). Vascular wilt of roselle (*Hibiscus Sabdariffa* L. var. sabdariffa) in the Humid forest region of south −western Nigeria. *Plant Pathology Journal, 4*(29), 122−125. https://doi.org/10.3923/ppj.2005.122.125

Ayipio, E., Abu, M., Agyare, R. Y., Azewongik, D. A., & Bonsu, S. K. (2018). Growth and yield performance of roselle accessions as influenced by intercropping with maize in the Guinea savannah ecology of Ghana. *International Journal of Agronomy*. https://doi.org/10.1155/2018/9821825. Article ID 9821825.

Azooz, M. M. (2009). Foliar application with riboflavin (Vitamin B2) enhancing the resistance of *Hibiscus sabdariffa* L. (Deep red sapplels variety) to salinity stress. *Journal of Biological Sciences, 9*(2), 109−118. https://doi.org/10.3923/jbs.2009.109.118

Babatunde, F. E., & Mofoke, A. (2006). Performance of roselle (*Hibiscus sabdariffa* L) as influenced by irrigation schedules. *Pakistan Journal of Nutrition, 5*(4), 363−367. https://doi.org/10.3923/pjn.2006.363.367

Beshir, A. A., & Babiker, S. A. (2009). Performance of Sudanese desert lambs fed graded levels of roselle (*Hibiscus sabdariffa*) seeds instead of groundnut cake. *Pakistan Journal of Nutrition, 8*(9), 1442−1445. https://doi.org/10.3923/pjn.2009.1442.1445

Carvajal-Zarrabal, O., Hayward-Jones, P. M., Orta-Flores, Z., Nolasco-Hipolito, Barradas-Dermitz, C. D. M., Aguilar-Uscanga, M. G., & Pedroza-Hernandez, M. F. (2009). Effect of *Hibiscus sabdariffa* L. dried calyx ethanol extract on fat absorption-excretion, and body weight implication in rats. *Journal of Biomedicine and Biotechnology*. https://doi.org/10.1155/2009/394592. Article ID 394592.

Changdee, T., Polthanee, A., Akkasaeng, C., & Morita, S. (2009). Effect of different waterlogging regimes on growth, some yield and roots development parameters in three fiber crops (*Hibiscus cannabinus* L., *Hibiscus sabdariffa* L. and *Corchorus olitorius* L.). *Asian Journal of Plant Science, 8*(8), 515−525. https://doi.org/10.3923/ajps.2009.515.525

Chauhan, A., & Kaith, B. (2011). Transforming waste biomass to novel grafted copolymer. *World Journal of Engineering, 8*(4), 347−356. https://doi.org/10.1260/1708-5284.8.4.347

Chinedu, P. A., Ijeoma, E., Olusoal, A., & Ayobami, O. A. (2011). Polyphenolic content and antioxidant activity of *Hibiscus sabdariffa* calyx. *Research Journal of Medicinal Plants, 5*(5), 557−566. https://doi.org/10.3923/rjmp.2011.557.566

Deyanira, O., Enrique, J.-F., Alejandro, Z., Armando, H.-A., Jaime, T., & Laura, A. (2009). Inhibition of angiotensin converting enzyme (ACE) activity by the anthocyanins delphinidin- and cyanidin-3-O-sambubiosides from *Hibiscus sabdariffa*. *Journal of Ethnopharmacology, 127*, 7−10. https://doi.org/10.1016/j.jep.2009.09.059

Dutt, D., Upadhyay, J. S., Singh, B., & Tyagi, C. H. (2009). Studies on Hibiscus cannabinus and *Hibiscus sabdariffa* as an alternative pulp blend for softwood: An optimization of kraft delignification process. *Industrial Crops and Products, 29*(1), 16−26. https://doi.org/10.1016/j.indcrop.2008.03.005

Eromosele, I. C., Ajayi, J. O., Kwasi, Njaprim, G., & Modibbo, U. (1999). Characterization of cellulosic fibers. *Journal of Applied Polymer Science, 73*, 2057−2060. https://doi.org/10.1002/(SICI)1097-4628(19990906)73:10<2057::AID-APP24>3.0.CO;2-E

Francisco, J. A.-A., Alejandro, Z., Dolores, P.-G., Julio, C. A.-P., Eunice, R.-N., Efrain, A. C.-S., Laura, I. V.-C., & Ruben, R.-R. (2007). Effect of *Hibiscus sabdariffa* on obesity in MSG mice. *Journal of Ethnopharmacology, 114*(2007), 66−71. https://doi.org/10.1016/j.jep.2007.07.020

Ganiyu, O. (2009). The neuroprotective potentials of sour (*Hibiscus sabdariffa* calyx) and green (camellia senensis) teas on some pro-oxidants induced oxidative stress in Brain. *Asian Journal of Clinical Nutrition, 1*(1), 40−49. https://doi.org/10.3923/ajcn.2009.40.49

Gasim, A. E., Yagoub, A., & Mohammed, M. A. (2008b). Furundu, a meat substitute from fermented roselle (*Hibiscus sabdariffa* L.) seed: Investigation on amino acids composition, protein fractions, minerals content and HCL-extractability and microbial growth. *Pakistan Journal of Nutrition, 7*(2), 352−358. https://doi.org/10.3923/pjn.2008.352.358

Gasim, A. E., Yagoub, A., Mohammed, M. A., & Baker, A. A. A. (2008a). Effect of soaking and cooking on chemical composition, bioavailability of minerals and in vitro protein digestibility of roselle (*Hibiscus sabdariffa* L.) seed. *Pakistan Journal of Nutrition, 7*(1), 50−56. https://doi.org/10.3923/pjn.2008.50.56

Glew, R. S., Amoako-Atta, B., Ankar-Brewoo, G., Presley, J., Chuang, L.-T., Millson, M., Smith, B. R., & Glew, R. H. (2010). Furthering an understanding of West African plant foods: Mineral, fatty acid and protein content of seven cultivated indigenous leafy vegetables of Ghana. *British Food Journal, 112*(10), 1102−1114. https://doi.org/10.1108/00070701011080230

Gomez-Leyva, J. F., Martinez, A. L. A., Lopez, M. I.,G., Silos, E. H., Ramirez-Cervantes, F., & Andrade-Gonzalez, I. (2008). Multiple shoot regeneration of roselle (*Hibiscus sabdariffa* L.) from a shoot apex culture system. *International Journal of Botany, 4*(3), 326−330. https://doi.org/10.3923/ijb.2008.326.330

Habibullah, S. A., Bilbis, L. S., Ladan, M. J., Ajagbonna, O. P., & Saidu, Y. (2007). Aqueous extracts of *Hibiscus sabdariffa* calyces reduces serum triglycerides but increases serum and egg yolk cholesterol of shika brown laying hens. *Asian Journal of Biochemistry, 2*(1), 42−49. https://doi.org/10.3923/ajb.2007.42.49

Hammadi, O. A., & Naji, N. I. (2014). Effect of acidic environment on the spectral properties of *Hibiscus sabdariffa* organic dye used in dye-sensitized solar cells use of roselle in industrial applications. *Iraq Journal of Applied Physics, 10*(2), 27−31. https://www.iasj.net/iasj/download/71f393c7544b6cd3.

Ibrahim, I. A., Badwi, A. M. A. E., Bakhiet, A. O., Gadir, W. S. A., & Adam, S. E. I. (2007). A 9 week feeding study of cuminum cyminum and *Hibiscus sabdariffa* in Bovans chicks. *Journal of Pharmacology and Toxicology, 2*(7), 666−671. https://doi.org/10.3923/jpt.2007.666.671

Inggrid, H. M., Jaka, H. M., & Santoso, H. (2017). Natural red dyes extraction on roselle petals. *IOP Conference Series: Materials Science and Engineering, 162*(2017). https://doi.org/10.1088/1757-899X/162/1/012029

Iyare, E. E., & Adegoke, O. A. (2008). Postnatal weight gain and onset of puberty in rats exposed to aqueous extract of *Hibiscus sabdariffa* in utero. *Pakistan Journal of Nutrition, 7*(1), 98–101. https://doi.org/10.3923/pjn.2008.98.101

Kayabas, N., Kızıl, S., & Toncer, O. (2001). An investigation on the colours obtained from roselle (*Hibiscus sabdariffa* L.) and their colour fastnesses in woolen carpet yarns. *Turkish Journal of Field Crops, 6*, 14–18. https://dergipark.org.tr/en/pub/tjfc/issue/17146/179269.

Lee, S. V., Halim, N. A., Arof, A. K., & Abidin, Z. H. Z. (2013b). Characterisation of poly(vinyl alcohol) coating mixed with anthocyanin dye extracted from roselle flower with different nitrate salt. *Pigment & Resin Technology, 42*(2), 146–151. https://doi.org/10.1108/03699421311301142

Lee, S. V., Vengadaesvaran, B., Arof, A. K., & Abidin, Z. H. Z. (2013a). Characterisation of poly(acrylamide-co-acrylic acid) mixed with anthocyanin pigment from *Hibiscus sabdariffa* l. *Pigment & Resin Technology, 42*(2), 103–110. https://doi.org/10.1108/03699421311301089

Liu, J.-Y., Chen, C.-C., Wang, W.-H., Hsu, J.-D., & Wang, C.-J. (2006). The protective effects of *Hibiscus sabdariffa* extract on CCl4-induced liver fibrosis in rats. *Food and Chemical Toxicology, 44*(3), 336–343. https://doi.org/10.1016/j.fct.2005.08.003

Manjula, G. S., Krishna, H. C., Reddy, M. C., Karan, M., & Kumar, M. M. (2018). Effect of storage temperature on various parameters of extracted pigment from roselle (*Hibiscus sabdariffa* L.) calyces for edible colour. *International Journal of Current Microbiology and Applied Sciences, 7*(1), 3382–3390. https://doi.org/10.20546/ijcmas.2018.701.400

Maqsudul, A. A. K., Nasimul, G. M., Rahman, M., Islam, M. R., Nuruzzaman, M., & Khandker, S. (2001). Effect of bast fiber cultivation on soil fertility. *Online Journal of Biological Sciences, 1*(12), 1127–1129. https://doi.org/10.3923/jbs.2001.1127.1129

Mati, E., & Hugo, H. D. (2011). Ethnobotany and trade of medicinal plants in the Qaysari market, Kurdish autonomous region, Iraq. *Journal of Ethnopharmacology, 133*, 490–510. https://doi.org/10.1016/j.jep.2010.10.023

Mazzio, E. A., & Soliman, K. F. A. (2009). *In vitro* screening for the tumoricidal properties of international medicinal herbs. *Phytotherapy Research, 23*(3), 385–398. https://doi.org/10.1002/ptr.2636

Mgaya-Kilima, B., Remberg, S. V., Chove, B. E., & Wicklund, T. (2015). Physiochemical and antioxidant properties of roselle-mango juice blends; effects of packaging material, storage temperature and time. *Food Science and Nutrition, 3*(2), 100–109. https://doi.org/10.1002/fsn3.174

Modibbo, U. U., Aliyu, B. A., Nkafamiya, I. I., & Manji, A. J. (2007). The effect of moisture imbibition on cellulosic bast fibers as industrial raw materials. *International Journal of the Physical Sciences, 2*(7), 163–168. https://doi.org/10.5897/IJPS.9000562

Monteiro, M. J. P., Costa, A. I. A., Fliedel, G., Cisse, M., Bechoff, A., Pallet, D., Tomlins, K. I., & Pintado, M. (2017). Chemical-sensory properties and consumer preference of hibiscus beverages produced by improved industrial processes. *Food Chemistry, 225*, 202–212. http://publications.cirad.fr/une_notice.php?dk=583995.

Muhammed, A., Ali, A., Kamal, M., & Mohammad, H. T. A. (2011). Ethnopharmacological survey of medicinal herbs in Jordan, the Northern Badia region. *Journal of Ethnopharmacology, 137*(1), 27–35. https://doi.org/10.1016/j.jep.2011.02.007 (article in press).

Mwasiagi, J. I., Yu, C. W., Phologolo, T., Waithaka, A., Kamalha, E., & Ochola, J. R. (2014). Characterization of the Kenyan *Hibiscus sabdariffa* L. (Roselle), bast fiber. *Fibres and Textiles in Eastern Europe, 22*(3(105)), 31–34. http://www.fibtex.lodz.pl/2014/3/31.pdf.

Nadembegaa, P., Boussimb, J. I., Nikiemac, J. B., Ferruccio Poli, F., & Fabiana Antognoni, F. (2011). Medicinal plants in Baskoure, Kourittenga Province, Burkina Faso: An ethnobotanical study. *Journal of Ethnopharmacology, 133*, 378–395. https://doi.org/10.1016/j.jep.2010.10.010

Nnebue, O. M., Ogoke, I. J., Obilo, O. P., Agu, C. M., Ihejirika, G. O., & Ojiako, F. O. (2014). Estimation of planting dates for roselle (*Hibiscus sabdariffa* L.) in the humid tropical environment of Owerri, south-eastern Nigeria. *Agrosearch, 14*(2), 168–178. https://doi.org/10.4314/agrosh.v14i2.7

Nurul, H. J., & Mikael, A. P. (2018). Physicochemical properties and oxidative storage stability of milled roselle (*Hibiscus sabdariffa* L.) seeds. *Molecules, 23*(385), 1–15. https://doi.org/10.3390/molecules23020385

Nwachukwu, D. C., Ejezie, F. E., AA Eze, A., Achukwu, P. U., & Nwadike, K. I. (2009). Toxicological effects of aqueous extract of *Hibiscus Sabdariffa* on the liver and kidney. *International Journal of Medicine and Health Development, 14*(1). https://www.ajol.info/index.php/jcm/article/view/44855.

Olayemi, F., Adedayo, R., Muhummad, R., & Bamishaiye, E. (2011). The nutritional quality of three varieties of Zobo (*Hibiscus sabdariffa*) subjected to the same preparation conditions. *American Journal of Food Technology, 6*(8), 705–708. https://doi.org/10.3923/ajft.2011.705.708

Ologundundu, A., & Obi, F. O. (2005). Prevention of 2,4 dinitrophenylhydrazine (DNPH)-induced tissue damage in rabbits by orally administered decoction of dried flower of *Hibiscus sabdariffa* L. *Journal of Medical Science, 5*(3), 208–211. https://doi.org/10.3923/jms.2005.208.211

Olotuah, O. F. (2006). Suitability of some local bast fibre plants in pulp paper making. *Journal of Biological Sciences, 6*(3), 635–637. https://doi.org/10.3923/jbs.2006.635.637

Oluwaniyi, O. O., Dosumu, O. O., Awoloal, G. V., & Abdulraheem, A. F. (2009). Nutritional analysis and stability studies of some natural and synthetic food colourants. *American Journal of Food Technology, 4*(5), 218–225. https://doi.org/10.3923/ajft.2009.218.225

Omonkhua, A. A., Adesunloro, C. A., Osaloni, O. O., & Olubodun, S. O. (2009). Evaluation of the effects of aqueous extracts of *Hibiscus sabdariffa* calyces on cadmium −induced oxidative damage in rats. *Journal of Biological Sciences*, 9(1), 68–72. https://doi.org/10.3923/jbs.2009.68.72

Orji, B. O., & Obi, F. O. (2017). Effects of concurrent administration of paracetamol and aqueous extract of *Hibiscus sabdariffa* L calyx on paracetamol hepatotoxicity in mice. *Journal of Pharmaceutical, Chemical and Biological Sciences, 5*(2), 108–117. https://www.jpcbs.info/2017_5_2_05_Blessing.pdf.

Ozougwu, S. U., & Anyakoha, E. U. (2016). Acceptability of cotton fabric treated with dye extracted from roselle (*Hibiscus sabdariffa*) calyces based on its phytochemical composition and evaluation of organoleptic attributes. *African Journal of Agricultural Research, 11*(33), 3074–3081. https://doi.org/10.5897/AJAR2014.9108

Plotto, A. (2004). *HIBISCUS: Post-production management for improved market access* (pp. 3–15). Rome: Food and Agriculture Organization of the United Nations (FAO). http://www.fao.org/fileadmin/user_upload/inpho/docs/Post_Harvest_Compendium_-_Hibiscus.pdf.

Preciado-Saldana, A. M., Domınguez-Avila, J. A., Ayala-Zavala, J. F., Villegas-Ochoa, M. A., Sayago-Ayerdi, S. G., Wall-Medrano, A., Gonzalez-Cordova, A. F., & Gonzalez-Aguilar, G. A. (2019). Formulation and characterization of an optimized functional beverage from hibiscus (*Hibiscus sabdariffa* L.) and green tea (*Camellia sinensis* L.). *Food Science and Technology International, 25*(7), 547–561. https://doi.org/10.1177/1082013219840463

Rasdhari, M., Parekh, T., Dave, N., Patel, V., & Subhash, R. (2008). Evaluation of various physio-chemical properties of *Hibiscus sabdariffa* and L. casei incorporated probiotic yoghurt. *Pakistan Journal of Biological Sciences, 11*(17), 2101–2108. https://doi.org/10.3923/pjbs.2008.2101.2108

Salvador, F.-A., Inmaculada, C. R.-M., Raul, B.–D., Federica, P. J. J., Vicente, M. A. S.-C., & Alberto, F.–G. (2011). Quantification of the polyphenolic fraction and in vitro antioxidant and in vivo anti-hyperlipemic activities of *Hibiscus sabdariffa* aqueous extract. *Food Research International, 44*(5), 1490–1495. https://doi.org/10.1016/j.foodres.2011.03.040

Singha, A. S., & Thakur, V. K. (2009). Physical, chemical and mechanical properties of *Hibiscus sabdariffa* fiber/polymer composite. *International Journal of Polymeric Materials, 58*(4), 217–228. https://doi.org/10.1080/00914030802639999

Singha, A. S., Thakur, V. K., Mehta, I. K., Shama, A., Khanna, A. J., Rana, R. K., & Rana, A. K. (2009). Surface-modified *Hibiscus sabdariffa* fibers: Physicochemical, thermal, and morphological properties evaluation. *International Journal of Polymer Analysis and Characterization, 14*(8), 695–711. https://doi.org/10.1080/10236660903325518

Suwitchayanona, Piyatida, P., Pukclaia, P., Ohnob, O., Kiyotake, K., & Kato-Noguchia, H. (2015). Isolation and identification of an allelopathic substance from *Hibiscus sabdariffa*. *Natural Product Communications, 10*(5), 765–766. https://doi.org/10.1177/1934578X1501000516

Syahnaz, A., Naquib, S., Devagi, K., & Hollena, N. (2019). Growth performance of roselle (*Hibiscus sabdariffa*) under application of food waste compost and Fe3O4 nanoparticle treatment. *International Journal of Recycling of Organic Waste in Agriculture, 8*(2009), 299–309. https://doi.org/10.1007/s40093-019-00302-x

Talukder, F. A. H., Islam, M. I. S., Chanda, C., Ahmad, I., & Zakaria Ahmed, Z. (2001). Phenology of jute, kenaf and roselle seed crops at different date of sowing. *Pakistan Journal of Biological Sciences, 4*(11), 1316–1318. https://doi.org/10.3923/pjbs.2001.1316.1318

Takuji, T., Takahiro, T., & Mayu, T. (2011). Potential cancer chemopreventive activity of protocatechuic acid. *Journal of Experimental and Clinical Medicine, 3*(1), 27–33. https://doi.org/10.1016/j.jecm.2010.12.005

Ukoha, U. U., Mbagwu, S. I., Ndukwe, G. U., & Obiagboso, C. (2015). Histological and biochemical evaluation of the kidney following chronic consumption of *Hibiscus sabdariffa*. *Advances in Biology*. https://doi.org/10.1155/2015/486510. Article ID 486510.

Usoh, I. F., Akpan, E. J., Etim, E. O., & Farombi, E. O. (2005). Antioxidant actions of dried flower extracts of *Hibiscus sabdariffa* L. on sodium arsenite –induced oxidative stress in rats. *Pakistan Journal of Nutrition, 4*(3), 135–141. https://doi.org/10.3923/pjn.2005.135.141

CHAPTER 4

Roselle (*Hibiscus sabdariffa* L.): Processing for Value Addition

M. AHIDUZZAMAN • TAHMINA SADIA JAMINI • A.K.M. AMINUL ISLAM

1 INTRODUCTION

Roselle (*Hibiscus sabdariffa* L.) is more than an eye-catching crop and is locally known by different names in different countries (Ismail et al., 2008). It belongs to the family Malvaceae; it originated from West Africa (Shoosh, 1993) and is commonly available in the tropics especially in the African countries (Abu-Tarboush et al., 1997). Tropical warm and humid climate is suitable for roselle cultivation and production. The temperature range within which roselle can grow is between 18 and 35°C, with an optimum of 25°C (Ansari et al., 2013). Therefore roselle can be easily grown and used to promote nutraceutical and pharmaceutical industries. Superior quality roselle is produced in Sudan, Jamaica, and Egypt. The best quality roselle is produced in Sudan (Mohamed et al., 2012). Senegal and Mali are also producing roselle for their domestic use, which is sold in the local market. China and Thailand are the largest producers in Asia. Malaysia also started roselle cultivation as a relatively new crop to create an industry (Mohamad et al., 2002). Roselle is popularly recognized as "*mesta*" or "*chukur*" in the Indian subcontinent including Bangladesh (Halimatul et al., 2007; Rao, 2008). It is commercially cultivated in different countries such as India, Indonesia, Malaysia, Sudan, Egypt, and Mexico (Patel, 2014). There is great market potential for roselle as a cash crop for farmers in warmer climates where it grows well. A single roselle plant may yield as many as 250 calyces or 1−1.5 kg fresh weight depending on the environmental conditions and management. Yield for leaves is about 10 tons per hectare. For centuries, roselle has been used in a number of dishes, beverages, and the conventional remedy of diseases. Roselle is also famous for its high nutritional and medicinal values. Nutritional analysis of the calyces of roselle showed that they are high in calcium, iron, niacin, and riboflavin. Roselle is also a source of antioxidants and anthocyanins, which act as free radical scavengers and inhibit lipid peroxidation (Islam et al., 2016). Roselle contains carbohydrates, crude fiber, ash content, and other constituents. It is also rich in minerals especially potassium and magnesium, and vitamins (ascorbic acid, niacin, and pyridoxine) are also present in appreciable amounts (Luvonga et al., 2010). The calyx is usually used to prepare jam, jelly, cakes, ice cream, preserves, and herbal beverage (Hirunpanich et al., 2006; Islam et al., 2016). Roselle contains a group of natural pigments, which exist in the dried calyces and exhibit antioxidant activity. It might be a potential food additive used to prevent contamination from a group of food spoilage-causing bacteria (Chao & Yin, 2009). Roselle is also used for the treatment of diseases, as it exhibits antibacterial activities against a group of food spoilage-causing bacteria (Olaleye, 2007). Besides food and medicinal use, it provides natural fibers replacing synthetic fibers in composite components. Therefore roselle has the potential to enable a better selection of materials for domestic and industrial uses.

2 DOMESTIC AND INDUSTRIAL USES

The roselle plant has multipurpose uses. Uses of different parts of roselle are many and varied both in food and in traditional medicine. Roselle leaves, fruits, roots, and seeds are used as food and as fiber in textiles in different parts of the world (Cobley, 1975; Duke et al., 2003). The young leaves and tender stems of roselle are consumed raw as a green vegetable; leaves are also used in salad. The calyces are the most important part of the roselle plant. Roselle calyx is attached to the seed capsule and is detached before use. The calyces may be merely chopped and added fresh to fruit salad (Mady, Manuel, Mama, Augustin, & Max, 2009; Mady, Manuel, Mama, & Ndiaye, 2009). Fleshy red calyces of roselle are also used for the production

Roselle. https://doi.org/10.1016/B978-0-323-85213-5.00005-6

of soft drinks. In Africa, they are cooked as a side dish and eaten with pulverized peanuts. For stewing as sauce, they may be left intact, if tender, and cooked with sugar. In North East India, leaves and calyces of roselle are cooked with chicken or fish. The poor people in Myanmar consume roselle leaves as an inexpensive vegetable (Dy Phon, 2000). In Bangladesh, roselle or mesta leaves are steamed with dried or fresh fish to make a paste with garlic, onion, and chilies. A delicious soup is also prepared from mesta leaves, along with prawn (Islam et al., 2016). Leaves are also used to make *pacchadi* by mixing with spices. Roselle seeds are a good source of protein, fat, and total sugars and are widely used in the diet in many African countries. The root has been reported as an aperient because of the presence of tartaric acid (Brennan, 2015). After harvesting the calyces, roselle plants are often used as fodder for livestock.

Dried calyces of roselle are commercially available, and it is appreciated to obtain concentrated extracts, which might be used in the food and pharmaceutical industries. The bast fiber is a good substitute for jute and it is used for making twine, cordage, rope, netting, and sacks. The bast fiber is also used in the paper industry. The bright red color with exceptional flavor and other organoleptic attributes of roselle calyces make them valuable food products (Islam et al., 2016) such as wine, syrup, ice cream, pies, snakes, tarts, and other desserts (Buletin et al., 1984; Eslaminejad & Zakaria, 2011). The red calyces are used to make a soft drink and are also used as food coloring. The dried red calyces are commonly used to prepare a tea, drunk hot or, more commonly, cold after adding sugar. The fresh drink of roselle is very popular from Senegal to Sudan and in Egypt and other North African countries where it is referred to as *karkadé*'. The calyces are also boiled down to make a syrup concentrate. Owing to their high inherent pectin content, roselle jams and jellies are also popular, particularly in Senegal, the Caribbean region, and southern Asia. The roselle seed oil is used for cooking, e.g., in Chad, Tanzania, and China. However, the seed oil is claimed to contain some toxic substances and may be better used in the soap and cosmetics industries (Al-Okbi et al., 2017). It is also reported to be a potential feedstock for biodiesel (Ahmed et al., 2016; Nakpong & Wootthikanokkhan, 2010). The oil is also used as an ingredient of paints. The ornamental value of roselle is of another interest, as a garden plant or cut flower (Ansari et al., 2013). The decorative red stalks with ripe red fruits are exported to Europe where they are used in flower arrangements (Alegbejo et al., 2003).

3 PHYSICOCHEMICAL PROPERTIES OF ROSELLE CALYCES

Roselle (*H. sabdariffa* L.) is primarily cultivated because of the consumption of its calyx and is commercially important in the food industry for the production of juices, jams, salads, pigments, and beverages (Borrás-Linares et al., 2015). The proximal compositions of roselle calyces have been reported, which include crude protein, crude fat, carbohydrates, total ash, ascorbic acid, total phenol, total flavonoid, total anthocyanin, and total soluble and insoluble fibers (Duarte-Valenzuela et al., 2016). On a dry basis, calyces contain proteins, fats, carbohydrates, raw fiber, and ashes (Adanlawo and Ajibade, 2006), as well as vitamins, organic acids, and phytosterols (Ismail et al., 2008). Roselle calyces are also an important source of minerals, including K, Ca, and Mg, as well as trace elements such as Fe, Mn, Zn, and Cu (Evans & Halliwell, 2001). They also contain ascorbic acid, β-carotene, and lycopene in the concentrations of 141.09 mg/100 g, 1.88 mg/100 g, and 164.34 μg/100 g, respectively (Hinojosa-Gómez et al., 2018). Roselle calyces corroborated with antioxidant compounds (e.g., phenols, anthocyanins, and antioxidant activity), in addition to their high content of total and insoluble fibers. Anthocyanins present in roselle are delphinidin-3-sambubioside, cyanidin-3 sambubioside, delphinidin-3-glucoside, and cyanidin-3-glucoside. They contribute benefits for health as a good source of antioxidants and as a natural food colorant (Kilima et al., 2014). The minerals play a role in health by functioning as antioxidants or as components of antioxidant enzymes. In addition to the benefits of the mineral content, roselle calyces confer pharmacologic, nutraceutical, and cosmetologic properties. Within this group are the polyphenols including delphinidin and cyanidin (Borrás-Linares et al., 2015; Jabeur et al., 2017), which show antioxidant activity in the human body (Wang et al., 2011).

4 DRYING OF ROSELLE CALYCES

Dehydration or drying is a method of food preservation, noted as the oldest and the most widely used method for food preservation that is practiced since human civilization (Midilli et al., 2002; Sacilik, 2007). It is a preservation method that removes water biologically available to microbes in order to reduce microorganism growth (Esper & Muhlbauer, 1998). It is widely adapted as the foremost postharvest operation for agricultural produce because it conserves the quality of the produce in dried form (Janjai & Tung, 2005). In addition, it facilitates efficient packaging, increased transportability, and longer periods of storage (Vengaiah & Pandey, 2007). However, there is some caution in drying of delicate crops like roselle. The roselle calyces are bright red (Fig. 4.1) comprising anthocyanins such as cyanidin-3-sambubioside, cyanidin-3-glucoside, and delphinidin-3-glucoside (Castañeda-Ovando et al., 2009), in addition to ascorbic acid. Anthocyanins exhibit many times more antioxidant activity than ascorbic acid (Wang et al., 1997). Hence anthocyanins ensure the antioxidant activity of roselle extract. However, the anthocyanins can be removed from food products when exposed to heat, oxygen, and light during processing (Chen et al., 2005). Rapid thermal degradation of anthocyanins in roselle extract occurs at temperatures above 100°C (Mazza & Miniati, 1993).

4.1 Sun Drying

The calyces can be dried in sunshine directly. However, the quality is of concern in this process. Sun drying of roselle is done in the following manner: after harvesting, it is dried naturally in sunlight when the sun shines, repeating the next day, and the process is carried out in an average of 3–4 days to reach an acceptable level of moisture content. The duration of the drying process depends on the strength of sunshine and the duration of exposure, as well as the environment (relative humidity and wind). The end product quality is often badly affected by open-sky exposure to pollution, leading to lower economic profits for producers (Hahn et al., 2011). Besides, owing to the exposure to the open-sky environment on floor or rooftops of houses, roselle gets adulterated by acquiring pollutants, bird wastage, debris, fungal attacks due to inadequate drying, etc. that lead to an inferior quality of the product, causing difficulties to market and providing low economic returns to producers. Therefore alternative drying methods are required to conserve the quality of the products. As a result, a variety of solar dryers are being offered to be attractive as a viable alternative to open-sky sun drying. In solar drying the drying process can be controlled by varying the drying parameters and quicker drying can be achieved and the produce is well protected during the process. Hence solar drying is discussed in the following section.

4.2 Solar Drying

The solar drying system is quite controllable to protect the product from deterioration during the drying operation. The system uses solar energy indirectly. The drying chamber could be a cabinet type or a tunnel where air heated by solar energy is blown. The air could be moved by natural draft where a chimney is integrated in the system or circulated with a blower in suction or forced mode. A typical schematic of a solar dryer is shown in Fig. 4.2 that facilitates producing quality dried products. The dryer has a cabinet where in the product is spread on tray for drying. Heat energy required for drying is collected through a solar collector. Heat is conserved by heating up water in a tank. There are two loops for water circulation: one connects the solar collector and the water tank that heats up the water in the tank and the other one connects the water tank and the heat exchanger that heats up the air before entering the dryer cabinet. A water pump is used to recirculate the water in both loops. The air blower pushes air into the heat exchanger so that the air gets warm and passes out the heat exchanger and enters the drying cabinet. The hot air passes through the trays loaded with roselle calyces and passes out through the exit port. Electric energy is needed to drive the water pumps and air blower. A solar photovoltaic (SPv) is integrated in the system to run the auxiliary devices.

Several authors reported their developed solar dryers and evaluated them by fitting different thin-layer drying models. Saeed (2010) demonstrated a cabinet type solar dryer for roselle drying. His dryer consists of a solar collector for increasing the intensity of the heating the water in the storage tank. The total

FIG. 4.1 Processing of roselle calyces in the Department of Agro-Processing, Bangabandhu Sheikh Mujibur Rahman Agricultural University.

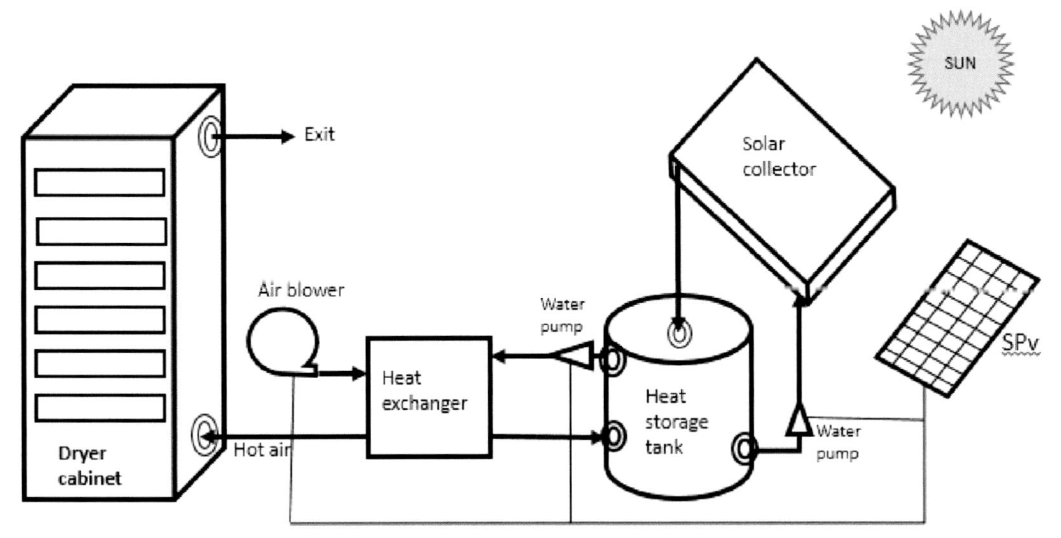

FIG. 4.2 Schematic diagram of an improved solar dryer. *SPv*, solar photovoltaic.

area of the collector was 9.86 m^2. This system integrated a silica-gel dehumidifier to increase the drying power of the hot air; there is also a regeneration unit to revive the silica gel for use in the next run. The dryer can stimulate the temperature of drying air up to 65°C. The dryer was evaluated by validating the logarithmic model of drying (Togrul & Pehlivan, 2002, 2003; Wang et al., 2007). Another study was conducted by Hahn et al. (2011) on the optimization of roselle drying time and the quality of the dried product. They used a tunnel type solar dryer. They constructed a tunnel covered by polyethylene plastic for drying roselle calyces. The performance of the dryer was evaluated by inspecting calyx color, drying rate, and crispness of dried products. The tunnel dryer was also evaluated by operating in three different drying operation conditions: (1) fan control, (2) an air recirculating system using silica gel, and (3) solar-biogas hybrid system by burning 1 kg of biogas for increasing air temperature to obtain 1 kg of dried product. The drying time varied from 27 to 4.5 h. The highest drying temperature raised to 70°C. Castañeda-Miranda et al. (2014) conducted another study on the continuous drying of roselle using solar cabinet dryer. They used solar energy by using a solar concentrator to heat up oil used as a thermal energy carrier fluid. The concentrator is a cylindric parabolic reflector that projects concentrated rays at the fluid inside the pipe. Hot oil is preserved in an insulated storage tank. The hot oil flows through the heat-exchanger coil in a drying prechamber wherein drying air flows through the heat exchanger and gets heated to a temperature at a satisfactory level before entering the cabinet. Then the hot air passes cross-currently though the dryer cabinet and dehydrates the calyces. The exhaust air, along with product humidity, is ejected from the drying chamber to the environment. The product is fed to the drying system at the top of the cabinet, and after passing through the dryer, the product is collected in a glass-protected chamber where its final humidity is measured for quality assurance purposes by a humidity sensor. The product is then ready for packaging and shipment.

4.3 Oven Drying

Oven drying is nothing but forced convection cabinet dryer. The heat energy is supplied from an electric heater. It could be operated on a laboratory scale or an industrial scale. Laboratory scale is often used by researchers to facilitate their work. A typical forced convection cabinet type oven dryer is shown in Fig. 4.3. The industrial type of oven drying requires a huge supply of raw materials to run a business economically. This type of drying method facilitates tuning up the quality of the dried product as per requirement.

4.4 Spray Drying

4.4.1 Spray-drying process

Spray dryer can dry a liquid product efficiently compared with other methods. It turns a solution

FIG. 4.3 Drying of roselle calyces in a forced convection cabinet dryer in the Department of Agro-Processing, Bangabandhu Sheikh Mujibur Rahman Agricultural University.

into a powder in a single step. A schematic of a spray dryer is shown in Fig. 4.4. The *spray dryer* consists of a high pressure feed pump, an atomizer for atomizing the feed liquid inside the drying chamber, an air heating unit to provide predefined temperature in air, an air dispenser, a drying chamber, a dried powder product collector, and an exhaust air cleansing system. This method of drying is very much suitable for preserving natural color, volatiles and phenolic compounds, etc. Heated air is used as the drying medium; however, if the liquid is flammable or oxygen sensitive then nitrogen is used. Spray drying parameters such as atomization pressure, feed flow rate, feed viscosity, feed surface tension, inlet temperature, drying airflow rate, outlet temperature, residence time inside drying chamber, and glass-transition temperature are need to be predefined before starting the process. The parameters could vary according to the droplet size and physicochemical properties of the feed liquid.

4.4.2 Preparation of roselle extract

Roselle calyces are collected after harvesting and washed properly with clean water. The calyces are then dried by a method described in the previous sections. The dried calyces are stored in a hermetically sealed container until used. Dried roselle calyces are soaked in water at 60°C for 50–60 min at a ratio of 1 g dried roselle calyces in 10 mL distilled water (Chumsri et al., 2008; Farimin & Nordin, 2009). Then the extract is filtered (Whatman filter paper No. 2) and concentrated on a rotary evaporator until the concentrate reaches 6 ± 0.2 g/100 mL (Díaz-Bandera et al., 2015). The concentrated extract is stored in an amber bottle at 4°C until next use.

4.4.3 Roselle-carrier mixing

Carrier materials are used to maximize powder recovery and retain volatiles and total phenolic content. Maltodextrin, pectin, gelatin, carboxymethyl cellulose, whey powder, carrageenan gum, and gum arabic are the common carrier materials are reported by several researchers (Díaz-Bandera et al., 2015; Farimin & Nordin, 2009; Serrano-Cruz et al., 2013). Different carrier materials show varied efficiency in the spray drying process (Table 4.1).

5 PROCESSING OF ROSELLE FOR VALUE ADDITION

Roselle contains anthocyanins and many functional food components that enable its use in food formulation and medicinal purposes. In Nigeria a popular fruit drink known as "zobo" is widely used (Bolade et al., 2009; Joseph & Adogbo, 2015). Beside drinks, roselle is used in the preparation of fruit jam, jelly, tea, etc. Drinks and other food products can be prepared from freshly harvested roselle or dried calyces. A few of the processed products from roselle are discussed in the following sections.

5.1 Roselle Tea

Roselle tea is the extract of its calyces and it contains huge health-beneficial phytochemicals. The quality of roselle tea greatly depends on various parameters followed during processing. The drying temperature greatly affects the retention of phytochemicals after drying. Roselle calyces dried at 80°C give the best color, aroma, and taste after infusion (Nguyen & Chuyen, 2020). The calyces are cleaned and dried carefully before processing for tea. The dried calyces are then

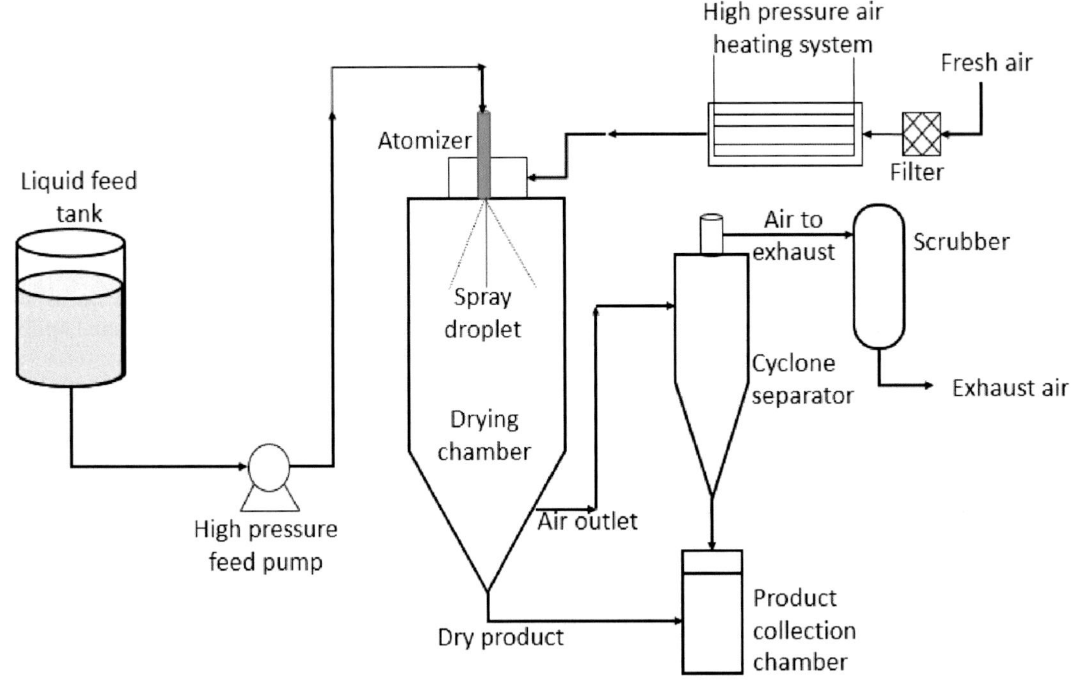

FIG. 4.4 Schematic representation of the spray drying system.

crushed. The expected particle size is passed through a mesh size of 200 μm (Joseph & Adogbo, 2015). These tea-size particles are then stored in an airtight vessel awaiting fortification with additives before finally packaging them. About 1−2.0 g of the dried calyx granules per tea bag is considered adequate for daily dosage (Adogbo & Bello, 2006). Infusion or brewing temperature also affect the quality of roselle tea. Brewing temperature varies from 60 to 100°C as reported by several researchers; however, 90°C temperature for 30 min gives the best taste and aroma and helps infuse maximum phytochemicals from the extract (Nguyen & Chuyen, 2020). The optimum processing parameters for roselle tea are presented in Table 4.2, and the roselle tea after infusion is shown in Fig. 4.5.

5.2 Roselle Juice and Drink

Roselle drink is a nonalcoholic nutritious beverage. It is popular in different regions of the world and is called by various names. For example, it has been named as "*zobo*" in Nigeria, "*sorrel*" in the Caribbean region, and "karkade" in Sudan. The drink is bright red and provides nutrients including vitamin C, calcium, iron, phosphorus, niacin, riboflavin, fiber, thiamine, carotene, etc. The drink can be prepared easily from dried calyces. About 20 g of clean dried calyces are added into 1 L of boiling water and left for 10 min for infusion. Then it is cooled, the juice extract is filtered and sweetened with 50 g of sugar, and it is ready to serve (Omemu et al., 2006). This drink can also be prepared from roselle extract blend with aloe vera gel or blended with pineapple juice. Fig. 4.6 shows the juice obtained from roselle.

5.3 Jam and Jelly

5.3.1 Roselle jam

One of the attractive and effective means of roselle utilization is jam preparation (Ashaye & Adeleke, 2009). Jam processing has been known since the 18th century. Roselle jam can easily be prepared from core-removed roselle calyces and sugar (Figs. 4.7 and 4.8). The jams made from roselle calyces are red and tangy. Roselle jam contains various nutrients especially vitamins (B_1, B_2, B_3, and C), minerals, and anthocyanins from calyces, which are good for health. Roselle jam and jelly are also industrially processed in different countries of the world and marketed in superstores (Mohamed et al., 2012; Morton, 1987).

5.3.2 Roselle mixed-fruit jam and jelly

Roselle mixed-fruit jam and jelly provides more variant nutrients and energy. It can easily be prepared like jam

TABLE 4.1
Roselle Powder Recovery From the Spray-Drying Process by Using Different Carrier Materials.

Carrier-Roselle Mixture	Powder Recovery (%)	Drying Efficiency (%)
Roselle (nonencapsulated)	69.77	76.38
Maltodextrin	85.98	79.41
Pectin	62.24	76.64
Gelatin	50.76	75.54
Carboxymethyl cellulose	54.12	74.42
Whey powder	50.44	76.81
Carrageenan gum	54.34	74.22
Gum arabic	63.05	76.34

Adapted from Díaz-Bandera, D., Villanueva-Carvajal, A., Dublán-García, O., Quintero-Salazar, B., & Dominguez-Lopez, A. (2015). Assessing release kinetics and dissolution of spray-dried Roselle (*Hibiscus sabdariffa* L.) extract encapsulated with different carrier agents. *LWT-Food Science and Technology, 64*(2), 693–698.

TABLE 4.2
Processing Parameters for Roselle Tea.

Parameters	Magnitude
Drying temperature of calyces	80°C
Particle mesh size	200 μm
Solid-liquid ratio (roselle/water)	1:10
Brewing temperature	90°C
Brewing time	30 min

Adapted from Adogbo, G. M., & Bello, T. K. (2006). Processing roselle (*Hibiscus sabdariffa*) calyx for high-content anthocyanins and ascorbic acid. *Annals of Nigerian Medicine, 2*(1), 37–42; Nguyen, Q. V., & Chuyen, H. V. (2020). Processing of herbal tea from roselle (*Hibiscus sabdariffa* L.): Effects of drying temperature and brewing conditions on total soluble solid, phenolic content, antioxidant capacity and sensory quality. *Beverages, 6*(1), 2.

from roselle, except there is addition of juice from other fruits. Roselle-pineapple mixed-fruit jam and jelly are prepared in the laboratory of Department of Agro-Processing, Bangabandhu Sheikh Mujibur Rahman Agricultural University, Gazipur, Bangladesh. The flowchart for mixed-fruit jam preparation is shown in Fig. 4.9. The roselle-pineapple mixed jelly is prepared as jam, except roselle pulp is replaced with roselle extract and pineapple pulp is replaced with pineapple juice. The mixed jam is dark red, whereas the mixed jelly is bright red (Fig. 4.10).

5.4 Pickles

Freshly harvested calyces are deseeded and cut into five pieces along with the crease of the five sepals. Then the calyces are washed with clean water and left them for draining. Next surface water-free calyces are soaked in brine solution for 2–3 days. After that the brine solution

FIG. 4.5 Roselle tea after infusion (Islam et al., 2016).

FIG. 4.6 **(A)** Roselle and aloe vera gel blend drink and **(B)** roselle, aloe vera, and pineapple juice blend drink prepared in the laboratory of Department of Agro-Processing, Bangabandhu Sheikh Mujibur Rahman Agricultural University. **(C)** Nigerian zobo (www.steemit.com) drink.

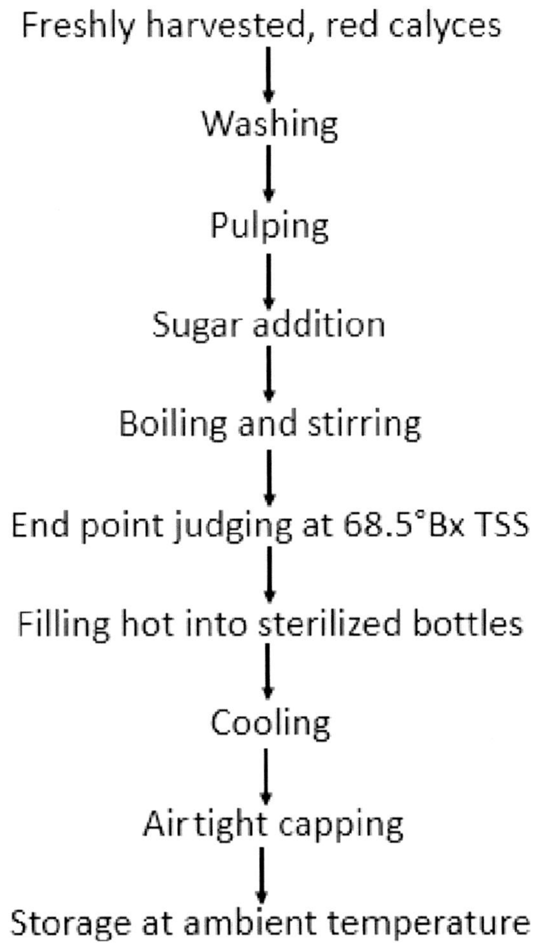

FIG. 4.7 Flowchart for the preparation of jam from roselle. *TSS*, total soluble solid.

is exhausted and the calyces are filled into a sterilized glass container leaving one-fourth blank. Then pickling solution is poured into the container leaving 1.5 cm headspace. Vinegar or sugar syrup (50 degrees Brix) can be used as the pickling solution. After filling with solution, it goes for pasteurization. The container is pasteurized in boiling water for 30 min until the temperature in the middle of the container reaches 74°C. The container is then closed with a cap, immersed in boiling water for 15 min, tempered in warm water for 30 s, and then cooled in running water. Finally, the container is tightly closed and left for 1 week until the pickle is ready to eat (Ibrahim et al., 2014).

6 ROSELLE FIBER

Roselle is well known for its bast fibers and fruit calyces. Roselle food products and their health benefits are already discussed in the previous sections. Roselle fiber is a natural fiber. Several researchers explored its competence as an element in composite materials. The chemical composition of roselle fibers includes cellulose (60%), hemicellulose (15%), and lignin (10%) (Thiruchitrambalam et al., 2010). Usually, mechanical strength properties (tensile strength and Young's modulus) of the fibers are higher in fibers containing a higher ratio of cellulose (Kalia et al., 2011). The lion's share of cellulose in roselle fiber attracts researchers to explore its use in composite materials. The bast fiber from roselle plant is silky, soft, and light-colored and has the same chemical and physical properties as jute (*Corchorus capsularis* L.) fiber; therefore it is a satisfactory replacement of synthetic fibers in industrial use. Fig. 4.11 shows roselle fiber and the fiber cloth woven from roselle. There are huge challenges that need to be overcome to use natural roselle fibers to a full

FIG. 4.8 Steps in the processing of roselle jam: **(A)** collect and clean fresh calyces, **(B)** boil fresh calyces, **(C)** blend boiled calyces, **(D)** heat blended calyces with sugar, and **(E)** ready jam preserved in jars (Islam et al., 2016).

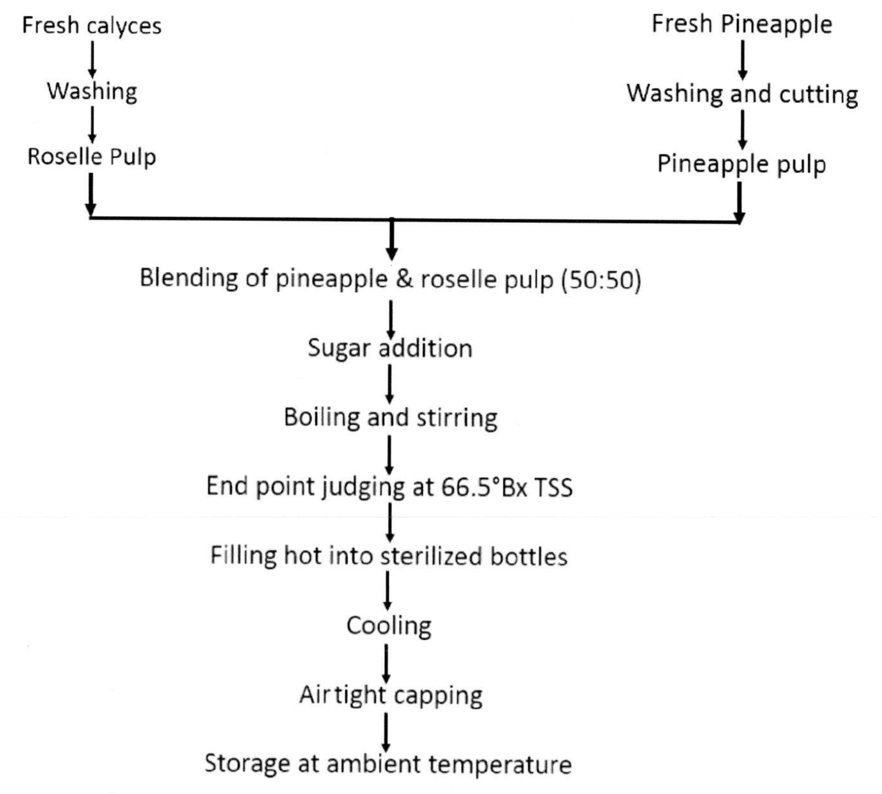

FIG. 4.9 Flowchart for the preparation of roselle and pineapple jam. *TSS*, total soluble solid.

range of biocomposite applications. The success in the future is dependent on both further advancement in technology and exploring a design approach in biocomposites that best fits the performance characteristics of the roselle fiber.

7 PROCESSING PROSPECT OF ROSELLE

Roselle is a multipurpose crop, as it provides food and fiber. As it is rich in numerous phytochemicals, anthocyanins, antioxidants, minerals, etc., long-term consumption of such type of diets offers strong defense against chronic diseases (Odigie et al., 2003; Wallstrom et al., 2000). To be benefitted from roselle food and fiber, we need processing technology. A number of technologies exist, such as juice, jam, jellies, mixed-fruit jam and jellies, etc. Proper marketing channels are also important for a viable economical industrial evolution. There are products available at various corners of the world. Also research is continuing to get new types of

FIG. 4.10 **(A)** Roselle and pineapple blended jam and **(B)** roselle and pineapple blended jelly prepared in the laboratory of Department of Agro-Processing, Bangabandhu Sheikh Mujibur Rahman Agricultural University.

technologies to work on the antioxidant activity of roselle extract and its medicinal uses to treat various diseases. Drying and preservation technologies are developed by various researchers. Dehydrated or dried calyces can be stored in a cool place year round to ensure sustainable supply. The worldwide business of roselle in raw, dried, and processed food form is increasing gradually. Germany and the United States import bulk amounts of roselle each year. The US import value is about $22 million for a volume of 5000 metric tons of dried roselle calyces. The dried calyces are mainly used for making herbal tea (Islam et al., 2016). The annual export of fresh calyces amounts to RM2.5 million in Malaysia. The domestic consumption amounts to 500 tons per year, and most of it is consumed in the form of juice and drinks (Mohammad et al., 2002). Production of roselle in many regions, local processing for consumption, and international trades make roselle crops have a very much promising future.

8 CONCLUSION

Roselle is an annual shrub popularly grown in tropical and subtropical countries for its fleshy and colorful calyces. The calyces are a good source of minerals, vitamins, and anthocyanins; therefore their consumption is helpful for the malnourished people in developing countries. Roselle calyces could be used for the preparation of different processed products and can be applied in a variety of food products as a good source of nutrients. Roselle production is season specific, so it should be dried to make it available round the year as an input for processed industries. This drying should be scientific to maintain the original quality of the calyces, although drying is an oldest method of food preservation and most widely used for food preservation since human civilization. Roselle calyces contain polyphenols and anthocyanins and these compounds can be removed from food products when exposed to heat, oxygen, and light during drying. Rapid thermal degradation of the anthocyanins of roselle calyces occurs at high temperatures. Proper drying of roselle calyces is important to retain total phenolic content, antioxidant capacity, and sensory quality of the dried roselle. This chapter describes the different drying processes of roselle calyces, their preservation, their physicochemical properties, and the various roselle food products for domestic and industrial use.

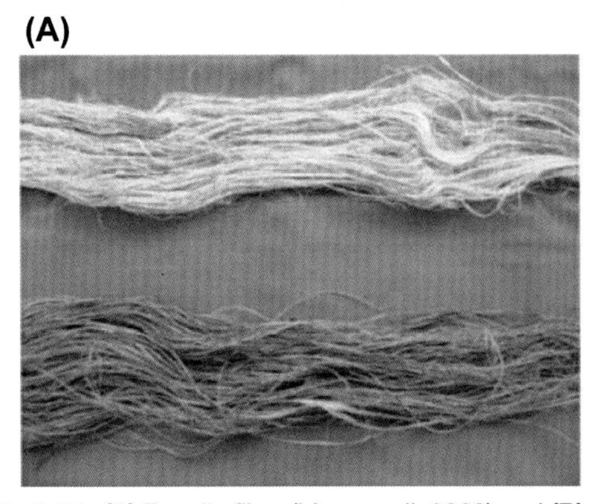

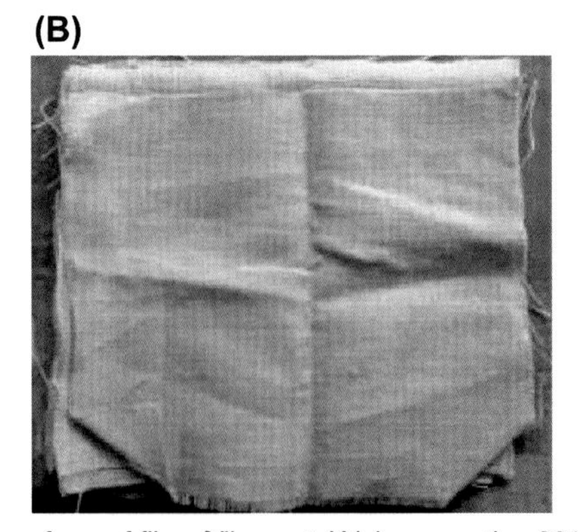

FIG. 4.11 **(A)** Roselle fiber (Managooli, 2009) and **(B)** woven form of fiber (Vijayan & Krishnamoorthy, 2018).

ACKNOWLEDGMENT

Authors are thankful to the authorities of Bangabandhu Sheikh Mujibur Rahman Agricultural University, Gazipur 1706, Bangladesh, for their support.

REFERENCES

Abu-Tarboush, H. M., Ahmed, A. A., & Kahtani, H. A. (1997). Some nutritional and functional properties of Karkade (*H. sabdariffa*) seed products. Cereal Chemistry, *74*, 352–355.

Adanlawo, I. G., & Ajibade, V. A. (2006). Nutritive value of the two varieties of roselle (*Hibiscus sabdariffa*) calyces soaked with wood ash. *Pakistan Journal of Nutrition, 5*, 555–557.

Adogbo, G. M., & Bello, T. K. (2006). Processing roselle (*Hibiscus sabdariffa*) calyx for high-content anthocyanins and ascorbic acid. *Annals of Nigerian Medicine, 2*(1), 37–42.

Ahmed, N. B., Abdalla, B. K., Elamin, I. H. M., & Elawad, Y. I. (2016). Biodiesel production from roselle oil seeds and determination the optimum reaction conditions for the transesterification process. *International Journal of Engineering Trends and Technology, 39*(2), 105–111.

Al-Okbi, S. Y., Abdel-Razek, A. G., Mohammed, S. E., & Ottai, M. E. S. (2017). Roselle seed as a potential new source of healthy edible oil. *Journal of Biological Sciences, 17*, 267–277. https://doi.org/10.3923/jbs.2017.267.277

Alegbejo, M. D., Abo, M. E., & Alegbejo, J. O. (2003). Current status and future potential of roselle production and utilization in Nigeria. *Journal of Sustainable Agriculture, 23*(2), 5–16. https://doi.org/10.1300/J064v23n02_03

Ansari, M., Eslaminejad, T., Sarhadynejad, Z., & Eslaminejad, T. (2013). An overview of the roselle plant with particular reference to its cultivation, diseases and usages. *European Journal of Medicinal Plants, 3*(1), 135–145.

Ashaye, O., & Adeleke, T. O. (2009). Quality attributes of stored Roselle jam. *International Food Research Journal, 16*(3), 363–371.

Bolade, M. K., Oluwalana, I. B., & Ojo, O. (2009). Commercial practice of roselle (*Hibiscus sabdariffa* L.) beverage production: Optimization of hot water extraction and sweetness level. *World Journal of Agricultural Sciences, 5*(1), 126–131.

Borrás-Linares, I., Fernández-Gutiérrez, A., Segura-Carretero, A., Arráez-Roman, D., Andrade-Gonzáles, I., Gómez-Leyva, J. F., ... Fernández-Arroyo, S. (2015). Characterization of phenolic compounds, anthocyanidin, antioxidant and antimicrobial activity of 25 varieties of Mexican Roselle (*Hibiscus sabdariffa*). *Industrial Crops and Products, 69*, 385–394.

Brennan, C. S. (2015). The importance of food science and technology in modern society. *International Journal of Food Science and Technology, 50*(1), 1–2.

Buletin, P. G. M., Duke, J. A., & Atchley, A. A. (1984). Proximate analysis. In BR Christie (Ed.), *The handbook of Plant Science in Agriculture*. Boca Raton, FL: CRC Press, Inc.

Castañeda-Miranda, A., Ríos-Moreno, J. G., Meza-Jiménez, J., Ortega-Moody, J. A., Herrera-Ruiz, G., & Trejo-Perea, M. (2014). A continuous production roselle (*Hibiscus sabdariffa* L.) dryer using solar energy. *Journal of Food Agriculture and Environment, 12*(1), 96–104.

Castañeda-Ovando, A., de Lourdes Pacheco-Hernández, M., Elena Páez-Hernández, M., Rodríguez, J. A., & Galán-Vidal, C. A. (2009). Chemical studies of anthocyanins: A review. *Food Chemistry, 113*, 859–871.

Chao, C. Y., & Yin, M. C. (2009). Antibacterial effects of roselle calyx extracts and protocatechuic acid in ground beef and apple juice. *Foodborne Pathogens and Disease, 6*(2), 201–206.

Chen, H. H., Tsai, P. J., Chen, S. H., Su, Y. M., Chung, C. C., & Huang, T. C. (2005). Grey relational analysis of dried roselle (*Hibiscus sabdariffa* L.). *Journal of Food Processing and Preservation, 29*(3-4), 228–245.

Chumsri, P., Sirichote, A., & Itharat, A. (2008). Studies on the optimum conditions for the extraction and concentration of roselle (*Hibiscus sabdariffa* Linn.) extract. *Songklanakarin Journal of Science and Technology, 30*(1), 133–139.

Cobley, L. S. (1975). *An introduction to Botany of tropical crops* (Vol. 11). Longman Group U.K.

Díaz-Bandera, D., Villanueva-Carvajal, A., Dublán-García, O., Quintero-Salazar, B., & Dominguez-Lopez, A. (2015). Assessing release kinetics and dissolution of spray-dried Roselle (*Hibiscus sabdariffa* L.) extract encapsulated with different carrier agents. *LWT-Food Science and Technology, 64*(2), 693–698.

Duarte-Valenzuela, Z. N., Zamora-Gasga, V. M., Montalvo-González, E., & Sáyago-Ayerdi, S. G. (2016). Caracterización nutricional de 20 variedades mejoradas de jamaica (*Hibiscus sabdariffa* L.) cultivadas en México. *Revista Fitotecnia Mexicana, 39*, 199–206.

Duke, A. J., Bogenschutz-Godwin, M. J., & Ducellier, J. (2003). *Handbook of medicinal spices* (pp. 186–187). Boca Raton: CRC Press LLC.

Dy Phon, P. (2000). *Dictionary of plants used in Cambodia* (1st ed., pp. 343–344). Cambodia: Imprimerie Olympic, Phnom Penh.

Esper, A., & Muhlbauer, W. (1998). Solar drying-an effective means of food preservation. *Renewable Energy, 15*, 95–100.

Evans, P., & Halliwell, B. (2001). Micronutrients: Oxidant/antioxidant status. *British Journal of Nutrition, 85*, S67–S74.

Eslaminejad, T., & Zakaria, M. (2011). Morphological characteristics and pathogenicity of fungi associated with Roselle (*Hibiscus sabdariffa*) diseases in Penang, Malaysia. *Microbial Pathogenesis, 5*(15), 325–327.

Farimin, A. O. A., & Nordin, E. (2009). Physical properties of powdered roselle-pineapple juice-effects of maltodextrin. In *National conference on postgraduate research* (pp. 90–97).

Hahn, F., Hernandez, G., Hernandez, J., Perez, C., & Vargas, J. M. (2011). Optimization of roselle drying time and drying quality. *Canadian Biosystems Engineering, 53*, 31–38.

Halimatul, S. M. N., Amin, I., Esa, M. N., Nawalyah, A. G., & Siti Muskinah, M. (2007). Protein quality of Roselle (*Hibiscus sabdariffa* L.) seeds. *ASEAN Food Journal, 14*(2), 131–140.

Hinojosa-Gómez, J., Martin-Hernández, C. S., Heredia, J. B., León-Félix, J., Osuna-Enciso, T., & Muy-Rangel, M. D. (2018). Roselle (*Hibiscus sabdariffa* L.) cultivars calyx produced hydroponically: Physicochemical and nutritional quality. *Chilean Journal of Agricultural Research, 78*(4), 478–485. https://doi.org/10.4067/S0718-58392018000400478

Hirunpanich, V., Utaipat, A., Morales, N. P., Bunyapraphatsara, N., Sato, H., Herunsale, A., & Suthisisang, C. (2006). Hypocholesterolemic and antioxidant effects of aqueous extracts from the dried calyx of *Hibiscus sabdariffa* L. in hypercholesterolemic rats. *Journal of Ethnopharmacology, 103*(2), 252–260.

Ibrahim, R., Adenan, M. F., & Lani, M. N. (2014). The potential sugar replacements in the development of low calorie Roselle (*Hibiscus sabdariffa* L.'UKMR-2') pickles. *IOSR Journal of Applied Chemistry, 7*(10), 01–07.

Islam, A. K. M. A., Jamini, T. S., Islam, A. K. M. M., & Sabina, Y. (2016). Roselle: A functional food with high nutritional and medicinal values. *Fundamental and Applied Agriculture, 1*(2), 44–49.

Ismail, A., Ikram, E. H. K., & Nazri, H. S. M. (2008). Roselle (*Hibiscus sabdariffa* L.) seeds-nutritional composition, protein quality and health benefits. *Food, 2*, 1–16.

Jabeur, I., Pereira, E., Barros, L., Calhelha, R. C., Sokovic, M., Oliveira, M. B. P., & Ferreira, I. C. (2017). *Hibiscus sabdariffa* L. as a sourse of nutrients, bioactive compounds and colouring agent. *Food Research International, 100*, 717–723.

Janjai, S., & Tung, P. (2005). Performance of a solar dryer using hot air from roof-integrated solar collectors for drying herbs and spices. *Renewable Energy, 30*, 2085–2095.

Joseph, A. D., & Adogbo, G. M. (2015). Processing and packaging of *Hibiscus sabdariffa* for preservation of nutritional constituents. *International Journal of Scientific Engineering and Research, 6*(4), 532–536.

Kalia, S., Kaith, B. S., & Kaur, I. (2011). *Cellulosic fibers: Bio- and nano-polymer composites*. New York: Springer.

Kilima, M. B., Remberg, S. F., Chove, B. E., & Wicklund, T. (2014). Physio-chemical, mineral composition and antioxidant properties of roselle (*Hibiscus sabdariffa* L.) extract blended with tropical fruit juices. *African Journal of Food, Agriculture, Nutrition and Development, 14*(3), 8963–8978.

Luvonga, W. A., Njoroge, M. S., Makokha, A., & Ngunjiri, P. W. (2010). Chemical characterisation of *Hibiscus sabdariffa* (Roselle) calyces and evaluation of its functional potential in the food industry. In *Jkuat annual scientific conference proceedings* (pp. 631–638).

Mady, C., Manuel, D., Mama, S., Augustin, N., & Max, R. (2009). The bissap (*Hibiscus sabdariffa* L.): Composition and principal uses. *Fruits, 64*, 179–193.

Mady, C., Manuel, D., Mama, S., & Ndiaye, A. (2009). The bissap (*Hibiscus sabdariffa*): Composition and principal uses. *Fruits, 64*, 179–193.

Managooli, V. A. (2009). *Dyeing mesta (Hibiscus sabdariffa) fibre with natural colourant*. Dharwad: Dharwad University of Agricultural Sciences.

Mazza, G., & Miniati, E. (1993). Introduction, chapter 1. In *Anthocyanins in fruit, vegetables and grains* (pp. 1–23). Boca Raton, Florida, USA: CRC Press.

Midilli, A., Kucuk, H., & Yapar, Z. (2002). A new model for single-layer drying. *Drying Technology, 20*, 1503–1513.

Mohamad, O., Nazir, B. M., Rahman, M. A., & Herman, S. (2002). Roselle: A new crop in Malaysia. *Bulletin Genetics Society Malaysia, 7*(1–2), 12–13.

Mohamed, B. B., Sulaiman, A. A., & Dahab, A. A. (2012). Roselle (*Hibiscus sabdariffa* L.) in Sudan, cultivation and their uses. *Bulletin of Environment, Pharmacology and Life Sciences, 1*, 48–54.

Morton, J. (1987). Roselle. In *Fruits of warm climates* (pp. 281–286). Miami, FL: Julia F. Morton.

Nakpong, P., & Wootthikanokkhan, S. (2010). Roselle (*Hibiscus sabdariffa* L.) oil as an alternative feedstock for biodiesel production in Thailand. *Fuel, 89*(8), 1806–1811. https://doi.org/10.1016/j.fuel.2009.11.040

Nguyen, Q. V., & Chuyen, H. V. (2020). Processing of herbal tea from roselle (*Hibiscus sabdariffa* L.): Effects of drying temperature and brewing conditions on total soluble solid, phenolic content, antioxidant capacity and sensory quality. *Beverages, 6*(1), 2.

Odigie, I. P., Ettarh, R. R., & Adigun, S. (2003). Chronic administration of aqueous extract of *Hibiscus sabdariffa* attenuates hypertension and reverses cardiac hypertrophy in 2K-1C hypertensive rats. *Journal of Ethnopharmacology, 86*, 181–185.

Olaleye, M. T. (2007). Cytotoxicity and antibacterial activity of methanolic extract of *Hibiscus sabdariffa*. *Journal of Medicinal Plants Research, 1*(1), 9–13.

Omemu, A. M., Edema, M. O., Atayese, A. O., & Obadina, A. O. (2006). "A survey of the microflora of *Hibiscus sabdariffa* (Roselle) and the resulting "Zobo" juice. *African Journal of Biotechnology, 5*(3), 254–259.

Patel, S. (2014). *Hibiscus sabdariffa*: An ideal yet under-exploited candidate for nutraceutical applications. *Biomedicine and Preventive Nutrition, 24*, 23–27.

Rao, P. U. (2008). Nutrient composition and biological evaluation of mesta (*Hibiscus sabdariffa*) seeds. *Plant Foods for Human Nutrition, 49*(1), 27–34.

Sacilik, K. (2007). Effect of drying methods on thin-layer drying characteristics of hull-less seed pumpkin (*Cucurbita pepo* L.). *Journal of Food Engineering, 79*, 23–30.

Saeed, I. E. (2010). Solar drying of roselle (*Hibiscus sabdariffa* L.): Effects of drying conditions on the drying constant and coefficients, and validation of the logarithmic model. *Agricultural Engineering International: CIGR Journal, 12*(1), 167–181.

Serrano-Cruz, M. R., Villanueva-Carvajal, A., Rosales, E. J. M., Dávila, J. F. R., & Dominguez-Lopez, A. (2013). Controlled release and antioxidant activity of Roselle (*Hibiscus sabdariffa* L.) extract encapsulated in mixtures of carboxymethyl cellulose, whey protein, and pectin. *LWT-Food Science and Technology, 50*(2), 554–561.

Shoosh, W. G. A. A. (1993). *Chemical composition of some roselle (Hibiscus sabdariffa) genotypes* (pp. 1–109). Sudan: Department of Food Science and Technology, Faculty of Agriculture, University of Khartoum.

Thiruchitrambalam, M., Athijayamani, A., & Sathiyamurthy, S. (2010). A review on the natural fiber-reinforced polymer composites for the development of roselle fiber-reinforced polyester composite. *Journal of Natural Fibers, 7*, 307–323.

Togrul, I. T., & Pehlivan, D. (2002). Mathematical modeling of solar drying of apricots in thin layers. *Journal of Food Engineering, 55*(1), 209–216.

Togrul, I. T., & Pehlivan, D. (2003). Modeling of drying kinetics of single apricot. *Journal of Food Engineering, 58*(1), 23–32.

Vengaiah, P. C., & Pandey, J. P. (2007). Dehydration kinetics of sweet pepper (*Capsicum annum* L.). *Journal of Food Engineering, 81*, 282–286.

Vijayan, R., & Krishnamoorthy, A. (2018). Experimental analysis of hybrid (roselle, aloe vera and glass) natural fiber-reinforced composite material. *International Journal of Mechanical and Production Engineering Research and Development, 8*(4), 303–314.

Wallstrom, P., Wirfalt, E., Janzon, L., Mattisson, I., Elmstahl, S., Johansson, U., & Berglund, G. (2000). Fruit and vegetable consumption in relation to risk factors for cancer: A report from the malmo diet and cancer study. *Public Health Nutrition, 3*, 263–271.

Wang, H., Cao, G., & Prior, R. L. (1997). Oxygen radical absorbing capacity of anthocyanins. *Journal of Agricultural and Food Chemistry, 45*(2), 304–309.

Wang, Z., Sun, J., Liao, X., Chen, F., Zhao, G., Wu, J., & Hu, X. (2007). Mathematical modeling on hot air drying of thin layer apple pomace. *Food Research International, 40*, 39–46.

Wang, S. C., Lee, S. F, Wang, C. J., Lee, C. H., Lee, W. C., & Lee, H. J. (2011). Aqueous extract from *Hibiscus sabdariffa* Linnaeus ameliorate diabetic nephropathy via regulating oxidative status and Akt/Bad/14-3-3γ in an experimental animal model. *Evidence-Based Complementary and Alternative Medicine*, 938126.

FURTHER READING

Grubben, G. J. H., & Ngwerume, F. C. (2004). *Cucurbita moschata* Duchesne. In G. J. H. Grubben, & O. A. Denton (Eds.), *PROTA 2: Vegetables/légumes. PROTA, Wageningen, Netherlands*.

Imad Eldin Saeed, I. E. (2010). Solar drying of roselle (*Hibiscus sabdariffa* L.) part II: Effects of drying conditions on the drying constant and coefficients, and validation of the logarithmic model. *Agricultural Engineering International: The CIGR Ejournal, XII*. Manuscript 1488.

Mgaya, B. M., Remberg, S. F., Chove, B. E., & Wicklund, T. (2014). Physiochemical, mineral composition and antioxidant properties of roselle (*Hibiscus sabdariffa* L.) extract blended with tropical fruit juices. *African Journal of Food, Agriculture, Nutrition and Development, 14*(3), 8963–8978.

Nadlene, R., Sapuan, S. M., Jawaid, M., Ishak, M. R., & Yusriah, L. (2016). A review on roselle fiber and its composites. *Journal of Natural Fibers, 13*(1), 10–41.

CHAPTER 5

Current Knowledge on Roselle Polyphenols: Content, Profile, and Bioaccessibility

Y. MARTÍNEZ-MEZA • R. REYNOSO-CAMACHO • J. PÉREZ-JIMÉNEZ

1 POLYPHENOLS AS BIOACTIVE DIETARY COMPONENTS

Polyphenols are secondary metabolites produced by plants in response to biotic and abiotic stress, as a defense mechanism against external conditions. They possess an aromatic ring with at least one hydroxyl group, and their structure can vary from that of simple molecules (monomers and oligomers) to polymeric structures. Based on their chemical structure, they are commonly classified into different families, with the most important being phenolic acids (hydroxybenzoic acids and hydroxycinnamic acids), flavonoids (flavonols [catechins and proanthocyanidins], flavones, isoflavones, flavanones, anthocyanidins), stilbenes, and lignans (Manach et al., 2004). Another classification of polyphenols is based on their solubility and presence in the food matrix. Thus extractable polyphenols (EPPs) have a low molecular weight, are free within the food matrix, and can be retrieved from the supernatants of aqueous or aqueous-organic extractions. In contrast, the nonextractable polyphenols (NEPPs), which are either high-molecular-weight compounds or low-molecular-weight compounds associated with macromolecules such as proteins or dietary fiber, remain in the residues of such extractions (Arranz et al., 2009).

The polyphenol content of food is affected by several factors, including annual climate variations, maturity, postharvest processing and storage, and production methods (Amarowicz et al., 2009). Nevertheless, many food items are known to be rich in specific polyphenols, as shown in Table 5.1. Using these data, it has been possible to estimate the average daily polyphenol intake in different populations, obtaining values of about 1 g per person when considering only EPPs (Pérez-Jiménez et al., 2011) and 2 g per person when including NEPPs (Arranz et al., 2010). We should also highlight the fact that interest in polyphenols has not only focused on food but also considered some alternative sources, such as agro-industrial by-products. Thus some studies indicate that agro-industrial by-products are potentially an important source of polyphenols and their consumption could have beneficial health effects, as well as offer economic and environmental advantages (Pérez-Jiménez and Saura-Calixto, 2018). An important aspect to bear in mind when assessing the benefits of polyphenols is that different classes exhibit important differences regarding their bioaccessibility and bioavailability. In general, polyphenols are extensively transformed after intake and two key organs are involved in their metabolic fate: the small intestine, where a certain fraction may be absorbed, and the colon, where bioactive absorbable metabolites may be generated via the action of the microbiota (Rodriguez-Mateos et al., 2014). For instance, it is known that relatively low-molecular-weight polyphenols, such as isoflavones and gallic acid, are the most easily absorbed, followed by catechins, flavanones, and quercetin glycosides, with different kinetic profiles. Meanwhile, larger polyphenols such as proanthocyanins are poorly absorbed in the small intestine and are partially transformed in the colon (Manach et al., 2005).

Although polyphenols are known to contribute to sensory aspects of foods, the main reason for the interest in these compounds over the recent decades has been their beneficial health effects. Different animal, human, and epidemiologic studies have shown that various polyphenols have antioxidant and antiinflammatory properties that could offer protection against

Roselle. https://doi.org/10.1016/B978-0-323-85213-5.00008-1

TABLE 5.1
Examples of Characteristic Polyphenols of the Main Polyphenol Classes and Richest Food Sources.

Class	Subclass	Compound	Richest Food Sources
Phenolic acids	Hydroxybenzoic acids	Gallic acid	Chestnut, dried cloves, green chicory
	Hydroxybenzoic acids	Ellagic acid	Chestnut, blackberry, black raspberry
	Hydroxycinnamic acids	Caffeic acid	Black chokeberry, sage, dried spearmint
	Hydroxycinnamic acids	Ferulic acid	Hard wheat, dark chocolate, date
Flavonoids	Anthocyanins	Cyanidin-3-O-glucoside	Black elderberry, blackberry, blackcurrant
	Anthocyanins	Delphinidin-3-O-glucoside	Blackcurrant, lowbush blueberry, black bean
	Flavanols	(+)-Epicatechin	Cocoa powder, dark chocolate, broad bean pod
	Flavanols	Procyanidin dimer B1	Cocoa powder, peach, nectarine
	Flavanones	Hesperidin	Dried peppermint, blond orange juice, blood orange juice
	Flavanones	Naringenin	Fresh rosemary, grapefruit juice, red wine
	Flavones	Apigenin	Dried marjoram, fresh Italian oregano, fresh sage
	Flavones	Luteolin	Dried Mexican oregano, globe artichoke heads, fresh thyme
	Flavonols	Kaempferol	Capers, cumin, cloves
	Flavonols	Quercetin	Black elderberry, dried Mexican oregano, capers
	Isoflavones	Daidzein	Tempe, soy paste (cheonggukjang), tofu
	Isoflavones	Genistein	Soy flour, soy paste (cheonggukjang), roasted soybean
Stilbenes		Pallidol	Red wine, rosé wine, white wine
		Resveratrol	Lingonberry, European cranberry, red currant
Lignans		Lariciresinol	Broccoli, kale, cashew nut
		Matairesinol	Sesame seed meal, flaxseed meal, kale
Other polyphenols	Curcuminoids	Curcumin	Dried turmeric, curry powder
	Phenolic terpenes	Rosmanol	Fresh rosemary
	Tyrosols	Hydroxytyrosol	Black olive, green olive, extra-virgin olive oil

Adapted from Neveu, V., Pérez-Jiménez, J., Vos, F., Crespy, V., du Chaffaut, L., Mennen, L., Knox, C., Eisner, R., Cruz, J., Wishart, D., & Scalbert, A. (2010). Phenol-explorer: An online comprehensive database on polyphenol contents in foods. Database, 2010. http://phenol-explorer.eu/.

diseases such as neurodegenerative disorders (Almeida et al., 2016), cardiovascular diseases (Mendonça et al., 2019; Yamagata et al., 2015), diabetes mellitus (Cao et al., 2019), or several types of cancer (Lambert et al., 2005; Zhou et al., 2016). This is due to a combination of biochemical mechanisms; while their antioxidant capacity is the most commonly considered characteristic, these compounds are also able to regulate several signaling pathways, inhibit different enzymes, and stimulate the growth of beneficial microbiota species (Boto-Ordóñez et al., 2014; Gobert et al., 2014; Mackenzie et al., 2009; Rodríguez-Ramiro et al., 2012; Schaffer & Halliwell, 2012; Shi and Williamson. 2016).

Thus polyphenols are a class of bioactive dietary compounds with health-related properties. In this context, it is pertinent to explore the current knowledge of the polyphenol content and profile of roselle. Also, as polyphenols need to be bioaccessible in biological fluids in order to exert their effects, we will also discuss in this chapter information concerning the metabolic fate of roselle polyphenols.

2 ROSELLE AS AN IMPORTANT SOURCE OF POLYPHENOLS

2.1 Polyphenols in Roselle: Content and Profile

Several studies have explored the presence of polyphenols in roselle extracts, evaluating total contents and establishing detailed polyphenol profiles. Spectrophotometric measurements have yielded a total polyphenol content of 14 mg/g in roselle calyces (Amaya-Cruz et al., 2017); this value was not affected by modifications in the particle size. In the same study, other spectrophotometric techniques revealed a flavonoid content of 10 mg/g and an anthocyanin content of about 5 mg/g. A different study obtained a crude extract with a polyphenol content of 22 mg/g, which was increased to 280 mg/g after a purification process (Herranz-López et al., 2012). Moreover, comparison between 25 roselle varieties provided values between 24 and 100 mg/g (Borrás-Linares et al., 2015).

Besides these general determinations, several studies have assessed the detailed polyphenol profile of roselle extracts. The most detailed studies apply high-performance liquid chromatography-mass spectrometry (HPLC-MS) techniques (Amaya-Cruz et al. 2017, 2019; Ezzat et al., 2016; Fernández-Arroyo et al., 2011; Herranz-López et al., 2012; Morales-Luna et al., 2019; Rodríguez-Medina et al., 2009), as HPLC analysis allows separation of the major phenolic compounds in roselle (Salazar-González et al., 2012). An overview of these findings is shown in Table 5.2. Overall, a total of 104 polyphenols have been identified, belonging to different classes: phenolic acids (hydroxybenzoic and hydroxycinnamic), flavonoids (flavonols, flavanols, flavones, flavanones, anthocyanins), and others. Anthocyanins, as characteristic red pigments, are not present in white roselle varieties (Morales-Luna et al., 2019); indeed, as roselle exhibits a wide range of colorations, a study suggested different equations that associate each anthocyanin with specific color parameters (Camelo-Méndez et al., 2016). It should be highlighted that roselle contains some specific flavonols, such as the flavonoid gossypetin (3,5,7,8,3′,4′-hexahydroxyflavone), that, although reported in other plant materials (Harborne, 1969), are rather scarce in edible plants. Other examples are hibiscitrin and hibiscetin-3-O-glucoside (3,5,7,8,3′,4′,5′-heptahydroxyflavone) isolated from roselle 70 years ago (Rao & Seshadri, 1942).

The individual compounds delphinidin-3-O-sambubioside and 3-caffeoylquinic acid constitute some 60% of the phenolic compounds in roselle (Piovesana et al., 2019). In the case of anthocyanins, one study reported that just two of them (cyanidin-3-O-rutinoside and delphinidin-3-O-sambubioside) accounted for more than 90% of the total anthocyanin content of roselle (Amaya-Cruz et al., 2017). However, other studies found the two main anthocyanins to be the sambubioside derivatives of cyanidin and delphinidin (Fernández-Arroyo et al., 2011; Salazar-González et al., 2012). This may be associated with varietal differences; indeed, the delphinidin/cyanidin ratio (from 1:1 to 4:1) was determined as a specific characteristic in the cluster analysis of 25 roselle varieties (Borrás-Linares et al., 2015).

Regarding other classes of polyphenols, quercetin, as both aglycone and glycoside (glucoside, rutinoside, or sambubioside), was the major flavonol, with a total concentration of about 100 mg/g, while chlorogenic acid isomers constituted 97% of the phenolic acid family content (Fernández-Arroyo et al., 2011). Interestingly, when a red roselle variety was compared with a white one, it was found that, excluding anthocyanins (absent in white varieties), the polyphenol content in the red variety was twofold higher than that in the white variety; however, it cannot be ruled out that this difference was due to agronomic conditions (Morales-Luna et al., 2019). Be that as it may, it is clear that some content data for individual compounds must be considered as approximate values because, owing to the lack of commercial standards, semiquantification based on related standards is commonly performed (Fernández-Arroyo et al., 2011). However, for the anthocyanin class a close association was found between the value obtained by spectrophotometric techniques and the sum of individual compounds measured by HPLC (Salazar-González et al., 2012).

Research over the past decade has advanced the characterization of roselle polyphenols, including the identification of some phenolic compounds not previously reported: syringetin-3-O-glucoside, luteolin-7-O-D-glucuronide methyl ester, 4′-O-acetylquercitrin, and two isomers of cleomiscosin, commonly named cleomiscosin A and B (Borrás-Linares et al., 2015). Some compounds, nevertheless, remain unidentified, as in the case of a delphinidin derivative with a molecular ion at *m/z* 613.1416, which corresponds to the formula $C_{26}H_{29}O_{17}$ (Ezzat et al., 2016), or the 10 unidentified phenolic compounds that appear in the analysis of 25 roselle varieties (Borrás-Linares et al., 2015). An important aspect to bear in mind is that some minor

TABLE 5.2
Polyphenols Identified in Calyces of Roselle.

ID Compound	Compound	Molecular Formula	Molecular Weight [M-H]⁻	References
PHENOLIC ACIDS				
Hydroxybenzoic acids				
1	Gallic acid	$C_7H_6O_5$	170.1195	Amaya-Cruz et al. (2019), Morales-Luna et al. (2019), Pérez-Ramírez et al. (2015)
2	Methyl digallate	$C_{15}H_{12}O_9$	336.0488	Herranz-López et al. (2012)
3	Protocatechuic acid	$C_7H_6O_4$	154.1201	Morales-Luna et al. (2019)
4	Vanillic acid	$C_8H_8O_4$	168.1467	Morales-Luna et al. (2019)
5	Isovanillic acid	$C_8H_8O_4$	168.1467	Morales-Luna et al. (2019)
6	4-Hydroxybenzoic acid	$C_7H_6O_3$	138.1207	Morales-Luna et al. (2019)
Hydroxycinnamic acids				
7	*p*-Coumaric acid	$C_9H_8O_3$	164.1580	Morales-Luna et al. (2019)
8	Coumaroylquinic acid isomer I	$C_{16}H_{18}O_8$	338.3093	Herranz-López et al. (2012) Pérez-Ramírez et al. (2015)
9	Coumaroylquinic acid isomer II	$C_{16}H_{18}O_8$	338.3093	Amaya-Cruz et al. (2019)
10	Coumaroylquinic acid isomer III	$C_{16}H_{18}O_8$	338.3093	Amaya-Cruz et al. (2019)
11	Caffeic acid	$C_9H_8O_4$	180.1574	Pérez-Ramírez et al. (2015)
12	Caffeoyl glucose	$C_{15}H_{18}O_9$	342.2980	Herranz-López et al. (2012)
13	Caffeic acid hexoside	$C_{15}H_{18}O_9$	342.0951	Morales-Luna et al. (2019), Amaya-Cruz et al. (2019)
14	Caffeoylquinic acid isomer I	$C_{16}H_{18}O_9$	354.0951	Fernández-Arroyo et al. (2011), Rodríguez-Medina et al. (2009), Amaya-Cruz et al. (2019), Borrás-Linares et al. (2015), Herranz-López et al. (2012), Morales-Luna et al. (2019)
15	Caffeoylquinic acid isomer II	$C_{16}H_{18}O_9$	354.0951	Fernández-Arroyo et al. (2011), Rodríguez-Medina et al. (2009), Amaya-Cruz et al. (2019), Borrás-Linares et al. (2015)
16	Caffeoylquinic acid isomer III	$C_{16}H_{18}O_9$	354.0951	Amaya-Cruz et al. (2019), Borrás-Linares et al. (2015), Fernández-Arroyo et al. (2011), Rodríguez-Medina et al. (2009)
17	Dicaffeoylquinic acid	$C_{25}H_{24}O_{12}$	516.1268	Amaya-Cruz et al. (2019)
18	Methyl chlorogenate isomer I	$C_{17}H_{20}O_9$	368.353	Borrás-Linares et al. (2015)
19	Methyl chlorogenate isomer II	$C_{17}H_{20}O_9$	368.3353	Borrás-Linares et al. (2015)
20	Methyl chlorogenate isomer III	$C_{17}H_{20}O_9$	368.3353	Borrás-Linares et al. (2015)

TABLE 5.2
Polyphenols Identified in Calyces of Roselle.—cont'd

ID Compound	Compound	Molecular Formula	Molecular Weight [M-H]⁻	References
21	Ethyl chlorogenate	$C_{17}H_{20}O_9$	382.1264	Borrás-Linares et al. (2015)
22	2-O-trans-caffeoyl-hydroxycitric acid	$C_{15}H_{14}O_{11}$	370.0536	Borrás-Linares et al. (2015), Herranz-López et al. (2012)
23	5-O-caffeoylshikimic acid	$C_{16}H_{16}O_8$	335.0766	Borrás-Linares et al. (2015), Ezzat et al. (2016), Fernández-Arroyo et al. (2011), Herranz-López et al. (2012), Rodríguez-Medina et al. (2009)
24	Ellagic acid	$C_{14}H_6O_8$	302.1926	Amaya-Cruz et al. (2019), Morales-Luna et al. (2019), Pérez-Ramírez et al. (2015)
25	Ferulic acid	$C_{10}H_{10}O_4$	194.1840	Amaya-Cruz et al. (2019), Morales-Luna et al. (2019)
26	Feruloylquinic acid isomer I	$C_{17}H_{20}O_9$	368.3353	Amaya-Cruz et al. (2019)
27	Feruloylquinic acid isomer II	$C_{17}H_{20}O_9$	368.3353	Amaya-Cruz et al. (2019)
28	Sinapic acid	$C_{11}H_{12}O_5$	224.0681	Morales-Luna et al. (2019), Pérez-Ramírez et al. (2015)
29	Rosmarinic acid	$C_{18}H_{16}O_8$	360.3148	Pérez-Ramírez et al. (2015)
FLAVONOIDS				
Flavonols				
30	Quercetin	$C_{15}H_{10}O_7$	302.2357	Amaya-Cruz et al. (2019), Borrás-Linares et al. (2015), Ezzat et al. (2016), Fernández-Arroyo et al. (2011), Herranz-López et al. (2012), Pérez-Ramírez et al. (2015), Rodríguez-Medina et al. (2009)
31	Quercetin-3-O-glucoside	$C_{21}H_{20}O_{12}$	464.3763	Amaya-Cruz et al. (2019), Borrás-Linares et al. (2015), Fernández-Arroyo et al. (2011), Herranz-López et al. (2012), Rodríguez-Medina et al. (2009)
32	Quercetin-3-O-sambubioside	$C_{26}H_{28}O_{16}$	596.1377	Ezzat et al. (2016), Fernández-Arroyo et al. (2011), Herranz-López et al. (2012), Rodríguez-Medina et al. (2009)
33	Quercetin-3-O-rutinoside (rutin)	$C_{27}H_{30}O_{16}$	610.5175	Fernández-Arroyo et al. (2011), Herranz-López et al. (2012), Morales-Luna et al. (2019), Pérez-Ramírez et al. (2015), Rodríguez-Medina et al. (2009)

Continued

TABLE 5.2
Polyphenols Identified in Calyces of Roselle.—cont'd

ID Compound	Compound	Molecular Formula	Molecular Weight [M-H]$^-$	References
34	Quercetin pentoside-rutinoside	$C_{32}H_{38}O_{20}$	743.2032	Amaya-Cruz et al. (2019)
35	4″-O-acetylquercitrin	$C_{23}H_{22}O_{12}$	490.1111	Borrás-Linares et al. (2015)
36	Kaempferol	$C_{15}H_{10}O_6$	286.2363	Amaya-Cruz et al. (2019), Morales-Luna et al. (2019)
37	Kaempferol-3-O-rutinoside	$C_{27}H_{30}O_{15}$	601.5101	Ezzat et al. (2016), Fernández-Arroyo et al. (2011), Herranz-López et al. (2012), Rodríguez-Medina et al. (2009)
38	Kaempferol hexoside	$C_{21}H_{20}O_{11}$	449.1081	Amaya-Cruz et al. (2019), Ezzat et al. (2016)
39	Kaempferol pentoside-hexoside	$C_{26}H_{28}O_{15}$	581.1504	Amaya-Cruz et al. (2019)
40	Kaempferol-3-O-sambubioside (leucoside)	$C_{26}H_{28}O_{15}$	580.1428	Herranz-López et al. (2012)
41	Kaempferol-3-O-(6″-coumarylglucoside)	$C_{30}H_{26}O_{13}$	594.5196	Fernández-Arroyo et al. (2011), Rodríguez-Medina et al. (2009)
42	Kaempferol-3-O-(6″-acetylglucoside)	$C_{23}H_{22}O_{12}$	490.1111	Borrás-Linares et al. (2015)
43	Kaempferol-3-O-glucuronide	$C_{21}H_{18}O_{12}$	462.0798	Borrás-Linares et al. (2015)
44	Kaempferol-3-O-glucuronic acid methyl ether isomer I	$C_{22}H_{20}O_{12}$	476.0958	Borrás-Linares et al. (2015)
45	Kaempferol-3-O-glucuronic acid methyl ether isomer II	$C_{22}H_{20}O_{12}$	476.0959	Borrás-Linares et al. (2015)
46	Myricetin	$C_{15}H_{10}O_8$	318.2351	Amaya-Cruz et al. (2019), Borrás-Linares et al. (2015), Ezzat et al. (2016), Herranz-López et al. (2012)
47	Myricetin-3-O-glucoside	$C_{21}H_{20}O_{13}$	480.3757	Amaya-Cruz et al. (2019), Herranz-López et al. (2012)
48	Myricetin-3-O-arabinogalactose	$C_{26}H_{27}O_{17}$	613.1407	Ezzat et al. (2016), Fernández-Arroyo et al. (2011), Herranz-López et al. (2012), Rodríguez-Medina et al. (2009)
49	Myricetin pentosylhexoside	$C_{26}H_{27}O_{17}$	612.1431	Ezzat et al. (2016)
50	Syringetin-3-O-glucoside	$C_{23}H_{24}O_{13}$	508.1217	Borrás-Linares et al. (2015)
Flavanols				
51	Catechin	$C_{15}H_{14}O_6$	290.2681	Morales-Luna et al. (2019)
52	Epicatechin	$C_{15}H_{14}O_6$	290.2681	Morales-Luna et al. (2019), Pérez-Ramírez et al. (2015)
53	Prodelphinidin B3	$C_{30}H_{26}O_{14}$	610.1322	Borrás-Linares et al. (2015), Herranz-López et al. (2012)
54	Methyl-epigallocatechin	$C_{16}H_{16}O_7$	320.0896	Borrás-Linares et al. (2015), Herranz-López et al. (2012)

TABLE 5.2
Polyphenols Identified in Calyces of Roselle.—cont'd

ID Compound	Compound	Molecular Formula	Molecular Weight [M-H]⁻	References
55	Gallocatechin gallate	$C_{22}H_{18}O_{11}$	158.3717	Pérez-Ramírez et al. (2015)
Flavones				
56	Gossypetin-3-O-glucoside (gossytrin)	$C_{21}H_{19}O_{13}$	479.0837	Ezzat et al. (2016)
57	3,5,7,8,3′,4′-hexahydroxy flavone (gossypetin)	$C_{15}H_{10}O_{8}$	318.0376	Rao and Seshadri (1942)
58	3,5,7,8,3′,4′,5′-heptahydroxyflavone (hibiscetin)	$C_{15}H_{10}O_{9}$	334.0376	Rao and Seshadri (1942)
59	Luteolin-7-O-D-glucuronidemethylester	$C_{27}H_{20}O_{12}$	475.0882	Borrás-Linares et al. (2015)
Flavanones				
60	Tetra-O-methyljeediflavanone	$C_{34}H_{30}O_{11}$	614.1788	Herranz-López et al. (2012)
61	Hesperidin	$C_{28}H_{34}O_{15}$	610.5606	Pérez-Ramírez et al. (2015)
62	Naringenin	$C_{15}H_{12}O_{5}$	272.2528	Pérez-Ramírez et al. (2015)
Anthocyanins				
63	Delphinidin	$C_{15}H_{11}O_{7}$	303.2436	Ezzat et al. (2016)
64	Delphinidin-3-O-glucoside	$C_{21}H_{21}O_{12}$	465.1033	Amaya-Cruz et al. (2019), Frank et al. (2005)
65	Delphinidin-3-O-arabinoside	$C_{20}H_{19}ClO_{11}$	435.0927	Amaya-Cruz et al. (2017)
66	Delphinidin-3-O-sambubioside	$C_{26}H_{29}ClO_{16}$	597.1456	Frank et al. (2005)
67	Delphinidin hexosyl pentosyl malonate	$C_{29}H_{31}O_{19}$	683.1437	Ezzat et al. (2016)
68	Delphinidin-3-O-sambubioside (hibiscin)	$C_{26}H_{29}O_{16}$	597.1456	Amaya-Cruz et al. (2017, 2019), Ezzat et al. (2016), Fernández-Arroyo et al. (2011), Herranz-López et al. (2012), Morales-Luna et al. (2019), Rodríguez-Medina et al. (2009)
69	Delphinidin-3-O-gentiobioside	$C_{27}H_{31}O_{17}$	627.1561	Ezzat et al. (2016)
70	Delphinidin-3-O-neohesperidoside	$C_{27}H_{31}O_{16}$	611.1612	Ezzat et al. (2016)
71	Delphinidin-3-O-galactoside	$C_{21}H_{21}O_{12}$	465.1033	Ezzat et al. (2016)
72	Delphinidin-3-O-feruloyl-glucoside	$C_{31}H_{29}O_{15}$	641.5530	Amaya-Cruz et al. (2017)
73	Delphinidin-3,5-O-dihexoside	$C_{27}H_{31}O_{17}$	627.1561	Amaya-Cruz et al. (2017, 2019)
74	Delphinidin rutinoside	$C_{27}H_{31}O_{16}$	611.1612	Amaya-Cruz et al. (2019)
75	Delphinidin-3-*O*-(6″-coumaroylglucoside)	$C_{30}H_{27}O_{14}$	611.1401	Amaya-Cruz et al. (2019), Morales-Luna et al. (2019)
76	Cyanidin	$C_{15}H_{11}O_{6}$	287.2442	Ezzat et al. (2016)
77	Cyanidin-3-O-glucoside	$C_{21}H_{21}O_{11}$	449.1084	Amaya-Cruz et al. (2017, 2019), Frank et al. (2005)
78	Cyanidin-3-O-(3″,6″-O-dimalonyl-glucoside)	$C_{27}H_{25}O_{17}$	621.1091	Amaya-Cruz et al. (2017)
79	Cyanidin malonyl-dihexoside	$C_{30}H_{33}O_{19}$	697.1616	Amaya-Cruz et al. (2019)
80	Cyanidin-3,5-O-dihexoside	$C_{27}H_{31}O_{16}$	611.1612	Amaya-Cruz et al. (2017, 2019)

Continued

TABLE 5.2
Polyphenols Identified in Calyces of Roselle.—cont'd

ID Compound	Compound	Molecular Formula	Molecular Weight [M-H]⁻	References
81	Cyanidin-3-*O*-(6″-acetylglucoside)	$C_{23}H_{23}O_{12}$	491.1190	Amaya-Cruz et al. (2019), Morales-Luna et al. (2019)
82	Cyanidin-3-O-(6″-dioxalil-glucoside)	$C_{25}H_{21}O_{17}$	593.0778	Amaya-Cruz et al. (2017)
83	Cyanidin-3-O-sambubioside	$C_{26}H_{30}O_{15}$	579.1493	Ezzat et al. (2016), Fernández-Arroyo et al. (2011), Frank et al. (2005), Herranz-López et al. (2012), Morales-Luna et al. (2019), Rodríguez-Medina et al. (2009)
84	Cyanidin-3-O-rutinoside	$C_{27}H_{31}O_{15}$	595.1663	Amaya-Cruz et al. (2017)
85	Cyanidin-3-O-(6″-succinyl-glucoside)	$C_{25}H_{25}O_{14}$	549.1244	Amaya-Cruz et al. (2017)
86	Cyanidin sambubioside-hexoside	$C_{32}H_{39}O_{20}$	743.2035	Amaya-Cruz et al. (2019)
87	Cyanidin-3-*O*-(6″-coumaroylglucoside)	$C_{30}H_{27}O_{13}$	595.1452	Morales-Luna et al. (2019), Amaya-Cruz et al. (2019)
88	Cyanidin hexoside-rutinoside	$C_{33}H_{41}O_{20}$	757.2191	Amaya-Cruz et al. (2019)
89	Pelargonidin-3-O-glucoside	$C_{21}H_{21}O_{10}$	433.1135	Amaya-Cruz et al. (2017)
90	Pelargonidin-3-O-sambubioside	$C_{26}H_{29}O_{14}$	565.1558	Amaya-Cruz et al. (2019)
91	Petunidin-3,5-O-diglucoside	$C_{28}H_{33}ClO_{17}$	676.1406	Amaya-Cruz et al. (2017)
92	Petunidin-3-O-glucoside	$C_{22}H_{23}O_{12}$	479.1190	Amaya-Cruz et al. (2017)
93	Petunidin-3-O-galactoside	$C_{22}H_{23}O_{12}$	479.1189	Amaya-Cruz et al. (2017)
94	Peonidin-3-O-glucoside	$C_{22}H_{23}O_{11}$	463.4114	Pérez-Ramírez et al. (2015)
95	Peonidin-3-O-acetylglucoside	$C_{24}H_{25}O_{12}$	505.1300	Morales-Luna et al. (2019)
96	Peonidin-3-*O*-(6″-coumaroylglucoside)	$C_{38}H_{41}O_{17}$	769.2343	Morales-Luna et al. (2019)
97	Malvidin-3-O-acetylglucoside	$C_{25}H_{27}O_{13}$	535.1452	Ezzat et al. (2016), Morales-Luna et al. (2019)
98	Malvidin-3-O-glucoside	$C_{23}H_{25}O_{12}$	493.1346	Pérez-Ramírez et al. (2015)
99	Malvidin-3-*O*-(6″-coumaroylglucoside)	$C_{21}H_{31}O_{14}$	639.1714	Morales-Luna et al. (2019)
OTHER POLYPHENOLS				
100	N-feruloyltyramine	$C_{18}H_{20}NO_4$	313.1314	Borrás-Linares et al. (2015), Ezzat et al. (2016), Fernández-Arroyo et al. (2011), Herranz-López et al. (2012), Rodríguez-Medina et al. (2009)
101	7-Hydroxycoumarin	$C_9H_6O_3$	162.0317	Fernández-Arroyo et al. (2011)
102	Cleomiscosin A	$C_{20}H_{18}O_8$	385.0919	(Borrás-Linares et al., 2015)
103	Cleomiscosin B	$C_{20}H_{18}O_8$	385.0929	Borrás-Linares et al. (2015)
104	Vanillin	$C_8H_8O_3$	152.1473	Pérez-Ramírez et al. (2015)

All references include HPLC-MS (high-performance liquid chromatography-mass spectrometry) analysis except reference Pérez-Ramírez et al. (2015), where HPLC-DAD (diode-array detector) analysis was performed.

compounds may not be detected in a general polyphenol extract but may need additional purification; such is the case of prodelphinidin B3, caffeoylglucose, myricetin 3-O-glucoside, and tetra-O-methyljeediflavanone (Herranz-López et al., 2012). This should be considered when comparing results between studies and particularly when apparent discrepancies are found.

As for all vegetal materials, phenolic content may be affected by several agronomic factors, and the best conditions for producing polyphenol-rich roselle plants are still to be established. Thus when total polyphenol content and anthocyanin content were evaluated in three roselle varieties at different moments of maturity (preflowering stage and days 3, 7, 21, and 35 after flowering), different tendencies were observed for each variety (Christian & Jackson, 2009). Another study showed that the use of poultry litter as organic fertilizer did not modify the phenolic compound content of roselle, thus constituting a potential use through which to valorize this residue (Formagio et al., 2015). Meanwhile, comparison of two soil media showed that the polyphenol content of roselle increased in a 2:1:1 (v/v/v) topsoil/organic matter/sand formulation, compared with the 2:1:2 formulation (Aishah, Rohana, et al., 2019).

A final interesting aspect regarding the polyphenol profile of roselle is the potential induction of synthesis of phenolic compounds as a result of different treatments. One study explored the use of plant growth regulators to stimulate the production of anthocyanins and phenolic compounds in callus and in vitro suspension cells of three varieties of roselle. The results showed that treatment with inorganic nitrogen increased the anthocyanin content through an activation of the transcriptional expression of genes encoding anthocyanins and flavone synthase 3′-hydroxylase (De Dios-López et al., 2011). Similarly, treatment of roselle with jasmonic acid 65 days after transplantation significantly increased total polyphenol content (Aishah, Huda, et al., 2019). Further studies in this area may help produce roselle enriched in specific components, which may affect its health benefits, as discussed in the next section.

Studies of roselle polyphenols have focused on calyces, as these are the plant part most commonly consumed via the production of a beverage. However, leaves are used in vegetable soups and salads in Africa (Lyu et al., 2020), so some authors have explored the phenolic profile of roselle leaves. In particular, two systematic studies evaluated 49 and 31 roselle varieties where the major flavonoid compounds in the roselle leaves were rutin (quercetin-3-rutinoside) and kaempferol-3-O-rutinoside. The major phenolic acid in roselle leaves is neochlorogenic acid (Lyu et al., 2020; Wang et al., 2016).

2.2 Strategies to Improve the Extraction of Bioactive Compounds From Roselle

Some studies have explored different solvent combinations for extracting polyphenols from roselle. Thus it was found that ethanol extraction provided a value for total polyphenols of 10 mg/g, whereas when ethanol/water (50:50, v/v) was used, this value increased to 24 mg/g (Salazar-González et al., 2012). Another study explored the use of aqueous two-phase extraction, comparing sodium citrate/polyethylene glycol 2000 and sodium citrate/acetone; the best results were obtained with the use of 15.83% of polymer and 25.16% of sodium citrate, which allowed the recovery of most of the phenolic compounds (95%) and interestingly the removal of 93% of the sugars (Rodríguez-Salazar & Valle-Guadarrama, 2019).

However, most research in this topic focuses on the use of novel technologies to obtain polyphenol-rich extracts from roselle. Ultrasound-assisted extraction (UAE) is a sustainable methodology for preparing extracts that are rich in bioactive compounds; it may improve the extraction efficiency by promoting mass transfer due to rupture of the cell wall and reduction of particle size due to the acoustic effect of cavitation. For roselle, the optimal UAE conditions were found to be 40 min, 40 kHz, 180 W, and ethanol as the solvent, except for tannins, in which case more sonication time was necessary, that is, 120 min under the same conditions of frequency and power (Peredo Pozos et al., 2020). Other authors applied UAE to roselle in order to extract anthocyanins to be used as natural food colors (Almahy et al., 2017; Pinela et al., 2019).

Another technology applied to the extraction of polyphenols from roselle is microwave-assisted extraction (MAE), which can optimize the extraction of phenolic compounds. However, different variables (microwave power, solvent/solid ratio, extraction temperature, and process time) can affect the polyphenolic profile of the extracts. According to Pimentel-Moral et al. (2018), a high temperature (164°C) for 12.5 min and an ethanol/water (80:20, v/v) mixture as the solvent applied to the system of 3 g of dried roselle powder per 30 mL of solvent were the ideal conditions to maximize the extraction yield of bioactive compounds from roselle. However, glycoside flavonoids can become degraded, maybe due to the temperature, because the application of a lower temperature (60°C) at 500 W in a relation of

14:1 solvent:sample, showed an increased extraction of polyphenols after 3 min, without degradation (Alara & Abdurahman, 2019). In addition, cyclic processes can be applied to decrease the process temperature (55°C), at 80 W for 15 min in a relation of 4:1 (water/material, v/v) and thus avoid the degradation of compounds due to the effects of high temperatures (Nguyen, 2020). Other authors have evaluated the combination of MAE and subsequent extraction and studied how the different variables affect total monomeric anthocyanins, total phenolic compounds, and antioxidant activity. The results indicated that the best extraction conditions were 700 W for 8 min with subsequent extraction in an acidic aqueous solution for 6 h. These conditions led to an extract with 1.63 mg/g of delphinidin-3-sambubioside equivalents for anthocyanins, 29.62 mg/g of gallic acid equivalents for total phenolic compound, and 133.25 μmol/g of Trolox equivalents for antioxidant capacity (Cassol et al., 2019).

Supercritical extraction has also been explored to obtain polyphenol-rich solutions from roselle. Thus several conditions with supercritical carbon dioxide were tested, with pressure ranging from 10 to 24 MPa and temperatures of 323, 333, and 343 K. The most successful extraction was obtained at 343 K and 24 MPa (Lukmanto et al., 2013). A related procedure, subcritical water extraction, was used to obtain polyphenol-rich extracts from defatted roselle seeds (Tran-Thi et al., 2012). Optimum conditions allowed the polyphenol content of the extract to be increased by up to 12 times, with ferulic acid and *p*-coumaric acid as the main constituents.

Additionally, although roselle beverages are traditionally prepared at home, in the past few years, several commercial beverages have been produced. A study evaluated the optimal conditions for polyphenol extraction during the preparation of this kind of beverages, which were found to be 95°C for 60 min (Pérez-Ramírez et al., 2015). Moreover, those authors evaluated the effect of the addition of stevia and citric acid to stabilize the polyphenol content of the final product and prevent the decrease in polyphenol levels that occurs during storage (due to the transformation of anthocyanins into chalcones, among other processes). They concluded that there was a preserving effect when stevia was added at 14−15 g/L and citric acid at 0.2−0.3 g/L, as evidenced by the determination of total polyphenols, individual polyphenols, and the browning index. Interestingly, the incorporation of stevia increased the stability of gallic acid, epigallocatechin gallate, rosmarinic acid, and quercetin, whereas some phenolic compounds, such as gallocatechin gallate, rutin, sinapic acid, vanillin, and ellagic acid, presented similar degradation when compared with the control. The stability of anthocyanins in stored freeze-dried roselle calyces has also been evaluated, finding that when the process was performed in the presence of an amorphous polysaccharide (pullulan) the color degradation decreased by 1.5−1.8 times, compared with the control (Gradinaru et al., 2003). A final aspect to be mentioned regarding polyphenol stability during roselle storage is the influence of water activity: it was established that when this value is above 0.3, there is an increase in polyphenol degradation due to matrix swelling and dissolution (Maldonado-Astudillo et al., 2019).

2.3 Beneficial Effects of Roselle Polyphenol Extracts

Other chapters in this book explore the different health effects associated with roselle consumption and derived from the combined effect of its constituents, both nutrients and phytochemicals. The fact that both white (without anthocyanins) and red varieties of roselle prevented body-weight gain and decreased adipocyte hyperplasia in rats fed a hypercaloric diet suggests that the biological effects of roselle are not exclusively due to anthocyanins (Morales-Luna et al., 2019).

Here, we will only consider studies performed with polyphenol-rich extracts from roselle. Some of them addressed the potential of these extracts to prevent complications associated with diabetes. A study in rats with induced type 1 diabetes found that supplementation with a polyphenol-rich extract from roselle improved the glucose and lipid profile, as well as cardiac function, as evidenced by modifications in ventricular pressure and histopathologic determinations (Yusof et al., 2018). Extracts of polyphenols from roselle have been shown to decrease serum triglyceride and total cholesterol levels in plasma in hamsters, in a dose-dependent manner. A decrease in cholesterol and triglyceride levels in the liver was also observed. These beneficial effects were attributed to the regulation of lipid synthesis and phosphorylation of AMP-activated protein kinase by polyphenols from roselle through a reduction of the expression of fatty acid synthase in liver cells (Yang et al., 2010). Another study focused on the injurious effects on the male reproductive organ in diabetic rats (Budin et al., 2018). Again, a polyphenol-rich extract from roselle was provided to rats in which diabetes had been induced by streptozotocin; compared with the control group, the treated animals showed fewer morphologic alterations of the testis as well as a decrease in oxidative injury in this organ. Another complication of diabetes attenuated by polyphenols from roselle is diabetic nephropathy through the inhibition of high glucose-induced angiotensin II receptor-1, thus attenuating renal epithelial-mesenchymal transition (Huang et al., 2016).

In connection with the aforementioned study of effects on cardiac function, a different study found that a polyphenol extract from roselle (containing 12 flavonoids and 7 phenolic acids), when directly perfused, lowered systolic heart function as well as heart rate, while simultaneously increasing maximal velocity of relaxation. After observing that the extract abolished inotropic responses elicited by pharmacologic agonists for L-type Ca^{2+} channel, ryanodine receptor, β-adrenergic receptor, and SERCA (sarcoplasmic/endoplasmic reticulum Ca^{2+}-ATPase) blocker, it was concluded that polyphenols in roselle show negative inotropic, negative chronotropic, and positive lusitropic responses, possibly by modulating calcium entry, release, and reuptake in the heart (Lim et al., 2016).

Other studies have focused on the potential hepatoprotective effects of polyphenol-rich roselle extracts. Thus in a preclinical study, rats had damage induced using carbon tetrachloride (Adetutu & Owoade, 2013) and supplementation with the extract decreased the levels of hepatic enzymes alanine aminotransferase and aspartate aminotransferase and the hepatic levels of malondialdehyde and increased the hepatic expression of antioxidant enzymes. Similarly, another preclinical study (Ezzat et al., 2016) prepared anthocyanin-rich extracts from roselle and evaluated their hepatoprotective activity by examining the hepatic, inflammatory, and oxidative stress markers and performing a histopathologic examination of rats with thioacetamide-induced hepatotoxicity. The supplementation significantly reduced the serum levels of alanine aminotransferase, aspartate aminotransferase, and hepatic malondialdehyde, as well as several hepatic inflammatory markers (tumor necrosis factor α, interleukin 6, and interferon gamma), at the same time as it increased the expression of endogenous antioxidant enzymes.

The beneficial effects of polyphenols from roselle on oxidative status were also observed in a study where rats received an acute dose of a polyphenol-rich extract prepared from roselle (Fernández-Arroyo et al., 2012). This intake significantly decreased lipid oxidation, as measured by malondialdehyde, as well as significantly increased antioxidant response, determined as the activity of the superoxide dismutase enzyme and total plasma antioxidant capacity. These effects were observed 20, 60, and 120 min after the intake (although the modification in superoxide dismutase was not significant after 120 min). The fact that significant differences were observed as early as 20 min after intake is related to the fast absorption of some polyphenols present in roselle, as discussed in the following.

It is known that oxidation is closely linked to inflammation, and roselle polyphenols have also shown antiinflammatory activities. In particular, the previously mentioned gossypetin activity was compared with that of the antiinflammatory drug ketorolac tromethamine in carrageenin-induced paw edema in rats. It was found that 100 mg/kg of gossypetin showed (52.05%) inhibition of inflammation in comparison with ketorolac tromethamine (56.73%) (Mounnissamy et al., 2002).

Regarding anticancer effects, anthocyanins in an extract from roselle calyces could inhibit tumor growth, lung metastasis, and tumor angiogenesis mediated via the PI3K/Akt and Ras/MAPK (mitogen-activated protein kinase) cascade pathways (Su et al., 2018). Lo et al. (2007) suggested that anthocyanins from *Hibiscus* induce apoptosis of proliferating smooth muscle cells via activation of the p38 MAP kinase and p53 pathways. In vivo models also demonstrated the anticancer effect of a polyphenol-enriched extract from *Hibiscus sabdariffa* through the suppression of the metastatic capacity of DLD-1 cells via downregulation of focal adhesion kinase(FAK)-associated signaling and CD44/c-MET signaling (Huang et al., 2018). Not only polyphenols in roselle calyces have shown anticancer effects but also leaf polyphenols have shown promising results in apoptotic and autophagic activities– an effect partially attributed to epicatechin gallate (Chiu et al., 2015).

Regarding other pathologic conditions, some polyphenols present in roselle, such as the widely distributed quercetin and the more specific gossypetin and hibiscetin, were tested in an in silico study to compare them to the antimalarial compound artemisinin (Nerdy, 2017). It is remarkable that these phenolic compounds showed greater inhibitory activity of plasmepsin 1 and 2 than artemisinin.

Another interesting aspect is that some beneficial effects of roselle polyphenols may be ascribed to some specific compounds. Thus in a study where 35 purified roselle polyphenol fractions were obtained, only three of them significantly inhibited in vitro adipogenesis at concentrations that may be used in a real scenario (Herranz-López et al., 2012). To complicate matters even further, these active fractions differed in their polyphenol composition: one was rich in delphinidin-3-O-sambubioside; another in cyanidin-3-O-sambubioside, chlorogenic acid, and tetra-O-methyljeediflavanone; and the last was especially rich in glycosylated flavonols, such as quercetin-3-O-sambubioside and myricetin 3-O-glucoside. Moreover, although most of the isolated fractions were not effective at preventing adipogenesis, the highest effect was observed when the original extract containing all the fractions was tested.

Another study involved partial least squares discriminant analysis (PLS-DA) of roselle phytochemicals on body weight, adipocyte diameter, and hepatic triglycerides in high-fat high-fructose diet-induced obese rats (Morales-Luna et al., 2019). The authors found that the phytochemicals responsible for the observed effects were organic acids (a nonphenolic group) and anthocyanins in the case of red varieties, while other flavonoids and phenolic acids did not contribute to them.

Moreover, regarding specific roselle compounds, it was established (Gulsheen et al., 2019), after combining a detailed polyphenol fractionation with animal studies, that gossypetin is the anxiolytic and antidepressant constituent of roselle calyces, traditionally consumed for nervous disorders.

Overall, these studies show that although there is evidence that polyphenols contribute to the health benefits of roselle, further research is needed to ascertain the exact contribution and role of these compounds in the biological effects of this material, especially considering that it is rich in other phytochemicals, such as characteristic organic acids.

Another important aspect is that although preclinical studies have shown promising results, there is a need to develop clinical trials with polyphenol-rich extracts from roselle. In this way, a randomized, double-blind placebo-controlled trial in which obese or overweight subjects consumed a commercial mixture of polyphenols from roselle and *Lippia citriodora* for 84 days found that this product significantly decreased body mass index and fat mass, as compared to the placebo (Marhuenda et al., 2020). Supplementation with this product combination to overweight subjects in another study improved anthropometric measurements, while decreasing blood pressure and heart rate (Boix-Castejón et al., 2018). Furthermore, the acute intake of 10 g of roselle extract significantly reduced the level of the inflammatory marker MCP-1 in healthy volunteers (Beltrán-Debón et al., 2010). Interestingly, the significant improvement in flow-mediated dilation observed after supplementation with roselle extract for 2 weeks to subjects with a 1%–10% cardiovascular risk was accompanied by a significant increase in the circulating levels of gallic acid and its methylated metabolites (Abubakar et al., 2019). Meanwhile, in another study that reported increased plasma antioxidant capacity and decreased plasma lipid oxidation in healthy subjects consuming an aqueous roselle extract, an increase in the urinary excretion of several roselle polyphenols and derived metabolites was also observed (Frank et al., 2012). Overall, these results are rather promising, but as the number of subjects was only about a dozen, they need to be confirmed in larger clinical trials.

2.4 Roselle By-Products as Another Polyphenol Source

Decoction residues generated as by-products can retain a significant amount of bioactive constituents. According to Sáyago-Ayerdi et al. (2013) the decoction residues from different cultivars of roselle are rich in dietary fiber (40%–45.7%) and protein (4.9%–12.3%), but their polyphenol content (5–13 mg/g) was particularly significant. This content does not seem to be affected by particle size as previously mentioned for calyces: it remains in a constant range when samples with particle sizes of 250–177, 177–150, and 150–74 μm were compared (Amaya-Cruz et al., 2017).

The specific polyphenol profile of a roselle by-product extract was evaluated by UPLC-ESI-QTOF MS (ultraperformance liquid chromatography/electrospray ionization/quadrupole time-of-flight/mass spectrometry) analysis (Amaya-Cruz et al., 2019). A total of 31 compounds were identified, including 10 hydroxycinnamic acids (ellagic acids and several derivatives of caffeic acid and ferulic acid), 1 hydroxybenzoic acid (gallic acid), 9 flavonols (the aglycons myricetin, quercetin, and kaempferol, as well as several glycosides of these compounds), and 11 anthocyanins (derivatives of cyanidin, pelargonidin, and delphinidin). Interestingly, the flavonol kaempferol hexoside-rhamnoside was detected in the by-product, but not in the calyx, probably because the concentration was below the limit of detection, while it was concentrated in the decoction residue. When anthocyanins were quantified in the by-product (Amaya-Cruz et al., 2017), it was found that two of them (cyanidin-3-O-rutinoside and delphinidin-3-O-sambubioside) corresponded to 95% of the total anthocyanin content, a value similar to that found in calyces. Also, when comparing the individual content of anthocyanins in calyces and by-products, the percentage decrease was 62% for cyanidin-3-O-rutinoside and 60% for delphinidin-3-O-sambubioside, the two major anthocyanins. However, the maximum decrease was observed in delphinidin-3-O-arabinoside (100%) and the smallest reduction was in petunidin-3-O-galactoside (14%). These variations in anthocyanin content, when comparing calyces with the decoction residue, are shown in Fig. 5.1.

Regarding the beneficial effects for the control of obesity and its complications, the roselle by-products

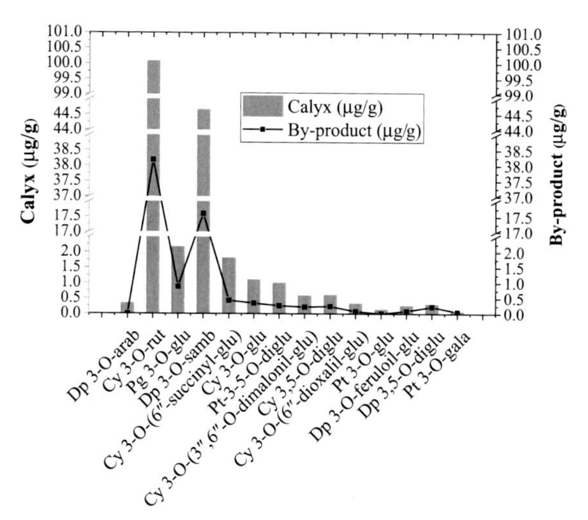

FIG. 5.1 Anthocyanin content in calyces and by-products of roselle. *arab*, arabinoside; *Cy*, cyanidin; *Dp*, delphinidin; *gala*, galactoside; *glu*, glucoside; *Pg*, pelargonidin; *Pt*, petunidin; *rut*, rutinoside; *samb*, sambubioside.

generated similar results to those of calyx powder in terms of the reduction in body weight gain (10% and 14%, respectively), adipocyte hypertrophy (17% and 13%, respectively), insulin resistance (48% and 59%, respectively), and hepatic steatosis (15% and 25%, respectively) in rats fed a high-caloric diet, as compared to the nonsupplemented control group (Amaya-Cruz et al., 2019). In contrast to the effects observed with polyphenol extracts derived exclusively from EPPs, the health effects of roselle by-products are due to a combination of the mechanisms of action of EPPs and NEPPs, as discussed in the next section.

3 THE SPECIFIC CASE OF NONEXTRACTABLE POLYPHENOLS IN ROSELLE

As previously indicated, NEPPs are compounds associated either with macromolecules such as proteins and carbohydrates that are present in cell walls—the class of hydrolyzable polyphenols (HPPs)—or with polyphenols with a high molecular weight—the class of nonextractable proanthocyanidins (NEPAs). These characteristics mean that the content of NEPs in roselle beverages is negligible, as they are in the complete matrix (calyces) or in the by-products generated after the calyx decoction process. Data on the NEPP content in roselle are still scarce, compared with those on EPPs. Among the studies of this topic, HPP content in whole dried calyces has been reported to be in the range of 6–14 mg/g, while that of NEPAs is in the range of 14–83 mg/g (Amaya-Cruz et al., 2019; Duarte-Valenzuela et al., 2016; Sáyago-Ayerdi et al., 2013, 2014). The high variability in the composition depends on genetic, climatic, and ecologic factors (Bakasso et al., 2013; Torres-Morán et al., 2011). Thus a study comparing 20 genetically improved roselle varieties grown in three Mexican states (Duarte-Valenzuela et al., 2016) found differences of up to twofold in HPP content and up to sixfold in NEPA content. Detailed HPLC-MS analysis on the NEPP profile in roselle-derived materials is still lacking.

Comparison between the by-products generated after calyx decoction and the original calyx showed that the latter possesses a higher NEPP content. Thus a study evaluating NEPP content in by-products and calyces found HPP values of 6.18 and 2.85 mg/g and NEPA values of 6.67 and 3.82 proanthocyanidin equivalents (PAE) mg/g, respectively (Amaya-Cruz et al., 2019). Another study reported the same tendency: HPPs in the range 8.5–13.1 g/kg in decoction residues and 6.4–8.8 g/kg in calyces and NEPAs in the range 14.4–22.5 g/kg in decoction residues and 10.3–18.5 g/kg in calyces (Sáyago-Ayerdi et al., 2013). In this case the difference in the values can partially be attributed to the decoction process because the conditions of the first study corresponded to 60 g of calyces added to 1 L of boiling water and heated for 15 min, whereas in the second study the conditions were 5 g of calyces added to 100 mL of boiling water for 5 min.

Taking into account the quantitative relevance of NEPPs for total polyphenol content in roselle, these compounds, for which several health-related properties have been reported (Pérez-Jiménez et al., 2013), must be considered when assessing the health benefits of roselle. However, studies on the biological effects of whole roselle calyces or the decoction residue are nearly inexistent. As mentioned earlier, in a study in rats, both a roselle by-product and calyx powder showed significant beneficial effects in body weight gain, adipocyte hypertrophy, insulin resistance, and hepatic steatosis; due to the low EPP content in the by-product, a relevant contribution of NEPPs to the observed effects was suggested (Amaya-Cruz et al., 2019). In another study, hypercholesterolemic rats were supplemented with agave dietary fiber (ADF) or ADF enriched with calyces of roselle (ADF-JC); the group fed with ADF-JC showed the lowest weight gain from week 3 until the end of the study as well as the lowest value in a marker of protein oxidation (Mda-Lys). As the ADF group was not supplemented with polyphenols, the beneficial effects were attributed to polyphenols from roselle (EPPs and

NEPPs) (Sáyago-Ayerdi et al., 2014). Finally, a study (Sáyago-Ayerdi et al., 2020) found that after dynamic in vitro fermentation of roselle calyces, significant modifications in the microbiota profile took place, with *Bifidobacterium* being the most abundant genus. Overall, these results are very promising, but many more studies are still needed in order to elucidate the exact contribution of NEPPs to the beneficial health effects of roselle.

4 CURRENT KNOWLEDGE OF THE BIOACCESSIBILITY AND BIOAVAILABILITY OF ROSELLE POLYPHENOLS

The beneficial effects associated with roselle polyphenols are related to two important factors: bioaccessibility and bioavailability. The former concept is understood as the potential for the release of polyphenols from the food matrix, which could then be absorbed, although it does not necessarily mean that they actually are absorbed. Some in vitro studies have explored how roselle polyphenols would be released during their passage through the gastrointestinal tract. Thus it was observed that in vitro digestion of whole calyces led to a significant increase in antioxidant capacity of the digestion fluids, derived from the release of polyphenols from the food matrix (Villanueva-Carvajal et al., 2013). A comparative digestion study between the decoction residue and the whole calyces (Mercado-Mercado et al., 2015) reported an overall bioaccessibility of 71% and 27%, respectively, highlighting the potential of the residue as a functional ingredient. In that study, the authors also evaluated the profile of the main phenolic compounds released during in vitro digestion, finding different profiles for the samples: the residue led to the release of caffeic acid, gallic acid, and chlorogenic acid, while the most abundant compounds after digestion of calyces were gallic acid, syringic acid, and caffeic acid.

Regarding bioavailability, as previously explained, polyphenols undergo several transformations after intake, and this applies to roselle polyphenols just as to all others, as shown in Fig. 5.2. It should be highlighted that research in this area is still scarce, although as indicated by Herranz-López et al. (2017), understanding the pharmacokinetics of compounds from roselle is essential because their absorption, metabolism, and distribution might determine the mode of action and molecular targets involved in their bioactivity.

Polyphenol metabolites detected after roselle ingestion in both preclinical and clinical studies are listed in Table 5.3. Some studies have focused on specific compounds, such as the bioavailability of gossypetin in a mouse model, where it was found that its bioavailability was 8% 1 h after intake and 24% after 6 h (Khan et al., 2015). Moreover, significant increases in gallic acid and methylated metabolite content were detected in a clinical trial (Abubakar et al., 2019). Similarly, pharmacokinetic parameters of anthocyanins following consumption of 150 mL of a roselle extract (147.4 mg of total anthocyanins) were determined in 6 healthy volunteers, with the maximum excretion rates at 1.5–2.0 h after intake being 0.018%–0.021% of urinary excretion of the administered doses of anthocyanins (Frank et al., 2005). This low bioavailability of anthocyanins was observed in another study of supplementation of healthy volunteers with roselle extract; however, the authors found a significant increase in plasma antioxidant capacity and a significant decrease in lipid oxidation (Frank et al., 2012). As it would be very difficult to ascribe these effects to compounds with such low bioavailability, it may be hypothesized that the role of roselle in oxidative status and, conversely, its beneficial effects are due to less abundant compounds or other unidentified metabolites. Indeed, it is known that most anthocyanins circulate in the human body as derived metabolites (Kay et al., 2005). In this way, a study in rats evaluated the whole profile of metabolites after roselle supplementation; a total of 14 phenolic compounds were detected, 9 of them corresponding to metabolites generated after intake (Fernández-Arroyo et al., 2012). Based on the different half-time elimination values for flavonoid conjugates and phenolic acids, the authors suggested that the latter would be linked to effects on membranes, while the former would be associated with more immediate effects. Nevertheless, an overview of the metabolic fate of roselle polyphenols in humans is still lacking. Moreover, microbial metabolites should be specifically assessed because although they have been reported after intake of several polyphenol-rich foods and their formation may be expected after roselle intake, they have not specifically been reported yet. Moreover, factors modifying polyphenol absorption, such as chemical structure (glycosylation), interaction with other components, or dose (Bohn, 2014), have not been studied in roselle.

Finally, to improve the bioavailability of phenolic compounds, different strategies have been adopted, such as encapsulation, in order to avoid degradation of compounds during processing and improve their physicochemical stability. For instance, Pimentel-Moral et al. (2019) developed nanostructured lipid

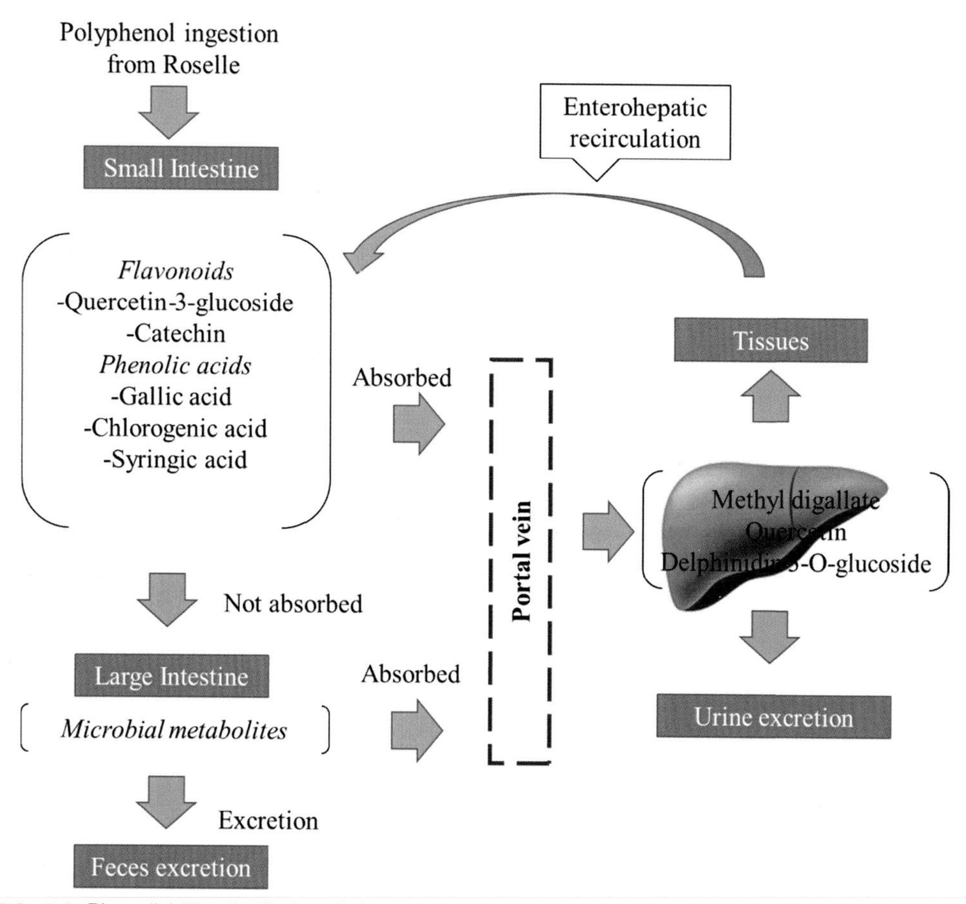

FIG. 5.2 Bioavailability of polyphenols from roselle. Note: Microbial metabolites, although expected to be formed from polyphenols present in roselle based on studies with other food sources, have not been reported in bioavailability studies with this material.

TABLE 5.3
Phenolic Metabolites Detected in Acute Preclinical and Clinical Studies After Supplementation With Roselle Extracts.[a]

ID[b]	Compound	Fluid[c]	Study Model	References
PHENOLIC ACIDS				
Hydroxybenzoic acids				
1	Gallic acid	P	Men with low cardiovascular risk	Abubakar et al. (2019)
105	3-O-methyl gallic acid	P	Men with low cardiovascular risk	Abubakar et al. (2019)
106	4-O-methyl gallic acid	P	Men with low cardiovascular risk	Abubakar et al. (2019)
2	Methyl digallate	P	Rats	Fernández-Arroyo et al. (2012)
107	Hippuric acid	P	Men with low cardiovascular risk	Abubakar et al. (2019)
		U	Healthy volunteers	Frank et al. (2012)

Continued

TABLE 5.3
Phenolic Metabolites Detected in Acute Preclinical and Clinical Studies After Supplementation With Roselle Extracts.[a]—cont'd

ID[b]	Compound	Fluid[c]	Study Model	References
Hydroxycinnamic acids				
14	Caffeoylquinic acid isomer I	P	Rats	Fernández-Arroyo et al. (2012)
FLAVONOIDS				
Flavonols				
30	Quercetin	P	Rats	Fernández-Arroyo et al. (2012)
108	Quercetin diglucuronide isomer I	P	Rats	Fernández-Arroyo et al. (2012)
109	Quercetin diglucuronide isomer II	P	Rats	Fernández-Arroyo et al. (2012)
110	Quercetin diglucuronide isomer III	P	Rats	Fernández-Arroyo et al. (2012)
111	Quercetin diglucuronide isomer IV	P	Rats	Fernández-Arroyo et al. (2012)
112	Quercetin glucuronide isomer I	P	Rats	Fernández-Arroyo et al. (2012)
113	Quercetin glucuronide isomer II	P	Rats	(Fernández-Arroyo et al., 2012)
114	Methylquercetin	P	Rats	Fernández-Arroyo et al. (2012)
36	Kaempferol	P	Rats	(Fernández-Arroyo et al., 2012)
115	Kaempferol glucuronide isomer I	P	Rats	Fernández-Arroyo et al. (2012)
116	Kaempferol glucuronide isomer II	P	Rats	Fernández-Arroyo et al. (2012)
Flavones				
57	3,5,7,8,3′,4′-hexahydroxy flavone (gossypetin)[a]	L	Mice	Khan et al. (2015)
Anthocyanins				
64	Delphinidin-3-O-glucoside	P,U	Healthy subjects	Frank et al. (2005, 2012)
66	Delphinidin-3-O-sambubioside	P,U	Healthy subjects	Frank et al. (2005, 2012)
117	Delphinidin monoglucuronide	U	Healthy subjects	Frank et al. (2012)
77	Cyanidin-3-O-glucoside	P,U	Healthy subjects	Frank et al. (2005)
83	Cyanidin-3-O-sambubioside	P,U	Healthy subjects	Frank et al. (2005)
118	Cyanidin monoglucuronide	U	Healthy subjects	Frank et al. (2012)
OTHER POLYPHENOLS				
100	N-feruloyltyramine	P	Rats	Fernández-Arroyo et al. (2012)

[a] In the study by Khan et al. (2015), the supplementation was performed with gossypetin.
[b] ID codes from Table 5.2 have been kept.
[c] *L*, liver; *P*, plasma; *U*, urine.

carriers as a strategy to improve the bioavailability of the bioactive compounds in roselle after increasing the content of these bioactive compounds in extracts by MAE or pressurized liquid extraction techniques. In another study, aqueous and ethanolic extracts of *H. sabdariffa* were spray-dried using maltodextrin and gum arabic as carrier agents (Navidad-Murrieta et al., 2020). However, in vivo studies of the modifications in bioavailability caused by these encapsulation techniques are still pending.

5 CONCLUSIONS

There is already accumulated evidence of the important polyphenol content of roselle, as well as incipient data on the metabolic fate of these compounds and promising results regarding their health effects. Nevertheless, detailed studies are still needed of the phenolic profile of the different roselle varieties as well as on the individual compounds in the NEPP fraction. Similarly, the bioavailability of roselle polyphenols, especially in humans, should be studied in depth. Finally, results concerning the by-products from roselle calyx decoction indicate its potential as a new food supplement or as an ingredient; this is a relevant aspect for material that is currently discarded.

REFERENCES

Abubakar, S. M., Ukeyima, M. T., Spencer, J. P. E., & Lovegrove, J. A. (2019). Acute effects of *Hibiscus sabdariffa* calyces on postprandial blood pressure, vascular function, blood lipids, biomarkers of insulin resistance and inflammation in humans. *Nutrients, 11*(2), 341. https://doi.org/10.3390/nu11020341

Adetutu, A., & Owoade, A. O. (2013). "Hepatoprotective and antioxidant effect of *Hibiscus* polyphenol rich extract (HPE) against carbon tetrachloride (CCl4)–induced damage in rats. *Journal of Advances in Medicine and Medical Research, 3*(4), 1574–1586. https://doi.org/10.9734/BJMMR/2013/3762

Aishah, M. S., Huda, A. N., Nadirah, C. M. C. N. A., & Jalifah, L. (2019). "Application of jasmonic acid: Effects on growth and phenolic constituents' production of roselle (*Hibiscus sabdariffa* var. UKM-2). *Journal of Physics: Conference Series, 1358*(1), 012004. https://doi.org/10.1088/1742-6596/1358/1/012004

Aishah, M. S., Rohana, T., Masni, M. A., & Jalifah, L. (2019). Growth and phenolic constituents production of roselle (*Hibiscus sabdariffa* var. UKM-2) in response to soil media. *Journal of Physics: Conference Series, 1358*(1), 012003. https://doi.org/10.1088/1742-6596/1358/1/012003

Alara, O. R., & Abdurahman, N. H. (2019). MIcrowave-assisted extraction of phenolics from Hibiscus sabadariffa calyces: kinetics modelling and porcess intensification. *Industrial Crops and Products, 137*, 528–535.

Almahy, H. A., Abdel-Razik, H. H., El-Badry, Y. A., & Ibrahim, A. M. (2017). Ultrasound-assisted extraction of anthocyanin pigments from *Hibiscus sabdariffa* (Rosella) and its phytochemical activity at Kingdom of Saudi Arabia. *International Journal Chemistry Science, 15*(4), 196.

Almeida, S., Alves, M. G., Sousa, M., Oliveira, P. F., & Silva, B. M. (2016). Are polyphenols strong dietary agents against neurotoxicity and neurodegeneration? *Neurotoxicity Research, 30*(3), 345–366. https://doi.org/10.1007/s12640-015-9590-4

Amarowicz, R., Carle, R., Dongowski, G., Durazzo, A., Galensa, R., Kammerer, D., Maiani, G., & Piskula, M. K. (2009). Influence of postharvest processing and storage on the content of phenolic acids and flavonoids in foods. *Molecular Nutrition and Food Research, 53*(S2), S151–S183. https://doi.org/10.1002/mnfr.200700486

Amaya-Cruz, D. M., Perez-Ramirez, I. F., Ortega-Diaz, D., Rodriguez-Garcia, M. E., & Reynoso-Camacho, R. (2017). Roselle (*Hibiscus sabdariffa*) by-product as functional ingredient: Effect of thermal processing and particle size reduction on bioactive constituents and functional, morphological, and structural properties. *Journal of Food Measurement and Characterization, 12*(1), 135–144. https://doi.org/10.1007/s11694-017-9624-0

Amaya-Cruz, D., Peréz-Ramírez, I. F., Pérez-Jiménez, J., Nava, G. M., & Reynoso Camacho, R. (2019). Comparison of the bioactive potential of roselle (*Hibiscus sabdariffa* L.) calyx and its by-product: Phenolic characterization by UPLC-QTOF MS^E and their anti-obesity effect in vivo. *Food Research International, 126*, 108589. https://doi.org/10.1016/j.foodres.2019.108589

Arranz, S., Saura-Calixto, F., Shaha, S., & Kroon, P. A. (2009). High contents of nonextractable polyphenols in fruits suggest that polyphenol contents of plant foods have been underestimated. *Journal of Agricultural and Food Chemistry, 57*(16), 7298–7303. https://doi.org/10.1021/jf9016652

Arranz, S., Silván, J. M., & Saura-Calixto, F. (2010). Nonextractable polyphenols, usually ignored, are the major part of dietary polyphenols: A study on the Spanish diet. *Molecular Nutrition and Food Research, 54*(11), 1646–1658. https://doi.org/10.1002/mnfr.200900580

Bakasso, Y., Zaman-Allah, M., Mariac, C., Billot, C., Vigouroux, Y., Zongo, J. D., & Saadou, M. (2013). Genetic diversity and population structure in a collection of Roselle (*Hibiscus sabdariffa* L.) from Niger. *Plant Genetic Resources, 12*(02), 207–214. https://doi.org/10.1017/s1479262113000531

Beltrán-Debón, R., Alonso-Villaverde, C., Aragonés, G., Rodríguez-Medina, I., Rull, A., Micol, V., Segura-Carretero, A., Fernández-Gutiérrez, A., Camps, J., & Joven, J. (2010). The aqueous extract of *Hibiscus sabdariffa* calices modulates the production of monocyte chemoattractant protein-1 in humans. *Phytomedicine, 17*(3–4), 186–191. https://doi.org/10.1016/j.phymed.2009.08.006

Bohn, T. (2014). Dietary factors affecting polyphenol bioavailability. *Nutrition Reviews, 72*(7), 429–452. https://doi:10.1111/nure.12114.

Boix-Castejón, M., Herranz-López, M., Pérez Gago, A., Olivares-Vicente, M., Caturla, N., Roche, E., & Micol, V. (2018). Hibiscus and lemon verbena polyphenols modulate appetite-related biomarkers in overweight subjects: A randomized controlled trial. *Food and Function, 9*(6), 3173–3184. https://doi.org/10.1039/c8fo00367j

Borrás-Linares, I., Fernández-Arroyo, S., Arráez-Roman, D., Palmeros-Suárez, P. A., Del Val-Díaz, R., Andrade-Gonzáles, I., Fernández-Gutiérrez, A., Gómez-Leyva, J. F., & Segura-Carretero, A. (2015). Characterization of phenolic compounds, anthocyanidin, antioxidant and antimicrobial activity of 25 varieties of Mexican Roselle (*Hibiscus sabdariffa*). *Industrial Crops and Products, 69*, 385–394. https://doi.org/10.1016/j.indcrop.2015.02.053

Boto-Ordóñez, M., Urpi-Sarda, M., Queipo-Ortuño, M. I., Tulipani, S., Tinahones, F. J., & Andres-Lacueva, C.

(2014). High levels of bifidobacteria are associated with increased levels of anthocyanin microbial metabolites: A randomized clinical trial. *Food and Function, 5*(8), 1932–1938. https://doi.org/10.1039/c4fo00029c

Budin, S. B., Rahman, W. Z., Jubaidi, F. F., Yusof, N. L., Taib, I. S., & Zainalabidin, S. (2018). Roselle (*Hibiscus sabdariffa*) polyphenol-rich extract prevents testicular damage of diabetic rats. *Journal of Applied Pharmaceutical Science, 8,* 65–70. https://doi.org/10.7324/JAPS.2018.8210

Camelo-Méndez, G. A., Jara-Palacios, M. J., Escudero-Gilete, M. L., Gordillo, B., Hernanz, D., Paredes-López, O., Vanegas-Espinoza, P. E., Del Villar-Martínez, A. A., & Heredia, F. J. (2016). Comparative study of phenolic profile, antioxidant capacity, and color-composition relation of Roselle cultivars with contrasting pigmentation. *Plant Foods for Human Nutrition, 71*(1), 109–114. https://doi.org/10.1007/s11130-015-0522-5

Cao, H., Ou, J., Chen, L., Zhang, Y., Szkudelski, T., Delmas, D., Daglia, M., & Xiao, J. (2019). Dietary polyphenols and type 2 diabetes: Human study and clinical trial. *Critical Reviews in Food Science and Nutrition, 59*(20), 3371–3379. https://doi.org/10.1080/10408398.2018.1492900

Cassol, L., Rodrigues, E., & Norea, C. P. Z. (2019). Extracting phenolic compounds form hibiscus sabadariffa L. calyx using microwave assisted extraction. *Industrial Crops and Products, 133,* 168–177.

Chiu, C. T., Hsuan, S. W., Lin, H. H., Hsu, C. C., Chou, F. P., & Chen, J. H. (2015). *Hibiscus sabdariffa* leaf polyphenolic extract induces human melanoma cell death, apoptosis, and autophagy. *Journal of Food Science, 80*(3), H649–H658. https://doi.org/10.1111/1750-3841.12790

Christian, K. R., & Jackson, J. C. (2009). Changes in total phenolic and monomeric anthocyanin composition and antioxidant activity of three varieties of sorrel (*Hibiscus sabdariffa*) during maturity. *Journal of Food Composition and Analysis, 22*(7–8), 663–667. https://doi.org/10.1016/j.jfca.2009.05.007

De Dios-López, A. D., Montalvo-González, E., Andrade-González, I., & Gómez-Leyva, J. F. (2011). Induction of anthocyanins and phenolic compounds in cell cultures of Roselle (*Hibiscus sabdariffa* L.) in vitro. *Revista Chapingo Serie Horticultura, 17*(2), 77–87.

Duarte-Valenzuela, Z. N., Zamora-Gasga, V. N., Montalvo-González, E., & Sáyago-Ayerdi, S. G. (2016). Nutritional composition in 20 improved Roselle (*Hibiscus sabdariffa* L.) varieties grown in México. *Revista Fitotecnica Mexicana, 39*(3), 199–206. https://doi.org/10.35196/rfm.2016.3.199-206

Ezzat, S. M., Salama, M. M., Seif El-Din, S. H., Saleh, S., El-Lakkany, N. M., Hammam, O. A., Salem, M. B., & Botros, S. S. (2016). Metabolic profile and hepatoprotective activity of the anthocyanin-rich extract of *Hibiscus sabdariffa* calyces. *Pharmaceutical Biology, 54*(12), 3172–3181. https://doi.org/10.1080/13880209.2016.1214739

Fernández-Arroyo, S., Herranz-López, M., Beltrán-Debón, R., Borrás-Linares, I., Barrajón-Catalán, E., Joven, J., Fernández-Gutiérrez, A., Segura-Carretero, A., & Micol, V. (2012). Bioavailability study of a polyphenol-enriched extract from *Hibiscus sabdariffa* in rats and associated antioxidant status. *Molecular Nutrition and Food Research, 56*(10), 1590–1595. https://doi.org/10.1002/mnfr.201200091

Fernández-Arroyo, S., Rodríguez-Medina, I. C., Beltrán-Debón, R., Pasini, F., Joven, J., Micol, V., Segura-Carretero, A., & Fernández-Gutiérrez, A. (2011). Quantification of the polyphenolic fraction and in vitro antioxidant and in vivo anti-hyperlipemic activities of *Hibiscus sabdariffa* aqueous extract. *Food Research International, 44*(5), 1490–1495. https://doi.org/10.1016/j.foodres.2011.03.040

Formagio, A. S., Ramos, D. D., Vieira, M. C., Ramalho, S. R., Silva, M. M., Zárate, N. A., Foglio, M. A., & Carvalho, J. E. (2015). Phenolic compounds of *Hibiscus sabdariffa* and influence of organic residues on its antioxidant and antitumoral properties. *Brazilian Journal of Biology, 75*(1), 69–76. https://doi.org/10.1590/1519-6984.07413

Frank, T., Janßen, M., Netzel, M., Straß, G., Kler, A., Kriesl, E., & Bitsch, I. (2005). Pharmacokinetics of anthocyanidin-3-glycosides following consumption of *Hibiscus sabdariffa* L. Extract. *The Journal of Clinical Pharmacology, 45*(2), 203–210. https://doi.org/10.1177/0091270004270561

Frank, T., Netzel, G., Kammerer, D. R., Carle, C., Kler, A., Kriesl, A., Bitsch, I., Bitsch, R., & Netzel, M. (2012). Consumption of *Hibiscus sabdariffa* L. aqueous extract and its impact on systemic antioxidant potential in healthy subjects. *Journal of the Science of Food and Agriculture, 92*(10), 2207–2218. https://doi.org/10.1002/jsfa.5615

Gobert, M., Rémond, D., Loonis, M., Buffière, C., Santé-Lhoutellier, V., & Dufour, C. (2014). Fruits, vegetables and their polyphenols protect dietary lipids from oxidation during gastric digestion. *Food and Function, 5*(9), 2166–2174. https://doi.org/10.1039/c4fo00269e

Gradinaru, G., Biliaderis, C. G., Kallithraka, S., Kefalas, P., & Garcia-Viguera, C. (2003). Thermal stability of *Hibiscus sabdariffa* L. Anthocyanins in solution and in solid state: Effects of copigmentation and glass transition. *Food Chemistry, 83*(3), 423–436. https://doi.org/10.1016/S0308-8146(03)00125-0

Gulsheen, P., Kumar, A., & Sharma, A. (2019). In vivo antianxiety and antidepressant activity of Hibiscus sabdariffa calyx extracts. *Journal of Biologically Active Products from Nature, 9*(3), 205–214.

Harborne, J. B. (1969). Gossypetin and herbacetin as taxonomic markers in higher plants. *Phytochemistry, 8*(1), 177–183. https://doi.org/10.1016/S0031-9422(00)85810-0

Herranz-López, M., Fernández-Arroyo, S., Pérez-Sanchez, A., Barrajón-Catalán, E., Beltrán-Debón, R., Menéndez, J. A., Alonso-Villaverde, C., Segura-Carretero, A., Joven, J., & Micol, V. (2012). Synergism of plant-derived polyphenols in adipogenesis: Perspectives and implications. *Phytomedicine, 19*(3–4), 253–261. https://doi.org/10.1016/j.phymed.2011.12.001

Herranz-López, M., Olivares-Vicente, M., Encinar, J. A., Barrajón-Catalán, E., Segura-Carretero, A., Joven, J., & Micol, V. (2017). Multi-targeted molecular effects of *Hibiscus sabdariffa* polyphenols: An opportunity for a global approach to obesity. *Nutrients, 9*(8), 907. https://doi.org/10.3390/nu9080907

Huang, C. C., Hung, C. H., Chen, C. C., Kao, S. H., & Wang, C. J. (2018). *Hibiscus sabdariffa* polyphenol-enriched extract inhibits colon carcinoma metastasis associating with FAK and CD44/c-MET signaling. *Journal of Functional Foods, 48*, 542–550. https://doi.org/10.1016/j.jff.2018.07.055

Huang, C. N., Wang, C. J., Yang, Y. S., Lin, C. L., & Peng, C. H. (2016). *Hibiscus sabdariffa* polyphenols prevent palmitate-induced renal epithelial mesenchymal transition by alleviating dipeptidyl peptidase-4-mediated insulin resistance. *Food and Function, 7*(1), 475–482. https://doi.org/10.1039/c5fo00464k

Kay, C. D., Mazza, G., & Holub, B. J. (2005). Anthocyanins exist in the circulation primarily as metabolites in adult men. *Journal of Nutrition, 135*(11), 2582–2588. https://doi.org/10.1093/jn/135.11.2582

Khan, A., Manna, K., Das, D. K., Kesh, S. B., Sinha, M., Das, U., Biswas, S., Sengupta, A., Sikder, K., Datta, S., Ghosh, M., Chakrabarty, A., Banerji, A., & Dey, S. (2015). Gossypetin ameliorates ionizing radiation-induced oxidative stress in mice liver—a molecular approach. *Free Radical Research, 49*(10), 1173–1186. https://doi:10.3109/10715762.2015.1053878.

Lambert, J. D., Hong, J., Yang, G. Y., Liao, J., & Yang, C. S. (2005). Inhibition of carcinogenesis by polyphenols: Evidence from laboratory investigations. *The American Journal of Clinical Nutrition, 81*(1), 284S–291S. https://doi.org/10.1093/ajcn/81.1.284S

Lim, Y. C., Budin, S. B., Othman, F., Latip, J., & Zainalabidin, S. (2016). Roselle polyphenols exert potent negative inotropic effects via modulation of intracellular calcium regulatory channels in isolated rat heart. *Cardiovascular Toxicology, 17*(3), 251–259. https://doi.org/10.1007/s12012-016-9379-6

Lo, C. W., Huang, H. P., Lin, H. M., Chien, C. T., & Wang, C. J. (2007). Effect of *Hibiscus* anthocyanins-rich extract induces apoptosis of proliferating smooth muscle cell via activation of P38 MAPK and p53 pathway. *Molecular Nutrition and Food Research, 51*(12), 1452–1460. https://doi.org/10.1002/mnfr.200700151

Lukmanto, S., Roesdiyono, N., Ju, Y. H., Indraswati, N., Soetaredjo, F. E., & Ismadji, S. (2013). Supercritical CO_2 extraction of phenolic compounds in Roselle (*Hibiscus sabdariffa* L.). *Chemical Engineering Communications, 200*(9), 1187–1196. https://doi.org/10.1080/00986445.2012.742433

Lyu, J. I., Kim, J. M., Kim, D. G., Kim, J. B., Kim, S. H., Ahn, J. W., Kang, S. Y., Ryu, J., & Kwon, S. J. (2020). Phenolic compound content of leaf extracts from different Roselle (*Hibiscus sabdariffa*) accessions. *Plant Breeding and Biotechnology, 8*(1), 1–10. https://doi.org/10.9787/PBB.2020.8.1.1

Mackenzie, G. G., Delfino, J. M., Keen, C. L., Fraga, C. G., & Oteiza, P. I. (2009). Dimeric procyanidins are inhibitors of NF-κB–DNA binding. *Biochemical Pharmacology, 78*(9), 1252–1262. https://doi.org/10.1016/j.bcp.2009.06.111

Maldonado-Astudillo, Y. I., Jiménez-Hernández, J., Arámbula-Villa, G., Flores-Casamayor, V., Álvarez-Fitz, P., Ramírez-Ruano, M., & Salazar, R. (2019). Effect of water activity on extractable polyphenols and some physical properties of *Hibiscus sabdariffa* L. calyces. *Food Measure, 13*, 687–696. https://doi.org/10.1007/s11694-018-9981-3

Manach, C., Scalbert, A., Morand, C., Rémésy, C., & Jiménez, L. (2004). Polyphenols: Food sources and bioavailability. *The American Journal of Clinical Nutrition, 79*(5), 727–747. https://doi.org/10.1093/ajcn/79.5.727

Manach, C., Williamson, G., Morand, C., Scalbert, A., & Rémésy, C. (2005). Bioavailability and bioefficacy of polyphenols in humans. I. Review of 97 bioavailability studies. *The American Journal of Clinical Nutrition, 81*(1), 230S–242S. https://doi.org/10.1093/ajcn/81.1.230S

Marhuenda, J., Perez, S., Victoria-Montesinos, D., Abellán, M. S., Caturla, N., Jones, J., & López-Román, J. (2020). A randomized, double-blind, placebo controlled trial to determine the effectiveness a polyphenolic extract (*Hibiscus sabdariffa* and Lippia citriodora) in the reduction of body fat mass in healthy subjects. *Foods, 9*(1), 55. https://doi.org/10.3390/foods9010055

Mendonça, R. D., Carvalho, N. C., Martin-Moreno, J. M., Pimenta, A. M., Lopes, A. C. S., Gea, A., Martinez-Gonzalez, M. A., & Bes-Rastrollo, M. (2019). Total polyphenol intake, polyphenol subtypes and incidence of cardiovascular disease: The SUN cohort study. *Nutrition, Metabolism, and Cardiovascular Diseases, 29*(1), 69–78. https://doi.org/10.1016/j.numecd.2018.09.012

Mercado-Mercado, G., Blancas-Benítez, F. J., Velderraín-Rodríguez, G. R., Montalvo-onzález, E., González-Aguilar, G. A., Álvarez-Parrilla, E., & Sáyago-Ayerdi, S. G. (2015). Bioaccessibility of polyphenols released and associated to dietary fibre in calyces and decoction residues of Roselle (*Hibiscus sabdariffa* L.). *Journal of Functional Foods, 18*, 171–181. https://doi.org/10.1016/j.jff.2015.07.001

Morales-Luna, E., Pérez-Ramírez, I. F., Salgado, L. M., Castaño-Tostado, E., Gómez-Aldapa, C. A., & Reynoso-Camacho, R. (2019). The main beneficial effect of roselle (*Hibiscus sabdariffa*) on obesity is not only related to its anthocyanin content. *Journal of the Science of Food and Agriculture, 99*(2), 596–605. https://doi.org/10.1002/jsfa.9220

Mounnissamy, V. M., Gopal, V., Gunasegaran, R., & Saraswathy, A. (2002). Antiinflammatory activity of gossypetin isolated from *Hibiscus sabdariffa*. *Indian Journal of Heterocyclic Chemistry, 12*(1), 85–86.

Navidad-Murrieta, M. S., Pérez-Larios, A., Sánchez-Burgos, J. A., Ragazzo-Sánchez, J. A., Luna-Bárcenas, G., & Sáyago-Ayerdi, S. G. (2020). Use of a taguchi design in *Hibiscus sabdariffa* extracts encapsulated by spray-drying. *Foods, 9*(2), 128. https://doi.org/10.3390/foods9020128

Nerdy, N. (2017). In silico docking Roselle (*Hibiscus sabdariffa* L.) calyces flavonoids as antimalarial against plasmepsin 1 and plasmepsin 2. *Asian Journal of Pharmaceutical and Clinical Research, 10*(10), 183–186. https://doi.org/10.22159/ajpcr.2017.v10i10.19770

Nguyen, M. P. (2020). Microwave-assisted extraction of phytochemical constituents in Roselle (*Hibiscus sabdariffa* L.).

Journal of Pharmaceutical Research International, 32(2), 1–12. https://doi.org/10.9734/JPRI/2020/v32i230397

Peredo Pozos, G. I., Ruiz López, M. A., Zamora Nátera, J. F., Álvarez Moya, C., Barrientos Ramírez, L., Reynoso Silva, M., & Vargas Radillo, J. J. (2020). Antioxidant capacity and antigenotoxic effect of *Hibiscus sabdariffa* L. extracts obtained with ultrasound-assisted extraction process. *Applied Sciences, 10*(2), 560. https://doi.org/10.3390/app10020560

Pérez-Jiménez, J., Díaz-Rubio, M. E., & Saura-Calixto, F. (2013). Non-extractable polyphenols, a major dietary antioxidant: Occurrence, metabolic fate and health effects. *Nutrition Research Reviews, 26*(2), 118–129. https://doi.org/10.1017/S0954422413000097

Pérez-Jiménez, J., Fezeu, L., Touvier, M., Arnault, N., Manach, C., Hercberg, S., Galan, P., & Scalbert, A. (2011). Dietary intake of 337 polyphenols in French adults. *The American Journal of Clinical Nutrition, 93*(6), 1220–1228. https://doi.org/10.3945/ajcn.110.007096

Pérez-Jiménez, J., & Saura-Calixto, F. (2018). Fruit peels as sources of non-extractable polyphenols or macromolecular antioxidants: Analysis and nutritional implications. *Food Research International, 111*, 148–152. https://doi.org/10.1016/j.foodres.2018.05.023

Pérez-Ramírez, I. F., Castaño-Tostado, E., Ramírez-de León, J. A., Rocha-Guzmán, N. E., & Reynoso-Camacho, R. (2015). Effect of stevia and citric acid on the stability of phenolic compounds and in vitro antioxidant and antidiabetic capacity of a roselle (*Hibiscus sabdariffa* L.) beverage. *Food Chemistry, 172*, 885–892. https://doi.org/10.1016/j.foodchem.2014.09.126

Pimentel-Moral, S., Borrás-Linares, I., Lozano-Sánchez, J., Arráez-Román, D., Martínez-Férez, A., & Segura-Carretero, A. (2018). Microwave-assisted extraction for *Hibiscus sabdariffa* bioactive compounds. *Journal of Pharmaceutical and Biomedical Analysis, 156*, 313–322. https://doi.org/10.1016/j.jpba.2018.04.050

Pimentel-Moral, S., Teixeira, M. C., Fernandes, A. R., Borrás-Linares, I., Arráez-Román, D., Martínez-Férez, A., Segura-Carretero, A., & Souto, E. B. (2019). Polyphenols-enriched *Hibiscus sabdariffa* extract-loaded nanostructured lipid carriers (NLC): Optimization by multi-response surface methodology. *Journal of Drug Delivery Science and Technology, 49*, 660–667. https://doi.org/10.1016/j.jddst.2018.12.023

Pinela, J., Prieto, M. A., Pereira, E., Jabeur, I., Barreiro, M. F., Barros, L., & Ferreira, I. (2019). Optimization of heat- and ultrasound-assisted extraction of anthocyanins from *Hibiscus sabdariffa* calyces for natural food colorants. *Food Chemistry, 275*, 309–321. https://doi.org/10.1016/j.foodchem.2018.09.118

Piovesana, A., Rodrigues, E., & Noreña, C. (2019). Composition analysis of carotenoids and phenolic compounds and antioxidant activity from *Hibiscus* calyces (*Hibiscus sabdariffa* L.) by HPLC-DAD-MS/MS. *Phytochemical Analysis, 30*(2), 208–217. https://doi.org/10.1002/pca.2806

Rao, P. S., & Seshadri, T. R. (1942). Isolation of hibiscitrin from the flowers of *Hibiscus sabdariffa*: Constitution of hibiscetin. *Proceedings of the Indiana Academy of Science, 15*, 148–153. https://doi.org/10.1007/BF03051846

Rodriguez-Mateos, A., Vauzour, D., Krueger, C. G., Shanmuganayagam, D., Reed, J., Calani, L., Mena, P., Del Rio, D., & Crozier, A. (2014). Bioavailability, bioactivity and impact on health of dietary flavonoids and related compounds: An update. *Archives of Toxicology, 88*(10), 1803–1853. https://doi.org/10.1007/s00204-014-1330-7

Rodríguez-Ramiro, I., Ramos, S., Bravo, L., Goya, L., & Martín, M.Á. (2012). Procyanidin B2 induces Nrf2 translocation and glutathione S-transferase P1 expression via ERKs and p38-MAPK pathways and protect human colonic cells against oxidative stress. *European Journal of Nutrition, 51*(7), 881–892. https://doi.org/10.1007/s00394-011-0269-1

Rodríguez-Salazar, N., & Valle-Guadarrama, S. (2019). Separation of phenolic compounds from Roselle (*Hibiscus sabdariffa*) calyces with aqueous two-phase extraction based on sodium citrate and polyethylene glycol or acetone. *Separation Science and Technology, 55*(13), 1–12. https://doi.org/10.1080/01496395.2019.1634730

Rodríguez-Medina, I. C., Beltrán-Debón, R., Molina, V. M., Alonso-Villaverde, C., Joven, J., Menéndez, J. A., Segura-Carretero, A., & Fernández-Gutiérrez, A. (2009). Direct characterization of aqueous extract of *Hibiscus sabdariffa* using HPLC with diode array detection coupled to ESI and ion trap MS. *Journal of Separation Science, 32*, 3441–3448. https://doi.org/10.1002/jssc.200900298

Salazar-González, C., Vergara-Balderas, F. T., Ortega-Regules, A. E., & Guerrero-Beltrán, J. A. (2012). Antioxidant properties and color of *Hibiscus sabdariffa* extracts. *Ciencia e Investigacian Agraria, 39*(1), 70–90. https://doi.org/10.4067/S0718-16202012000100006

Sáyago-Ayerdi, S. G., Mateos, R., Ortiz-Basurto, R. I., Largo, C., Serrano, J., Granado-Serrano, A. B., Sarriá, B., Bravo, L., & Tabernero, M. (2014). Effects of consuming diets containing *Agave tequilana* dietary fibre and Jamaica calyces on body weight gain and redox status in hypercholesterolemic rats. *Food Chemistry, 148*, 54–59. https://doi.org/10.1016/j.foodchem.2013.10.004

Sáyago-Ayerdi, S. G., Velázquez-López, C., Montalvo-González, E., & Goñi, I. (2013). "By-product from decoction process of *Hibiscus sabdariffa* L. calyces as a source of polyphenols and dietary fiber. *Journal of the Science of Food and Agriculture, 94*(5), 898–904. https://doi.org/10.1002/jsfa.6333

Sáyago-Ayerdi, S. G., Zamora-Gasga, V. M., & Venema, K. (2020). Changes in gut microbiota in predigested *Hibiscus sabdariffa* L calyces and agave (*Agave tequilana* weber) fructans assessed in a dynamic *in vitro* model (TIM-2) of the human colon. *Food Research International, 132*, 109036. https://doi.org/10.1016/j.foodres.2020.109036

Schaffer, S., & Halliwell, B. (2012). Do polyphenols enter the brain and does it matter? Some theoretical and practical considerations. *Genes and Nutrition, 7*(2), 99–109. https://doi.org/10.1007/s12263-011-0255-5

Shi, Y., & Williamson, G. (2016). Quercetin lowers plasma uric acid in pre-hyperuricaemic males: A randomised, double-blinded, placebo-controlled, cross-over trial. *British Journal of Nutrition, 115*(5), 800–806. https://doi.org/10.1017/S0007114515005310

Su, C. C., Wang, C. J., Huang, K. H., Lee, Y. J., Chan, W. M., & Chang, Y. C. (2018). Anthocyanins from *Hibiscus sabdariffa* calyx attenuate in vitro and in vivo melanoma cancer metastasis. *Journal of Functional Foods, 48*, 614–631. https://doi.org/10.1016/j.jff.2018.07.032

Torres-Morán, M. I., Escoto-Delgadillo, M., Ron-Parra, J., Parra-Tovar, G., Mena-Munguía, S., Rodríguez-García, A., & Rodríguez-Sahagún, A. (2011). Relationships among twelve genotypes of Roselle (*Hibiscus sabdariffa* L.) cultivated in Western Mexico. *Industrial Crops and Products, 34*(1), 1079–1083. https://doi.org/10.1016/j.indcrop.2011.03.020

Tran-Thi, N. Y., Yuliana, M., Kasim, N. S., Le, N. T. H., Lee, D. Y., & Ju, Y. H. (2012). Subcritical water extraction of phenolic-rich product from defatted Roselle seed. *Journal of Chemical Engineering of Japan, 45*(11), 911–916. https://doi.org/10.1252/jcej.12we136

Villanueva-Carvajal, A., Bernal-Martínez, L. R., García-Gasca, M. T., & Domínguez-López, A. (2013). In vitro gastrointestinal digestion of *Hibiscus sabdariffa* L.: The use of its natural matrix to improve the concentration of phenolic compounds in gut. *LWT-Food Science and Technology, 51*(1), 260–265. https://doi.org/10.1016/j.lwt.2012.10.007

Wang, J., Cao, X., Ferchaud, V., Qi, Y., Jiang, H., Tang, F., Yue, Y., & Chin, K. L. (2016). Variations in chemical fingerprints and major flavonoid contents from the leaves of thirty-one accessions of *Hibiscus sabdariffa* L. *Biomedical Chromatography, 30*(6), 880–887. https://doi.org/10.1002/bmc.3623

Yamagata, K., Tagami, M., & Yamori, Y. (2015). Dietary polyphenols regulate endothelial function and prevent cardiovascular disease. *Nutrition, 31*(1), 28–37. https://doi.org/10.1016/j.nut.2014.04.011

Yang, M. Y., Peng, C. H., Chan, K. C., Yang, Y. S., Huang, C. N., & Wang, C. J. (2010). The hypolipidemic effect of *Hibiscus sabdariffa* polyphenols via inhibiting lipogenesis and promoting hepatic lipid clearance. *Journal of Agricultural and Food Chemistry, 58*(2), 850–859. https://doi.org/10.1021/jf903209w

Yusof, M. N. L., Zainalabidin, S., Mohd Fauzi, N. M., & Budin, S. B. (2018). *Hibiscus sabdariffa* (Roselle) polyphenol-rich extract averts cardiac functional and structural abnormalities in type 1 diabetic rats. *Applied Physiology Nutrition and Metabolism, 43*(12), 1224–1232. https://doi.org/10.1139/apnm-2018-0084

Zhou, Y., Zheng, J., Li, Y., Xu, D. P., Li, S., Chen, Y. M., & Li, H. B. (2016). Natural polyphenols for prevention and treatment of cancer. *Nutrients, 8*(8), 515. https://doi.org/10.3390/nu8080515

CHAPTER 6

Modifications and Physicomechanical Behaviors of Roselle Fiber-HDPE Biocomposites for Biomedical Uses

TAOFIK OLADIMEJI AZEEZ • RABBONI MIKE GOVERNMENT • INNOCENT OCHIAGHA EZE • SAMUEL CHIDI IWUJI

1 INTRODUCTION

Roselle, known as *Hibiscus sabdariffa*, belong to the family of Malvaceae that can be found in many parts of the World including Nigeria (Singh, 2012). There are two distinct types identify in Nigeria, especially the Southwestern part (Yoruba land) out of about 300 species: whitish green and red (Fig. 6.1). Calyx of the red type is used for the production of a beverage called hibiscus tea/roselle tea, which attributed to their nutritional values (Builders et al.,. 2013; Ismail et al., 2008; Padmakumari et al., 2011). The beverage extracted from the red fruit calyx using water and alcoholic extraction have been reported to possess medicinal potentials such as antioxidants (Ismail et al., 2008; Okasha et al., 2008), wound healing (Builders et al., 2013), and improvements in circulating reproductive hormones (Omotuyi et al., 2010; Padmakumari et al., 2011) to mention a few due to the presence of bioactive agents like ascorbic acid, flavonoids, phenolic compounds, and chelating metallic ions of copper, zinc, manganese, and iron (Abou-arab et al., 2011; Builders et al., 2013). The leaves and fruit calyx of whitish green type is used as vegetable soup and for treatment of anemia, jaundice, malaria, painful and irregular menstruation, and urethutis due to high protein, fiber, carbohydrates, and low fat contents, which justify their nutritional importance in human daily diet with the presence of terpenoids, alkaloids, philobatanins, steroids, flavonoids, cardiac glycosides, and reducing sugar (Arowosegbe et al., 2015; Okasha et al., 2008; Omotuyi

(a)

(b)

FIG. 6.1 Roselle plant **(A)** red type **(B)** whitish green type.

Roselle. https://doi.org/10.1016/B978-0-323-85213-5.00013-5

et al., 2010). Upon removal of the calyces after harvest, the bast become agrowaste which threat the environs (Azeez et al., 2020, 2018b; Azeez & Onukwuli, 2016a; Durowaye et al., 2019). The waste from agricultural residue can be potentially used as reinforcement for polymer composites (Ilyas & Sapuan, 2020a, 2020b; Ilyas et al., 2018, 2021; Sabaruddin et al., 2020; Omran et al., 2021). Literature have revealed the use of roselle fiber for reinforcement of polymers such as phenolformaldehyde (Singha & Thakur, 2008), acrylic denture base material (Okeke et al., 2018) where mechanical properties such as compressive, tensile, flexural, impact strengths, and wear resistance were characterized. HDPE disposal becomes alarming and threat to the environment as a result of its usefulness for packaging of products (Azeez, 2019). However, the use of HDPE for making bone tissue, grafts, suture, and wound dressing as biomedical applications has been an interesting area of study, but the use of roselle fiber to improve the performance of HDPE matrix are scarce in the literature, which may be attributed to technical and economic implications. Since roselle fibers are biocompatible due to their composition, its uses in making HDPE composites may serves a better utilization in biomedical applications provided it meet the technical feasibility based on mechanical properties and environmental factors that may be influenced by physical properties. In this study, the physical and mechanical properties of roselle fiber-reinforced HDPE composites were investigated.

2 MATERIALS AND METHODS

2.1 Roselle Fiber Extraction

Water retting technique was used for fiber extraction for 21–27 days, then washed with deionized water and sun dried for 28days.

2.2 Modifications of Roselle Fiber

Several chemical techniques have been used for modification of roselle fiber for the purpose of enhancing its physical and mechanical behaviors. The effect of techniques such as acetic anhydride (AC), mercerization (NaOH), sodium lauryl sulfate (SLS), and ethylene diamine tetraacetic acid (EDTA) treatments on physico-mechanical properties of roselle fiber in composite applications have not been reported. The modification of roselle fiber using AC and EDTA at optimum conditions of 3.0% and 4.3% concentration, 125, and 40 min was carried out.

2.3 Physical and Tensile Properties of Roselle Fiber

The density, aspect ratio, and water absorption of fiber and density were studied with the use of water-solid displacement method and fiber length to diameter ratio of Eqs. (6.1) and (6.2). The water absorption parameters of the fiber and composites was evaluated using percentage variation and power's law as stated in Eqs. (6.3) and (6.4), respectively. The variation in physical behaviors (density, aspect ratio, and water absorption) when chemically modified with AC and EDTA at optimum conditions of 3.0% and 4.3% concentration, 125 and 40 min.

$$\rho_f = \frac{M}{V} \tag{6.1}$$

$$A_f = \frac{L_f}{d_f} \tag{6.2}$$

$$W_i = \frac{M_t - M_0}{M_0} x100\% \tag{6.3}$$

$$\frac{M_t}{M_m} = kt^n \tag{6.4}$$

Where A_f, P_f, L_f, d_f, M, and V are aspect ratio, density, length, thickness, quantity, and displaced volume of water of roselle fiber, respectively. W_i, M_t, and M_m are water absorption, water content at any time t and saturated water content of roselle fiber, and k and n represent water absorption rate constant and sorption index, respectively.

2.4 Characterization of Mechanical Properties of Fiber and HDPE Composites

Instron Universal Testing Machine model 3369 was reported to be used for estimation of tensile properties of roselle fiber on 100 mm length as well as tensile, flexural, and impact properties, while Rockwell hardness equipment was used for evaluation of hardness parameter.

2.5 Microstructural Behaviors of Roselle Fiber and Its Composites

SEM ASPEX 3020 model coupled with EDS 4000 model was carried out on surface fracture of fiber and composites mounted on stubs of silver plate at 20 KeV and 5.0 × 10−5 torr upon vacuum evacuation of thin film platinum (Azeez & Onukwuli, 2018a, b) to study the morphology and elemental composition of the sample.

3 RESULTS AND DISCUSSION

3.1 Extraction and Yield of Roselle

The extraction techniques employed are the retting method with or without the heat applied and resulted in extraction with residue (solid content) after separation (Azeez & Onukwuli, 2018a, b; Okasha et al., 2008; Omemu et al., 2006). Both the extract and residue of roselle have been utilized for health benefits such as beverages, nutriceuticals, and food such as vegetables (Atta et al., 2011). The extract of red roselle flower, leaf, seed, fruit and root have been reported to compose of ash, carbohydrates, fats and oils, fiber, minerals, proteins, vitamins such as alkaloids, flavonoids, niacin, saponin, thiamine, β-carotene, riboflavin (Mahadevan & Kamboj, 2009) with hepatoprotective and protective potentials against cancer, hypertension, hyperlipidemias and oxidation, inhibits microbial infections as well as haste wound healing (Ismail et al., 2008; Mahadevan & Kamboj, 2009; Omemu et al., 2006). The extract of red type of roselle using water and acidified water yields of a range of 10%–14% of the plant (Abou-arab et al., 2011; Cilsse et al., 2009). This indicated variability in the extraction methods and composition which might be due to sources, soil and environmental conditions of propagation roselle plant. However, the reports in literature as regards the use of solid content (residue) of red type of roselle with large content for biomedical uses are limited. The compositions of the extract of whitish green roselle type for flower, fruit, leaf, seed, and root are scarce but it has been used traditionally as vegetable. The composition of yields of solid content called fiber from bast of stem of green roselle type has been reported to made up of ash, cellulose, hemicellulose, lignin, moisture, pectin, water soluble, and wax with percentage constituents of about 1.54%, 60.93%, 13.32%, 12.70%, 2.04%, 3.88%, 3.77%, and 1.82%, respectively which composed of 79.6% of dry matter (Azeez & Onukwuli, 2016a; Azeez & Onukwuli, 2018a, b). This means roselle fiber is a biopolymer composite and its usefulness as biopolymer materials in biomedical applications are found scarce in literature.

3.2 Density

The density of roselle fiber was reduced by 46.5% and 37.74% when modified with AC and EDTA, respectively, as presented in Table 6.1. The results show a better reduction in density compared with when roselle fiber treated with sodium hydroxide and SLS as reported by Azeez and Onukwuli (2018a, b). This is an indication of degradation of some amorphous constituents of modified fiber compared with unmodified fiber. The use of modified roselle fiber reveals the lighter of modified fiber in biological system like suture application in tissue engineering and other biomedical applications such as nanostructural materials for scaffold formation and improve tissue regeneration (Banigo et al., 2018).

3.3 Aspect Ratio

The improvement in aspect ratio was observed when modified with AC and EDTA (see Table 6.1). This is as a result of removal of some constituents that soluble in AC and EDTA, which shrunk the thickness of the fiber, thereby increasing the strength of fiber. This influenced by the concentration and treatment time of fiber (Azeez et al., 2020). The use of modified roselle fiber may improve the strength of fiber in holding the tissue together and fast reabsorption into the tissue as well as compatibility of the tissue.

3.4 Water Absorption of Fiber

The change in water absorption of unmodified and modified with AC and EDTA are presented in Fig. 6.2. Water absorption depicts the durability and compatibility of materials in design. There exists an exponential increase in water absorption of both modified and unmodified fiber for the first 5 min., then gradual increase with increasing time for up to 25 min. and later reaches stationary stage where saturation point reaches and there is no longer increase in weight as observed. The model plots of modified roselle fibers were found lower compared with unmodified roselle fiber. This depicts reduction in water absorption of roselle fiber when modified with AC and EDTA. In design of orthopedic,

TABLE 6.1
Physical Parameters of Unmodified and Modified Roselle Fiber at Optimum Conditions.

Fiber Sample	C (%)	t (mins)	d_f (mm)	L_f (mm)	A_f	ρ_f (g/cm^3)	W_f (%)	n	k
B	0	0	0.021	100	4761.9	0.514	366.67	0.0946	0.4785
B_{AC}	3.0	125.59	0.019	100	5263.16	0.275	381.82	0.2811	0.1111
B_{EDTA}	4.33	41	0.016	100	6250.00	0.320	331.25	0.3209	0.7878

B represents roselle fiber, Subscripts, *f*, AC and EDTA are fiber, modification agents of roselle fiber, subscript f represent fiber. *C* is the concentration [illegible]

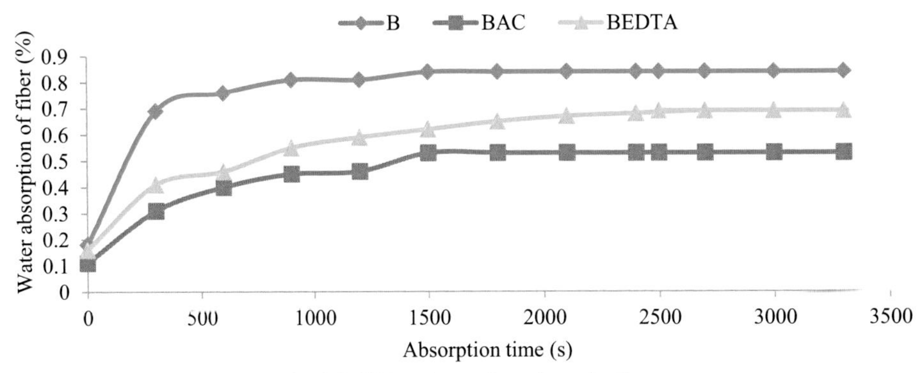

FIG. 6.2 Water absorption of roselle fiber.

suture, grafts, wound dressing and bone tissue engineering applications, the high water absorption of HDPE and its composites may not only cause defect and quality of the products but causes postoperation infections. Roselle fiber modified with EDTA shows the lowest water absorption which is 35.42% lower than unmodified fiber while the one modified with AC increased by 15.15% (Table 6.1). The use of EDTA for modification of roselle has proved to be better in enhancing the mechanical properties than mercerization as reported by Azeez and Onukwuli (2018a, b) and Okeke et al. (2018a, b). The use of modified roselle fiber modified with EDTA may be considered as better products due to its lowest water or moisture absorption and be used in biomedical applications.

The power law of Eq. (6.3) was used to evaluate the kinetics parameters as shown in Fig. 6.3. The report shows that unmodified and modified (AC and EDTA) have been reported to be water diffusion controlled of less Fickian behavior since $n < 0.5$ (Azeez, 2017) with reduction in the sorption rate constant (k) of roselle fibers which in contrast with Gierszewska-Drużyńska & Ostrowska-Czubenko (2012) report. The less Fickian behavior may be attributed to fiber shrinkage, reduction in pore creation, source and nature of the fibers, and reduction in hemicellulose and lignin content which make the roselle fiber to be more hydrophobic. This is in contrast with Gierszewska-Drużyńska & Ostrowska-Czubenko (2012) report.

3.5 Tensile Properties of Fiber

A tensile test with the Instron Universal Testing Machine of model 3369 was used to measure the tensile strength and modulus on five strands of fiber of length 140 mm and gauge length of 100 mm with average thickness of 0.003 ± 0.0005 mm. The modifications of roselle fiber with AC and EDTA, respectively, improved the tensile strength by 114.81% and 321.5% of unmodified fiber as well as tensile modulus by 231.52% and 149.10% of unmodified roselle fiber. The improvement in tensile strength and modulus of roselle fiber are in agreement and justified some reports (Azeez & Onukwuli, 2018a, b; Azeez & Onukwuli, 2017; Kalia et al., 2009; Azeez & Okechukwu, 2016a; Supri & Lim, 2009). The result obtained for modifications of roselle fiber using AC and EDTA was found to be higher than mercerization (see Table 6.2) as reported in the literature (Azeez & Onukwuli, 2016a; Azeez & Onukwuli, 2018a, b).

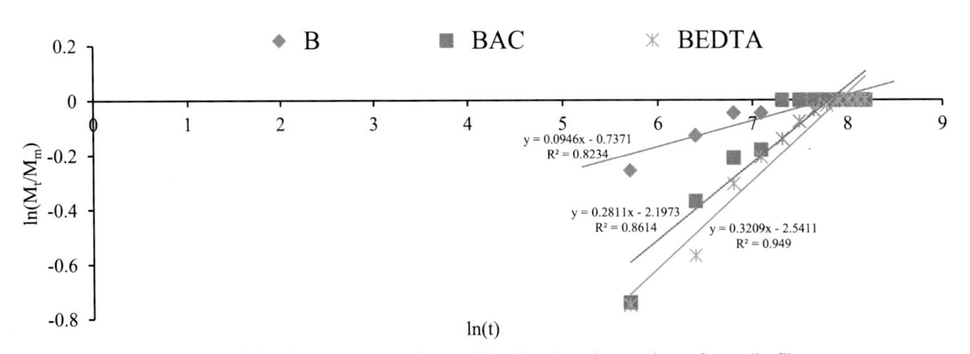

FIG. 6.3 Kinetic parameters (*n* and *k*) of water absorption of roselle fiber.

TABLE 6.2
Tensile Properties of Unmodified and Modified Roselle Fiber.

Fiber Sample	C (%)	t (mins)	T_{sa} (MPa)	T_{ma} (MPa)
B	0	0	43.5326	2294.6
B_{AC}	3.0	125.59	93.5138	7604.66
B_{EDTA}	4.33	41	183.49	5715.9

4 HIGH-DENSITY POLYETHYLENE (HDPE) MATRIX AND ITS COMPOSITES

A composite refers to combination of two or more materials (resin matrix, reinforcement fibers/fillers, etc.) differs in composition with enhanced material properties when imbedded but retain their constituents. This implies the synergy of the individual constituents of composites indicate the quality and the performance of the composites. Constituents of polymer composites include polymers called resin/matrix, a weaker material and a reinforcement, a strong load-carrying material such as fibers and fillers (synthetic or natural). HDPE has wide applications like industrial and domestic uses like plastic bag, cup, jug, corrosion protection of steel pipelines, water pipes, chair and tables, storage of food and fuels, electrical and pumping box as well as biomedical applications (plastic surgery for facial and skeletal remolding, plastic lumber reconstruction, cell lines, infrared lenses to mention few). However, the defects of HDPE include the tensile and flexural moduli, environmental conditions of moisture and water absorption and impact load (Lin et al., 2015; Rizov, 2017). The influence of modification of roselle fiber hybridized HDPE matrix on the physical (density, water absorption, and its kinetics) and mechanical (tensile, flexural, hardness, and impact properties) performance of roselle fiber-HDPE composites was examined. The tensile strength and modulus, flexural strength and modulus, hardness and impact strength of HDPE have been reported to be 27.628 and 792.59 MPa, 34.519 MPa, and 1339.7 MPa, 24 HR, and 0.9628Jmm^{-2}, respectively (Azeez, 2019; Azeez et al., 2020; Azeez & Onukwuli, 2017). The properties of composites depend on their constituent which governs by different models such as Halpin-Tsai. Mori-Tanaka, Chamis, and Brintrup models as reported in literatures (Cao et al., 2013; Facca et al., 2006; Kim et al., 2010), but differs in experimental evaluation of parameters with high degree of correlation.

4.1 Mechanical Properties of HDPE Composites

In this work, Instron Universal Testing Machine model 3369 model was reported to be used for estimation of tensile, flexural and impact properties while Rockwell hardness equipment was used for evaluation of hardness parameter.

4.1.1 Tensile properties

Figs. 6.4 and 6.5 depict the tensile strength and modulus of unmodified and modified roselle fiber-reinforced HDPE composites. The use of unmodified, AC, and EDTA-modified roselle fiber reduces the tensile strength of HDPE matrix at fiber loading of 2.5–3.75w/w%. The reduction in tensile strength is as a result of uneven distribution of fiber. The tensile strength increased by 0.88% and 3.38% of HDPE matrix when modified with EDTA and AC at fiber loading of 5% and 6.25%, respectively. The increased in tensile strength may be attributed to even distribution, enhanced interfacial adhesion and compatibility of the fiber with HDPE matrix. This is an indication for improvement in performance of HDPE composites. More so, the tensile modulus of HDPE matrix was found to be improved by the imbedded roselle fiber for unmodified, and AC and EDTA-modified roselle fiber at 5, 7.5, and 2.5w/w %, respectively. The mechanism of improvement was supported by the report of researchers (Bonnia et al., 2010; Luyt, 2009; Troëdec et al., 2007).

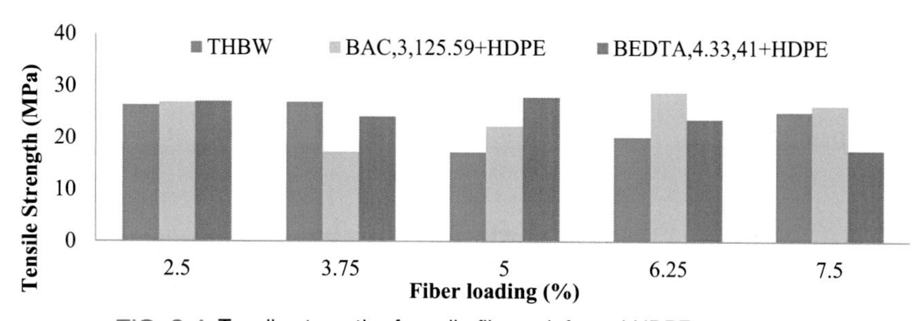

FIG. 6.4 Tensile strength of roselle fiber-reinforced HDPE composites.

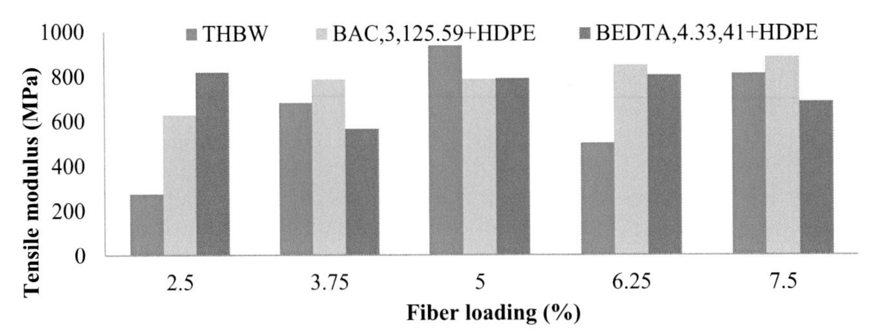

FIG. 6.5 Tensile modulus of roselle fiber-reinforced HDPE composites.

4.1.2 Flexural properties

The flexural strength and modulus of roselle fiber reinforce HDPE composites for unmodified and modified with AC and EDTA are presented in Figs. 6.6 and 6.7. In Fig. 6.6, the flexural strength of unmodified and AC-modified roselle fiber-reinforced HDPE composites were found to be higher compared to HDPE matrix at fiber loading of 2.5w/w% while increases in fiber load reduced the flexural strength of the composites. The unmodified roselle fiber yields the highest flexural strength at fiber loading of 6.25w/w%. The flexural strength of AC-modified roselle fiber-reinforced HDPE composites found to be unevenly sinusoidal with increased fiber loading. In case of EDTA-modified roselle fiber-reinforced HDPE composites. There is a reduction in flexural strength at fiber loading of 2.5w/w% and later increases with increased fiber loading up to 6.25w/w %. The reduction in flexural strength may be due to uneven distribution of fiber loading, while the increase in flexural strength of HDPE composites of roselle fiber is as a result of good dispersion of fiber in HDPE matrix (Razak et al., 2014). More so, the effect of roselle fiber

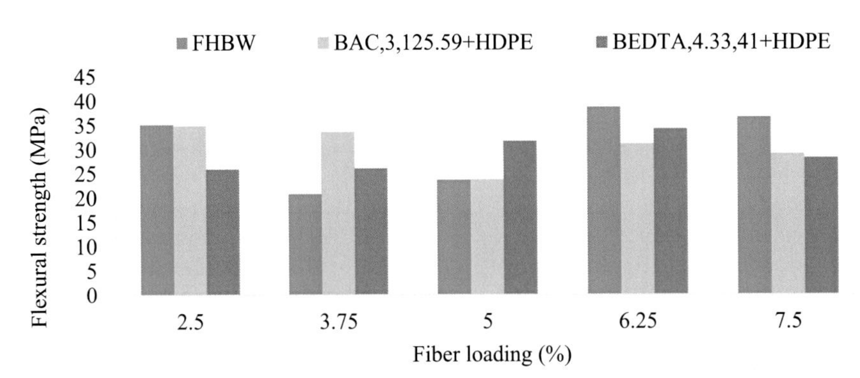

FIG. 6.6 Flexural strength of roselle fiber-reinforced HDPE composites.

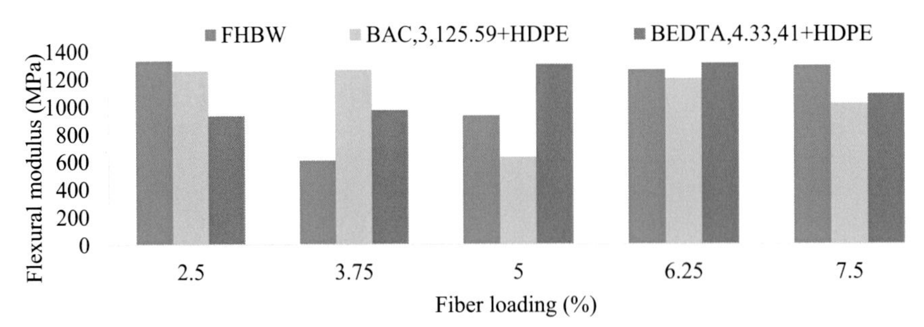

FIG. 6.7 Flexural modulus of roselle fiber-reinforced HDPE composites.

seems to reduce the flexural modulus of HDPE composites in unevenly progression and later increases with increased fiber loading (see Fig. 6.7). The modification of roselle fiber with AC and EDTA follows the similar pattern as against the report of Azeez et al., 2020, 2018b. There is need for optimization of the HDPE composites of roselle fiber based on constituents and modification for optimum performance and compatibility in terms of flexibility when used in biomedical applications.

4.1.3 Hardness

In Fig. 6.8, there is increase in hardness of HDPE composites with increased fiber loading up to 3.75, 6.25 and 7.5w/w% for unmodified, AC, and EDTA-modified roselle fiber to reach maximum level. This indicated that AC and EDTA modifications improved the hardness of roselle fiber-reinforced HDPE composites. The improvement is associated to fiber aggregation and orientation. This is supported by reporting in the literature (Azeez et al., 2020, 2018b; Ishidi et al., 2011). More so, further increase in unmodified roselle fiber loading leads to reduction in hardness due to pore formation in the composites, which weaken the toughness of the HDPE composites. The hardness of composites should be considered an important mechanical parameter that may hinder the compatibility of tissue and bone model especially in regeneration and rehabilitation of human impairment.

4.1.4 Impact strength

Roselle fiber including AC and EDTA modification reduced the impact strength of HDPE composites evaluated using Eq. (6.5) and seen in Fig. 6.9. This explains the poor wettability of roselle fiber incorporation in HDPE matrix and results in weak interfacial adhesion between the fiber and polymer, thereby creates an intercalation region. This weak interfacial force that exists between HDPE and hydrophilic roselle fiber usually causes reduction in mobility of HDPE resin (Dedeepya et al., 2012; Raju et al., 2012). Although there is variation in the extent of reduction in impact strength based on modifications, fiber loading, orientation, and even distribution of roselle fiber.

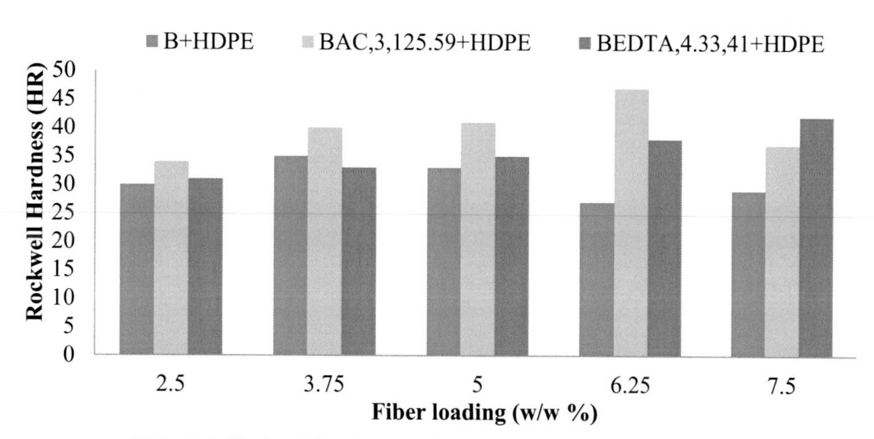

FIG. 6.8 Rockwell hardness of fiber-reinforced HDPE composites.

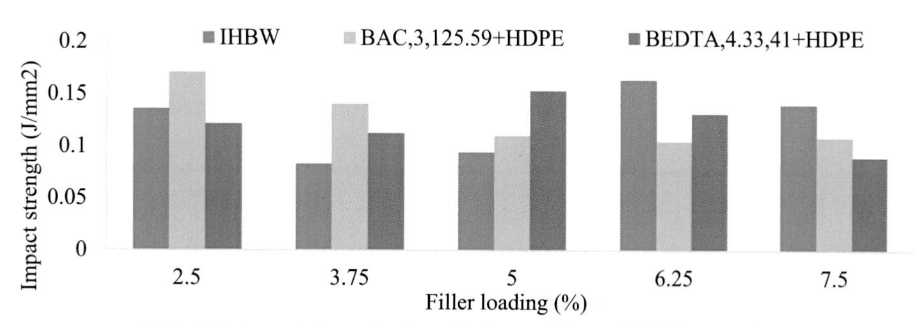

FIG. 6.9 Impact strength of roselle fiber rerinforced HDPE compsites.

$$I_s = \frac{E}{A} \quad (6.5)$$

Optimization of HDPE composites of roselle fiber was achieved using central composite design of response surface methodology in design of experiment software 6.0.8 version. The optimal mechanical properties were obtained based on roselle fiber and HDPE loading as presented in Table 6.3. It can be observed that the HDPE composites of unmodified roselle fiber reduced the tensile strength, flexural modulus, and impact strength but increased the tensile modulus, flexural strength, and hardness. Modification of roselle fiber with AC and EDTA improved tensile strength and hardness of the composites by 0.46% and 7.57%, 15.63%, and 12.5%, respectively, but reduces the tensile strength, flexural strength, and impact strength. The improvement in tensile strength was found to be more than when mercerized and modified with SLS, which attributed interfacial adhesion between the fiber and HDPE matrix as reported by Azeez and Onukwuli (2017).

4.2 Microstructural Analysis of HDPE Composites

In Figs. 6.10 and 6.11, The strength of HDPE composites of unmodified roselle fiber reduced compared with HDPE matrix which corroborated with crystalline carbon counts while the modulus depends on pores formation in the composites as seen in SEM with EDX of Fig. 6.10B compared with Fig. 6.10A which in also revealed incorporation of unmodified roselle fiber by Fig. 6.11A in HDPE matrix. In Fig. 6.11A, there is a smooth surface of fiber, which indicates the presence of amorphous constituents of fiber but disappeared when treated with AC and EDTA, thereby causing rough fiber surfaces and shrinks in fiber as shown in Fig. 6.11B and C. More so, the enhancement in strength can be observed in Fig. 6.10C and D which imbedded with AC and EDTA modified roselle fiber characterized with rough surfaces, shrinkage of fiber surfaces, and improved aspect ratio, as shown in Fig. 6.11B and C, respectively. This may also justify removal some amorphous constituents (Azeez & Onukwuli, 2016b) The presence of calcium, sodium, bromine, zinc, and chlorine arise from environmental contamination may contribute to improvement in strength of the composites as a result of compatibility while oxygen content alters the magnitude of modulus of the HDPE composites. These environmental contaminants may not be toxic to the human system and advance effect on tissue and bone regeneration and rehabilitation applications of the HDPE composites due to their contents.

4.3 Physical Properties of HDPE Composites

4.3.1 Density

The property of the composites is a dependent of properties of individual constituents, which may be governed by the mixture rule (Eq. 6.6) and density of the composites can be evaluated using Eq. (6.7) as reported in literature (Cao et al., 2013; Facca et al., 2006), which are not in agreement with the experimental data as presented in Table 6.4. It can be seen that the density of the HDPE matrix lesser than the HDPE composites of unmodified roselle fiber while the density of AC and EDTA-modified roselle fiber lesser than that of unmodified roselle fiber-reinforced HDPE composites

TABLE 6.3
Optimal Data of Mechanical Properties of Roselle Fiber - HDPE Composites.

Composite Parameters	H	H_B	$H_{B,AC}$	$H_{B,EDTA}$
W_f	0	7.5	7.5	4.95
W_m	100	92.5	92.5	92.5
T_{sa}	27.628	26.1001	26.22	28.076[a]
T_{ma}	792.59	906.89[b]	885.59	842.847
F_{sa}	34.519	37.6319[c]	28.986	34.0867
F_{ma}	1390.7	1289.92	1017.9	1382.81[d]
H_a	24	32	37[e]	36
I_s	0.9628	0.136806	0.108577	0.14579[f]

Superscript a,b,c,d,e,f represent maximum tensile strength, tensile modulus, flexural strength, flexural modulus, hardness and impact strength.

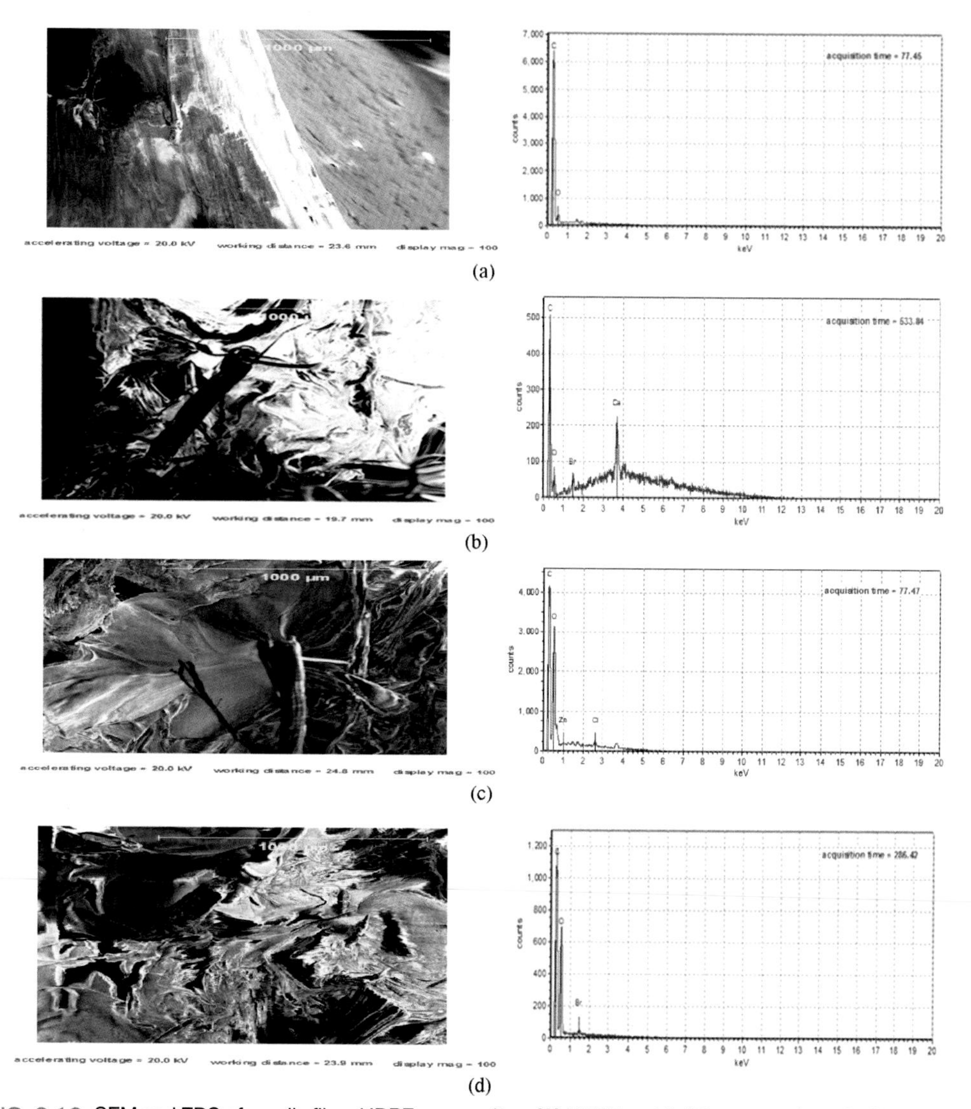

FIG. 6.10 SEM and EDS of roselle fiber-HDPE composites: **(A)** HDPE matrix **(B)** untreated roselle fiber-HDPE **(C)** AC treated roselle fiber-HDPE (**D**) EDTA treated roselle fiber-HDPE.

(Table 6.4). This is attributed to removal of amorphous constituents (pectin, lignin, and hemicellulose) from roselle fiber. This indicated that the use of modified roselle fiber-reinforced HDPE composites may be lighter and easily carry by human system as against the use of metal as an implant, regenerative and rehabilitation materials.

$$V_f + V_m = 1 \tag{6.6}$$

$$\rho_c = \rho_m V_m + \rho_f V_f \tag{6.7}$$

Where V_f and V_m are fiber volume fraction and matrix volume fraction. ρ_c, ρ_f, and ρ_m represent the density of composite, fiber, and matrix.

4.3.2 Water absorption

Environmental factors such as moisture, water, and contaminants influence the performance, durability, and quality of designed composite products. This can be influenced through the kinetics of penetration,

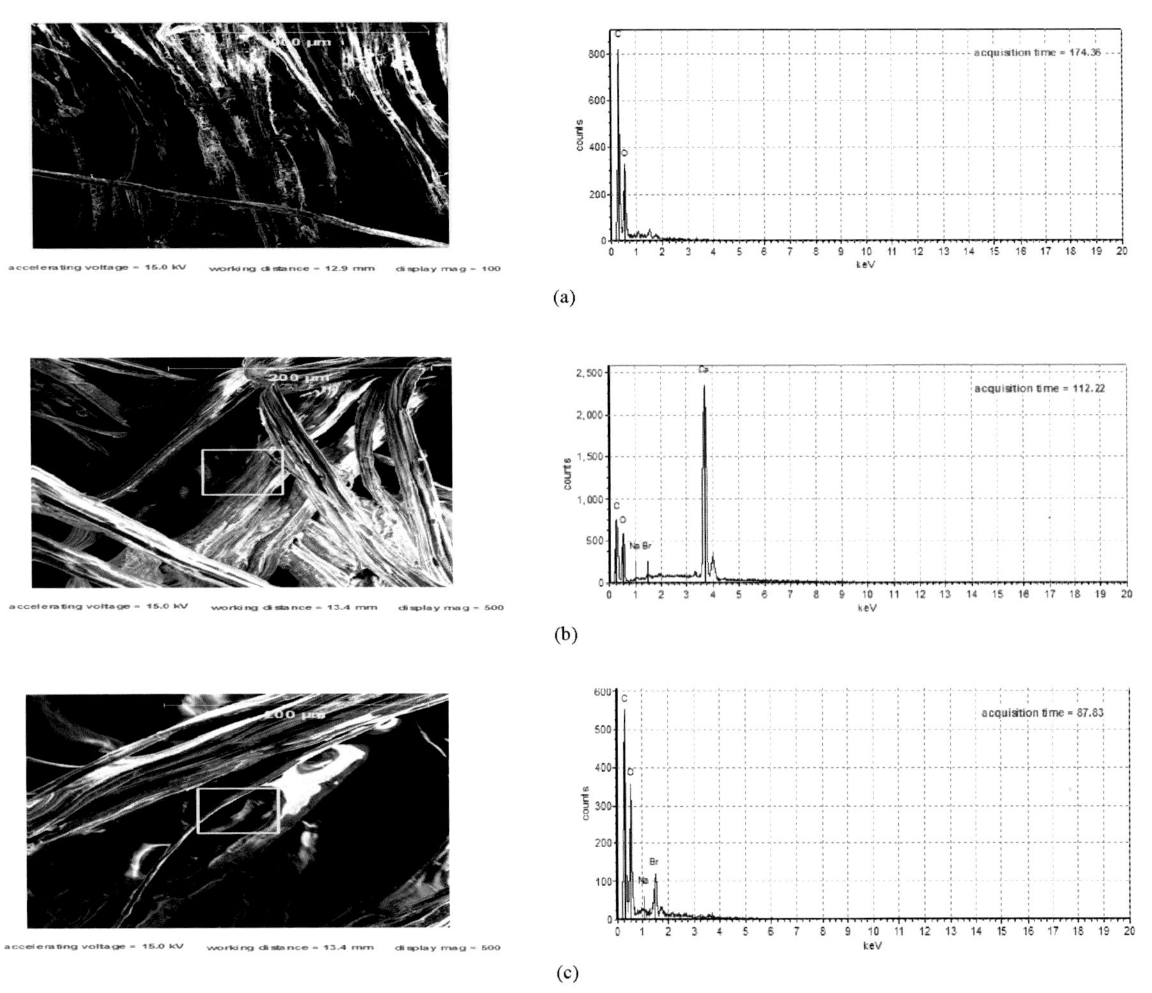

FIG. 6.11 SEM and EDS of roselle fibers **(A)** unmodified **(B)** modified with AC **(C)** Modified with EDTA.

TABLE 6.4
Density and Water Absorption Kinetics of Fiber Reinforce HDPE Composites.

Composite Sample	ρ (g/cm^3)	h_c (mm)	n (1/s)	k	M_m (%)
H	0.9225	3.0	0.8803	0.000788	167.6259
H_B	1.0467	3.0	0.8567	0.000854	99.0625
$H_{B,AC}$	1.0444	3.0	0.663	0.004436	117.2727
$H_{B,EDTA}$	0.8167	3.0	0.8466	0.000972	152.1739

rheological properties and decline in adhesive force between the constituents. In this work, the water absorption and its kinetics on use of roselle fiber for HDPE composites were examined as seen in Figs. 6.12 and 6.13. It can be deduced from Fig. 6.12 that the water absorption of HDPE matrix, unmodified, AC and EDTA-modified roselle fiber-reinforced HDPE composites progressively increases until it reaches saturated region when it can no longer absorb water or moisture with increased time. The saturated time for HDPE and

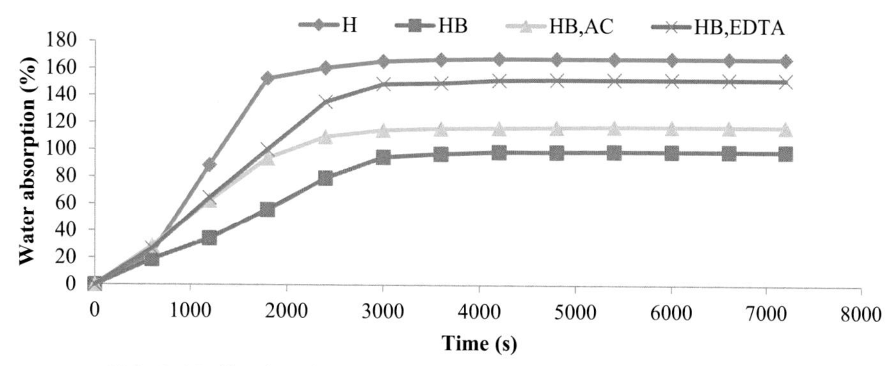

FIG. 6.12 Kinetics of water absorption of roselle fiber-HDPE composites.

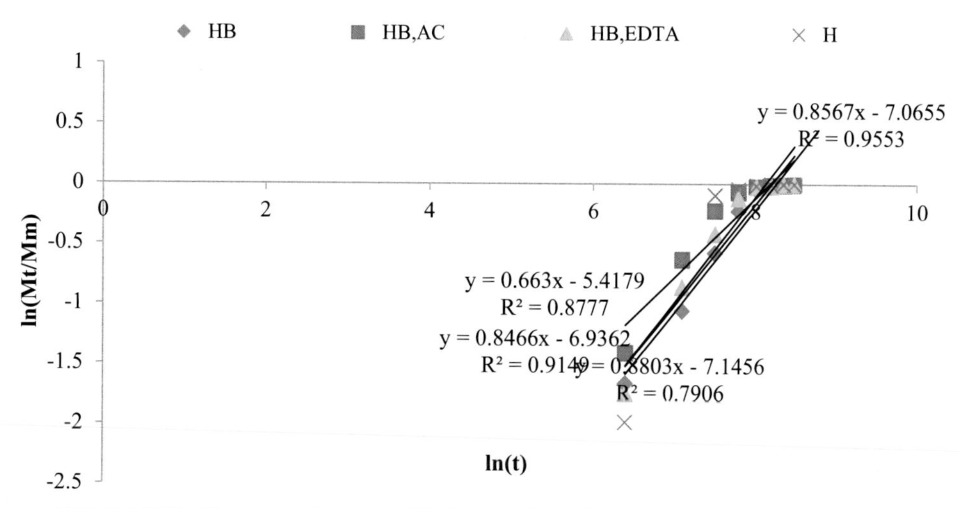

FIG. 6.13 Kinetics parameters (*n* and *k*) of water absorption roselle fiber-HDPE composites.

its composites has been found to be 4200s. The saturation water absorption of HDPE matrix, HDPE composites of unmodified, AC and EDTA-modified roselle fiber are 167.53%, 99.06%, 117.27%, and 152.17%, respectively. This shows that AC and EDTA reduced the water absorption of the composites more than mercerization and SLS treatment reported by Azeez and Onukwuli (2017). Thus, AC and EDTA improved the durability and biodegradation of the composites.

The water absorption mechanism reveals that water flows through the HDPE matrix and its composites with roselle fiber are by non-Fickian behaviors since $0.5 < n < 1.0$, irrespective of fiber modifications. This condition is anomalous because water diffusion rate is relatively comparable with HDPE relaxation rate due to hydrophilic nature of fibers. The magnitude of *n* reduced when imbedded with modified roselle fiber (Gierszewska-Druzyńska & Ostrowska-Czubenko, 2012). The modification of roselle fiber with AC and EDTA, respectively reduced the values of *n*. This means AC and EDTA modifications improved hydrophobicity of HDPE composites of roselle fiber and increased value of water absorption rate constant, *k* which associated with structural composition and orientation of roselle fiber in the HDPE composites. It can also be deduced that the higher *k* value, the lower reduction in *n* value for non-Fickian behaviors.

5 CONCLUSION

The following can be drawn from this study that acetic anhydride and EDTA modifications at optimum processes:

1. Acetic anhydride and EDTA reduced the density and water absorption of roselle fiber with improved aspect ratio of roselle fiber.

2. The use of acetic anhydride and EDTA improved the tensile strength and hardness of roselle fiber-reinforced HDPE composites with reduction in flexural and impact properties strength.
3. The density and water absorption of the HDPE composites also improved as a result of modifications using acetic anhydride and EDTA as a result of degradation of amorphous consttuents.
4. The variation in mechanical properties of HDPE composites of roselle fiber corroborates with the morphology and elemental contents of EDX.
5. Treatment of roselle fiber gave superior and favorable enhanced performance in terms of tensile strength, density, and water absorption of the roselle fiber-reinforced HDPE composites when optimally modified with EDTA.
6. The favorable behaviors of HDPE composites based on modifications enhanced its quality and performance. Hence, it enhances its usefulness in biomedical applications.

LIST OF ABBREVIATIONS

AC	Acetic anhydride
B	Roselle fiber
EDTA	Ethylene diamine acetic acid
FHBW	Flexural property of HDPE-Roselle fiber without modification
HB	hardness of roselle HDPE-roselle fiber
HDPE	High density polyethylene matrix
IHBW	Impact property of HDPE-roselle fiber without modification
THBW	Tensile property of HDPE-roselle fiber without modification

REFERENCES

Abou-arab, A. A., Abu-salem, F. M., & Abou-arab, E. A. (2011). Physico- chemical properties of natural pigments (anthocyanin) extracted from roselle calyces (*Hibiscus subdariffa*). *Journal of American Science, 7*(7), 445–456.

Arowosegbe, S., Oyeyemi, S. D., & Alo, O. (2015). Investigation on the medicinal and nutritional potentials of some vegetables consumed in Ekiti state, Nigeria. *International Research Journal of Natural Sciences, 3*(1), 16–30.

Atta, S., Seyni, H. H., Bakasso, Y., Sarr, B., Lona, I., & Saadou, M. (2011). Yield character variability in roselle (*Hibiscus sabdariffa* L.). *African Journal of Agricultural Research, 6*(6), 1371–1377. http://www.academicjournals.org/AJAR.

Azeez, T. O. (2017). *Evaluation of the potentials of chemically treated roselle, food gum and kapok fibers as reinforcement in polymer bio – composites* (Ph.D. thesis). Awka: Nnamidi Azikiwe University.

Azeez, T. O. (2019). Thermoplastic recycling: Properties, modifications, and applications. In *Thermosoftening plastics* (pp. 1–19). IntechOpen. https://doi.org/10.5772/intechopen.81614.

Azeez, T. O., & Onukwuli, O. D. (2016a). Effect of chemical agents on morphology , tensile properties and water diffusion behaviour of Hibiscus sabdariffa fibers. *Chemical and Process Engineering Research, 42*, 76–83.

Azeez, T. O., & Onukwuli, O. D. (2016b). Modified food gum (*Cissus populnea*) fibers: Microstructural behaviour, physico - mechanical properties and kinetics of water absorption *ARPN Journal of Engineering and Applied Sciences, 11*(17), 10655–10663.

Azeez, T. O., & Onukwuli, O. D. (2017). Effect of chemically modified *Cissus populnea* fibers on mechanical , microstructural and physical properties of *Cissus populnea*/high density polyethylene composites. *Engineering Journal, 21*(2), 25–42. https://doi.org/10.4186/ej.2017.21.2.25

Azeez, T. O., & Onukwuli, O. D. (2018a). Tensile responses of treated *Cissus populnea* fibers. *Nigerian Journal of Technology (NIJOTECH), 37*(1), 173–183.

Azeez, T. O., & Onukwuli, D. O. (2018b). Properties of white roselle (*Hibiscus sabdariffa*) fibers. *Journal of Scientific & Industrial Research, 77*(9), 525–532.

Azeez, T. O., Onukwuli, D. O., Nwabanne, J. T., & Banigo, A. T. (2020). Cissus populnea fiber - unsaturated polyester composites: Mechanical properties and interfacial adhesion *Cissus populnea* fiber - unsaturated polyester composites. *Journal of Natural Fibers, 17*(9), 1281–1294. https://doi.org/10.1080/15440478.2018.1558159

Azeez, T. O., Onukwuli, D. O., Walter, P. E., & Menkiti, M. C. (2018). Influence of chemical surface modifications on mechanical properties of combretum dolichopetalum fibre - high density polyethylene (HDPE) composites. *Pakistan Journal of Scientific and Industrial Research Series A: Physical Sciences, 61A*(1), 28–34.

Banigo, A. T., Azeez, T. O., Ejeta, K. O., & Egbuchulam, M. (2018). Application of nanotechnology in biomedical engineering and medicine: A review. *Science Focus, 23*(2), 55–65. https://doi.org/10.36293/sfj.2019.0007

Bonnia, N. N., Ahmad, S. H., Zainol, I., Mamun, A. A., Beg, M. D. H., & Bledzki, A. K. (2010). Mechanical properties and environmental stress cracking resistance of rubber toughened polyester/kenaf composite. *Express Polymer Letters, 4*(2), 4–10. https://doi.org/10.3144/expresspolymlett.2010

Builders, P., Kabele-Toge, B., Builders, M., Chindo, B., Anwunobi, P., & Isimi, Y. (2013). Wound healing potential of formulated extract from *Hibiscus sabdariffa* calyx. *Indian Journal of Pharmaceutical Sciences, 75*(1), 45–52. https://doi.org/10.4103/0250-474X.113549

Cao, Y., Wang, W., Wang, Q., & Wang, H. (2013). Application of mechanical models to flax fiber/wood fiber/plastic composites. *Bioresources, 8*(3), 3276–3288.

CiIsse, M., Cornier, M., Sakho, M., NDiaye, A., Reynes, M., & Sock, O. (2009). Article de Synthèse Le Bissap (*Hibiscus Sabdariffa* L.): Composition et Principales Utilisations. *Fruits, 64*(3), 179–193. https://doi.org/10.1051/fruits/2009013

Dedeepya, M., Raju, T. D., & Kumar, T. J. (2012). Effect of alkaline treatment on mechanical and thermal properties oftypha angustifolia fiber reinforced composites. *International Journal of Mechanical and Industrial Engineering, 1*(4), 12–14.

Durowaye, S., Sekunowo, O., Kuforiji, C., Nwafor, C., & Ekwueme, C. (2019). Effect of agro waste particles on the mechanical properties of hybrid unsaturated polyester resin matrix composites. *Journal of the Institute of Engineering, 15*(2), 123–132. https://doi.org/10.3126/jie.v15i2.27656

Facca, A. G., Kortschot, M. T., & Yan, N. (2006). Predicting the elastic modulus of natural fibre reinforced thermoplastics. *Composites Part A: Applied Science and Manufacturing, 37*, 1660–1671. https://doi.org/10.1016/j.compositesa.2005.10.006

Gierszewska-Drużyńska, M., & Ostrowska-Czubenko, J. (2012). Mechanism of water diffusion into noncrosslinked and ionically crosslinked chitosan membranes. *Progress in Chemistry and Application of Chitin and Its Derivatives, XVII*, 59–66.

Ilyas, R. A., & Sapuan, S. M. (2020a). The preparation methods and processing of natural fibre bio-polymer composites. *Current Organic Synthesis, 16*(8), 1068–1070. https://doi.org/10.2174/157017941608200120105616

Ilyas, R. A., & Sapuan, S. M. (2020b). Biopolymers and biocomposites: Chemistry and technology. *Current Analytical Chemistry, 16*(5), 500–503. https://doi.org/10.2174/157341101605200603095311

Ilyas, R. A., Sapuan, S. M., Atikah, M. S. N., Asyraf, M. R. M., Rafiqah, S. A., Aisyah, H. A., Nurazzi, N. M., & Norrahim, M. N. F. (2021). Effect of hydrolysis time on the morphological, physical, chemical, and thermal behavior of sugar palm nanocrystalline cellulose (*Arenga pinnata* (Wurmb.) Merr). *Textile Research Journal, 91*(1–2), 152–167. https://doi.org/10.1177/0040517520932393

Ilyas, R. A., Sapuan, S. M., Ishak, M. R., & Zainudin, E. S. (2018). Development and characterization of sugar palm nanocrystalline cellulose reinforced sugar palm starch bionanocomposites. *Carbohydrate Polymers, 202*(December), 186–202. https://doi.org/10.1016/j.carbpol.2018.09.002

Ishidi, E. Y., Kolawale, E. G., Sunmonu, K. O., Yakubu, M. K., Adamu, I. K., & Obele, C. M. (2011). Study of physiomechanical properties of high density polyethylene (HDPE) – palm kernel nut shell (*Elaeis guineasis*). *Journal of Emerging Trends in Engineering and Applied Sciences, 2*(6), 1073–1078. jeteas.scholarlinkresearch.org.

Ismail, A., Ikram, H. E. K., & Mohd Nazri, H. (2008). Roselle (*Hibiscus sabdariffa* L.) seeds – nutritional composition , protein quality and health benefits. *Global Science Books, 2*(1), 1–16.

Kalia, S., Kaith, B. S., & Kaur, I. (2009). Pretreatments of natural fibers and their application as reinforcing material in polymer composites — a review. *Polymer Engineering and Science*, 1253–1272. https://doi.org/10.1002/pen

Kim, M., Mirza, F. A., & Song, J. I. (2010). Micromechanics modeling for the stiffness and strength properties of glass fibers/CNTs/epoxy composites. *WIT Transactions on the Built Environment, 112*, 279–290. https://doi.org/10.2495/HPSM100261

Lin, J.-H., Pan, Y.-J., Liu, C.-F., Huang, C.-L., & Hsieh, C.-T. (2015). Preparation and compatibility evaluation of polypropylene/high density polyethylene polyblends. *Materials, 8*, 8850–8859. https://doi.org/10.3390/ma8125496

Luyt, A. S. (2009). Editorial corner – a personal view natural fibre reinforced polymer composites – are short natural fibres really reinforcements or just fillers? *Express Polymer Letters, 3*(6), 3144. https://doi.org/10.3144/expresspolymlett.2009.41

Mahadevan, N., & Kamboj, P. (2009). *Hibiscus sabdariffa* Linn. – an overview. *Natural Product Radiance, 8*(1), 77–83.

Okasha, M. A. M., Abubakar, M. S., & Bako, I. G. (2008). Study of the effect of aqueous *Hibiscus sabdariffa* Linn seed extract on serum prolactin level of lactating female albino rats. *European Journal of Scientific Research, 22*(4), 575–583.

Okeke, K. N., Onwubu, S. C., Iwueke, G. C., & Arukalam, I. O. (2018). Characterization of *Hibiscus sabdariffa* fiber as potential reinforcement for denture acrylic resins. *Revista Materia, 23*(4). https://doi.org/10.1590/s1517-707620180004.0550

Okeke, K. N., Vahed, A., & Singh, S. (2018). Improving the strength properties of denture base acrylic resins using *Hibiscus sabdariffa* natural fiber. *Journal of International Dental and Medical Research, 11*(1), 248–254.

Omemu, A. M., Edema, M. O., Atayese, A. O., & Obadina, A. O. (2006). A survey of the microflora of *Hibiscus sabdariffa* (roselle) and the resulting 'Zobo' juice. *African Journal of Biotechnology, 5*, 254–259.

Omotuyi, I. O., Ologundudu, A., Onwubiko, V. O., Wogu, M. D., & Obi, F. O. (2010). *Hibiscus sabdariffa* Linn anthocyanins alter circulating reproductive hormones in rabbits (*Oryctolagus cuniculus*). *Journal of Diabetes and Endocrinology, 1*, 36–45.

Omran, A. A. B., Abdulrahman, A., Mohammed, B. A., Sapuan, S. M., Ilyas, R. A., Asyraf, M. R. M., Rahimian Koloor, S. S., & Petrů, M. (2021). Micro- and nanocellulose in polymer composite materials: A review. *Polymers, 13*(2), 231. https://doi.org/10.3390/polym13020231

Padmakumari, P., Anupama, C., Abbulu, K., & Pratyusha, A. P. (2011). Evaluation of fruit calyces mucilage of *Hibiscus sabdariffa* Linn as. *International Journal of Research in Pharmaceutical and Biomedical Sciences, 2*(8), 516–519.

Raju, G. U., Gaitonde, V. N., & Kumarappa, S. (2012). Experimental study on optimization of thermal properties of groundnut shell particle reinforced polymer composites. *International Journal of Emerging Sciences, 2*(3), 433–454.

Razak, I. N. A., Ibrahim, N. A., Zainuddin, N., Rayung, M., & Saad, W. Z. (2014). The influence of chemical surface modification of kenaf fiber using hydrogen peroxide on the mechanical properties of biodegradable kenaf fiber/poly(lactic acid) composites. *Molecules, 19*, 2957–2968. https://doi.org/10.3390/molecules19032957

Rizov, V. (2017). Delamination fracture analysis of an elastic-plastic functionally graded multilayered beam. *International Journal of Mechanical and Materials Engineering, 12*(4), 1–12. https://doi.org/10.1186/s40712-017-0073-7

Sabaruddin, F. A., Tahir, P. M., Sapuan, S. M., Ilyas, R. A., Lee, S. H., Abdan, K., Mazlan, N., Roseley, A. S. M., & Abdul Khalil, H. P. S. (2020). The effects of unbleached and bleached nanocellulose on the thermal and flammability of polypropylene-reinforced kenaf core hybrid polymer bionanocomposites. *Polymers, 13*(1), 116. https://doi.org/10.3390/polym13010116

Singh, D. P. (2012). Mesta (*Hibiscus cannabinus* & *Hibiscus sabdariffa*). In *Central research institute for jute & allied fibres barrackpore 743, 24-Parganas (North) West Bengal, India* (pp. 1–28).

Singha, A. S., & Thakur, V. K. (2008). Fabrication and study of lignocellulose *Hibscus sabdariffa* fiber reinforced polymer composites. *Bioresources, 3*(4), 1173–1186.

Supri, A. G., & Lim, B. Y. (2009). Effect of treated and untreated filler loading on the mechanical, morphological, and water absorption properties of water hyacinth fibers- low density polyethylene composites. *Journal of Physical Science, 20*(2), 85–96.

Troëdec, M. L., Peyratout, C. S., Chotard, T., Bonnet, J.-P., Smith, A., & Guinebretière, R. (2007). Physico-chemical modifications of the interactions between hemp fibres and a lime mineral matrix: Impacts on mechanical properties of mortars. In J. G. Heinrich, & C. Aneziris (Eds.), *10th international conference of the European ceramic society* (pp. 451–456) (Berlin, Germany).

FURTHER READING

Agbeboh, N. I., Olajide, J. L., Oladele, I. O., & Babarinsa, S. O. (2019). Kinetics of moisture sorption and improved tribological performance of keratinous fiber-reinforced orthophthalic polyester biocomposites. *Journal of Natural Fibers, 16*(5), 744–754. https://doi.org/10.1080/15440478.2018.1434849

CHAPTER 7

Roselle (*Hibiscus sabdariffa* L.): Nutraceutical and Pharmaceutical Significance

TAHMINA SADIA JAMINI • A.K.M. AMINUL ISLAM

1 INTRODUCTION

Roselle (*Hibiscus sabdariffa* L.) is a tropical tetraploid (2n = 72) annual species. It belongs to the family of Malvaceae and originated from West Africa (Shoosh, 1993). It is popularly used as a vegetable. But it has some medicinal importance as well. Roselle is locally known by different names in different countries (Ismail et al., 2008). Roselle is commonly known as bissap, chukur, mesta, sorrel, karkade, lemon bush, jelly plant, jelly okra, wonjo, asam susur, and saril. In Bangladesh, it is popularly recognized as 'mesta' or 'chukur." Roselle is commonly available in the tropics, especially in the African countries (Abu-Tarboush et al., 1997). It is widely cultivated in tropical Africa, Sudan, Egypt, Ethiopia, Mali, Nigeria, Chad, India, Indonesia, Philippines, Malaysia, Brazil, Australia, Mexico, and Hawaii and Florida of the United States. The world largest producer is Thailand and China but the highest quality is from Sudan (Food and Agriculture Organization). Roselle from Egypt and Sudan is highly paid in the United States and Germany as compared with the price of Chinese Roselle. The quality of roselle from China and Thailand is comparatively low because of excessive precipitation during production (Islam et al., 2016).

Roselle is an annual or perennial herb or woody-based subshrub that requires around 6 months for completing its production cycle, growing to a height of 2.0–3.5 m. Roselle is a miracle plant with various uses (Crane, 1949). The leaves and calyx are used as vegetable in many countries of the tropics. There are three different color groups (Fig. 7.1): green, red, and dark red are available in the tropics (Purseglove, 1974). The calyx of red and dark red types is used to extract juice for fresh drink after sweetened and the leaves of green types are used as vegetables (Babalola, 2000). Two botanical types of roselle, namely, *H. sabdariffa* var *sabdariffa* and *H. sabdariffa* var *altissima*, have been recognized by Purseglove (1974). The altissima variety is grown for fiber in India, Java, and the Philippines.

2 DOMESTIC APPLICATIONS

The tender leaves and stalks of the roselle plant are eaten as a leafy vegetable just like spinach. Roselle leaves are also used in salad. The red calyx is the most important part of the roselle plant. Roselle fruits are best prepared for use by washing, then making an incision around the tough base of the calyx below the bracts to free and remove it with the seed capsule attached. The calyces are then ready for immediate use. They may be merely chopped and added to fruit salads (Mady et al., 2009). In Africa, they are cooked as a side dish eaten with pulverized peanuts. For stewing as sauce or as a filling for tarts or pies, they may be left intact, if tender, and cooked with sugar. In Assam (India), leaves of mesta are also cooked along with chicken or fish. The leaves are widely consumed as an affordable vegetable in Myanmar by poor people (Dy Phon, 2000, pp. 343–344). In Bangladesh, roselle or mesta leaves are steamed with dried or fresh fish to make a paste with garlic, onion, and chilies or cooked with fish. A popular soup or dish is also prepared from mesta leaves along with prawn stock (Islam et al., 2016). The young leaves of roselle are consumed as dal after steaming with lentils in India. Leaves are also used to make pacchadi (pesto) by mixing with spices. The root has been reported as aperients due to the presence of tartaric acid (Mclean, 1973). The root has been reported to be

Roselle. https://doi.org/10.1016/B978-0-323-85213-5.00001-9

FIG. 7.1 Different types of roselle (dark red, red, green).

used as an aperient due to the presence of tartaric acid (Mclean, 1973). After harvesting of the calyces, roselle plants are often used as fodder for livestock.

3 INDUSTRIAL APPLICATIONS

Roselle is an underutilized industrial crop. It is used for numerous purposes from health to other different uses in making wine, cakes, ice cream, and flavors and is also dried and brewed into tea (Rao, 1996; Tsai et al., 2002). In this section, we discuss the different uses of roselle as an industrial crop. The plant parts such as calyx, leaves, roots, fruits, and seeds are used for different purposes:

3.1 Uses of Calyces

The red calyces are used to make a fruity drink and are also used as food colorings in jams, jelly, marmalade, ice cream, sorbets, butter, pies, sauces, tarts, and other deserts. Besides this, calyces are also helpful in decreasing blood viscosity, reducing hypertension, and helping form healthy bone and teeth, as they contain a high amount of ascorbic acid or vitamin C (Fleck and Munves, 1962; Siong et al., 1988; Wong et al., 2002). The bright red color coupled with exceptional flavor and other organoleptic attributes makes them valuable food products (El-Adawy & Khalil, 1994) such as in wine, syrup, ice cream, pies, snakes, tarts, and other desserts (Duke & Atchley, 1984; Eslaminejad and Zakaria, 2011) (Fig. 7.2).

3.1.1 Roselle calyces as tea

Roselle calyces are rich in ascorbic acid. Roselle tea made from calyces has tons of health benefits. It is a caffeine-free herbal tea. Roselle tea reduces blood sugar level and prevents hypertension. The tea is made out of the dried calyces. Roselle tea is a popular sugary herbal tea in Africa. It has also spread in Italy during the first decades of the 20th century and is quite common everywhere. In Thailand, roselle tea is believed to reduce cholesterol. Roselle tea is also consumed in Jamaica by adding more flavor with ginger (Mohamed et al., 2012).

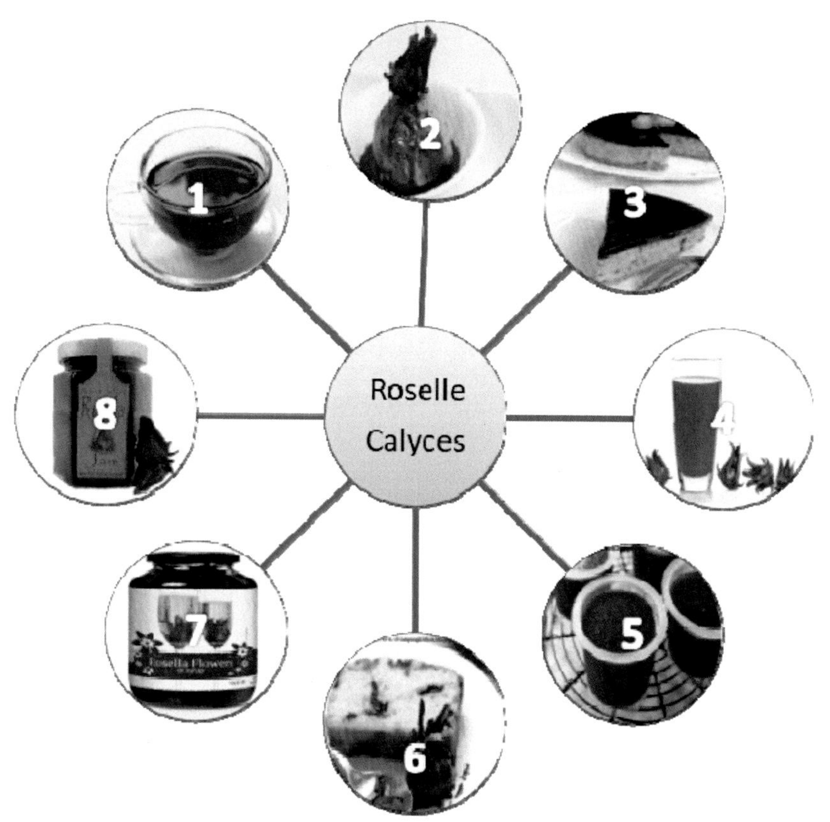

FIG. 7.2 Industrial application of roselle calyces: 1. roselle tea, 2. ice cream, 3. pudding, 4. juice, 5. jelly, 6. cake, 7. syrup, and 8. jam.

3.1.2 Roselle calyces as ice cream, yogurts, and puddings

Roselle calyces are used for making ice cream, yogurts, and puddings. Roselle yogurts that are prepared from roselle juice have shown fabulous health benefits to improve gastrointestinal functions (Heyman, 2000), which include lactose digestion and lactose intolerance symptoms among maldigesters. It also lowers cholesterol levels and reduces risk from hypertension (Taylor and Williams, 1998) and helps maintain the microfloral population in the intestine (Boudraa et al., 1990; Iwalokun and Shittu, 2007).

3.1.3 Roselle calyces as juice/beverage

The production of a nonalcoholic beverage from red roselle calyces is very popular. It can be an option to the industrially produced soft drinks in terms of cost-effective choice. Either fresh or dried roselle calyces are used to prepare drinks. Roselle juice is a popular drink in many social events of African countries such as Guyana, Antigua, Barbados, Dominica, Grenada, Jamaica, Trinidad, Tobago, Mali, Senegal, Gambia, Burkina Faso, Sudan, and Benin. The dried calyces and readymade drinks are widely available in groceries throughout the United Kingdom (Fellows and Axtell, 2014). In Senegal, roselle juice is used to treat conjunctivitis and, when pulverized, soothes sores and ulcers.

3.1.4 Roselle calyces as jam and jelly

In West Indies and elsewhere in the tropics, roselle calyces are used for making jam and jelly. Roselle jam and jelly are also manufactured in different countries of the world and are available in supershops (Mohamed et al., 2012; Morton, 1987). One of the attractive and effective means of roselle utilization is jam processing (Desrosier 1970). Jam processing has been known since the 18th century. Roselle jam is easy to make with only roselle calyces and sugar. The jam made from roselle calyces is red and tangy. Calyx of roselle is rich with various nutrients, especially vitamins (B_1, B_2, B_3, and

C), minerals, and antioxidants. Antioxidants present in roselle calyces are good for our heart and health. Roselle jam has been made since the colonial period and is still available in the community stalls of Australia. It is commonly sold as preserved fruits or jams in Myanmar (Islam et al., 2016).

3.2 Uses of Roselle Seeds

Roselle seed contains a high amount of oil that has good physical and chemical characteristics. It could be one of the promising sources of vegetable oil production. Oil extracted from seeds of *H. sabdariffa* has been shown to have an in vitro inhibitory effect on *Bacillus anthracis* and *Staphylococcus albus* (Gangrade et al., 1979). The physical and chemical properties of roselle seed oil suggest that it could have several important industrial applications and justify its added value for cultivation (Mohamed et al., 2007). Roselle seeds are a source of vital nutrients and bioactive compounds and are responsible for essential biological activities for overall well-being. Minerals are essential elements that activate enzymes performing metabolic functions of human body processes and their deficiency may develop various chronic diseases. Fortification of bakery products with a more economical nutrient source such as roselle seed flour could improve their dietary properties, as cereal flours in baking industry are usually deficient in some vital mineral elements (Rimamcwe et al., 2016).

3.3 Uses of Roselle Flowers

Roselle flowers are also widely used to make herbal tea (Mohamed et al., 2012). Flowers are axillary, solitary, predominantly yellow and red, regular, and pedicellate; epicalyx has about 12 reddish bracts, is connate at base, and 10–12 mm long; calyx is 15–30 mm long (enlarging in fruit to 40 mm) and has 5 reddish, large, accrescent, fleshy sepals, with all sepals fused at the base. The ornamental value of roselle is of another interest, as a garden plant or cut flower (Crane, 1949; Morton, 1987). The decorative red stalks with ripe red fruits are exported to Europe where they are used in flower arrangements (Alegbejo et al., 2003).

4 NUTRITIONAL IMPORTANCE

Roselle also has some medicinal properties. It contains high levels of antioxidants that reduce body fat, assist in weight loss, reduce high blood pressure and blood sugar levels, and help in digestion (Patel, 2014). Nutritional properties of roselle calyces were previously reported. Duke (1983) and Mat et al. (1985) found that 100 g of fresh roselle calyces contain 84.5% of moisture. The calyces contain 49 calories, 1.9 g protein, 0.1 g fat, 12.3 g total carbohydrate, 2.3 g fiber, and 1.2 g ash (Duke & Atchley, 1984). The nutritional properties of roselle are different in different areas. The results differ probably because of the different varieties, genetics, environment, ecology, and harvesting conditions of the plant (Atta 2003). Another experiment with roselle suggests that daily consumption of hibiscus tea, in an amount readily incorporated into the diet, lowers blood pressure in prehypertensive and mildly hypertensive adults (Vangalapati et al., 2014).

4.1 Proximate and Mineral Composition

4.1.1 Calorie

Calories provide energy to our bodies in the form of heat so that our bodies can function properly. The human body needs calories to survive and run the body function properly. Calories are the fuel to our body. According to Duke and Atchley (1984), 100 g fresh roselle calyces contain 49 calories.

4.1.2 Carbohydrates

Carbohydrates are required for several key functions in our body. They provide energy for daily work and are the primary fuel source for our brain's high energy demands. Fiber is a special type of carbohydrate that helps promote good digestive health. The pectin content in roselle had been shown by Riaz (1969). Galactose, galacturonic acid, and rhamnose were detected in roselle calyx by Hamidi et al. (1996). Ibrahhim et al. (1971) also detected two other free sugars, glucose and arabinose. Babalola (2000) and Ojokoh et al. (2003) stated that the roselle calyces contain high carbohydrate contents. Besides leaves, seeds and flowers of Roselle also contain Carbohydrate. Adanlawo and Ajibade (2006) showed that Roselle varieties soaked in wood ash have increased carbohydrate content. Amoasah et al. (2018) found the carbohydrate content of roselle calyces was high in oven-drying method. Carbohydrate content of different parts of roselle is clearly stated in Table 7.1.

4.1.3 Protein and fat

Protein is the building blocks of the body. Adults need to eat 60 g of protein per day. Children also need more protein because they are in the growing stage. Protein is not mainly used for energy. The body needs protein to maintain and replace tissues and to function and grow. If the body is not getting enough calories from other nutrients or from the fat stored in the body then protein is used for energy. The body breaks the protein down and

TABLE 7.1
Carbohydrate Content (%) of Roselle.

Plants Parts	Amount (%)	References
Calyces	68.75 (red calyces) 71.56 (green calyces)	Adanlawo and Ajibade (2006)
	60.51 (sundried) 62.46 (oven dried) 58.77 (solar dried)	Amoasah et al. (2018)
	79.68 (red calyces soaked in wood ash) 78.91 (green calyces soaked in wood ash)	Adanlawo and Ajibade (2006)
Seeds	21.7	El-Deab and Ghamry (2017)
Fresh leaves	8.7	Singh et al. (2007)
Flowers	4.38	Sayago-ayerdi et al. (2007)

stores its components as fat, if more protein is consumed than that is needed. Fats are complex molecules composed of fatty acids and glycerol. The body needs fats for growth and energy. Fats keep skin healthy. They also synthesize hormones and other substances needed for the body's activities.

Some research showed that roselle plant parts contain protein and fat, especially roselle seeds contain a high amount of protein and fats. According to El-Deab and Ghamry (2017), roselle seeds contain relatively high amounts of protein and fat. The relatively high fat and protein contents indicate that roselle seeds could become an excellent economic source for edible oil. Previous studies also showed that the seeds can be used as a potential source of proteins (Al-Wandawi et al., 1984; El-Adawy & Khalil, 1994). Roselle seeds vary in nutritional composition depending on the location and environmental conditions during cultivation. Protein from roselle seeds could be used as a supplement of food mixture for poor lysine sources (Elneairy, 2014). Balarabe (2019) found that in the dried form, roselle leaves have high protein content than the calyces. The crude protein content was higher in solar-dried calyces (5.86%) than oven-dried calyces (5.76%). But drying roselle calyces in an oven could be more desirable, as the crude fat content is preserved and consequently flavor and fat-soluble vitamins (Amoasah et al., 2018). Fat and protein content of roselle are given in Table 7.2.

4.1.4 Vitamins

Roselle is a medicinal plant. Roselle contains a high amount of vitamin C and thiamine in both fresh and dry forms. Vitamin C is one of the important antioxidant nutrients that plays a vital role in preventing losses and providing a decent level of antioxidants to our body. It also plays a crucial role in fighting infections and producing collagen and hormone and protecting our body against different diseases. Vitamin C is needed in our body in many different reactions. Vitamin C helps regulate blood pressure, contributes to reduce cholesterol levels, and aids in the removal of cholesterol deposits from arterial walls, thus preventing arteriosclerosis. The easily destroyed nutrient protects us from the effects of free radicals, dangerous unpaired oxygen fragments that are produced in huge numbers as a normal by-product of human metabolic processes (Addo, 2004). Fresh roselle leaves contain a higher amount of ascorbic acid than dried roselle leaves. So vegetables should be cooked fresh to retain their vitamin contents (Sarkiyayi & Ikioda, 2009). The vitamin content of roselle plants can be significantly influenced by variety, location, and environmental conditions. Green calyces were significantly higher than other types in ascorbic acid content (Babalola et al., 2001). The chemical composition of vitamins (vitamin C, thiamine) of roselle is given in Table 7.3.

4.1.5 Minerals (Zn, Fe, Mg, Na, K, and Mn)

Minerals are essential in preventing deficiency diseases (Cissouma et al., 2013) and could play a vital role in the normal functions of the body's biochemical processes. For example, phosphorus is essential for the development of teeth, especially in young children, and it plays an essential role in the formation and utilization of high-energy phosphate compounds (phosphagens). Phosphate is required for the formation of phospholipids, phosphoproteins, and nucleic acids (DNA and RNA) (Gatti et al., 2010). Calcium is a mineral that is necessary for life. In addition to building

TABLE 7.2
Fat and Protein Content of Different Parts of Roselle.

Plant Parts		Protein Content (%)	Fat (%)	References
Calyces	Red calyces	4.71	2.01	Adanlawo and Ajibade (2006)
	Green calyces	6.45	2.17	
	Red calyces soaked in wood ash	6.64	0.89	Adanlawo and Ajibade (2006)
	Green calyces soaked in wood ash	5.57	0.85	
	Oven drying method	5.76	2.88	Amoasah et al. (2018)
	Solar drying method	5.86	2.16	
	Sun drying method	5.65	1.25	
	Kjeldahl method	4.10	ND	Balarabe (2019)
Leaves	Dried	5.37	ND	Balarabe (2019)
Seeds	Fresh	23.52	19.36	El-Deab and Ghamry (2017)
	Dried	29.9	ND	Halimatul (2007)
	Boiled	32.7		
	Egyptian seed	31.02	21.60	Elneairy (2014)
	Libyan seed	28.67	16.94	
Flower		9.87	0.59	Sayago-ayerdi et al. (2007)

ND, not determinate.

TABLE 7.3
Vitamin Composition of Roselle.

Plant Parts	Vitamin C (mg/g)	Thiamine (mg/g)	References
Leaves	0.0736 (dried) 0.1802 (fresh)	0.00194 (dried) 0.75 (fresh)	Sarkiyayi and Ikioda (2009)
Calyces	141.09 (fresh) 86.5 (green) 63.5 (red) 54.8 (dark red)		Wong et al. (2002) Babalola et al. (2001)

bones and keeping them healthy, calcium helps prevent blood clots, nerves send messages, and muscles contract (Nof, 2016). Magnesium plays a very critical role in energy synthesis and storage, as phosphate and magnesium ion interaction makes magnesium essential to the basic nucleic acid chemistry of all cells of all known living organisms; magnesium boosts over 300 enzyme actions including all enzymes using or synthesizing ATP and those that use other nucleotides to synthesize DNA and RNA, as ATP molecules are normally found in a chelate with a magnesium ions (Romani, 2013). It is essential for the normal function of the heart, kidneys, arteries, and bones (Sleelig, 1980) and the neuromuscular system (Durlach, 1988, p. 360). Potassium was found to be the most abundant mineral present in both red and green varieties of roselle (Adanlawo and Ajibade, 2006). Potassium is not only for the chief electrolytes but also essential for the nervous system, maintenance of fluid volume in the body, contractile mechanism of muscles, maintenance of correct rhythm of heart beat, and blood clotting (Shahnaz et al., 2003). Mineral composition of dark red roselle calyces, including calcium, magnesium, and zinc, was very high but did not vary significantly from that of the red type. This trend indicates that the consumption of the highlighted calyces will have an active role in good bone and teeth formation. The compositions of minerals in roselle are shown in Table 7.4.

4.2 Amino Acids

A total of 19 amino acids were detected in the roselle seed analysis (Shaheen et al., 2012). These are leucine, methionine, tryptophan, valine, isoleucine, phenylalanine, threonine, cystine, tyrosine, lysine, aspartic acid, glutamic acid, arginine, proline, serine, histidine, alanine, glycine, and serine. The biological activity of

TABLE 7.4
The Mineral Composition of Roselle.

Type	Mineral Content	Fresh Red Calyces (mg/100 g)	Fresh Green Calyces (mg/100 g)	Seeds (mg/100 g)	Leaves (mg/g)
Macroelements	Calcium (Ca)	1583	1209	320.45	1.4
	Magnesium (Mg)	316.6	235	464.36	1.35
	Potassium (K)	2060	1850	1925.67	ND
	Phosphorus (P)	36.30	15.05	590.14	5.00
	Sodium (Na)	6.6	9.5	ND	ND
Microelements	Iron (Fe)	37.8	32.8	11.45	ND
	Zinc (Zn)	6.4	5.8	17.43	ND
	Manganese (Mn)	2.39	5.61	7.57	ND
	Nickle	1.78	3.57	ND	ND

ND, not determinate.
Rimamcwe, K. B., Chavan, U. D., Kotecha, P. M., & Lande, S. B. (2016). Physical properties and nutritional potentials of indian roselle (*Hibiscus sabdariffa* L.) seeds. *International Journal of Current Research*, *8*(9), 38644–38648, Kilima, B. M., Remberg, S. F., Chove, B. E., & Wicklund, T. (2014). Physio-chemical, mineral composition and antioxidant properties of Roselle (*Hibiscus Sabdariffa* L.) extract blended with tropical fruit juices. *African Journal of Food Agricultural, Nutrition and Development, 14*(3), 8963–8978, Babalola, S. O., Babalola, A. O., & Aworh, O. C. (2001). Compositional attributes of the calyces of roselle (*Hibiscus sabdariffa*). *Journal of Food Technology in Africa, 6*(4), 133–134, Adanlawo, I. G., & Ajibade, V. A. (2006). Nutritive value of the two varieties of roselle calyces soaked with wood ash. *Pakistan Journal of Nutrition, 5*(6), 555–557, Balarabe, M. A. (2019). Nutritional analysis of *Hibiscus sabdariffa* L. (roselle) leaves and calyces. *Plant*, 7(4), 62–65.

protein is more related to its amino acid makeup. In order to appreciate the physicochemical properties of the hydrolysates, amino acid composition analysis was carried out. The hydrolysates contained a good proportion of all essential amino acids as reported by Sathivel et al. (2003). Glutamic acid was the major amino acid in both hydrolysates. In general, arginine, aspartic acid, and glutamic acid were predominant in all the samples (Tounkara et al., 2014). Elneairy (2014) found that amino acid content of Egyptian seeds varies from that of Libyan seeds. He stated that roselle seeds vary in nutritional composition depending on the location and environmental conditions during cultivation. According to Atanda et al. (2018), the amino acid composition of roselle seeds boiled for 60 min revealed the overall suitability of roselle seeds as an alternative feed ingredient in poultry nutrition and thus they may serve as a potential source of functional ingredients. Roselle seeds could be used as supplement food or as diet enrichment, especially in low-protein diets, because of the increased histidine, threonine, serine, glutamic acid, and proline composition of roselle seeds. The roselle seeds boiled up to 60 min could be a biomarker for alternative feed ingredients in poultry diet. Chemical compositions of amino acids in roselle are given in Table 7.5.

4.3 Fatty Acid Composition

The fat quality is usually valued according to the content of essential fatty acids such as linoleic acid, linolenic acid, and arachidonic acid. Humans require some of these fatty acids in their diet to prevent fatty acid deficiency diseases including skin lesions, poor hair growth, and low growth rate (Kinsella, 1987). These qualities are well presented in *H. sabdariffa* L. fatty acids (Aldeen et al., 2015). Reports from different studies have shown that the seed contains about 20% edible health-promoting oil. Roselle seed oil could be a good source of essential fatty acids (Nyam et al., 2009). According to Shaheen et al. (2012), a high content of oleic and linoleic acids is found in roselle seed oil. Linoleic acid, which has the beneficial effects of reducing blood cholesterol levels and blood pressure, is determined as the major fatty acid in roselle seed oil (Nasrabadi et al., 2018). A high content of unsaturated fatty acid constituents in the roselle seeds and the ratio between saturated and unsaturated fatty acids have made roselle seeds as a source of edible oil, which has similar properties with cotton seed oil. Roselle seed oil is a rich source of α-tocopherol. The α-tocopherol is the second major component in roselle seed oil. Tocopherols are well known as biological antioxidants that can prevent or retard the oxidation of body lipids, which include polyunsaturated fatty acids and lipid components of cells and organelle membranes (Nyam et al., 2009). The total fatty acid composition of roselle is given in Table 7.6.

4.4 Total Anthocyanin Content

Roselle (*H. sabdariffa* L.) is rich in anthocyanin pigment. The calyx contains two main anthocyanins—delphinidin-3-sambubioside or delphinidin-3-

TABLE 7.5
Chemical Composition of Amino Acids in Roselle.

Amino Acid Content		ROSELLE SEED (G/100 G OF PROTEIN)					ROSELLE CALYCES (MG/G DRY MATTER)	
		TOUNKARA ET AL. (2014)			ELNEAIRY (2014)			
		RSPH 1.5	RSPH 3	FAO/WHO Child (Adults)	Egyptian Seed Powder	Libyan Seed Powder	Morton (1987)	Glew et al., 1997
Essential amino acids	Lysine	4.48	4.04	5.5(1.6)	5.37	3.02	3.90	2.77
	Histidine	2.34	2.28	1.6 (1.5)	2.97	5.43	1.50	1.19
	Leucine	7.90	6.26	6 (5.9)	7.32	6.93	5.00	4.21
	Isoleucine	3.82	3.93	4 (3)	3.24	4.65	3.00	2.70
	Phenylalanine	5.41	4.31	6.0	5.09	4.77	3.20	2.32
	Phenylalanine + tyrosine	8.19	6.33	4.1 (3.8)				
	Methionine	1.79	2.26	3.5	1.13	0.71	1.00	0.65
	Methionine + cysteine	3.04	3.88	2.3 (1.6)				
	Valine	5.11	5.61	5.0 (3.9)	3.26	5.33	3.80	3.33
	Threonine	3.25	3.76	4.0 (2.3)	4.86	2.64	3.00	2.36
Nonessential amino acids	Glycine	4.03	4.59		4.27	8.20	3.80	2.47
	Cysteine	1.25	1.62		2.64	2.61	1.30	o.87
	Aspartic acid	10.37	11.69		10.73	10.26		
	Glutamic acid	23.48	24.78		21.30	19.87	7.20	8.85
	Serine	4.43	4.48		4.40	2.88	3.50	2.65
	Arginine	10.37	11.69		10.58	9.63	3.60	4.48
	Alanine	4.45	4.74		4.69	5.41	3.70	3.46
	Tyrosine	2.78	2.02		3.64	6.31	2.20	1.44
	Proline	4.73	5.56		4.14	0.15	5.60	5.82

FAO, Food and Agriculture Organization; *RSPH*, roselle seed protein hydrolysate.

xylosylglucoside or hibiscin and cyanidin-3-sambubioside or cyanidin-3-xylosylglucoside or gossypicyanin—and two minor anthocyanins—delphinidin-3-glucoside and cyanidin-3-glucoside. Delphinidin-3-sambubioside represents 70% of the total content of anthocyanins. The anthocyanins represent the largest group of water-soluble pigments in the plant. They are highly valued in the food industry for their coloring properties, which can give food various hues of red and violet (Francis, 1990; Wang et al., 2000). The anthocyanin pigment can be used as a natural colorant and antioxidant. An antioxidant is an organic compound that has the ability to inhibit free radical reactions in the human body (Inggrid et al., 2017). Anthocyanin improves overall visual acuity, blood flow, and blood pressure; decreases cholesterol levels; and fights against oxidative stress (Lila, 2004; Zhu et al., 2016). Anthocyanin has pharmacologic properties such as reduction of coronary disease and anticancer, antitumor, antiinflammatory, and antidiabetic effects; in addition, they improve cognitive behavior. The pharmacologic effects of anthocyanins are related to their antioxidant activity (Hopkins et al., 2013; Kim et al., 2009). Tsai et al. (2002) also suggest that anthocyanin is the major source of antioxidant capacity in roselle extract.

Mollah et al. (2020) found that the total anthocyanin content of roselle calyces was 0.17 mg/100 g. Similar results were also reported by Juliani et al. (2009). But these findings differ from those of Jamini et al. (2019). In my research, the range of anthocyanin content of roselle calyces (fresh) was 2.1–87.7 μg/g. Different extraction procedures (modified protocol), genotypes, planting time, climate conditions, fertilizer doses, planting methods, and soil types may account for this variation. Mollah et al. (2020) also measured the total proanthocyanidin content (284.27 mg/100 g) of roselle calyces. Proanthocyanidins have much

TABLE 7.6
Fatty Acid Composition of Roselle.

Fatty Acids	Seed Oil (%)	References
Lauric acid (C12:0)	0.01	Rimamcwe et al. (2016)
Myristic acid (C14:0)	2.25	El-Deab and Ghamry (2017)
Palmitic acid (C16:0)	18.96	El-Deab and Ghamry (2017)
Palmitoleic acid (C16:1)	1.78	El-Deab and Ghamry (2017)
Heptadecanoic acid (C17:0)	0.12	Mohamed et al. (2007)
Margaroleic acid (C17:1)	0.14	Mohamed et al. (2007)
Heptadecadienoic acid (C17:2)	0.17	Mohamed et al. (2007)
Stearic acid (C18:0)	4.06	El-Deab and Ghamry (2017)
Oleic acid (C18:1)	31.94	El-Deab and Ghamry (2017)
Linoleic acid (C18:2)	40.10	Mohamed et al. (2007)
Linolenic acid (C18:3)	3.46	El-Deab and Ghamry (2017)
Arachidic acid (C20:0)	0.59	Rimamcwe et al. (2016)
Eicosenoic acid (C20:0)	0.01	Rimamcwe et al. (2016)
Behenic acid (C22:0) 0.25	0.25	Rimamcwe et al. (2016)
Lignoceric acid (C24:0)	0.14	Rimamcwe et al. (2016)
Saturated fatty acids	25.27	El-Deab and Ghamry (2017)
Unsaturated fatty acids	74.73	El-Deab and Ghamry (2017)
Unsat./Sat.	2.96:1	El-Deab and Ghamry (2017)

TABLE 7.7
Comparison of Anthocyanin Content of Roselle With Different Fruits.

Fruits	Anthocyanin Content (mg/g)
Strawberry (*Fragaria* spp.)	450–700
Grape (*Vitis vinifera* L.)	30–750
Blackberry (*Rubus fruticosus* L.)	67–230
Blood orange (*Citrus sinensis* L.)	70–100
Sweet cherry (*Prunus cerasus* L. var *Montmorency*)	35–82
Roselle (*Hibiscus sabdariffa*)	150

Mazza, G., & Miniati, E. (2000). *Anthocyanin in fruits, vegetables and grains*. CRC Press.

stronger antioxidant activities than vitamin C or vitamin E (Ariga, 2008). According to Inggrid et al. (2017), the highest total anthocyanin content in the dried roselle calyces extract was 80.4 mg/L and it was obtained at 5°C and pH 2. The researchers also added that temperature and pH significantly affect the total anthocyanin content and antioxidant activity. In addition, temperature and ultraviolet rays affect the stability of red color in roselle calyces. However, anthocyanins of *H. sabdariffa* are known for their instability (Chen et al., 2005; Esselen & Sammy, 1975; Mazza & Miniati, 2000; Tsai & Ou, 1996). One of the characteristics of *H. sabdariffa* is its high anthocyanin content that can reach 1.5 g/kg of dry calyx. This content is comparable to that of blackberry and superior to most other edible plants (Table 7.7).

TABLE 7.8
Total Phenolic Content of Roselle.

Plant Parts		Mean ± SD (%)	Source
Roselle calyces (mg/100 g)		521.46 ± 15.26	Mollah et al. (2020)
Roselle seed	Seed oil (mg/g)	23.65 ± 0.24	El-Deab and Ghamry (2017)
	Seed powder (mg/100 g)	18.8 ± 1.86	Nyam et al. (2014)

SD, standard deviation.

4.5 Total Polyphenol Content

Phenols play an important role in plant constituents because they contribute to the overall antioxidant activity (Khattak et al. 2008). In my research, the total phenol content of roselle calyces ranged from 488.98 to 869.45 μg/g. Mollah et al. (2020) found 521.46 mg/100 g phenol content in roselle calyces. Kilima et al. (2014) reported that the total phenolic content of roselle calyx extract ranged from 108 to 546 μg/g. Luvonga et al. (2009) also found that the total phenolic content of roselle extracts ranged from 582 to 606 μg/g. By investigating various studies, it is clearly seen that the phenolic content of roselle calyx is less variable among various genotypes of roselle. Besides calyces, roselle seeds also contain phenolic compounds. According to EL-Deab and Ghamry (2017), the phenolic content of roselle seed oil was 23.65 mg/g. The total phenolic content of roselle seed powder in methanolic extract was 18.8 mg gallic acid equivalents (GAE)/100 g seed (Nyam et al., 2014). The total phenolic content of roselle is summarized in Table 7.8.

4.6 Total Flavonoid Content

Flavonoid content was found in roselle. Flavonoids act as anticancerous agents, so regular intake of flavonoids may reduce the risk of cancer in humans. In my research, total flavonoid content of fresh roselle calyces ranged from 9.31 to 404.40 μg/g; Mollah et al. (2020) also found the same result (959.53 mg/100 mg).

5 PHARMACEUTICAL IMPORTANCE

5.1 Antioxidant Activity

The protective property of a compound to inhibit the oxidative mechanisms by scavenging reactive oxygen and free radicals is known as antioxidant activity. This activity protects lining organelles from premature cell damage and reduces aging. A large number of in vitro and in vivo studies have shown that roselle calyces contain potent antioxidants. According to Augustine et al. (2011), both the whole aqueous and anthocyanin-rich extracts of roselle are effective antioxidants. Studies have also highlighted that polyphenolic acid, flavonoids, and anthocyanins, which are found in roselle, are potent antioxidants (Crawford et al., 1998).

Akim et al. (2011) also evaluated the antioxidant capacity of commercialized roselle juice at three storage periods and its antiproliferative effect on breast (MCF-7 and MDA-MB-231), ovarian (Caov-3), and cervical (HeLa) cancer cell lines. The study showed that commercialized roselle juice has strong antioxidant capacity and antiproliferative activity on the four cancer cell lines despite different storage periods.

Ghosh et al. (2015) estimated the antigenotoxic property of *H. sabdariffa* L. (roselle) calyx extract, which is presumably attributed to its antioxidant properties. The results revealed that the calyx extract of roselle inhibited the DNA damage induced by sodium arsenite in a dose-dependent manner. The presence of phytochemical constituents such as polyphenols and flavonoids was ascribed to the observed changes. The antioxidant efficacy was substantiated by applying ferric reducing antioxidant power (FRAP) and 2,2-di(4-tert-octylphenyl)-1-picrylhydrazyl (DPPH) assays.

5.2 Antinutritional Factor

Atanda et al. (2018) conducted an experiment on the nutritive evaluation of boiling duration on the antinutrient composition of roselle seeds to examine how cooking roselle seeds (*H. sabdariffa* L.) at 100°C for different lengths of time (from 10 to 60 min) affects the antinutrient composition and to select the most optimal cooking time span for the seeds. The result of the laboratory trial showed that there were significant ($P \leq .05$) differences in the antinutritional factors of roselle seeds boiled at different durations. The chemical composition, micro- and macrominerals, and amino acid composition of roselle seeds boiled for 60 min revealed the overall suitability of roselle seeds as an alternative feed ingredient in poultry nutrition. It is therefore recommended that roselle seeds boiled for 60 min can be used to replace maize in the diets of broiler birds.

5.3 Antidiabetic Activity

Peng et al. (2011) extracted the polyphenolic components of roselle and studied their effects in a type II diabetic rat model (high-fat-diet model). Studies revealed anti-insulin resistance properties of the extract at a dose level of 200 mg/kg and reduction in hyperglycmia and hyperinsulinemia. The extract was found effective in lowering serum cholesterol levels, triacylglycerol levels, the ratio of low-density lipoprotein/high-density protein (LDL/HDL), and AGE formation and lipid peroxidation. Intestinal α-glycosidase and pancreatic α-amylase help in the digestion of complex carbohydrates present in the food into bioavailable monosaccharides and play an important role in postprandial hyperglycemia; therefore inhibition of these enzymes has been reported as an effective mechanism for the control of postprandial hyperglycemia. Hibiscus acid (hibiscus-type (2S,3R)-hydroxycitric acid lactone) has been shown to be a potent inhibitor of pancreatic α-amylase and intestinal α-glucosidase activity(Yamada et al., 2007). Adisakwattana et al. (2012) conducted an in vitro study and reported roselle extract as an effective inhibitor of pancreatic α-amylase.

Rosemary et al. (2014) estimated the antidiabetic effect of roselle calyx extract in streptozotocin-induced mice. Research on the antidiabetic activity of n-hexane, ethyl acetate, and ethanol extracts of roselle calyces (*H. sabdariffa* L.) has been conducted in order to improve the utilization of herbs and natural antidiabetics, and the information can be useful in extracting roselle calyces to obtain the optimal antidiabetic effect. The results showed that the ethanol extract of roselle calyces reduced the blood glucose levels in diabetic mice, whereas the *n*-hexane and ethyl acetate extracts did not. Chemical constituents of the ethanol extract of roselle calyces are likely efficacious in lowering blood glucose levels in diabetic mice.

5.4 Antihypertensive and Prehypertensive Effects

Hopkins et al. (2013) assessed roselle in the treatment of hypertension. The effectiveness of roselle in the treatment of risk factors associated with cardiovascular disease is assessed in this review by taking a comprehensive approach to interpret the randomized clinical trial (RCT) results in the context of the available ethnomedical, phytochemical, pharmacologic, and safety and toxicity information. Roselle decoctions and infusions of calyces, and on occasion leaves, are used in at least 10 countries worldwide in the treatment of hypertension and hyperlipidemia, with no reported adverse events or side effects. Roselle extracts have a low degree of toxicity with an LD_{50} (median lethal dose) ranging from 2000 to over 5000 mg/kg/day. There is no evidence of hepatic or renal toxicity as the result of roselle extract consumption, except for possible adverse hepatic effects at high doses. There is evidence that roselle acts as a diuretic; however, in most cases the extract did not significantly influence electrolyte levels. Animal studies have consistently shown that consumption of roselle extract reduces blood pressure in a dose-dependent manner. In RCTs, the daily consumption of tea or extract produced from roselle calyces significantly lowered systolic blood pressure and diastolic blood pressure in adults with pre- to moderate essential hypertension and type 2 diabetes. In addition, roselle tea was as effective at lowering blood pressure as the commonly used blood pressure medication captopril, but less effective than lisinopril. Total cholesterol, low-density lipoprotein cholesterol (LDL-C), and triglyceride levels were lowered in the majority of normolipidemic, hypolipidemic, and diabetic animal models, whereas high-density lipoprotein cholesterol (HDL-C) levels were generally not affected by the consumption of roselle extract. Over half of the RCTs showed that daily consumption of roselle tea or extracts had favorable influence on lipid profiles, including reduced total cholesterol, LDL-C, triglyceride levels, as well as increased HDL-C levels. Anthocyanins found in abundance in roselle calyces are generally considered the phytochemicals responsible for the antihypertensive and hypocholesterolemic effects; however, evidence has also been provided for the role of polyphenols and hibiscus acid. A number of potential mechanisms have been proposed to explain the hypotensive and anticholesterol effects, but the most common explanation is the antioxidant effects of the anthocyanin inhibition of LDL-C oxidation, which impedes atherosclerosis, an important cardiovascular risk factor. This comprehensive body of evidence suggests that extracts of hibiscuss sabdariffa (HS) are promising as a treatment of hypertension and hyperlipidemia; however, more high-quality animal and human studies informed by actual therapeutic practices are needed to provide recommendations for use that have the potential for widespread public health benefit. According to Herrera-Arellano et al. (2004), the effectiveness of an aqueous extract of roselle on mild to moderate hypertension was investigated in many researches. The aqueous extract of roselle was as effective as captopril in treating mild to moderate hypertension and there is no adverse effect with the treatment, confirming the effectiveness and safety of the extract. According to Haji Faraji and Haji Tarkhani (1999), the possible mechanism(s) of action of roselle extract is not investigated, daily consumption of the aqueous roselle extract resulted in decrease in systolic and diastolic blood pressure.

5.5 Anticancer Activities

Prabhakaran et al. (2017) conducted an experiment to investigate the anticancer activity of the flower of *H. sabdariffa* L. (roselle) against human hepatoma cell line (HepG2) in India. In vitro anticancer activity was carried out to screen the potency of cytotoxicity of the solid obtained from the ethyl acetate fraction from *H. sabdariffa* L. flower extract at different concentrations against HepG2 cell line. The study confirmed that *H. sabdariffa* L. is a potential plant with anticancer activity. The isolation of the pure compounds and determination of the bioactivity of individual compounds will be further performed.

Gheller et al. (2017) measured the antimutagenic effect of *H. sabdariffa* L. aqueous extract on rats treated with monosodium glutamate. This study evaluated the effects of *H. sabdariffa* aqueous extract against cyclophosphamide (CPA, 25 mg/kg)-induced damage to DNA in male Wistar rats by the micronucleus test. Under the conditions tested, *H. sabdariffa* L. presented a protective effect to CPA-induced damage to the DNA of the treated animals, and it is a potential candidate as a chemopreventive agent against carcinogenesis.

5.6 Antiaging Activity

Hutagaol (2017) formulated an antiaging gel from roselle calyx extract. Roselle is known to have many active compounds that function as antioxidants, such as flavonoids, polyphenols, and vitamin C. High antioxidant content has a high antiaging effect. His study aims to create a cream dosage form that has antiaging properties with active ingredients from the roselle calyx ethanolic extract that meets the requirements of physical evaluation and stability evaluation and has antiaging activity before and after use. The ethanol extract of roselle calyx is made by maceration. The results showed that the antiaging cream of roselle calyx ethanolic extract has good physical properties and good stability properties for 1 month of storage. All creams have significantly different antiaging activities and depend on the roselle calyx ethanolic extract concentration. The higher the concentration of roselle calyx ethanolic extract, the higher the antiaging activity. Antiaging activity with a concentration of roselle calyx ethanolic extract had no significant difference with vitamin C.

5.7 Immunoprotective Effects

Liu et al. (2006) measured the protective effects of roselle extract on CCl(4). Dried flower roselle extracts, a local soft drink material and medicinal herb, were studied for their protective effects against liver fibrosis induced by carbon tetrachloride (CCl(4)) in rats. The results suggested that roselle may protect the liver against CCl(4)-induced fibrosis. This protective effect is due to the antioxidant properties of roselle.

5.8 Antinociceptive Activity

Khatun et al. (2011) designed an experiment to investigate the CNS depressant and antinociceptive activity of methanol (85%) extracts of roselle fruit in Bangladesh. The CNS depressant was evaluated by observing the reduction of locomotor and exploratory activities in the open field and hole cross tests at doses of 250 and 500 mg/kg body weight, while the analgesic activity was examined using acetic acid-induced writhing test and formalin test in rat models at 100 and 200 mg/kg body weight. The results of the statistical analysis showed that the plant extract had significant ($P < .01$) dose-dependent CNS depressant and antinociceptive activities. In the acetic acid-induced writhing model, the extract showed better analgesic effects at higher doses characterized by a reduction in the number of writhings when compared with the control. The extract, at the dose of 200 mg/kg, exerted a maximum of 66.464% inhibition of writhing response, which is more potent to the reference drug indomethacin (58.8%). Altogether, these results suggested that the fruit extract of roselle possesses remarkable CNS depressant and analgesic properties.

Ali et al. (2011) also evaluated the antinociceptive activity of the roselle extract by using the acetic acid-induced writhing test. The findings indicate that the calyx extract of *H. sabdariffa* possesses significant antinociceptive activities that support its uses in traditional medicine.

5.9 Antiinflammatory Activity

In Nigeria, Meraiyebu et al. (2013) investigated the antiinflammatory activity of the methanolic extract of *H. sabdariffa* L. in adult Wistar rats. The results showed that the methanolic extract of *H. sabdariffa* L. possess antiinflammatory properties against carrageenan-induced inflammation similar to the action of diclofenac. According to the authors, the therapeutic effect of this extract will encourage its use in the treatment of inflammation.

In Malaysia, Nyam et al. (2015) also carried out an experiment to measure the antiinflammatory effects of roselle seeds. They demonstrated that roselle seed oil and roselle seed extract contain significant levels of phytochemicals such as phenol, flavonoid, saponin, terpenoid, and alkaloid. The study also showed that roselle seed oil and roselle seed extract exhibited antiinflammatory effects in edema-induced rats. Hence it was concluded that the phytochemical content may affect the antiinflammatory activity of the samples studied.

In Indonesia, Mardiah et al. (2015) conducted another experiment to assess antiinflammatory property of purple roselle extract in rats with diabetes induced by streptozotocin. They stated that the pathogenesis of diabetes mellitus involves a low-level inflammatory process due to the increase in blood glucose levels. Antioxidant compounds contained in roselle can suppress or reduce the levels of free radicals and lead to decreased levels of tumor necrosis factor α. This result showed that roselle extract tends to reduce the level of inflammation so that the impact of continuous severity due to inflammation can be suppressed.

Ali et al. (2011) also tested the antiinflammatory effect of roselle extract by using the xylene-induced ear edema mouse model. The findings indicate that the calyx extract of *H. sabdariffa* possesses significant antiinflammatory activities that support its uses in traditional medicine.

5.10 Antidiarrheal Effect

Ali et al. (2011) evaluated the antinociceptive, antiinflammatory, and antidiarrheal activities of the ethanolic calyx extract of *H. sabdariffa* L. in mice. In this study, the dried calyces of *H. sabdariffa* were subjected to extraction with 95% ethanol and the extract was used to investigate the possible activities. Castor oil-induced diarrheal mouse model was used to evaluate the antidiarrheal activity of the extract. The findings indicate that the calyx extract of *H. sabdariffa* possesses significant antidiarrheal activities and can be recommended to be used in traditional medicine.

5.11 Immunomodulatory Effects

Fakeye et al. (2008) conducted an experiment on the immunomodulatory effect of extracts of *H. sabdariffa* L. in a mouse model. This study established the immunity-enhancing properties of the extracts of this plant, confirming that the immunomodulatory activity is cell mediated and humoral. The insoluble fraction could find use as an immunostimulatory agent in humans. Hussein et al. (2019) conducted an experiment to evaluate the immunomodulatory and antiinflammatory activities of the chemically characterized defatted alcoholic extract and mucilage of *H. sabdariffa* L. leaves. The findings highlight the beneficial effect of *H. sabdariffa* L. leaves on immune response and inflammation, encouraging further investigation of this underutilized part.

5.12 Other Pharmacologic Effects

Roselle has been reported to possess lactogenic activity. Okasha et al. (2008) observed enhancement in the serum prolactin levels of lactating female albino rats on administration of the seed extract of roselle. Bako et al. (2014) studied the lactogenic effect of the ethyl acetate fraction of *H. sabdariffa* from 3 to 17 days of lactation. The results showed an increase in serum prolactin levels and milk production in lactating female albino rats, which confirms the lactogenic property of *H. sabdariffa*. Studies have shown that roselle tea contains an enzyme inhibitor that blocks the production of amylase, and it is possible that drinking a cup of hibiscus tea after meals can reduce the absorption of dietary carbohydrates and assist in weight loss. It was also reported that roselle is considered as a possible antiobesity agent. According to a study conducted among hypercholesterolemic patients, two capsules of roselle extract (1 g), given three times a day (for a total of 3 g/day), significantly lowered serum cholesterol levels. Another scientific study also confirmed that the ethanolic extract from the leaves of roselle significantly exhibits hypolipidemic effect. The roselle extract was also studied among subjects, some with and some without metabolic syndrome. Subjects with metabolic syndrome receiving ethanolic extract of roselle had significantly reduced glucose, total cholesterol, and LDL levels, as well as increased HDL levels.

6 CONCLUSIONS

Roselle is a miracle plant with various utilizations in nutrition and folk medicine. It is considered as a medicinal plant having a variety of medically important compounds called phytochemicals and is also well known for its nutritional properties. All the plant parts, namely, leaves, roots, fruit calyces, and seeds, are used as food and traditional medicine. Extracts from leaves and fruit calyces play a significant role in hypertension, cardiovascular diseases, diabetes, cancer, antiaging, obesity, etc., but further investigation is needed to find out the exact mechanism and active compound to formulate products from roselle.

ACKNOWLEDGMENTS

Authors are thankful to the authorities of the Bangabandhu Sheikh Mujibur Rahman Agricultural University, Gazipur 1706, Bangladesh for their support.

REFERENCES

Abu-Tarboush, H. M., Ahmed, A. A., & Kahtani, H. A. (1997). Some nutritional and functional properties of Karkade *(H. sabdariffa)* seed products. *Cereal Chemistry, 74*, 352–355.

Adanlawo, I. G., & Ajibade, V. A. (2006). Nutritive value of the two varieties of roselle calyces soaked with wood ash. *Pakistan Journal of Nutrition, 5*(6), 555–557.

Addo, A. A. (2004). *Seasonal availability of dietary ascorbic acid and incidence of scurvy in northern state of Nigeria*. PhD thesis. ABU Zaria: Department of Biochemistry.

Al-Wandawi, H., Al-Shaikhly, K., & Abdul-Rahman, M. (1984). Roselle seed: A new protein source. *Journal of Agricultural and Food Chemistry, 32*, 510–512.

Adisakwattana, S., Ruengsamran, T., Kampa, P., & Sompong, W. (2012). *In vitro* inhibitory effects of plant-based foods and their combinations on intestinal α-glucosidase and pancreatic α-amylase. *BMC Complementary and Alternative Medicine, 2*, 110. https://doi.org/10.1186/1472-6882-12-110

Akim, A., Lim, C. H., Asmah, R., & Zanainaul, A. Z. (2011). Antioxidant and anti-proliferative activities of Roselle juice on Caov3, MCF-7, MDA-MB-231 and Hela cancer cell lines. *Journal of Pharmacy and Pharmacology, 5*(7), 957–965.

Aldeen, B., Almontaser, M., Ahmed, M., Ahmed, M., & Elshafie, M. (2015). Chemical and physicochemical characteristics of roselle seeds oil (*Hibiscus sabdariffa* L.). B.Sc. thesis paper. Sudan University of Science and Technology College of Agricultural Studies, Department of Food Science and Technology.

Alegbejo, M. D., Abo, M. E., & Alegbejo, J. O. (2003). Current status and future potential of roselle production and utilization in Nigeria. *Journal of Sustainable Agriculture, 23*(2), 5–16. https://doi.org/10.1300/J064v23n02_03

Ali, K. M., Ashraf, A., Biswas, N. N., Karmakar, U., & Afroz, S. (2011). Antinociceptive, anti-inflammatory and antidiarrheal activities of ethanolic calyx extract of *Hibiscus sabdariffa* Linn. (Malvaceae) in mice. *Journal of Chinese Integrative Medicine, 9*(6), 626–631.

Amoasah, B., Appiah, F., & Kumah, P. (2018). Effects of different drying methods on proximate composition of three accessions of roselle (*Hibiscus sabdariffa*) calyces. *International Journal of Plant & Soil Science, 21*(1), 1–8.

Ariga, N. (2008). Environmental Impact of Organic Compounding Ingredients and Future Problems. *International Polymer Science and Technology, 35*(7), 170–175.

Atanda, A., Shuaibu Bawa, G., Duru, S., & Ojariafe Abeke, F. (2018). Nutritive evaluation of boiling duration on the anti-nutrients and amino acids composition of roselle seeds. *Asian Journal of Advances in Agricultural Research, 6*(3), 1–7.

Atta, M. B. (2003). Some characteristics of nigella *(Nigella sativa L.)* seed cultivated in Egypt and its lipid profile. *Food Chemistry, 83*(1), 63–68. https://doi.org/10.1016/S0308-8146(03)00038-4

Augustine, L. F., Shahnaz, V., Fernandez, R. S., Rao, V. V. M., Laxmaiah, A., & Nair, K. M. (2011). Perceived stress, life events & coping among higher secondary students of Hyderabad, India: A pilot study. *Indian Journal of Medical Research, 134*(1), 61–68.

Babalola, S. O. (2000). Chemical analysis of roselle leaf (*Hibiscus sabdariffa*). In *Proceeding of 24th annual conference of NIFST* (pp. 228–229).

Babalola, S. O., Babalola, A. O., & Aworh, O. C. (2001). Compositional attributes of the calyces of roselle (*Hibiscus sabdariffa*). *Journal of Food Technology in Africa, 6*(4), 133–134.

Bako, I. G., Abubakar, M. S., Mabrouk, M. A., & Mohammed, A. (2014). Lactogenic study of the effect of ethyl-acetate fraction of *Hibiscus sabdariffa* L. (Malvaceae) seed on serum prolactin level in lactating albino rats. *Advance Journal of Food Science and Technology, 6*(3), 292–296.

Balarabe, M. A. (2019). Nutritional analysis of *Hibiscus sabdariffa* L. (roselle) leaves and calyces. *Plant, 7*(4), 62–65.

Boudraa, G., Touhami, M., Pochart, P., Soltana, R., Mary, J. Y., & Desjeux, J. F. (1990). Effect of feeding yogurt versus milk in children with persistent diarrhea. *Journal of Pediatric Gastroenterology and Nutrition, 11*, 509–512.

Chen, H., Tsai, P. J., Chen, S. H., Su, Y. M., Chung, C. C., & Huang, T. T. C. (2005). Grey relational analysis of dried roselle (*Hibiscus sabdariffa* L.). *Journal of Food Processing and Preservation, 29*(3-4), 228–245. https://doi.org/10.1111/j.1745-4549.2005.00025.x

Cissouma, A. I., Tounkara, F., Nikoo, M., Yang, N., & Xu, X. (2013). Phyico chemical properties and antioxidant activity of roselle seed extracts. *Advance Journal of Food Science and Technology, 5*(11), 1483–1489.

Crane, J. C. (1949). Roselle—a potentially important plant fiber. *Economic Botany, 3*, 89–103.

Crawford, J. M., Harden, N., Leung, T., Lim, L., & Kiehart, D. P. (1998). Cellularization in Drosophila melanogaster is disrupted by the inhibition of rho activity and the activation of cdc42 function. *Dev. Biol., 204*(1), 151–164.

Desrosier, N. W. (1970). *The Technology of Food Preservation* (pp. 47–50). West Port Connecticut: AV Publ. Co Inc.

Duke, J. (1983). *Hibiscus sabdariffa L. Handbook of energy crops, Center for New Crops & Plants Products*. Purdue University, West Lafayette, Indiana.

Duke, J. A., & Atchley, A. A. (1984). Proximate analysis. In B. R. Christie (Ed.), *The handbook of plant science in agriculture*. Boca Raton, FL: CRC Press, Inc.

Durlach, J. (1988). *Magnesium in clinical practice*. London, Paris: Libbey Eurotext.

Dy Phon, P. (2000). *Dictionary of plants used in Cambodia* (1st ed.). Phnom Penh, Cambodia: Imprimerie Olympic.

El-Adawy, T. A., & Khalil, A. H. (1994). Characteristics of roselle seeds as a new source of protein and lipid. *Journal of Agricultural and Food Chemistry, 42*, 1896–1900.

El-Deab, S. M., & Ghamry, H. E. (2017). Nutritional evaluation of roselle seeds oil and production of mayonnaise. *International Journal of Food Science and Nutrition Engineering, 7*(2), 32–37.

Elneairy, N. A. (2014). Comparative studies on Egyptian and Libyan roselle seeds as a source of lipid and protein. *Food and Nutrition Sciences, 5*, 2237–2245.

Eslaminejad, T., & Zakaria, M. (2011). Morphological characteristics and pathogenicity of fungi associated with roselle *(Hibiscus Sabdariffa)* diseases in Penang, Malaysia. *Microbial Pathogenesis, 51*(5), 325–337.

Esselen, W. B., & Sammy, G. M. (1975). Applications for roselle as a red food colorant. *Food Production and Development, 9*(8), 37–40.

Fakeye, T. O., Adisa, R., & Musa, I. E. (2008). Attitude and use of herbal medicines among pregnant women in Nigeria. *BMC Complementary and Alternative Medicine, 9*(1), 53. https://doi.org/10.1186/1472-6882-9-53

Fellows, P. J., & Axtell, B. (Eds.). (2014). *Opportunities in Food Processing: A handbook for setting up and running a smallscale business producing high-value foods*. ACP-EU Technical Centre for Agricultural and Rural Cooperation (CTA) (p. 454). Wageningen, The Netherlands.

Fleck, H., & Munves, E. (1962). *Introduction to nutrition*. New York: Macmilan Publishing Co. Inc.

Francis, F. J. (1990). Colour analysis. In N. N. Nielsen (Ed.), *Food analysis* (pp. 599–612). Gaithersburg, MD, USA: Aspen Publ.

Gangrade, H., Mishra, S. H., & Kaushal, R. (1979). Antimicrobial activity of the oil and unsaponifiable matter of red roselle. *Indian Drugs, 16*, 147–148.

Gatti, M., Tokalty, I., & Rubio, A. (2010). Sodium A charge transfer insulator at high pressures. *Physical Review Letters, 104*(21), 404–414.

Gheller, A. C. G. V., Kerkhoff, J., Júnior, G. M. V., Eduardo de Campos, K., & Sugui, M. M. (2017). Antimutagenic effect of *Hibiscus sabdariffa* L. aqueous extract on rats treated with monosodium glutamate. *Scientific World Journal*, 1–8. https://doi.org/10.1155/2017/9392532

Ghosh, S., Singh, A., Mandal, S., & Mandal, L. (2015). Active hematopoietic hubs in Drosophila adults generate hemocytes and contribute to immune response. *Developmental Cell, 33*(4), 478–488.

Glew, R. H., VanderJagt, D. J., Lockett, C., Grivetti, L. E., Smith, G. C., Pastuszyn, A., & Millson, M. (1997). Amino acid, fatty acid, and mineral composition of 24 indigenous plants of Burkina Faso. *Journal of Food Composition and Analysis, 10*, 205–217.

Haji Faraji, M., & Haji Tarkhani, A. (1999). The effect of sour tea (*Hibiscus sabdariffa*) on essential hypertension. *Journal of Ethnopharmacology, 65*, 231–236.

Halimatul, S. M. N. (2007). Protein quality of roselle (*Hibiscus sabdariffa* L.) seeds. *ASEAN Food Journal, 14*(2), 131–140.

Hamidi, A. E. I., Saleh, M., & Ahmed, S. S. (1966). Investigation of *Hibiscus sabdariffa*. *Journal of Chemistry. UAR, 9*(1), 127.

Herrera-Arellano, A., Flores-Romero, S., Chávez-Soto, M. A., & Tortoriello, J. (2004). Effectiveness and tolerability of a standardized extract from *Hibiscus Sabdariffa* in patients with mild to moderate hypertension: A controlled and randomized clinical trial. *Phytomedicine, 11*(5), 375–382. https://doi.org/10.1016/j.phymed.2004.04.001

Heyman, M. (2000). Effect of lactic acid bacteria on diarrheal diseases. *Journal of the American College of Nutrition, 19*, 137S–146S.

Hopkins, A. L., Lamm, M. G., Funk, J. L., & Ritenbaugh, C. (2013). *Hibiscus sabdariffa* L. in the treatment of hypertension and hyperlipidemia: A comprehensive review of animal and human studies. *Fitoterapia, 85*, 84–94.

Hussein, M. A., Mohammed, A. A., & Atiya, M. A. (2019). Application of emulsion and Pickering emulsion liquid membrane technique for wastewater treatment: an overview. *Environmental Science and Pollution Research, 26*, 36184–36204. https://doi.org/10.1007/s11356-019-06652-3

Hutagaol, E. V. (2017). *Peningkatan Kualitas Hidup Pada Penderita Gagal Ginjal Kronik Yang Menjalani Terapi Hemodialisa Melalui Psychological Intervention Di Unit Hemodialisa Rs Royal Prima Medan Tahun 2016. Jurnal Universitas Prima Indonesia Meda, 2*, 18.

Ibrahhim, M. E. H., Karamalla, K. A., & Khattab, A. H. (1971). Biochemical studies on Karkade (roselle). (*Hibiscus sabdariffa*). *Journal of Food Science and Technology, 3*(1), 37–39.

Inggrid, H. M., Jaka, & Santoso, H. (2017). Natural red dyes extraction on roselle petals. *IOP Conference Series: Materials Science and Engineering, 162*.

Islam, A. K. M. A., Jamini, T. S., Islam, A. K. M. M., & Yeasmin, S. (2016). Roselle: A functional food with high nutritional and medicinal values. *Fundamental and Applied Agriculture, 1*(2), 44–49.

Ismail, A., Ikram, E. H. K., & Nazri, H. S. M. (2008). Roselle (*Hibiscus sabdariffa* L.) seeds-nutritional composition, protein quality and health benefits. *Food, 2*, 1–16.

Iwalokun, B. A., & Shittu, M. O (2007). Effect of *Hibiscus Sabdariffa* (Calyce) Extract on Biochemical and Organoleptic Properties of Yogurt. *Pakistan Journal of Nutrition, 6*(2), 172–182.

Jamini, T. S., Aminul Islam, A. K. M., Saikat, M. M. H., & Mohi-ud-Din, M. (2019). Phytochemical composition of calyx extract roselle (*Hibiscus sabdariffa* L) genotypes. *Journal of Food Technology and Food Chemistry, 2*, 102. https://doi.org/10.4314/jafs.v16i1.2

Juliani, H. R., Welch, C. R., Wu, Q., Diouf, B., Malainy, D., & Simon, J. E. (2009). Chemistry and quality of hibiscus (*Hibiscus sabdariffa*) for developing the natural product industry in Senegal. *Journal of Food Science, 74*, 113–121.

Khattak, K. F., Simpson, T. J., & Ihasnullah, F. (2008). Effect of gamma irradiation on the extraction yield, total phenolic content and free radical scavenging activity of Nigella sativa seed. *Food Chemistry, 110*, 967–972.

Khatun, A., Rashid, M. H., & Rahmatullah, M. (2011). Scientific validation of eight medicinal plants used in traditional medicinal systems of Malaysia: A review. *American-Eurasian Journal of Sustainable Agriculture, 5*(1), 67–75.

Kilima, B. M., Remberg, S. F., Chove, B. E., & Wicklund, T. (2014) Physio chemical, mineral [illegible] and antioxidant properties of roselle (*Hibiscus Sabdariffa* L.) extract blended with tropical fruit juices. *African Journal of Food, Agriculture, Nutrition and Development, 14*(3), 8963–8978.

Kim, S. H., Joo, M. H., & Yoo, S. H. (2009). Structural identification and antioxidant properties of major anthocyanins extracted from omija (*Schizandra chinensis*) fruit. *Journal of Food Science, 74*(2), 134–140.

Kinsella, J. E. (1987). *Sea foods and human health and diseases*. New York and Basel: Marcel Dekker Inc.

Lila, M. A. (2004). Anthocyanins and human health: An in vitro investigative approach. *Journal of Biomedicine and Biotechnology, 5*, 306–313.

Liu, J. Y., Chen, C.-C., Wang, W.-H., Hsu, J.-D., Yang, M.-Y., & Wang, C.-J. (2006). The protective effects of *Hibiscus sabdariffa* extract on CCl4-induced liver fibrosis in rats. *Food and Chemical Toxicology, 44*(3), 336–343.

Luvonga, W. A., Njoroge, M. S., Mokokha, A., & Ngunjiri, P. W. (2009). Chemical characterization of *Hibiscus sabdariffa* (roselle) calyces and evaluation of its functional potential in the food industry. Nairobi, Kenya: Jomo Kenyatta University of Agriculture and Technology.

Mady, C., Manuel, D., Mama, S., Augustin, N., & Max, R. (2009). The bissap (*Hibiscus sabdariffa* L.): Composition and principal uses. *Fruits, 64*, 179–193.

Mardiah, M., Zakaria, F. R., Prangdimurti, E., & Damanik, R. (2015). Anti-inflammatory of Purple Roselle Extract in Diabetic Rats Induced by Streptozotocin. *Procedia Food Science, 3*, 182–189. https://doi.org/10.1016/j.profoo.2015.01.020

Mat Isa, A., Isa, P. M. M., & Aziz, R. A. (1985). Chemical analysis and roselle processing (*Hibiscus sabdariffa* L.). *MARDI Research Bulletin, 13*, 68–74.

Mazza, G., & Miniati, E. (2000). *Anthocyanin in fruits, vegetables and grains*. Boca Raton, FL, USA: CRC Press.

Mclean, K. (1973). Roselle (*Hibiscus sabdariffa* L.) as a cultivated edible plant. In *UNDP/FAO project SUD/70/543*. Sudan Food Research Centre Khartoum North.

Meraiyebu, A. B., Olaniyan, O. T., Eneze, C., Anjorin, Y. D., & Dare, J. B. (2013). Anti-inflammatory activity of methanolic extract of *Hibiscus sabdariffa* on carrageenan induced inflammation in wistar rat. *International Journal of Pharmaceutical Science Invention, 2*(3), 22–24.

Mohamed, B. B., Sulaiman, A. A., & Dahab, A. A. (2012). Roselle (*Hibiscus sabdariffa* L.) in Sudan, Cultivation and Their Uses. *Bulletin of Environment pharmacology and life sciences, 1*(6), 48–54.

Mohamed, R., Fernandez, J., Pineda, M., & Aguilar, M. (2007). Roselle (*Hibiscus sabdariffa*) seed oil is a rich source of γ –tocopherol. *Journal of Food Science, 72*(3), S207–S211.

Mollah, M. A. F., Tareq, M. Z., Bashar, K. K., Zahidul Hoque, A. B. M., Karim, M. M., & Rafiq, M. Z. A. (2020). Antioxidant properties of BJRI vegetable mesta-1 (*Hibiscus sabdariffa* L.). *Plant Science Today, 7*(2), 154–156.

Morton, J. (1987). Roselle. In *Fruits of Warm Climates* (pp. 281–286). Miami, FL: Web Publications Purdue University.

Nasrabadi, Z. M., Zarringhalami, S., & Ganjloo, A. (2018). Evaluation of chemical, nutritional and antioxidant characteristics of roselle (*Hibiscus sabdariffa* L.) seed. *Nutrition and Food Sciences Research, 5*(1), 41–46.

Nof. 2016. National Osteoporosis Foundation, 251 18th Street S, Suite 630, Arlington, VA, 22202-1 (800): 231-4222.

Nyam, K. L., Leao, S. Y., Tan, C. P., & Long, K. (2014). Functional properties of roselle (*Hibiscus sabdariffa* L.) seed and its application as bakery product. *Journal of Food Science and Technology, 51*, 3830–3837. https://doi.org/10.1007/s13197-012-0902-x

Nyam, K. L., Tan, C. P., Lai, O. M., Long, K., & Che Man, Y. B. (2009). Some physicochemical properties and bioactive compounds of seed oils. *LWT, 42*, 1396–1403.

Nyam, K. V., Sin, L. N., & Kamariah, L. (2015). Phytochemical analysis and anti-inflammatory effect of kenaf and roselle seeds. *Malaysian Journal of Nutrition, 22*(2), 245–254.

Ojokoh, A. O., Adetuye, F. A., Akiuyosoye, E., & Oyetayo, V. O. (2003). Fermentation studies on roselle (*Hibiscus sabdariffa*) calyces neutralized with trona. In *Proceeding of 16th annual conference of Biotechnology society of Nigeria* (pp. 90–92).

Okasha, M. A. M., Abubakar, M. S., & Bako, I. G. (2008). Study of the effect aqueous *Hibiscus sabdariffa* L. seed extract on serum prolactin level in lactating albino rats. *European Journal of Scientific Research, 22*(4), 575–583.

Patel, S. (2014). Hibiscus sabdariffa: an ideal yet under-exploited candidate for nutraceutical applications. *Biomedicine and Preventive Nutrition, 4*(1), 23–27.

Peng, C. H., Chyau, C. C., Chan, K. C., Chan, T. H., Wang, C. J., & Huang, C. N. (2011). *Hibiscus sabdariffa* polyphenolic extract inhibits hyperglycemia, hyperlipidemia, and glycation-oxidative stress while improving insulin resistance. *Journal of Agricultural and Food Chemistry, 59*(18), 9901–9909.

Prabhakaran, G., Bhore, S. J., & Ravichandran, M. (2017). Development and Evaluation of Poly Herbal Molluscicidal Extracts for Control of Apple Snail *(Pomacea maculata)*. *Agriculture, 7*(3), 22. https://doi.org/10.3390/agriculture7030022

Purseglove, J. W. (1974). *Tropical crops: dicotyledons* (pp. 242–246). London: Longman.

Rao, P. U. (1996). Nutrient composition and biological evaluation of mesta (*Hibiscus sabdariffa*) seeds. *Plant Foods for Human Nutrition, 49*, 27–34.

Riaz, R. A. (1969). Peel in extinction from roselle sepals (*Hibiscus sabdariffa* L.). *Science and Industry, 5*(3), 435–441.

Rimamcwe, K. B., Chavan, U. D., Kotecha, P. M., & Lande, S. B. (2016). Physical properties and nutritional potentials of indian roselle (*Hibiscus sabdariffa* L.) seeds. *International Journal of Current Research, 8*(9), 38644–38648.

Romani, A. M. P. (2013). Magnesium in health and disease. In Astrid Sigel, Helmut Sigel, & Roland K. O. Sigel (Eds.), *Metal ions in life sciences: Vol. 13. Interrelations between essential metal ions and human diseases* (pp. 49–79).

Rosemary, R., & Haro, G. (2014). Antidiabetic effect of roselle calyces extract (*Hibiscus sabdariffa* L.) in streptozotocin induced mice. *International Journal of PharmTech Research, 6*(5), 1703–1711.

Sarkiyayi, S., & Ikioda, H. (2009). Estimation of thiamin and ascorbic acid contents in fresh and dried *Hibiscus sabdariffa* (roselle) and *Lactuca sativa* (tettuce). *Advance Journal of Food Science and Technology, 2*(1), 47–49.

Sathivel, S., Bechtel, P. J., Babbitt, J., Smiley, S., Crapo, C., & Reppond, K. D. (2003). Biochemical and functional properties of herring (*clupea harengus*) byproduct hydrolysates. *Journal of Food Science, 68*, 2196–2200.

Sayago-ayerdi, S. G., Arranz, S., Serrano, J., & Goni, I. (2007). Dietary fiber content and associated antioxidant compounds in roselle flower (*Hibiscus sabdariffa* L.) beverage. *Journal of Agricultural and Food Chemistry, 55*(19), 7886–7890.

Shaheen, M. A., El-Nakhlawy, F. S., & Al-Shareef, A. R. (2012). Roselle (*Hibiscus sabdariffa* L.) seeds as unconventional nutritional source. *African Journal of Biotechnology, 11*(41), 9821–9824.

Shahnaz, A., Atiq-Ur-Rahman, Qadiraddin, M., & Shanim, Q. (2003). Elemental analysis of Calendula officinalis plant and its probable therapeutic roles in health. *Pakistan Journal of Scientific and Industrial Research, 46*, 283–287.

Shoosh, W. G. A. A. (1993). Chemical composition of some Roselle *(Hibiscus sabdariffa L.)* genotypes (pp. 1–109). Sudan: Department of Food Science and Technology, Faculty of Agriculture, University of Khartoum.

Singh, P., Khan, M., & Hailemariam, H. (2017). Nutritional and health importance of *Hibiscus sabdariffa*: A review and indication for research needs. *Journal of Nutritional Health & Food Engineering, 6*(5), 125–128.

Siong, T. E., Noor, M. I., Azudin, M. N. and Idris, K. 1988. Nutrient composition of Malaysian foods. 299

Sleelig, G. F. (1980). Noncatalytic subunits of human blood plasma coagulation factor XIII. Preparation and partial characterization of modified forms. *Journal of Biological Chemistry, 255918*, 8881–8886.

Taylor, G. R. J., & Williams, C. M. (1998). Effects of probiotics and prebiotics on blood lipids. *British Journal of Nutrition, 80*, S225–S230.

Tounkara, F., Sodio, B., Chamba, M. V. M., Le, G., & Shi, Y. (2014). Nutritional and functional properties of roselle (*Hibiscus sabdariffa* L.) seed protein hydrolysates. *Emirates Journal of Food and Agriculture, 26*(5), 409–417.

Tsai, P. J., Mclntosh, J., Pearce, P., & Camben, B. (2002). Anthocyanin and antioxidant capacity in roselle (*Hibiscus sabdariffa* L.) extract. *Food Research International, 35*(4), 351–356.

Tsai, J., & Ou, M. (1996). Colour degradation of dried roselle during storage. *Food Science, 23*, 629–640.

Vangalapati, B., Poornima, M., & Anupama, H. (2014). Total phenolic content and free radical scavenging activity of Pterocarpus marsupium heartwood & Tribulus terrestris dry fruits: An in vitro comparative study. *Journal of Pharmacy Research, 8*(5), 610–613.

Wang, C. J., Wang, J. M., Lin, W. L., Chu, C. Y., Chou, F. P., & Tseng, T. H. (2000). Protective effect of Hibiscus anthocyanins against tert-butyl hydroperoxide-induced hepatic toxicity in rats. *Food and Chemical Toxicology, 38*, 411–416.

Wong, P. K., Salmah, Y., Ghazali, H. M., & Che, M. (2002). Physico-chemical characteristics of roselle (*Hibiscus sabdariffa* L.). *Nutrition & Food Science, 32*(2), 68–73.

Yamada, T., Hida, H., & Yamada, Y. (2007). Chemistry, physiological properties, and microbial production of hydroxycitric acid. *Applied Microbiology and Biotechnology, 75*(5), 977–982.

Zhu, Y., Bo, Y., Wang, X., Lu, W., & Wang, X. (2016). The effect of anthocyanins on blood pressure: A PRISMA-compliant meta-analysis of randomized clinical trials. *Medicine, 95*, e3380.

CHAPTER 8

Roselle (*Hibiscus sabdariffa* L.) in Sudan: Production and Uses

BAHAELDEEN BABIKER MOHAMED

1 INTRODUCTION

Roselle belongs to the family Malvaceae and is an important perennial shrub that grows in tropical and subtropical regions. In Sudan, where it is considered as an annual shrub, it is grown in rain-fed areas in the western states in large scale and is cultivated in small scale in irrigated areas (Abbas et al., 2019). It is an annual crop grown successfully in tropical and subtropical climates (Copley, 1975). The commercially important part of the plant is the fleshy calyx (sepals) surrounding the fruit (capsules). The whole plant can be used as beverage, or the dried calyces can be soaked in water to prepare a colorful cold drink or may be boiled in water and taken as a hot drink (N.B.A.P, 1999). The seeds contain 17.8%–21% nonedible oil (Ahmed, 1980) and 20% protein, and they are sometimes used for animal feed (Ahmed & Nour, 1981). Roselle is a flexible plant with a number of uses. It is intercropped with crop staples such as sorghum and sesame or planted along field margins. It requires little care. Its leaves, seeds, capsules, and stems are used in traditional medicines. In rural areas, women are usually responsible for growing roselle. They add value to the crop by developing products for market (McClintock, 2004). Mclean (1973) and Wilson and Menzel (1964) reported that *Hibiscus sabdariffa* is a tetraploid ($2n = 4x = 72$), whose chromosomes are related to the diploid ($2n = 2x = 36$) *Hibiscus cannabinus*. The two botanical types of roselle documented by Purseglove (1974) are *H. sabdariffa* var *sabdariffa*, grown for its fleshy, shiny red calyx, and *H. sabdariffa* var *altissima*, grown for its phloem fiber. Despite its potential economic importance, karkade has received little attention and there is a lack of information regarding its genetics, breeding, and production, particularly under rain-fed conditions.

Roselle may have been domesticated in western Sudan before 4000 BCE (Wilson & Menzel, 1964). It was first recorded in Europe in CE 1576. It seems to have been carried from Africa to the New World by slaves for use as a food plant. Roselle was called Jamaican sorrel in 1707 in Jamaica, where the regular use of the calyces as food seems to have been first practiced. The use of the plant as "greens" was known in Java as early as 1658 (Lainbourne, 1913; Wester, 1912). Taken to the New World, roselle was cultivated in Mexico, parts of Central America, the West Indies, and southern Florida, Texas, and California in the late 19th century. It is now grown for culinary purposes in much of the tropical world. The use of *H. sabdariffa* for fiber seems to have developed in regions other than Africa (Wilson & Menzel, 1964).

2 ROSELLE CLASSIFICATION

H. sabdariffa L. belongs to the family Malvaceae. It is an annual plant of 64–429 cm average height (see Fig. 8.1), having more than 300 species with the basic chromosomal number $x = 18$ ($4n = 72$) (Mohamed et al., 2012). It originated in Sudan (Africa), growing mostly in tropical and semitropical regions. Roselle is tetraploid, the chromosomes of *H. sabdariffa* are associated with the diploid, i.e., $2n = 2x = 36$, *H. cannabinus* (Wilson & Menzel, 1964). The two distinct botanical varieties of roselle are *H. sabdariffa* var *sabdariffa*, which is grown for its shiny, fleshy red calyx, and *H. sabdariffa* var *altissima* (Purseglove, 1968).

3 ROSELLE COMMON NAMES

H. sabdariffa L. is known in different countries by various common names (English name: red sorrel, roselle; Arabic name: karkade). In English-speaking countries, it is known as roselle, Jamaican sorrel, red sorrel, Indian sorrel, rozelle hemp, natal sorrel, and rosella. The Japanese name is rohzelu, sabdriqa or lalambari

Roselle. https://doi.org/10.1016/B978-0-323-85213-5.00018-4

FIG. 8.1 Roselle plant.

in Urdu, and lal-ambari, patwa, or laalambaar in Hindi (Kays, 2011). It is also known as roselle, razelle, sorrel, red sorrel, Jamaican sorrel, Indian sorrel, Guinea sorrel, sour-sour, and Queensland jelly plant (Mahadevan & Shivali, 2009; Morton, 1987a,b). In French, roselle is also the word for the red-winged thrush. In Switzerland, the edible calyx is called karkadé. The roselle fiber is called India rosella hemp, rosella fiber, rosella hemp, or Pusa hemp. Vernacular names for roselle include rozelle, jelly okra, lemon bush, and Florida cranberry (Small, 1997).

4 USES

Roselle is a multi use plant, and its outer leaves (calyx), also known as natal sorrel (Ageless, 1999), are frequently used in the production of jelly, jam, juice, wine, syrup, gelatin, pudding, cake, ice cream, and flavoring agents. Its brilliant red color and unique flavor make it a valuable food product (Tsai and Ou (1996). Roselle is an annual crop used in food, animal feed, nutraceuticals, cosmeceuticals, and pharmaceuticals. The calyces, stems, and leaves are acidic in flavor. The juice from the calyces is claimed to be a health-enhancing drink because of its high content of vitamin C, anthocyanins, and other antioxidants Mohamad et al. (2002).

Roselle is grown mainly for its fleshy calyx (sepals), which is a good source of natural antioxidants (anthocyanin and protocatechuic acid) that protect cell membranes from damage by free radicals and lipid peroxidation. Roselle is used to cure a number of diseases such as high blood pressure, liver disorders, fever, urinary tract infection, and muscular pain. Other plant parts are used as raw material for various foods. Roselle contains essential amino acids and anthocyanins (see Table 8.1).

In the countryside, accountability goes to women for caring and growing roselle. The crop is valued by them for developing market products (McClintock, 2004).

4.1 Roselle Drink

Dry calyces are used to produce a flavorsome and healthy drink (see Fig. 8.2) and are also used in making tea, jelly, marmalade, ices, ice cream, sorbets, butter, pies, sauces, tarts, and other desserts (Duke & Ayensu, 1985). The seeds have also been used as an aphrodisiac coffee substitute.

TABLE 8.1
Concentration of Essential Amino Acids and Anthocyanin Percentage (%) in *Hibiscus sabdariffa* L. Grown in Shambat, Sudan.

Essential Amino Acids	Concentration (μg/mL)	Anthocyanins	Percentage (%)
Threonine	38.96	Malvidin	23.45
Valine	21.65	Pelargonidin	23.97
Methionine	07.49		
Leucine	15.70		
Isoleucine	40.50		
Phenylalanine	35.05		
Histidine	32.90		
Lycine	24.73		
Argnine	20.15		

Source: Abbas et al. (2019).

4.2 Roselle Tea

Tea produced from dry calyces by boiling them in water is widely used in Sudan. Roselle mostly grown under rain fed in sudan and known as an organic product and is highly valued for its beneficial effects (see Table 8.1). Tea made from roselle flowers is widely exported for use in making a sweet herbal tea and is also commonly sold on the domestic market. The dried flowers are common in some countries. The drink made from the calyces has a mild diuretic and purgative effect, among many other effects. The drink is said to be a folk remedy for cancer. Restored roselle drink has no bacterial isolate (El-Sherif & Sarwat, 2007). Ageless (1999) reported that the benefits of taking herbal tea include soothing colds, clearing a blocked nose, clearing mucus, as an astringent, promoting kidney function, aiding digestion, as a general tonic, as a diuretic, and helping reduce fever (Fig. 8.3).

FIG. 8.2 Roselle drink.

FIG. 8.3 Roselle tea (https://elnasrltd.com).

5 MEDICINAL AND INDUSTRIAL APPLICATIONS

Many medicinal applications of the roselle plant have been developed around the world. Roselle is used to treat hypertension, pyrexia, and liver damage, and in ayurvedic medicine, it has been reported that the aqueous extract of *H. sabdariffa* L. attenuates hypertension and reverses cardiac hypertrophy (Odigie et al., 2003). The calyx extract has been used as an effective treatment against leukemia owing to its high content of polyphenols, particularly protocatechuic acid (Tseng et al., 2000). Roselle seeds, which until now have not had any commercial applications, are a source of a vegetable oil that is low in cholesterol and rich in other phytosterols and tocopherols, particularly β-sitosterol and γ-tocopherol. The overall characteristics of roselle seed oil allow for important industrial applications and represent added value for roselle cultivation (Mohamed et al., 2007). The alcoholic extract of *H. sabdariffa* L. has an anti-hyperammonemic and antioxidant effect on brain tissues (Essa et al., 2007).

FIG. 8.4 Roselle flowers.

6 ROSELLE DESCRIPTION

Roselle is an annual erect shrub that takes 5 months from planting to harvesting; it can also be regarded as a perennial (Bailey & Bailey, 1976). Species grown for their fiber are tall, with fewer branches, sometimes growing to more than 3–5 m in height. Culinary varieties are many-branched, bushy, and generally 1–2 m tall (Wester, 1920). Stems may be green or red, depending on the seed source. Roselle has a strong taproot. The young plants have leaves that are unlobed, but as the plant grows the later-developing leaves are shallowly to deeply palmate and three- or five-parted (sometimes seven-parted) (see Fig. 8.4). The large flowers have pale yellow petals (sometimes suffused with pink or red) and a dark red eye (Cooper, 1993).

The flowers are usually borne singly in the leaf axils. The sepals at the base of the large flowers and fruit vary from dark purple to bright red (sometimes white) at maturity and are quite fleshy. The calyx increases from 1 to 2 cm in length before the flower is fertilized, then to about 5.5 cm (occasionally longer) at maturity. Some forms of roselle contain a pigment that gives a brilliant red color to culinary products made from the plant; other forms are completely green. Edible types of roselle are usually succulent, have well-developed lateral branches, and lack a hairy covering (Wilson & Menzel, 1964).

Flowering is induced as the days become shorter and the light intensity decreases, beginning in September or later depending on the country. Flowers are red to yellow, with a dark center containing short peduncles (Fig. 8.5), and have both male and female organs. The seed pods begin ripening near the bottom and proceed to the top (Fig. 8.6). In Sudan, growers sometimes allow the seed to completely ripen and let the leaves drop prior to harvest (Plotto et al., 2004).

7 CLIMATE AND PLANTING

Roselle requires a monthly rainfall ranging from 130 to 250 mm in the first 3–4 months of growth. Dry weather is well tolerated and is desirable in the later months of growth. Rain or high humidity at harvest and drying times can downgrade the quality of the calyces and reduce the yield.

Roselle is very sensitive to changes in the length of day. This photoperiodism requires the planting time to be set according to the length of the day rather than rainfall requirements. It is a deep-rooted crop; therefore deep plowing is recommended in preparing the seedbed. Seeds are planted at a rate of 6–8 kg/ha and approximately 2.5 cm deep. Seeds are usually planted at the beginning of the rainy season, with 60 cm–1 m space between rows and 45–60 cm apart. Reduced planting rate produces a larger calyx. Sowing is done by hand or using a modern grain drill. A good alternative tool would be a corn planter small enough to accommodate the hibiscus seeds. Thinning is also

FIG. 8.5 Mature flower.

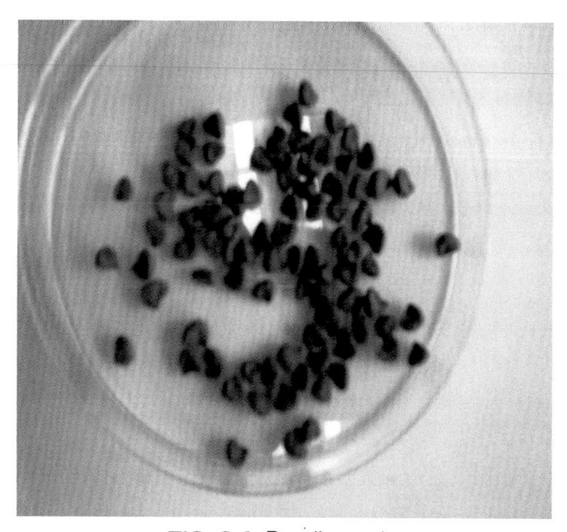

FIG. 8.6 Roselle seeds.

done by hand. Plotto et al. (2004) reported that there are over 100 commercial cultivars of roselle grown mainly in China, Thailand, Mexico, Sudan, Senegal, and Mali.

8 HARVEST AND STORAGE

Roselle is harvested from late November onward. The harvest is timed according to the ripeness of the seed. The fleshy calyces are harvested after the flower has dropped but before the seed pod has dried and opened. The longer the capsule remains on the plant after the seeds begin to ripen, the more susceptible the calyx is to disease and sun cracking (Plotto et al., 2004).

The calyces ripen about 3 weeks after the start of flowering, which is 100–160 days after the plants are transplanted outdoors (Duke & duCellier, 1993). The fruit ripens progressively from the bottom of the plant to the top. Harvesting is carried out by intensive hand labor, with the calyces being picked singly at the appropriate stage. The fruit may be harvested when fully grown but still tender, when they can be easily snapped off by hand; later harvesting requires clippers (Morton, 1987a, b). The fruit is easier to break off in the morning than at the end of the day. On average, each fruit yields about 7–10 g of sepals (Duke, 1993). Drying is the traditional method for preserving foods. Roselle drying is done in one of the two ways: by harvesting the fresh fruits and then sundrying the calyces or by leaving the fruit to partially dry on the plants and harvesting the dried fruit, keeping the crop well protected during the process. Dehydration depends on the two fundamental processes of heat transfer (heat is transferred into the fruit) and mass transfer (subsequent removal of moisture from it) (Potter & Hotchkiss, 1995). In Sudan, the fully developed fleshy calyx is peeled from the fruit by hand and dried naturally in shade (see Fig. 8.7).

9 PRODUCTION

Roselle is cultivated in several parts of Sudan, mostly in the western (Darfur and Kordofan) states. Traditional farmers cultivate it as one of the cash crops under rainy conditions, where bulk quantities are produced for both export and local consumption. The total cultivated area was estimated at 290,000 feddans (around 121,800 ha) in the 2000/2001 season as compared with 22,300–78,444 feddans (around 9370–32,950 ha) in the 1970s and 47,998–59,882 feddans (approximately 20,160–25,160 ha) in the 1980s (Mohamed et al., 2012).

Most breeding of roselle has been for its fiber yield (Duke, 1993). Sudan is presently the major producer of roselle; however, farmers regard it as a famine food. When drought is expected, farmers prefer to cultivate roselle rather than cereals because of its hardiness under adverse conditions (Mohamad et al., 2002). Roselle is

FIG. 8.7 **(A)** Manual harvester. **(B)** Dry calyces. (Source: Karkadi forum 2004.)

grown for its calyces, which are exported from Sudan, China, and Thailand, and it is also grown for its calyces in Mexico. In Sudan, it is collected by goat-herding nomadic tribes, but the product is frequently inferior because of poor processing conditions. Nevertheless, the Sudanese product is attractively bright red, very acidic, and extremely popular in Germany, which imports most of the crop.

Export price for the 1992–93 season for Sudanese, Chinese, and Thai roselle was of the order of $US1700.00/t, as reported by Duke and duCellier (1993). Karkade is grown in various parts of Sudan, particularly Kordofan and Darfur. It is one of the cash crops cultivated by traditional farmers in Kordofan and Darfur states under rain-fed conditions, where large quantities are produced for both local consumption and export. The total area under cultivation was estimated at 290,000 feddans (approximately 121,800 ha) in the 2000/2001 season, compared with 22,300–78,444 feddans (approximately 9370–32,950 ha) in the 1970s and 47,998–59,882 feddans (approximately 20,160–25,160 ha) in the 1980s. The increased area raised production from 454 tons in the 1960s to 26,000 tons in the 1999/2000 season (El-Awad, 2001).

In Sudan, especially in the North Kordofan state, the crop is grown mainly by traditional farming methods, exclusively under rain-fed conditions (El Naim & Ahmed, 2010). China and Thailand are also major producers and control much of the world's supply. Thailand has invested heavily in roselle production and its product is of superior quality, whereas China's product, with less stringent quality control practices, is less reliable and reputable. The world's best roselle comes from Sudan, but the quantity is low, and poor processing hampers quality. Mexico, Egypt, Senegal, Tanzania, Mali, and Jamaica are also important suppliers but production is mostly used domestically (Mohamad et al., 2002).

10 PEST CONTROL AND WEEDS

Major diseases of hibiscus are stem rot and root rot. Prevention techniques include monitoring the water content in an irrigated field and avoiding the planting of other crops that are also prone to these diseases. Insect damage is minor, but it does exist; pests include stem borer, flea beetles, abutilon moth, cotton bollworm, and cutworm. Mealy bugs and leafhoppers cause minor concerns, as is the cotton stainer. Plant enemies usually do not compete in a cultivated field (Plotto et al., 2004).

REFERENCES

Abbas, K. R., Fatma, S. E., & Awatif, A. M. (2019). Effect of maturity stages and cultivars on chemical constituents of Hibiscus sabdariffa (Roselle) grown in Sudan. *Elixir Bio Chem., 128*, 52801–52804.

Ageless The Trusted Herbal Anti-Aging. (1999). *Herbal remedies using Roselle (*Hibiscus sabdariffa*)*. http://www.ageless.co.za/rosella.htm.

Ahmed, A. K. (1980). *Karkade (*Hibiscus sabdariffa *L.) seed as new oilseed and a source of edible oil* (Ph.D. thesis). England: University of Reading.

Ahmed, A. H. R., & Nour, A. M. (1981). *Promising karkade seed derivatives: Edible oil and karkade*. Annual Report. Shambat, Sudan: Food Research Centre.

Bailey, L. H., & Bailey, E. Z. (1976). Revised by Staff of L.H. Bailey Hortorium. *Hortus third: A concise dictionary of plants cultivated in the United States and Canada* (Vol. 1). New York: MacMillan Publishing Co., 290 pp..

Cooper, B. (1993). The delightful *Hibiscus sabdariffa*. *Tea and Coffee Trade Journal, 165*(1), 100–102.

Copley, L. S. (1975). *An introduction to the botany of tropical crops*. U.K: Longman Group.

Duke, J. A. (1993). Medicinal plants and the pharmaceutical industry. In J. Janick, & J. E. Simon (Eds.), *New crops* (pp. 664–669). New York: John Wiley and Sons, Inc.

Duke, J. A., & Ayensu, E. S. (1985). *Medicinal plants of China* (Vol. 2). Algonac, MI, USA: Reference Publications, Inc.

Duke, J. A., & duCellier, J. L. (1993). *CRC handbook of alternative cash crops*. Boca Raton, FL, USA: CRC Press, 536 pp.

El Naim, A. M., & Ahmed, S. E. (2010). Effect of weeding frequencies on growth and yield of two roselle (*Hibiscus sabdariffa* L) varieties under rain fed. *Australian Journal of Basic and Applied Sciences, 4*(9), 4250–4255. ISSN 1991-8178.

El-Awad, H. O. (2001). *Roselle*. El-Obeid Research Station (in Arabic).

El-Sherif, M. H., & Sarwat, M. I. (2007). Physiological and chemical variations in producing roselle plant (*Hibiscus sabdariffa* L.) by using some organic farmyard manure. *World Journal of Agricultural Sciences, 3*(5), 609–616.

Essa, M. M., Subramanian, P., Suthakar, G., Manivasagam, T., Dakshayani, K. B., & Sivaperumal, R. (2007). *Hibiscus sabdariffa* L.(gungura) affect ammonium chloride- induced hyperammonemic rates. *Journal of Evidence-Based Integrative Medicine, 4*(3), 321–325.

Kays, S. J. (2011). *Cultivated vegetables of the world: A multilingual onomasticon* (p. 184). The Netherlands: University of Georgia. Wageningen Academic Publishers.

Lainbourne, J. (1913). The roselle. *Agricultural Bulletin, Straits and Federated Malay States, 2*, 59–67.

Mahadevan, N., & Shivali, K. P. (2009). *Hibiscus sabdariffa* Linn: An overview. *Natural Product Radiance, 8*, 77–83.

McClintock, N. (2004). *Roselle in Senegal and Mali. LEISA, Magazine on low external input and sustainable agriculture* (Vol. 20 (1)).

Mclean, K. (1973). *Roselle (*Hibiscus sabdariffa *L.), or karkade, as cultivated edible plants*. AG. S. SUD/70/543, Project Working Paper. Rome: FAO.

Mohamad, O., Mohd Nazir, B., Abdul Rahman, M., & Herman, S. (2002). *Roselle: A new crop in Malaysia. Bio Malaysia: A grand international biotechnology event*. Kuala Lumpur: Bulletin PGM.

Mohamed, B. B., Sulaiman, A. A., & Dahab, A. A. (2012). Roselle (*Hibiscus sabdariffa* L.) in Sudan, cultivation and their uses. *Bull. Environ. Pharmacol. Life Science, 1*, 48–54.

Mohamed, R., Fernadez, J., Pineda, M., & Aguilar, M. (2007). Roselle (*Hibiscus sabdariffa*) seed oil is a rich source of γ-tocopherol. *Journal of Food Science, 72*, 207–211.

Morton, F. J. (1987a). Roselle (*Hibiscus sabdariffa* L.). In C. F. Dowling, Jr. (Ed.), *Fruits of warm climates*. Miami, Florida: Creative Resources Systems, Inc.

Morton, J. F. (1987b). Roselle. In C. F. Dowling, Jr. (Ed.), *Fruits of warm climates* (pp. 281–286). Greensboro, NC, USA: Media Inc.

National Biodiversity Action Plan (N.B.A.P.). (1999). *Biodiversity in Kordofan region* (pp. 41–43). Sudan: El-obeid Agricultural Research Station. Report SUD/97/G31.

Odigie, I. P., Ettarh, R. R., & Adigun, S. (2003). Chronic administration of aqueous extract of *Hibiscus sabdariffa* attenuates hypertension and reverses cardiac hypertrophy in 2K-1C hypertensive rats. *Journal of Ethnopharmacology, 86*, 181–185.

Plotto, A., Mazaud, F., Röttger, A., & Steffel, K. (2004). *Hibiscus: Post-production management for improved market access organisation*. Food and Agriculture Organization of the United Nations (FAO), AGST.

Potter, N. N., & Hotchkiss, J. H. (1995). *Food science* (5th ed., pp. 24–68). New York: Chapman and Hall.

Purseglove, J. W. (1974). *Tropical crops: Dicotyledons*. London: Longman.

Small, E. (1997). *Culinary herbs* (p. 274). Ottawa, Ontario: NRC Research Press.

Tsai, J., & Ou, M. (1996). Colour degradation of dried roselle during storage. *Food Science, 23*, 629–640.

Tseng, T., Kao, T., Chu, C., Chou, F., Lin, W., & Wang, C. (2000). Induction of apoptosis by hibiscus protocatechuic acid in human leukemia cells via reduction of retinoblastoma (RB) phosphorylation and Bcl-2 expression. *Biochemical Pharmacology, 60*, 307–315.

Wester, P. J. (1912). Roselle, its cultivation and uses. *Philippine Agricultural Review, 5*, 123–132.

Wester, P. J. (1920). The cultivation and uses of roselle. *Philippine Agricultural Review, 13*, 89–99.

Wilson, F. D., & Menzel, M. Y. (1964). Kenaf (*Hibiscus canabinus*), roselle (*Hibiscus sabdariffa*). *Economic Botany, 18*(1), 80–90.

CHAPTER 9

Development and Characterization of Roselle Anthocyanins in Food Packaging

JIYONG SHI • JUNJUN ZHANG • ZHIHUA LI • XIAODONG ZHAI • XIAOWEI HUANG • SULAFA HASSAN • XIAOBO ZOU

1 INTRODUCTION

Roselle (*Hibiscus sabdariffa* L.), an annual shrub, is commonly used to make jellies, jams, and beverages. Its brilliant red color and unique flavor make it a valuable food product. The natural anthocyanins, which confer roselle serves as alternative coloring agents for a wide variety of foods (Fang, Zhao, Warner, & Johnson, 2017; Galanakis, Eleftheria, & Vassilis, 2013). The biological activities of anthocyanins, such as their antioxidant and anticarcinogenic activities, have been the issues of recent researches (Alfaro, Rivera, & Pérez, 2009). These activities have been reported to be several times stronger in roselle anthocyanins (RACNs) than the other plants; thus RACNs can act as multifunctional ingredients for food products (Galanakis, Eleftheria, & Vassilis, 2013). This chapter explores the bioactive properties of roselle in terms of antioxidant and antimicrobial activities of its calyces. The results highlight the multiple functional applications of this plant species, namely, as a source of bioactive compounds due to its bioactive activities and as a natural pigment for food packaging. Roselle's multifunctional properties (colorant and bioactive properties) can be explored with respect to the plant's ability to serve as a source of natural ingredients to be incorporated into intelligent packaging, edible coatings, and active packaging in the food packaging industry. In the modern food industry, natural, edible, and eco-friendly materials are always desirable, and smart food packaging is no exception. Recently, anthocyanins have been used to develop intelligent food packaging, edible coating, and active packaging (Ghaani et al., 2016; Emad, 2016; Zhang, Mao, & Zheng, 2014). For example, a colorimetric film was developed through the adsorption of anthocyanins extracted from *Bauhinia blakeana Dunn* into chitosan film (Piñeiro, Marrufo-Curtido, Vela, & Palma, 2017). This film was placed onto the headspace of a sealed tray with pork or fish, and the results showed that the film exhibited changes in color (from red to green) with the meat samples changed from fresh to spoiled. The aim of intelligent food packaging is to monitor the quality of individually packaged foods through indicators or sensors, whereas active food packaging can prolong the food shelf life of foods by absorbing food-derived chemicals from food or environment within the food packaging (Dainelli et al., 2008; Ghaani et al., 2016; Fang et al., 2017).

2 BIOACTIVE PROPERTIES OF ROSELLE ANTHOCYANINS

Anthocyanins are natural and nontoxic pigments that can exhibit visible color changes in response to changes in pH, such as by appearing yellow under alkaline conditions (Yusoff & Leo, 2017). Roselle extract is rich in bioactive compounds with antioxidant and antimicrobial properties (Alshami & Alharbi, 2014).

2.1 Extraction Methods of Roselle Anthocyanins

Methods of extracting anthocyanins generally include solvent extraction, ultrasound-assisted extraction (UAE), microwave-assisted extraction (MAE), ultrasonic extraction, enzymatic hydrolysis, and supercritical fluid extraction. Supercritical extraction technology exerts no pollutant effects, leaves no chemical residue, and does not destroy anthocyanin activity, but few researchers have focused on its use for anthocyanin (Du & Francis, 1973). Enzymatic hydrolysis has a relatively mild reaction condition, but it causes relatively severe defects that

[illegible]

can easily lead to the destruction of the anthocyanin structure (Charalampos & Komaitis, 2008), which limits its use in the extraction of RACNs. Therefore this chapter only describes solvent extraction, ultrasonic extraction, and MAE methods (Fig. 9.1).

2.1.1 Solvent extraction methods

Solvent extraction is the most commonly reported method because of its low cost and simple process and is the most commonly used method for anthocyanin extraction (Costa et al., 2010). Solvent type and temperature are two key parameters for solvent extraction. Water, ethanol, methanol, acetone, or a mixture of these is usually used as the solvent for the extraction process. However, because of the instability of anthocyanin under high-temperature and light conditions, the extraction process is generally performed at a low temperature, and a small amount of acid is usually added to the extraction solvent to form an anthocyanin cation for improving its stability. Sindi et al., 2014 quantitatively studied the effect of different solvents on the extraction rate of anthocyanin by using high-performance liquid chromatography (HPLC). The results indicated that the solvents were ranked as follows according to their extraction capacity: water > methanol > ethyl acetate > hexane. The experimental results demonstrated that the extraction capacity of the solvent had a marked influence on the polarity of the extraction solvent. The polarity of the solvent increased the extraction concentration, thus improving the extraction rate. An example of solvent extraction using an ethanol-water mixture as the solvent is described as follows: First, the roselle calyx is dried and ground into a powder (by being passed through an 80-mesh sieve). Subsequently, 1 g of roselle powder is added into ethanol at a material-to-liquid ratio of 1:20 (m/V), and the pH value is adjusted to 2.0 using Hcl (1 M). The sample is then heated at 50°C for 1 h and centrifuged at 3000 r/min for 6 min to obtain the extractions.

2.1.2 Ultrasound-assisted extraction methods

UAE, as a type of "high technology," has limited application for anthocyanin extraction because of the need for expensive equipment (Kazibwe et al., 2017). The capacity of UAE depends largely on the settings of parameters such as temperature, frequency, power, and time (Teh & Birch, 2014). The solid-to-liquid ratio and extraction temperature are also key parameters of UAE (Dzah, Courage, Duan, & Zhang, 2020). Ultrasound can be used for the extraction of bioactive compounds because the generated energy produces cavitation, which leads to a higher mass transfer rate during the extraction process than other extraction methods. Fig. 9.2 illustrates the UAE process for RACNs, which is described as follows. UAE is conducted using an ultrasonic homogenizer and a pulse mode. The solvent-to-feed ratio remains constant for all extractions, and the temperature is controlled using a water bath. The following experimental conditions can be evaluated: extraction time, ultrasound amplitude, and extraction temperature (Pinela et al., 2019).

2.1.3 Microwave-assisted extraction methods

During MAE, polar molecules in the electromagnetic field, which originally exhibit a random distribution, align according to the polarity of the electric field

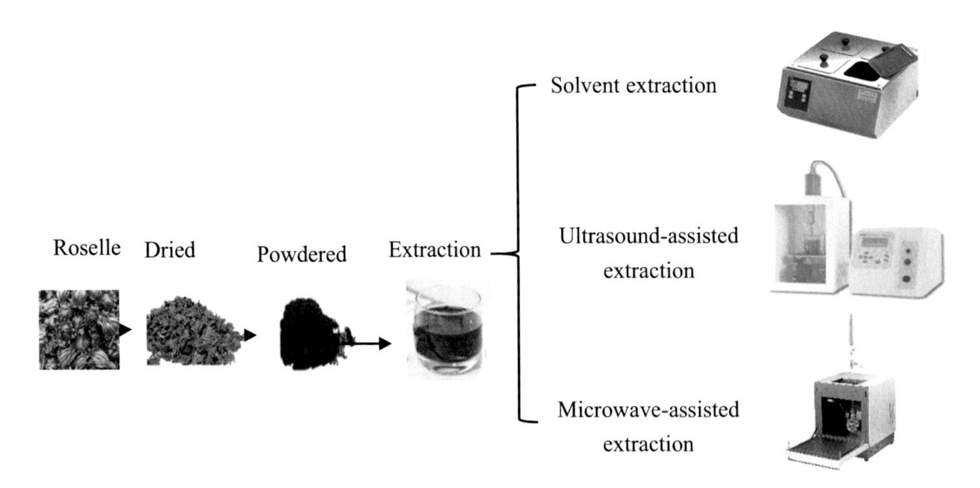

FIG. 9.1 The process of extraction of roselle anthocyanins.

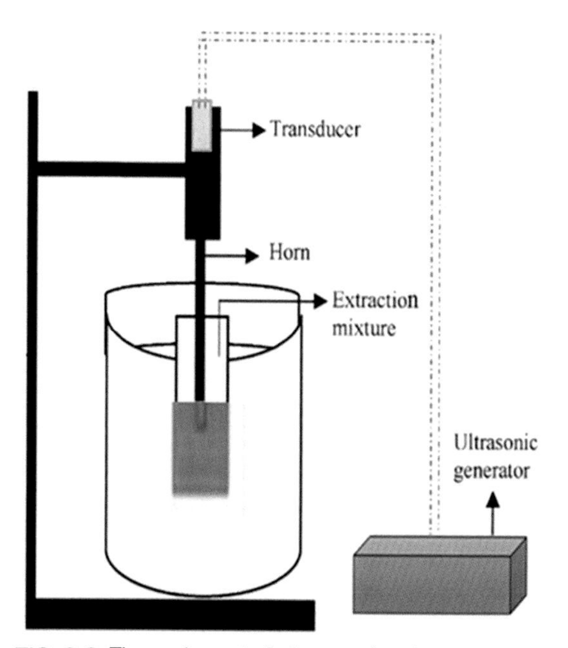

FIG. 9.2 The equipment of ultrasound-assisted extraction.

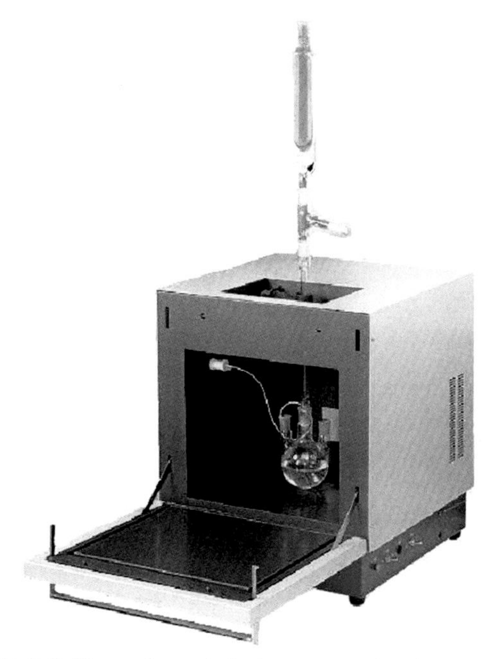

FIG. 9.3 The equipment of microwave-assisted extraction.

orientation and the frequency of the alternating electromagnetic field (Charalampos & Komaitis, 2008). The extraction solvent and material produce a thermal effect and nonthermal effect—which speed up the materials of infiltration and keep the solvent extract—thus enhancing the extraction effect (Alfaro, Rivera, & Pérez, 2009). The traditional solvent extraction method, with its long extraction time and low efficiency, degrades anthocyanins and reduces their biological activity (Galanakis, Eleftheria, & Vassilis, 2013) whereas MAE has the advantages of being less time consuming and offering high selectivity and low energy consumption (Piñeiro, Marrufo-Curtido, Vela, & Palma, 2017). Microwave technology has been widely used in the extraction of various bioactive components from plants. The efficiency of the MAE method can vary according to the parameter settings adopted for the extraction process, such as the solid-to-liquid ratio, solvent type, temperature, microwave power, and extraction time. As displayed in Fig. 9.3, MAE was performed using microwave experiment equipment with adjustable power. The device was equipped with one closed polytetrafluoroethylene (PTFE) vessel, a power sensor, a temperature sensor, a temperature controller, and a cooling system. The material was ground and placed into a PTFE vessel (Yusoff & Leo, 2017). After extraction, the vessel was cooled at room temperature and the material was filtered and freeze-dried to obtain crude extracts.

2.2 Qualitative Detection of Roselle Anthocyanin Components

RACNs have attracted the attention of scientists worldwide, and studies have demonstrated that they mainly take the form of cyanin and delphinine-like anthocyanins (Fig. 9.4). The main pigments found in the calyx of roselle are delphinidin-3-glucoside, cyanidin-3-glucoside, delphinidin-3-sambubioside, and cyanidin-3-sambubioside, and these compounds are responsible for roselle's red color (Martins et al., 2016). HPLC-mass spectrometry (HPLC-MS) was performed using the Agilent Technologies 1200 Series liquid chromatography system equipped with a diode-array detector to characterize the RACN extract. C18 columns (LiChrospher 5-C18, 150 mm × 4.6 mm × 5 μm) were used to perform HPLC-MS (quantitative analysis) at a temperature of 30°C and flow rate of 1 mL/min. Binary gradient elution of mobile phase A (100% acetonitrile) and mobile phase B (5% formic acid in deionized water) was performed at a flow rate of 1 mL/min for anthocyanin separation and the wavelength was set to 530 nm. The procedure was applied according to the following linear gradient: 10% A + 90% B for 0 min, 25% A + 75% B for 10 min, 60% A + 40% B for 20 min, 70% A + 30% B for 30 min, and 70% A + 30% B for 40 min. The capillary voltage was 3.0 kV for the positive (ESI+) mode.

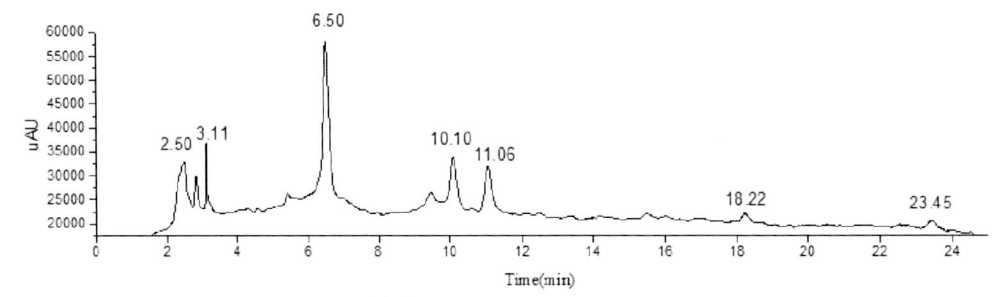

FIG. 9.4 High-performance liquid chromatogram of roselle anthocyanins recorded at 530 nm.

HPLC-MS has long been applied for the structural identification of anthocyanins, and it can also reveal their relative molecular weight. Combined with the HPLC-MS images of the anthocyanins, the fragment ions provide information regarding the parent nucleus and substituents in the molecules; subsequently, the approximate structural information of the anthocyanins can be inferred (Guzmán-Figueroa et al., 2016). Table 9.1 presents the qualitative analysis results of RACNs based on the chromatogram displayed in Fig. 9.4.

TABLE 9.1
Composition of Roselle Anthocyanins According to High-Performance Liquid Chromatography/Mass Spectrometry.

Retention Time (min)	Compound Identity	Parent and Product
2.5	Delphinidin-3-*O*-glucoside	466.4, 217.33, 261.43
3.11	Delphinidin-3-*O*-glucoside	466.4, 217.33, 261.43
6.50	Delphinine-3-elderberry glucoside	627.5, 261.43, 305.52
10.10	Cyanidin-3-elderberry glucoside	610.5, 349.25, 260.25
11.06	Cyanidin-3-elderberry glucoside	610.5, 260.4, 349.1
18.22	/	/
23.45	/	/

2.3 Quantitative Detection of Roselle Anthocyanins

2.3.1 Quantitative detection of total anthocyanin content

The total anthocyanin content was determined using differential quantitative detection with an ultraviolet-visible (UV-vis) spectrophotometer (Agilent CARY 100, Varian Corporation, United States; Hashemi and Shahani (2019). Accordingly, 100 μL of roselle extract was diluted in potassium chloride solution (pH = 1.0) and sodium acetate buffer (pH = 4.5), which were then incubated away from light at 23°C for 15 min. The samples were examined at 510 (λ_{max}) and 700 nm using UV-vis spectrophotometry. The absorbance variation (*A*) was calculated to be ($A = [A_{510}\,\text{nm} - A_{700}\,\text{nm}]$ pH 1.0 to $[A_{510}\,\text{nm} - A_{700}\,\text{nm}]$ pH 4.5), and the total anthocyanin content of cyanidin-3-*O*-glucoside (molecular weight, 449.2 g/mol) was determined using the following equation:

$$\text{TAC(mg/g)} = \frac{A \times W \times DF \times V}{\varepsilon \times L \times m} \tag{9.1}$$

where *V* is the volume of the solution (mL), *DF* is the dilution factor, *ε* is the molar extinction coefficient of cyanidin-3-*O*-glucoside (26,900 L/(mol.cm), *L* is the length of the optical path (1 cm), and *m* is the mass of the dry residue. The total anthocyanin content (TAC) is expressed in mg/g.

2.3.2 Quantitative detection of roselle anthocyanin components

The quantitative detection of RACN can also be achieved using the above mentioned HPLC method to detect the standard curve for each component. Finally, the components of the RACN can be determined.

2.4 The pH Sensitivity of Roselle Anthocyanins

The colors of RACNs were evaluated in different pH buffer solutions. As shown in Fig. 9.5A, RACNs became

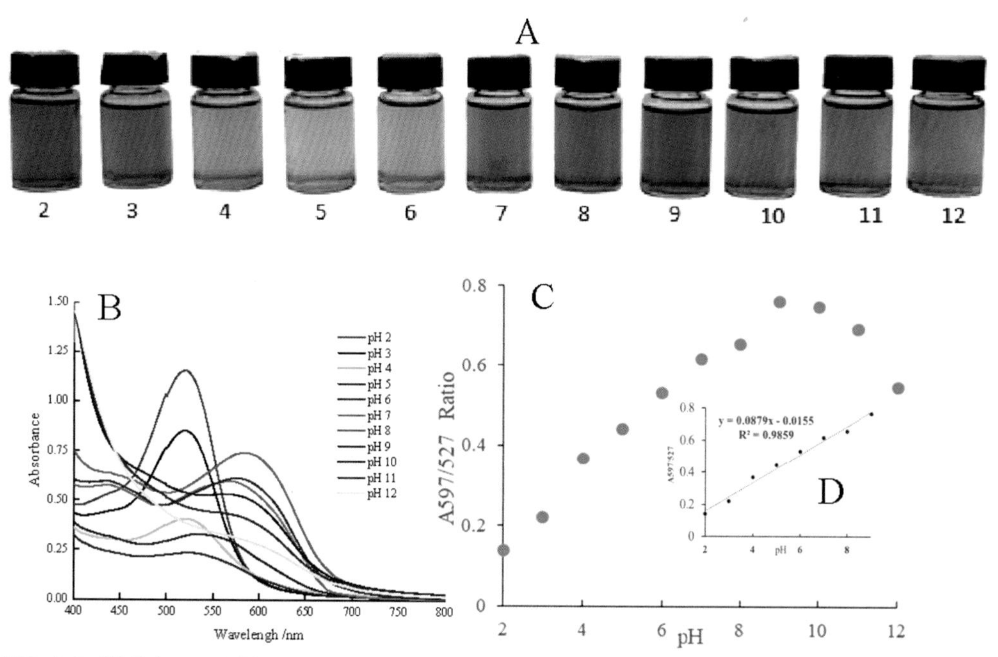

FIG. 9.5 **(A)** Colors and **(B)** UV-vis spectra of RACNs at pH 2–12, **(C)** absorbance ratio at 593 versus 529 nm of roselle solutions at pH 2–12, and **(D)** at pH 2–9.

red at pH 2.0–4.0, pink at pH 5.0–6.0, purple at pH 8.0–9.0, and yellow at pH 11.0–12.0. These color changes are attributable to the transformation of the chemical structure of the RACNs (Calhau et al., 2017; Grajeda-Iglesias et al., 2016). At pH 2.0–3.0, the RACNs mainly presented in the form of yellow salt ions, causing the solution to appear red. As the pH increased, their structures gradually transformed into quinoids at pH 4.0–6.0 and then into a colorless pseudobase at pH 7.0–9.0 and the color gradually turned blue. Finally, when the pH value was greater than 9.0, the RACNs degraded to become strongly alkaline and their color changed to yellow-green (Zhang et al., 2019).

Fig. 9.5B presents the corresponding absorption spectra of the RACNs in different pH buffer solutions. The maximum absorption peak appeared at approximately 527 nm at pH 2, and the absorbance decreased with the increase at pH 2–4. At pH levels higher than 5, the maximum absorption peak was at 548 nm and the absorbance at 597 nm in the alkaline environment (Zhang et al., 2019).

2.5 Antioxidant Properties of Roselle Anthocyanins

RACNs are polyphenols, which contain numerous phenolic hydroxyl groups, especially catechol or pyrogallol, which are easily oxidized to form a stable quinone structure and forming phenoxy groups (Formagio et al., 2015). Anthocyanins are capable of not only providing protons and oxygen free radicals but also combining with macromolecular substances such as protein. Furthermore, RACNs can be used as a catalyst for oxidation and the metal ion chelating reaction then for subsequently cutting off the chain reaction of lipid oxidation (Daniel et al., 2012). Numerous studies have demonstrated that anthocyanins play an antioxidant role in removing reactive oxygen species before directly chelating metal ion with proteins (Daniel et al., 2013). The antioxidant activity of RACNs is affected by the hydroxyl group on its B ring and the methoxide base (Kerch, 2015). The stronger antioxidant activity is due to the number of higher hydroxyl group and affected by the number and structure of its sugar ligand and acyl group.

The antioxidant activity of anthocyanins was determined using the 2,2-diphenyl-1-picryl-hydrazyl-hydrate (DPPH) free radical scavenging activity method reported by Tahir (2017). The RACN extract was added to a 0.06 mmol/L DPPH methanol solution (A_1) and left to stand for 30 min. With methanol as the blank background (A_0) and distilled water as the reference, the absorbance of the samples was measured at 517 nm, and the antioxidant activity of the samples was calculated according to the inhibition percentage of the DPPH radical using the following formula:

$$\text{DPPH scavenging rate (\%)} = \frac{A_0 - A_1}{A_0} \quad (9.2)$$

The antioxidant activity of roselle extract was periodically measured using the ferric ion reducing antioxidant power method reported by Haroon (Tahir, 2017). The diluted roselle extract solution was mixed with phosphate buffer (pH 6.6) and then mixed with potassium ferricyanide (10%). Subsequently, trichloroacetic acid was added, followed by continuous vortex mixing, and then the supernatant was diluted with distilled water and $FeCl_3$ (0.1%) solution. The absorbance was measured at 700 nm, and ascorbic acid (0–10 μg/mL) was used as the standard solution ($R^2 = 0.99$).

2.6 Antibacterial Activity of Roselle Anthocyanins

RACNs have a typical flavonoid structure with more phenolic hydroxyl groups. The RACNs of functional groups can be combined with proteins or enzymes through hydrogen bonding, leading to the destruction of the protein molecular structure, to its consequential inactivation, and finally to cytoplasm pyknosis (Emad, 2016). Anthocyanins can affect the stability of the cell membrane and directly affect the metabolism of microorganisms. The antimicrobial activity is affected by the absorption of required nutrients and exhibits an influence on microbial growth and the adhesion of bacteria to epithelial cells. The yellow ketone extracted from RACNs was reported as having different degrees of inhibition for *Escherichia coli*, *Staphylococcus aureus*, and *Bacillus subtilis* (Gutiérrez-Alcántara et al., 2016). The extraction of RACNs influences the membrane protein and nucleic acid content in *E. coli* and *S. aureus*, subsequently preventing normal DNA from being replicated or transcribed; this reduces the nucleic acid content and protein synthesis, eventually leading to loss of the bacteria's biological functions (He et al., 2016).

The antibacterial activity of RACNs was characterized using the antimicrobial circle method with filter paper, where the size of the antibacterial circle can serve as the basis for determining the antimicrobial effect (Liu et al., 2010). The bacteria that most commonly cause food spoilage are *E. coli*, *B. subtilis*, and *S. aureus*. These bacteria were injected into a sterile liquid medium and then placed in a shaker at 37°C for 24 h. Part of the activated strain solution was absorbed and injected into sterile water. After full oscillation, the initial strain solution of 10^6–10^7 CFU/mL was prepared and a sterilized glass-coated stick was used to disperse the bacterial solution evenly. Sterile filter paper with a diameter of 6 mm was then impregnated in the RACN extract solution for 20 min. The filter paper was removed using tweezers, drained on an ultraclean bench, and then pasted on the culture medium containing the tested bacteria. After culturing, the antimicrobial activity was characterized through measurement of the diameter of the circle that had formed on the filter paper.

Roselle has antibacterial properties because it is rich in polyphenols. The antibacterial mechanism may be due to the following reasons:

1. The polyphenols in RACNs can change the permeability of the bacterial cell membrane. The destruction of cell membranes by plant polyphenols is the antibacterial basis of RACNs. Hydroxyl groups in polyphenols can interact with bacterial cell membranes through hydrogen bonds. The functional structure of the cell membrane may be destroyed, resulting in extravasation of the contents. The potential of the cell membrane may change from a normal polarization state to a polarization or even a superstate, which interferes with proton kinetics and reduces the pH and ATP dynamic rate of the cell membrane. These changes can cause the death of microorganisms due to disorders of physiologic activities.
2. The polyphenols in RACNs can affect the expression of enzymes in bacterial cells. Because the behavior of polyphenols penetrating the cell membrane may inhibit the expression of intracellular enzymes such as ATP synthase and DNA gyrase, they interfere with the normal energy metabolism in bacterial cells. This also leads to the inability to produce DNA and protein and other biological macromolecules that help sustain the life of bacteria. So the polyphenols in RACNs can achieve the purpose of inhibiting bacteria.
3. The polyphenols in RACNs can destroy the cell walls of bacteria. Polyphenols can chelate with divalent cations on the outer membrane of bacteria to release lipopolysaccharides out of the cell wall. Then the integrity of the cell wall is destroyed, causing the cell contents to leak and the bacteria to die (Chibane et al., 2017).

3 APPLICATIONS OF ROSELLE ANTHOCYANINS

As mentioned, RACNs possess the properties of pH sensitivity, antioxidant activity, and antibacterial activity. Therefore they can be used to develop pH sensors that indicate the quality of food, which are closely related to the pH changes indicators used for intelligent packaging. RACNs can also be embedded in packaging materials to functionalize the materials with antioxidant and antibacterial activities for active packaging (Fang et al., 2017; Pereira et al., 2015)

3.1 Intelligent Films Based on pH Sensitivity for Freshness Monitoring

Intelligent food packaging can be achieved with anthocyanin-based colorimetric indicators (Choi et al., 2017). Colorimetric indicators have received considerable attention because they are easily perceptible to the naked eye. Most colorimetric sensors are gas sensors. They can react with volatile gases generated by food (e.g., amines and CO_2) or react with gases (e.g., O_2) in the package atmosphere during gas exchange between the inside and outside of the package. Various volatile gases can be generated by food during storage. For example, protein-rich foods and fermented foods can produce amines and CO_2 (Meng et al., 2015), respectively. These specific gases are quality indicators of foods during storage. Colorimetric gas sensors integrated into food packaging can display visible color changes by reacting with these specific gases and thus provide information regarding the quality of packaged foods.

1. As indicated in Fig. 9.6, the first colorimetric sensor array was developed using nine natural pigments (mainly containing anthocyanins) to indicate pork spoilage. The color of the sensor array changes according to the freshness level of the pork stored at 5°C. Moreover, a partial least squares prediction model can effectively reveal correlations of color changes with the biogenic amine index and total viable count (TVC) in pork. The study indicated that anthocyanins, as natural pigments, are promising materials for use in the development of colorimetric sensors. Similar anthocyanin-based colorimetric sensor arrays have also been used to indicate the quality of Chinese traditional salted pork (Huang et al., 2015).

   Consumers generally want to have real-time knowledge regarding the quality of packaged foods being sold, which could help them decide whether to buy the food. Thus colorimetric sensors that provide in situ and real-time monitoring of food quality are desirable. To offer these capabilities, colorimetric sensors must have strong self-stability, in addition to their sensitivity to target gases. Because anthocyanins have intrinsic instability when exposed to factors such as UV light, extreme temperatures, and oxygen, they are generally incorporated with macromolecules to improve their stability.
2. RACNs have also been used to develop colorimetric films (Zhai et al., 2017). As shown in Fig. 9.7, to develop these films, RACNs extractions were

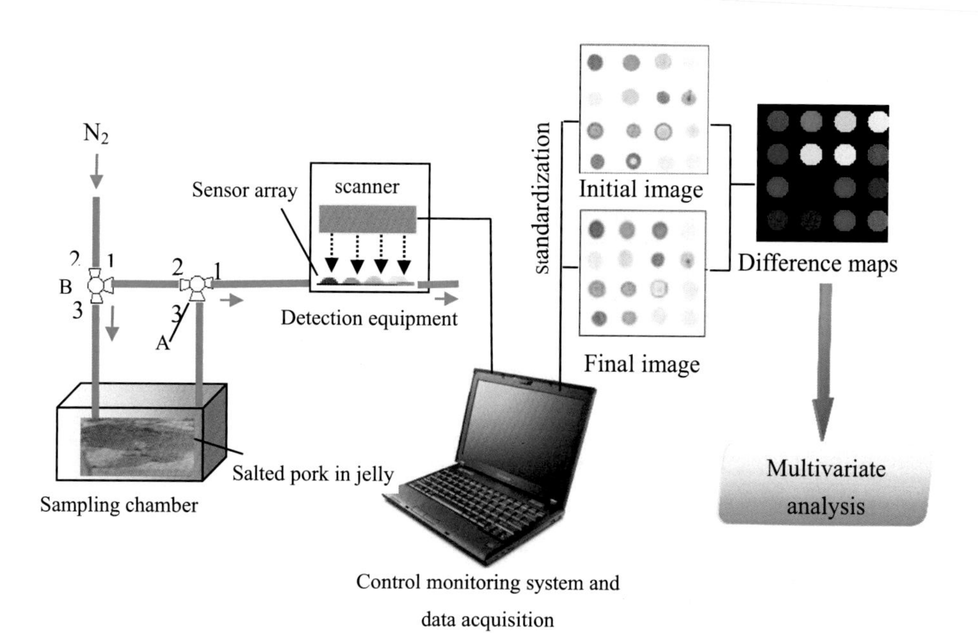

FIG. 9.6 Colorimetric sensor array for monitoring pork freshness.

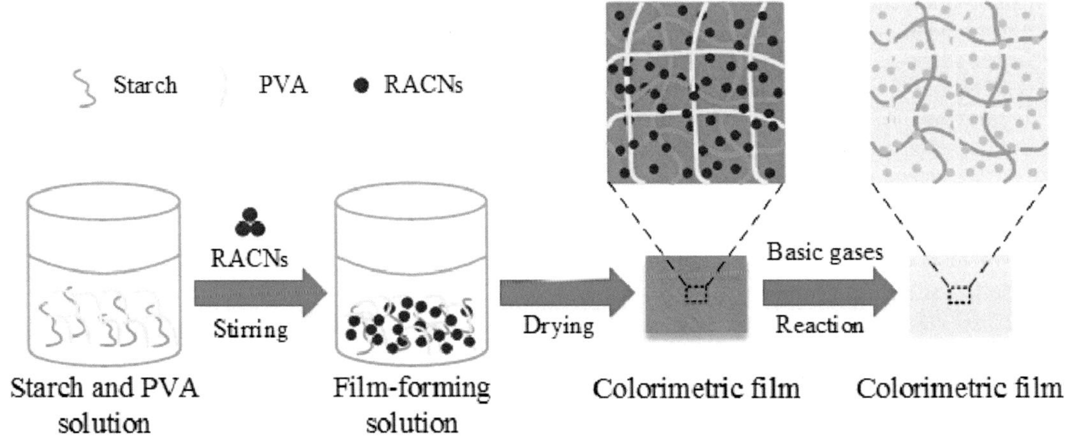

FIG. 9.7 Preparation of the colorimetric films. *PVA*, polyvinyl alcohol; *RACNs*, roselle anthocyanins.

imbedded at different concentrations in polyvinyl alcohol/starch (PSE) composite film using a casting/ solvent evaporation method.

The effects of RACN concentration on self-stability and gas (NH_3) sensitivity were investigated, and the results indicated that the composite films with higher RACN content possessed greater self-stability but lower sensitivity to NH_3. The film with moderate RACN content exhibited compromised performance. When the film was used to monitor fish freshness, it exhibited a red-to-green color change in response to volatile amines.

The PSE-30 film exhibited the highest color change rate, followed by the PSE-60 film and then the PSE-120 film (Fig. 9.8), consistent with the result that the colorimetric films with a lower RACN content had higher color change rates.

Another colorimetric film was also developed using RACNs in which were embedded in two of the following three substances: starch, polyvinyl alcohol, and chitosan. The effects of these polymers on the self-stability and gas sensitivity of the films were investigated. The results revealed that the PSE composite film had the best color stability and NH_3 sensitivity. As shown in Fig. 9.9, When it was used to monitor pork freshness, it also exhibited color changes from red to green and finally to yellow (Fig. 9.10, Zhang et al., 2019).

3. Research has also led to improvements in other applications of intelligent packaging for food

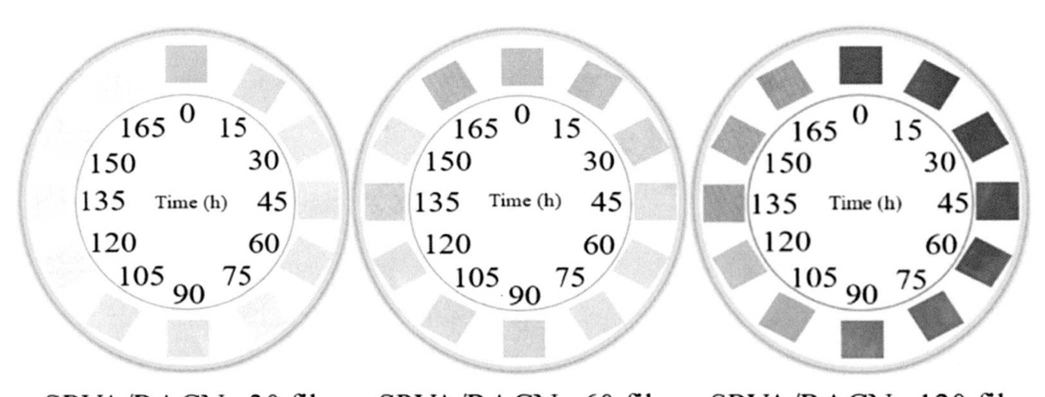

FIG. 9.8 Color change of colorimetric films during the storage of fish sample. *RACNs*, roselle anthocyanins; *SPVA*, starch/polyvinyl alcohol.

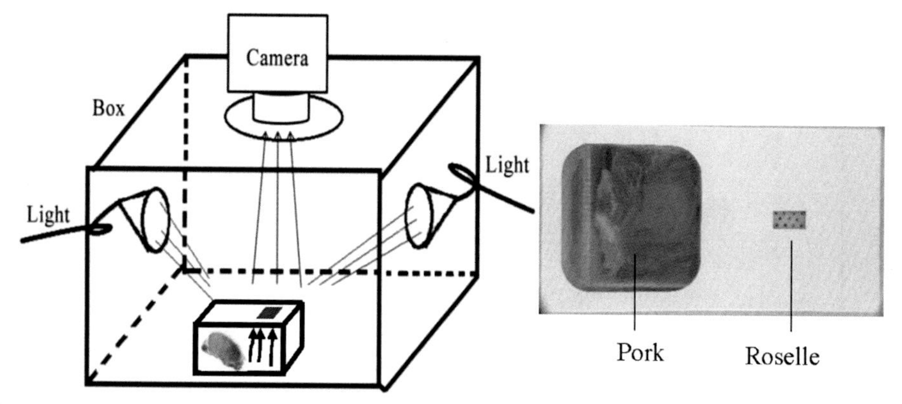

FIG. 9.9 Schematic of the process of acquiring composite film color images.

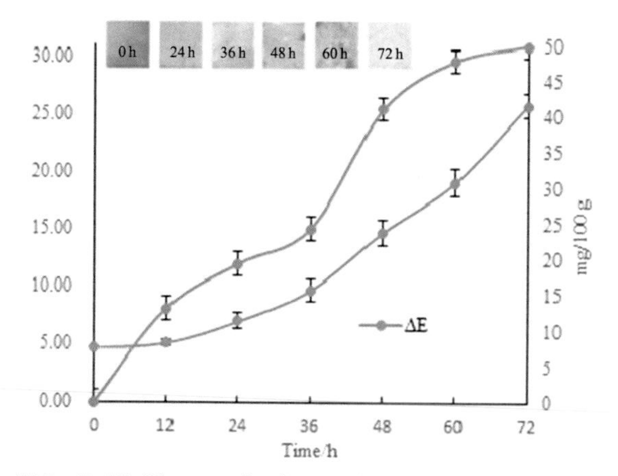

FIG. 9.10 Changes in the total volatile basic nitrogen content of pork samples and the corresponding ΔE.

freshness monitoring. As shown in Fig. 9.11, amine-responsive bilayer films were developed using anthocyanins incorporated into other materials for the visual monitoring of meat spoilage. One layer functions as a sensing layer to volatile amines, while the other layer serves as a light barrier layer to improve illumination stability (Zhai et al., 2020).

3.2 Edible Coatings Based on the Antioxidant Properties of Anthocyanins for Extending Food Shelf Life

Edible coatings with antioxidant properties constitute a new type of packaging that combines antioxidants with packaging materials. They represent an alternative method of preservation based on their ability to reduce the transpiration of fresh fruits and vegetables (Kerch, 2015). Some bioactive compounds can be used as a natural additive in food production to improve the preservative functions of edible coating and thereby enhance the properties of edible coatings, including their antioxidant and color-changing properties (Peretto et al., 2017).

RACNs have been used to develop edible coatings for monitoring the postharvest quality of blueberry fruits during storage at $4 \pm 0.5°C$. Herein, a coating solution was prepared and blueberry fruits were dipped in a gum-based formulation for 4 min and the excess coating materials were allowed to drip off. The coated fruits were placed on a polyethylene sheet and air-dried at $25 \pm 1°C$. Physicochemical, phytochemical, antioxidant capability, percentage decay, and microbiologic analyses of coated and uncoated blueberries were performed. The effects of RACNs in the coating film on polyphenol oxidase (PPO) and guaiacol peroxidase (POD) activities were also measured in the previous study. The results demonstrated that the RACN-coated fruits exhibited significantly lower PPO and POD activities than the uncoated fruits ($P < 0.05$) after a given storage time. The results also revealed that the PPO activity in a strawberry may be inhibited by RACN coatings because of the synergetic effects of the bioactive compounds. Therefore coating films containing RACNs have beneficial effects conferred by the antioxidant properties of RACNs for extending the shelf life of blueberries (Yang et al., 2017).

3.3 Active Films Based on the Antibacterial Properties for Extending Food Shelf Life

Few researchers have focused on the application of antibacterial films in food packaging. Antioxidant additives

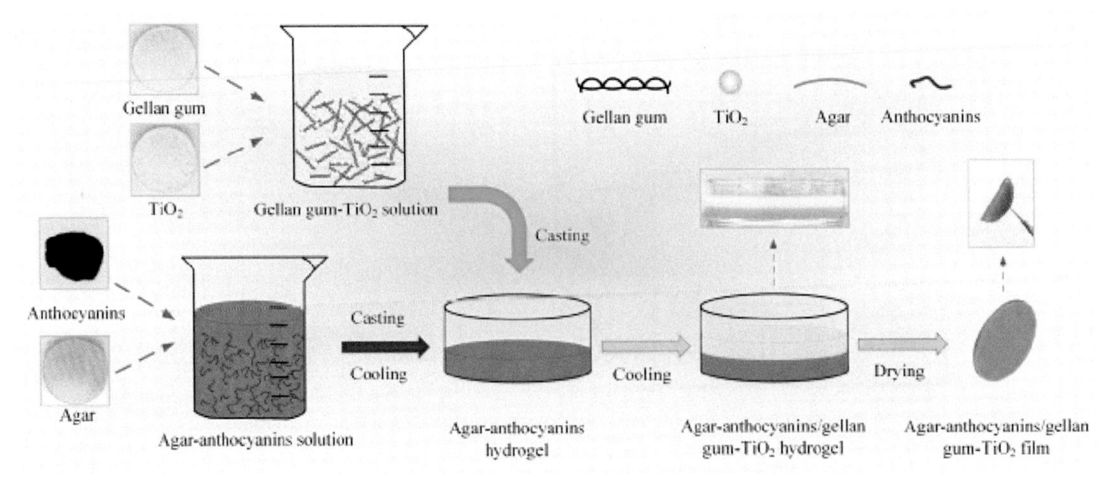

FIG. 9.11 Preparation of the bilayer colorimetric film.

can be slowly released from the inside of the coating material and diffused into the internal packaging space or into the food. Active films were developed based on the antibacterial properties of RACNs for extending the shelf life of pork. As shown in Fig. 9.12, fresh pork was wrapped in active film composed of biodegradable materials imbedded with different concentrations of RACNs (Lin et al., 2013), and the active film was placed in a refrigerator at 4°C in a polyethylene self-sealing bag to protect it from the effects of the external environment. The TVC can reflect the degree of microbial contamination in pork products.

As indicated in Fig. 9.13, the TVC of pork packaged with active films was significantly lower than that of the control pork ($P < .05$) because the polyvinyl alcohol/starch (PS) film inhibited the growth of aerobic microorganisms during the early stage of storage. On the fourth day, the TVC of the control pork reached 6.43 log CFU/g, indicating that the pork was no longer fresh (>6.00 log CFU/g (Huang et al., 2014), whereas the TVCs of the pork wrapped in 0.12% and 0.24% RACN active films were 5.66 and 5.55 log CFU/g, respectively, which were still at the subfresh level. The results indicated that during the storage period, the pork wrapped in active films had significantly lower TVCs than the control pork ($P < .05$). Additionally, with increases in RACN concentration, the total number of colonies exhibited a decreasing tendency. Therefore active films with RACNs extend the shelf life of pork due to the antibacterial properties of RACNs.

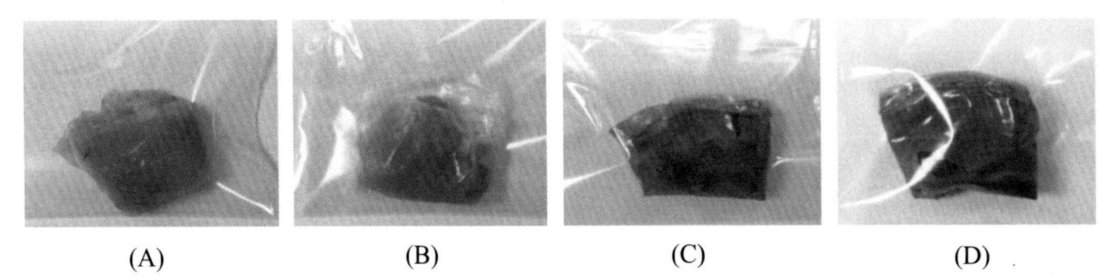

FIG. 9.12 Fresh pork packaged using composite films incorporated with roselle extract: **(A)** blank, **(B)** PS, **(C)** PSE/12%, and **(D)** PSE/24%. *PS*, polyvinyl alcohol/starch, *PSE*, polyvinyl alcohol/starch/RACNs extraction.

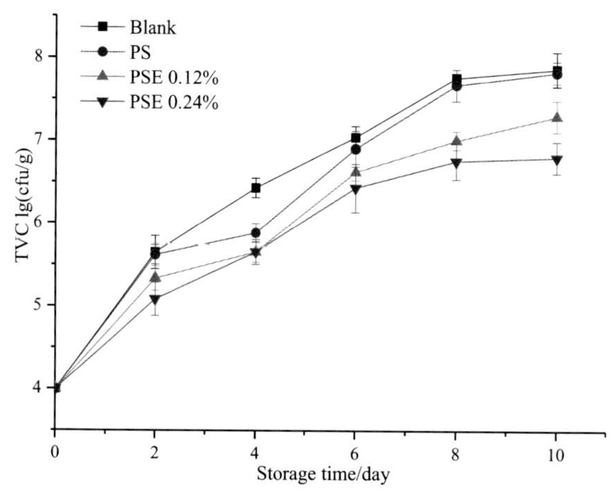

FIG. 9.13 Changes in the total viable count (TVC) of pork wrapped in different packaging materials during storage. *PS*, polyvinyl alcohol/starch; *PSE*, polyvinyl alcohol/starch/RACNs extraction.

4 CONCLUSION

RACNs can be obtained from a wide range of sources and are safe, nontoxic, and easy to extract. Moreover, they have some special functional effects, including pH sensitivity and antioxidant and antimicrobial activities, which may help stabilize and extend the shelf life of food products. RACNs have a good effect and can extend the shelf life of food products, thanks to their antioxidant properties. The antibacterial properties of RACNs can effectively inhibit the growth and reproduction of microorganisms that extend the shelf life of food products. Therefore anthocyanin-rich natural pigment extracts are ideal materials for the development of active intelligent packaging films, which can be used to prepare multipurpose packaging films with food-preservation and freshness-detection functions by the naked eye.

REFERENCES

Alfaro, A., Rivera, A., Pérez, R., et al. (2009). Integral valorization of two legumes by autohydrolysis and organosolv delignification. *Bioresource Technology, 100*(1), 440–445. https://doi.org/10.1016/j.biortech.2008.05.003.

Alshami, I., & Alharbi, A. E. (2014). Antimicrobial activity of *Hibiscus sabdariffa* extract against uropathogenic strains isolated from recurrent urinary tract infections. *Asian Pacific Journal of Tropical Disease, 4*(4), 317–322. https://doi.org/10.1016/S2222-1808(14)60581-8

Calhau, C., Silva, S., Morais, R. M., et al. (2017). Characterization of phenolic compounds, anthocyanidin, antioxidant and antimicrobial activity of 25 varieties of Mexican roselle (*Hibiscus sabdariffa*). *Industrial Crops and Products, 69*, 385–394. https://doi.org/10.1080/10408398.2015.1007963

Charalampos, P., & Komaitis, M. (2008). Application of microwave-assisted extraction to the fast extraction of plant phenolic compounds. *LWT - Food Science and Technology*. https://doi.org/10.1016/j.lwt.2007.04.013

Chibane, L. B., Degraeve, P., Ferhout, H., Bouajila, J., & Oulahal, N. (2017). Plant antimicrobial polyphenols as potential natural food preservatives. *Journal of the Science of Food and Agriculture, 99*, 1457–1474. https://doi.org/10.1002/jsfa.9357

Choi, I., Lee, J. Y., Lacroix, M., & Han, J. (2017). Intelligent pH indicator film composed of agar/potato starch and anthocyanin extracts from purple sweet potato. *Food Chemistry, 218*, 122–128. https://doi.org/10.1016/j.foodchem.2016.09.050

Costa, M. J., Maciel, L. C., Teixeira, J. A., Vicente, A. A., & Cerqueira, M. A. (2018). Use of edible films and coatings in cheese preservation: Opportunities and challenges. *Food Research International, 107*, 84–92.

Dainelli, D., Gontard, N., Spyropoulos, D., Zondervan Van Den Beuken, E., & Tobback, P. (2008). Active and intelligent food packaging: Legal aspects and safety concerns. *Trends in Food Science & Technology, 19*(Suppl. 1), S103–S112. https://doi.org/10.1016/j.tifs.2008.09.011

Daniel, D. E., Barragan Huerta, B. E., Vizcarra Mendoza, M. G., & Anaya Sosa, I. (2013). Effect of drying conditions on the retention of phenolic compounds, anthocyanins and antioxidant activity of roselle (*Hibiscus sabdariffa* L.) added to yogurt. *International Journal of Food Science & Technology, 48*(11), 2283–2291. https://doi.org/10.1111/ijfs.12215

Daniel, D. L., Huerta, B. E. B., Sosa, I. A., & Mendoza, M. G. V. (2012). Effect of fixed bed drying on the retention of phenolic compounds, anthocyanins and antioxidant activity of roselle (*Hibiscus sabdariffa* L.). *Industrial Crops and Products, 40*, 268–276. https://doi.org/10.1016/j.indcrop.2012.03.015

Du, C. T., & Francis, F. J. (1973). Anthocyanins of roselle (*Hibiscus sabdariffa*, L.). *Journal of Food Science, 38*(5), 810–812. https://doi.org/10.1111/j.1365-2621.1973.tb02081.x

Dzah, Courage, S., Duan, Y., Zhang, H., et al. (2020). The effects of ultrasound assisted extraction on yield, antioxidant, anticancer and antimicrobial activity of polyphenol extracts: a review. *Food Bioence*, Article 100547. https://doi.org/10.1016/j.fbio.2020.100547.

Emad, A. (2016). Antibacterial efficiency of the Sudanese roselle (*Hibiscus sabdariffa* L.), a famous beverage from Sudanese folk medicine. *Journal of Intercultural Ethnopharmacology, 5*(2), 186–190. https://doi.org/10.5455/jice.20160320022623

Fang, Z., Zhao, Y., Warner, R. D., & Johnson, S. K. (2017). Active and intelligent packaging in meat industry. *Trends in Food Science & Technology, 61*, 60–71. https://doi.org/10.1016/j.tifs.2017.01.002

Formagio, A. S. N., Ramos, D. D., Vieira, M. C., Ramalho, S. R., Silva, M. M., Zárate, N. A. H., Foglio, M. A., & Carvalho, J. E. (2015). Phenolic compounds of *Hibiscus sabdariffa* and influence of organic residues on its antioxidant and antitumoral properties. *Brazilian Journal of Biology, 75*(1), 69–76. https://doi.org/10.1590/1519-6984.07413

Galanakis, C. M., Eleftheria, M., & Vassilis, G. (2013). Recovery and fractionation of different phenolic classes from winery sludge using ultrafiltration. *Separation & Purification Technology, 107*, 245–251. https://doi.org/10.1016/j.seppur.2013.01.034.

Ghaani, M., Cozzolino, C. A., Castelli, G., & Farris, S. (2016). An overview of the intelligent packaging technologies in the food sector. *Trends in Food Science and Technology, 51*, 1–11. https://doi.org/10.1016/j.tifs.2016.02.008

Grajeda-Iglesias, Figueroa-Espinoza, M. C., Barea, B., & Fernandes. (2016). Isolation and characterization of anthocyanins from *Hibiscus sabdariffa* flowers. *Journal of Natural Products*. https://doi.org/10.1021/acs.jnatprod.5b00958

Gutiérrez-Alcántara, E. J., Rangel-Vargas, E., Gómez-Aldapa, C. A., Falfan-Cortes, R. N., Rodríguez-Marín, M. L., Godínez-Oviedo, A., Cortes-López, H., & Castro-Rosas, J. (2016). Antibacterial effect of roselle extracts (*Hibiscus sabdariffa*), sodium hypochlorite and acetic acid against multidrug-resistant Salmonella strains isolated from tomatoes. *Letters in Applied Microbiology, 62*(2), 177–184. https://doi.org/10.1111/lam.12528

Guzmán-Figueroa, M. P., Ortega-Regules, A. E., Bautista-Ortín, A. B., Gómez-Plaza, E., & Anaya-Berríos, C. (2016). New pyranoanthocyanins synthesized from roselle (*Hibiscus sabdariffa* L.) anthocyanins. *Journal of the Mexican Chemical Society, 60*(1), 13–18. https://doi.org/10.29356/jmcs.v60i1.65

Hashemi, A., & Shahani, A. (2019). Effects of salt stress on the morphological characteristics, total phenol and total anthocyanin contents of roselle (*Hibiscus sabdariffa* L.). *Plant Physiology Reports*. https://doi.org/10.1007/s40502-019-00446-y

He, Z., Xu, M., Zeng, M., Qin, F., & Chen, J. (2016). Interactions of milk A- and B-casein with malvidin-3-O-glucoside and their effects on the stability of grape skin anthocyanin extracts. *Food Chemistry, 199*, 314–322. https://doi.org/10.1016/j.foodchem.2015.12.035

Huang, X. W., Zou, X. B., Shi, J. Y., Guo, Y., Zhao, J. W., Zhang, J., & Hao, L. (2014). Determination of pork spoilage by colorimetric gas sensor array based on natural pigments. *Food Chemistry, 145*, 549–554. https://doi.org/10.1016/j.foodchem.2013.08.101

Huang, X., Zou, X., Zhao, J., Shi, J., Li, Z., & Shen, T. (2015). Monitoring the biogenic amines in Chinese traditional salted pork in jelly (Yao-Meat) by colorimetric sensor array based on nine natural pigments. *International Journal of Food Science & Technology, 50*. https://doi.org/10.1111/ijfs.12620

Kazibwe, Z., Kim, D.-H., Chun, S., & Gopal, J. (2017). Ultrasonication assisted ultrafast extraction of *Tagetes erecta* in water: Cannonading antimicrobial, antioxidant components. *Journal of Molecular Liquids, 229*, 453–458. https://doi.org/10.1016/j.molliq.2016.12.044

Kerch, G. (2015). Chitosan films and coatings prevent losses of fresh fruit nutritional quality: a review. *Trends in Food Science Technology*. https://doi.org/10.1016/j.tifs.2015.10.010

Lin, H., Zhao, J., Chen, Q., & Zhang, Y. (2013). Rapid detection of total viable count (TVC) in pork meat by hyperspectral imaging. *Food Research International, 54*(1), 821–828. https://doi.org/10.1016/j.foodres.2013.08.011

Liu, K.-S., Tsao, S.-M., & Yin, M.-C. (2010). In vitro antibacterial activity of roselle calyx and protocatechuic acid. *Phytotherapy Research, 19*(11), 942–945. https://doi.org/10.1002/ptr.1760

Martins, N., Roriz, C. L., Morales, P., Barros, L., & Ferreira, I. C. F. R. (2016). Food colorants: Challenges, opportunities and current desires of agro-industries to ensure consumer expectations and regulatory practices. *Trends in Food Science & Technology*, 1–15. https://doi.org/10.1039/C5RA03963K

Meng, X., Lee, K., Kang, T. Y., & Ko, S. (2015). An irreversible ripeness indicator to monitor the Co_2 concentration in headspace of packaged kimchi during storage. *Food Science & Biotechnology, 24*(1), 91–97. https://doi.org/10.1007/s10068-015-0014-2

Pereira, V. A., Jr., De Arruda, I. N. Q., & Stefani, Ricardo (2015). Active chitosan/PVA films with anthocyanins from *Brassica oleraceae* (red cabbage) as time–temperature indicators for application in intelligent food packaging. *Food Hydrocolloids, 43*(1), 180–188. https://doi.org/10.1016/j.foodhyd.2014.05.014

Peretto, G., Du, W. X., Avena-Bustillos, R. J., Berrios Jose, D. J., Sambo, P., & Mchugh, T. H. (2017). Electrostatic and conventional spraying of alginate-based edible coating with natural antimicrobials for preserving fresh strawberry quality. *Food Bioprocess Technology, 10*(1), 165–174. https://doi.org/10.1007/s11947-016-1808-9

Piñeiro, Z., Marrufo-Curtido, C., Vela, & Palma, M. (2017). Microwave-assisted extraction of stilbenes from woody vine material. *Food & Bioproducts Processing*. https://doi.org/10.1016/j.fbp.2017.02.006.

Pinela, J., Prieto, M. A., Pereira, E., Jabeur, I., Barreiro, M. F., Barros, L., & Ferreira, I. C. F. R. (2019). Optimization of heat- and ultrasound-assisted extraction of anthocyanins from *Hibiscus sabdariffa* calyces for natural food colorants. *Food Chemistry, 275*, 309–321. https://doi.org/10.1016/j.foodchem.2018.09.118

Sindi, H. A., Marshall, L. J., & Morgan, M. R. A. (2014). Comparative chemical and biochemical analysis of extracts of *Hibiscus sabdariffa*. *Food Chemistry, 164*(3), 23–29. https://doi.org/10.1016/j.foodchem.2014.04.097

Tahir, H. E. O. (2017). Assessment of antioxidant properties, instrumental and sensory aroma profile of red and white karkade/roselle (*Hibiscus sabdariffa* L.). *Journal of Food Measurement & Characterization*, (1), 1–10. https://doi.org/10.1007/s11694-017-9535-0

Teh, S. S., & Birch, E. J. (2014). Effect of ultrasonic treatment on the polyphenol content and antioxidant capacity of extract from defatted hemp, flax and canola seed cakes. *Ultrasonics Sonochemistry, 21*(1), 346–353. https://doi.org/10.1016/j.ultsonch.2013.08.002

Yang, Z., Zou, X., Li, Z., Huang, X., Zhai, X., Zhang, W., Shi, J., & Tahir, H. E. (2017). Improved postharvest quality of cold stored blueberry by edible coating based on composite gum Arabic/roselle extract. *Food and Bioprocess Technology*,

12(9), 1537–1547. https://doi.org/10.1007/s11947-019-02312-z

Yusoff, N. I., & Leo, C. P. (2017). Microwave assisted extraction of defatted roselle (Hibiscus sabdariffa L.) seed at subcritical conditions with statistical analysis. *Journal of Food Quality, 1*, 1–10. https://doi.org/10.1155/2017/5232458.

Zhai, X., Shi, J., Zou, X., Wang, S., Jiang, C., Zhang, J., Huang, X., Zhang, W., & Holmes, M. (2017). Novel colorimetric films based on starch/polyvinyl alcohol incorporated with roselle anthocyanins for fish freshness monitoring. *Food Hydrocolloids, 69*, 308–317. https://doi.org/10.1016/j.foodhyd.2017.02.014

Zhai, X., Zou, X., Shi, J., Huang, X., Sun, Z., Li, Z., Sun, Y., et al. (2020). Amine-responsive bilayer films with improved illumination stability and electrochemical writing property for visual monitoring of meat spoilage. *Sensors and Actuators B: Chemical, 302*, 127130. https://doi.org/10.1016/j.snb.2019.127130

Zhang, B., Mao, G., Zheng, D., et al. (2014). Separation, identification, antioxidant, and anti-tumor activities of hibiscus sabdariffa L. Extracts. *Separation Science & Technology, 49*(9), 1379–1388. https://doi.org/10.1080/01496395.2013.877037.

Zhang, J., Zou, X., Zhai, X., Huang, X. W., Jiang, C., & Holmes, M. (2019). Preparation of an intelligent pH film based on biodegradable polymers and roselle anthocyanins for monitoring pork freshness. *Food Chemistry, 272*, 306–312. https://doi.org/10.1016/j.foodchem.2018.08.041

CHAPTER 10

Aroma, Aroma-Active, and Phenolic Compounds of Roselle

S. SELLI • G. GUCLU • O. SEVINDIK • H. KELEBEK

1 INTRODUCTION

Edible flowers have been actively used for human consumption in various cultures for ages. The applications of their use include both as a flavoring agent and as a garnish in the culinary profession, as well as being a beneficial additive, having health-promoting properties in alternative medicine. Previous studies elucidated the chemical properties of edible flowers, showing the presence of valuable bioactive and nutraceutical components including phenolics, antioxidants, carotenoids, prebiotics/probiotics, fatty acids, organic acids, volatiles, dietary fibers, sterols, polyols, vitamins, mineral elements, amino acids, and carbohydrates (Fernandes et al., 2017; Junsathian et al., 2018; Kaisoon et al., 2011; Sáyago-Ayerdi et al., 2007; Takahashi et al., 2020). Within a broad array of edible flower species, one attractive plant, known as *Hibiscus sabdariffa* L., is sought out for its delicacy and medicinal properties. This plant has attracted a great deal of attention recently.

H. sabdariffa, with striking red calyxes and flowers, belongs to the Malvaceae family and is widely grown in many developing countries, as the plant can thrive well in tropical and subtropical regions. These areas include Benin, India, Saudi Arabia, Australia, Mexico, China, and Africa (Ismail et al., 2008; Yagoub et al., 2004; Zannou et al., 2020). It has various local names such as roselle in Australia, sinko in Benin, zobo in Nigeria, karkade in Sudan, and flor de Jamaica in Mexico (Hopkins et al., 2013). Also this plant was reported to be one of the highest volume specialty botanical products in international commerce (Plotto et al., 2004). It is an annual herb with a reddish cylindric stem, nearly glabrous. Its flowers are red, solitary, and axial and consist of calyx segments. The calyxes are bright red, thick, fleshy, and cuplike (Ross, 2003). The dainty aroma and alluring color of roselle flowers and calyxes attract humans and help widen their use.

The history of roselle use dates back to thousands of years, spreading across many continents and civilizations, and every part of the roselle plant, from its roots to leaves, have been used as foodstuffs or for medicinal purposes since then. Roselle plants are used to produce several food products such as hibiscus tea (called as zobo, bissap, or sour tea), jams, and juices. They are also incorporated into jellied confectionaries, ice creams, chocolates, puddings, cakes, and dairy products as a flavoring agent (Farag et al., 2015).

The extensive use of roselle in a wide variety of products stemmed from the rich nutritional value it has. Humans are greatly interested in foods that have good nutrition and medicinal properties. In addition to the attractive red color of this plant, its nutritional composition, including many bioactive and health-beneficial components, gives cause for the great interest from all over the world. It has been revealed that roselle extracts obtained from its flower, calyxes, and seeds have bioactive properties that may play an important role in the prevention of chronic diseases such as hypertension, hepatic disease, cardiovascular disease, atherosclerosis, and diabetes (Agoreyo et al., 2008; Ali et al., 2005; Chen et al., 2004; Maganha et al., 2010; McKay et al., 2010; Nzikou et al., 2011; Prenesti et al., 2007). Extracts have been related to reduce high cholesterol levels and can also work as anticancer, antimicrobial, antimutagenic, and antiproliferative agents (Hainida et al., 2008). Furthermore, roselle extract can be used to stabilize body temperature, treat drunkenness, soothe sore throats and coughing, and treat genital troubles, kidney disease, and bladder stones (Maganha et al., 2010; Neuwinger, 2000). The decoction from the seeds can relieve pain during urination and treat indigestion, and it can be used to enhance or induce lactation in cases of poor milk production. Further uses of the decoction include poor letdown and decreases in maternal mortality, as well as the treatment of stomach, liver, and high blood pressure problems (Gaya et al., 2009).

[illegible]

Roselle, having earned its reputation, is one of the rare plants that can manage to contribute to the color, aroma, and taste of a food product. This plant, along with the already present nutrients such as carbohydrates, fats, minerals, and vitamins, has a considerable number of secondary metabolites comprising the organoleptic properties, in addition to the pharmaceutical properties of the plant. These metabolites enclose volatile compounds such as furans, alcohols, aldehydes, ketones, and phenolic compounds, such as anthocyanins and flavonoids. Volatile compounds are responsible for the unique and dainty aroma of roselle and play a key role in the preference of consumers (Topi, 2020). The information about the aroma of the different parts of roselle is still very scarce in literature, yet researchers have managed to detect and identify them in a broad array of volatiles in the groups of ketones, aldehydes, furans, and alcohols originating from different pathways including fermentation, lipid oxidation, and Maillard reactions (Chen et al., 1998; Farag et al., 2015; Juhari et al., 2018; Ningrum et al., 2019; Pino et al., 2006; Ramirez-Rodrigues et al., 2011). Various aroma compounds have been discovered in the roselle plant; however, the unique composition of each volatile compound and how it contributes to the overall aroma is different in each specimen. Aroma-active compounds, also known as key odorants, are the compounds that create the olfactory perception that a person can detect (Contis et al., 1998; Kesen, 2020; Selli et al., 2006, 2008). Gas chromatography-mass spectrometry (GC-MS)-olfactometry is a useful method during the aroma studies for directly characterizing the key odorants in food stuffs by human nose. This technique can be used to detect aroma-active compounds present in very small amounts that have a concentration above a specific threshold (Acree et al., 1997). In the extant literature, there are only a limited number of studies investigating the aroma-active profile of roselle and its products (Juhari et al., 2018; Ramirez-Rodrigues et al., 2011; Zannou et al., 2020).

Phenolic compounds are the other crucial components of roselle, as they form the attractive color and also contribute to the bioactivity of the plant. These secondary metabolites have been selected as the subject of many scientific studies, and the innovative food products that are enhanced with phenolics have gained popularity in the recent decades, as they possess numerous proven health benefits, particularly against diabetes, cancer, neurodegenerative diseases, and cardiovascular diseases (Kelebek et al., 2020; Pandey & Rizvi, 2009; Sen & Sonmezdag, 2020). In that manner, roselle has gained attention with its great phenolic potential due to its anthocyanin and flavonoid content and bright red color. In the case of the phenolic compounds of roselle, the most promoting parts of the plant are its calyxes and petals (Tsai et al., 2002). Owing to its phenolic composition, roselle calyxes are an excellent source of natural antioxidants, providing even higher levels than well-known sources such as raspberries and blueberries (Carvajal-Zarrabal et al., 2005; Juliani et al., 2009).

Roselle, as mentioned earlier, includes a wide variety of compounds and all have different contributions to the physiology of the plant. This chapter aims to document the detailed information on the aroma, aroma-active, and phenolic compounds of the roselle (*H. sabdariffa* L.) plant and its products. It will also focus on the antioxidant properties of this plant.

2 AROMA COMPOSITION OF ROSELLE AND ROSELLE PRODUCTS

It is a known fact that the main parameters affecting the consumer preferences and the quality of a food product are the appearance, aroma, and taste (Yang et al., 2013). Among these features, aroma is a prevailing factor, which has an important role in the quality of roselle and its products. These compounds are composed of a wide variety of chemical groups including alcohols, norisoprenoids, volatile phenols, aldehydes, esters, terpenes, ketones, furans, and so on.

Although the roselle plant is emerging as a very competitive target for phytochemical studies, very little is known about its volatile composition. Such knowledge can be suspected to be relevant for understanding its olfactory and taste properties. The studies investigating the volatiles in roselle and its by-products are still very scarce. The dainty and delicate aroma of the roselle plant was reported to be generally derived from fatty acid and sugar compounds. Although a limited number of studies have been conducted on the aroma of roselle, it has been reported that furans, alcohols, aldehydes, and ketones play a major role in the aroma of this plant and its products (Chen et al., 1998; Pino et al., 2006). The summary of detected volatiles in roselle calyxes, flowers, seeds, and teas is presented in Table 10.1.

The consumption of roselle as a type of herbal tea is very common all over the world, and hence the studies investigating aroma compounds of this plant are generally focused on roselle teas. One of the earliest studies examining the volatiles of roselle tea was conducted by Chen et al. (1998). The authors applied the Likens-Nickerson steam distillation procedure for isolating volatiles and these compounds were analyzed by GC-MS.

TABLE 10.1
Prominent Volatile Composition of Roselle and Roselle Products Reported.

Product/Country	Prominent Volatile Compounds	Volatile Extraction Method	References
Dried roselle calyxes (Mexico)	Geraniol Menthol Benzaldehyde Linalool γ-Undecalactone Ethyl methyl phenylglycidate	SPME	Camelo-Méndez et al. (2013)
Fresh roselle calyxes (Taiwan)	Hexanol (*E*)-2-Hexenal (*Z*)-3-Hexenol 2-Hexanol Eugenol α-Terpineol Limonene Linalool oxide	Likens-Nickerson steam distillation	Chen et al. (1998)
Dried roselle flowers (Sudan)	Furfural Hexanal Ethyl acetate 2-Methylpropyl acetate 3-Methyl butanol acetate 1-Hexanol Phenylethyl alcohol 2-Ethyl-1-hexanol 1-Nonanol 6-Methyl-5-hepten-2-one 2,4-Di-tert-butylphenol	SPME	Tahir et al. (2017)
Dried roselle tea (Taiwan)	Furfural 5-Methyl-2-furfural Eugenol α-Terpineol Linalool oxide Acetic acid	Likens-Nickerson steam distillation	Chen et al. (1998)
Dried roselle tea (Cuba)	α-Terpineol Linalool Linalool oxide Furfural 5-Methyl-2-furfural Hexanal	Likens-Nickerson steam distillation	Pino et al. (2006)
Fresh and dried roselle tea (Mexico)	Hexanal Heptanal, Limonene Octanal 6-Methyl-5-hepten-2-one Nonanal 1-Octen-3-ol Acetic acid Decanal Bornylene (*E*)-2-Nonenal 1-Octanol [illegible]	SPME	Ramirez-Rodrigues et al. (2011)

Continued

TABLE 10.1
Prominent Volatile Composition of Roselle and Roselle Products Reported.—cont'd

Product/Country	Prominent Volatile Compounds	Volatile Extraction Method	References
Dried roselle tea (Benin)	Furfural 3-Penten-2-ol 2-Methyl-3-buten-2-ol Benzyl alcohol Acetic acid Nonanal Hexanal Benzaldehyde 3-Hydroxy-2-butanone Eugenol DL-Limonene Methyl-2-furoate	Liquid-liquid extraction	Zannou et al. (2020)
Roselle seeds (Malaysia and China)	2-Methylpropanal 2-Methylbutanal 3-Methylbutanal α-Phellandrene Hexanal 2-Ethyl-1-propanol Sabinene 4-Methyl-2-hexanone β-Phellandrene 3-Methylbutanol 2-Pentylfuran 1-Pentanol *p*-Cymene 1-Hexanol 2-Ethyl-5-methylphenol	Dynamic headspace sampling	Juhari and Petersen (2017)

SPME, solid-phase microextraction.

Up to 37 aroma compounds were characterized in the study and classified into four groups such as fatty acid derivatives, sugar derivatives, phenolic derivatives, and terpenes. Six-carbon alcohols and aldehydes formed as fatty acid derivatives including hexanol, (*E*)-2-hexenal, (*Z*)-3-hexenol, and 2-hexanol are reported to dominate the volatile composition of fresh roselle calyxes, while a small amount of these compounds were found in dried and frozen roselle teas. The formation of these compounds was reported to be the result of the lipoxygenase action on unsaturated fatty acids, especially during the disruption of cell structure in the presence of oxygen or the thermal decomposition of the fatty acids (Kesen et al., 2013; Schlozhauer et al., 1996). As for sugar derivative volatiles, furfural and 5-methyl-2-furfural were found in high amounts in dried roselle tea (Ramirez-Rodrigues et al., 2011; Zannou et al., 2020). It was proposed that these compounds were formed mainly during the drying process as a result of sugar degradation. Eugenol, as a phenolic derivative, originating from phenylalanine by enzymatic synthesis was reported to be one of the most abundant volatile compounds in both fresh roselle and its tea. This compound makes a significant contribution to the odor of foods because of its low detection threshold (6 μg/L) value (Selli, 2007). In the final group of terpenes, α-terpineol, limonene, and linalool oxide were found to be the three dominant compounds in fresh roselle calyxes. The drying process was reported to cause a drastic decrease in the amount of these compounds, as they were found in much lower concentrations in roselle teas.

The following research studied the volatile composition of roselle teas obtained from Cuba with the same extraction procedure as the previous study by Pino et al. (2006). A total of 81 aroma compounds were identified and quantified from the roselle tea samples. Among these, terpenoids such as linalool and α-terpineol were found to be the main volatile compounds. Fatty acids, alcohols, furans, aldehydes, esters, and volatile phenols were detected in minor amounts. Similar to Taiwanese roselle tea (Chen et al., 1998), the sugar derivatives furfural and 5-methyl-2-furfural were also detected from the degradation of ascorbic acid and sugars.

The volatile composition of roselle teas may differ according to the method of preparation and the conditions during these processes. Ramirez-Rodrigues et al. (2011) studied teas obtained from hot (98°C for 16 min) and cold (22°C for 240 min) infusions of fresh and dried roselle calyxes in terms of volatile aroma compounds by using the solid-phase microextraction (SPME)/GC-MS method. This study also investigates the contribution of each volatile to the overall aroma of roselle teas. Researchers identified a total of 32 aroma compounds, classified as aldehydes, alcohols, terpenes, and acids. Aldehydes comprised the most abundant amounts in overall composition of the infusions. This outcome could result from the fact that aldehydes are polar compounds and would be expected to be highly extracted by the polar water. Nonanal, decanal, octanal, and 1-octen-3-ol were reported as the major volatile compounds commonly found in all roselle beverages. Researchers also indicated that the hot infusion process increased the quantity and intensity of aroma compounds in dried roselle tea, showing that the hot infusion had the highest number of total volatiles. 1-Octen-3-one and 6-methyl-5-hepten-2-one were the newly identified ketones in roselle teas and they had the importance of contributing to the overall aroma, along with other low-threshold-valued compounds. Additionally, two furan aldehydes, furfural and 5-methyl-2-furfural, were present only in dried and hot-infused teas. As a result of the study, it was proven that extraction temperatures influence the aroma profiles of roselle teas by changing the relative proportions of all these volatiles.

The most recent study about roselle tea volatiles was conducted by Zannou et al. (2020). In this study, the aroma and aroma-active compounds in Beninese roselle infusions obtained from three different hot and cold infusion methods (by the application of aroma extract dilution analysis (AEDA) and GC-MS-olfactometric analyses of the representative aromatics) were elucidated. According to the results, a total of 39 aroma compounds were identified in the Beninese roselle teas, mainly including the groups of furans, alcohols, aldehydes, acids, ketones, volatile phenols, lactones, a terpene, and an ester. Cold-infused teas had a higher number of volatiles, as the brewing time was longer (24 h). This outcome was probably caused by the longer time of contact with water during brewing, resulting in the release of compounds more efficiently, according to their hydrophilic capacities. Regarding the volatiles, the most abundant compounds were found to be furan compounds, followed by alcohols and acids. Furfural and 5-hydroxymethylfurfural formed due to the Maillard reaction during the drying of roselle and had the highest concentrations in the overall aroma profile. 3-Penten-2-ol, detected for the first time in roselle tea, and 2-methyl-3-buten-2-ol, were the most abundant alcohols in the samples. The hot brewing process resulted in a decrease in their concentrations, as was expected. The other high concentration compounds were acetic acid, hexanal, nonanal, and benzaldehyde. Additionally, the applied principle component analysis managed to distinguish clearly between the tea samples, based on their aroma profiles as influenced by different infusion processes.

There are also few studies investigating the volatile composition of roselle calyxes, as they are the most attractive part of the plant that is rich in both volatiles and anthocyanins. The calyxes also have a desired sour taste (Ali et al., 2005; McKay et al., 2010). In one of those studies, four different varieties of dried Mexican roselle calyxes, namely, Rosa, Negra, Sudan, and Blanca, were compared in relation to their levels of volatiles and anthocyanins (Camelo-Méndez et al., 2013). While comparing the cultivars, the researchers tried different solvents (water and ethanol) during extraction and ethanol was reported to exhibit more efficacies in retaining both anthocyanin and volatile compounds. According to GC analysis, geraniol had the highest concentration levels of volatile composition in all the varieties. A total of nine aroma compounds, including geraniol, menthol, benzaldehyde, linalool, γ-undecalactone, and ethyl methyl phenylglycidate, were reported to be present in all roselle samples.

Another study concerning the volatiles of roselle calyxes was conducted by Tahir et al. (2017). They investigated the aroma compounds of red and white dried roselle calyxes using SPME and GC-MS. A total of 19 volatile compounds were determined to be present in white and red roselle calyxes and were classified as aldehydes, alcohols, ketones, esters, and volatile phenols. Furfural, hexanal, ethyl acetate, 2-methylpropyl acetate, 3-methyl butanol acetate, 1-hexanol, phenylethyl alcohol, 2-ethyl-1-hexanol, 1-nonanol, 6-methyl-5-hepten-2-one, and 2,4-di-tert-butylphenol were found as the main contributors of the roselle calyxes aroma profile, as their concentrations were the highest. A high concentration of furfural was detected in the samples. These authors stated that furfural was formed in the samples mostly during the drying process, similar to previous studies (Farag et al., 2015; Ramirez-Rodrigues et al., 2011; Zannou et al., 2020).

The flowers of the roselle plant are also quite attractive, with white or pink/red petals and unique odor properties. Researchers tried to determine the volatile composition of roselle flowers. Farag et al. (2015) used SPME coupled with GC-MS to extract and profile the aroma compounds and primary metabolites in the flowers of viz. Aswan and Sudan-1 roselle cultivars. Aswan *cv.* had higher concentrations of volatiles, with 74 compounds, when compared with the Sudan variety, which had 71 compounds. The authors detected that these two varieties had very similar aroma patterns qualitatively, with having a difference quantitatively. The main volatile classes were reported to be sugar- and fatty acid-derived compounds in both cultivars. They also identified the terpenoids (α-terpinene, β-phellandrene, and (*Z*)-geranylacetone), phenylpropanoids, aromatic derivatives ((*E*)-cinnamaldehyde, *p*-cresol, and eugenol), and amino acid-derived volatiles. Within all these compounds, furfural, (*E*)-cinnamaldehyde, acetic acid, and 1-octen-3-ol appeared to be the most important compounds accounting for the differences between the two cultivars. Similarly, it was reported that the drying temperatures of the roselle flower samples could also have a significant effect on the aroma of the product, favoring the preservation of short-chain alcohols and aldehydes or the accumulation of Maillard degradation products.

In another study, dynamic headspace sampling (DHS) was performed for the extraction of volatiles of roselle flowers obtained from different preparation methods (Juhari et al., 2018). Grinding, blending with water, heating, and grinding with water were used as preparation methods for the samples. A total of 125 aroma compounds were revealed and classified, including a wide variety of groups: terpenes, aldehydes, esters, ketones, alcohols, furans, acids, sulfur compounds, lactones, and others. Within these, terpenes (limonene, α-terpineol, and 1,8-cineole) and furan aldehydes (furfural) were reported to have the highest concentrations in the samples. The greatest amount of aroma compounds were discovered in the heated and ground roselle flower samples.

As mentioned earlier, the information detailing the composition of volatiles in roselle calyxes and flowers can be ascertained from existing literature. However, the composition properties of roselle seeds have rarely been studied, as the uses of roselle seeds are a lot less attractive than the flowers and calyxes. In one related study, researchers compared the volatile composition of Malaysian and Chinese roselle seeds to emphasize the effects of geographic origin (Juhari & Petersen, 2017). The roselle seeds from Malaysia were reported to be characterized by alcohols, aldehydes, and ketones, whereas samples from China were rich in mostly terpene compounds (Table 10.1). After the application of the DHS extraction method, these volatiles were reported in abundance: 2-methylbutanal, 3-methylbutanal, and hexanal in the aldehyde group; 1-hexanol, 2-methyl-1-propanol, and 1-pentanol as alcohols; α-phellandrene, β-phellandrene, sabinene, and *p*-cymene as terpenes.

Up to this point, an enhanced summary of the volatiles contained in the various roselle parts and roselle products have been discussed. As it can be noticed, even if the used raw materials are the same, there can be many differences in the volatile composition of the roselle plant. These differences in identified compounds and concentrations might be attributed to a number of parameters, including the geographic origin of the plant, its maturity state, the climate of the thriving area, cultivar change, preharvest and postharvest handling methods, and the processing of the plant. Additionally, the extraction procedure, including solvent polarity and the extraction conditions (time, solute concentration, and temperature), plays a substantial role in obtaining an accurate measurement of the roselle volatiles (Gonzalez-Palomares et al., 2009; Hidalgo-Villatoro et al., 2009).

3 KEY ODORANTS OF ROSELLE AND ROSELLE PRODUCTS

Every part of roselle plants, especially their striking calyxes and alluring flowers, has a unique and dainty aroma, attracting both insects and humans (Juhari et al., 2018). Although the commercial value of this plant and its consumption around the world are very high, the information about its key odorants and their production pathways is still very scarce. There are only three studies that have investigated the key odorants in roselle tea and contributed to the understanding of the volatile composition that forms the odor of roselle teas (Juhari et al., 2018; Ramirez-Rodrigues et al., 2011; Zannou et al., 2020).

Aroma-active compounds, or key odorants, are one or many volatiles that can create an olfactory perception in humans and are responsible for the odor of foods, event at low concentrations (Acree et al., 1997). Several research studies have been conducted to determine these compounds, as their contribution to the odor profile cannot be determined with only the application of GC. Therefore a combined system, namely, gas chromatography-olfactometry (GC-O), was developed for the implementation of aroma-active compound analysis. This system combines chromatography with the human nose to be used as a detector in order to define the quality and intensity of the olfactory sensation. GC-O has been suggested for use for the first time by Fuller et al. (1964), yet the system they had proposed was difficult to be reproduced because of the malaise caused by sniffing hot dry effluent gases and the sensitivity of volatile compounds. Later, Dravnieks and O'Donnell (1971) partially fixed the problem by incorporating humidified air to the GC-O system, as seen in Fig. 10.1. For each single aroma-active compound from the GC, the human nose is capable of detecting the activity, quality, and intensity of an odor. Since the first time GC-O had been used in an experiment, numerous techniques have been employed to gather and provide GC-O data and to evaluate the sensory contribution of single potent aroma-active compounds, which can be categorized in the following three categories: dilution analysis, detection frequency method, and time-intensity method (TIM) (Linssen et al., 1993; Pollien et al., 1997; Serot et al., 2001).

Dilution analysis is the most commonly used technique for the detection of aroma-active compounds. This technique consists of the determination of relative power, regarding the presence of the odor of aroma compounds in an extract (Van Ruth, 2001). Two research groups have investigated the development of this technique. The first group developed the Charm analysis (Acree et al., 1984) and the second group established AEDA (Grosch, 1994). During Charm analysis, a series of dilutions of aromatic extracts are randomly sniffed and graphed according to flavor dilution (FD) factor, retention time, or indices. During AEDA, the concentrated extract is diluted step by step and each diluted extract is injected and sniffed until the odor is not perceived, and the maximum dilution is reported as a FD value. AEDA is the most preferred technique, as it is easy to use and the obtained results reflect the samples aroma well. However, the only disadvantage of this technique is that it takes a lot of time (Delahunty et al., 2006).

The application of the AEDA procedure to detect the key odorants contained in roselle samples was determined in only a recent study (Zannou et al., 2020). In this study, the effects of hot and cold brewing techniques on the key odorants of Beninese roselle teas were investigated. Liquid-liquid extraction (LLE) was applied to obtain volatile concentration levels, and following the performance of AEDA by trained panelists made it possible to fully understand the aroma of roselle tea. A total of 23 different key odorants were detected in cold and hot infusions after a GC-MS-olfactometric analysis. The influential key odorants, with regard to FD factors in all samples, were prevailingly furans, alcohols, and aldehydes. Aroma compounds such as furfural (FD:4096, caramel, bready), 5-methyl-2-furfural (FD:1024, caramel), 3-penten-2-ol (FD:128, oily, herbal), and 3-methyl-3-buten-2-one (FD:128, fruity) demonstrated higher FD factors than other aroma-active compounds, indicating that the contribution of those compounds to the overall aroma of roselle teas was greater. Within all these compounds, two furan compounds, i.e., furfural and 5-methyl-2-furfural, with characteristic caramel, bready odor notes became

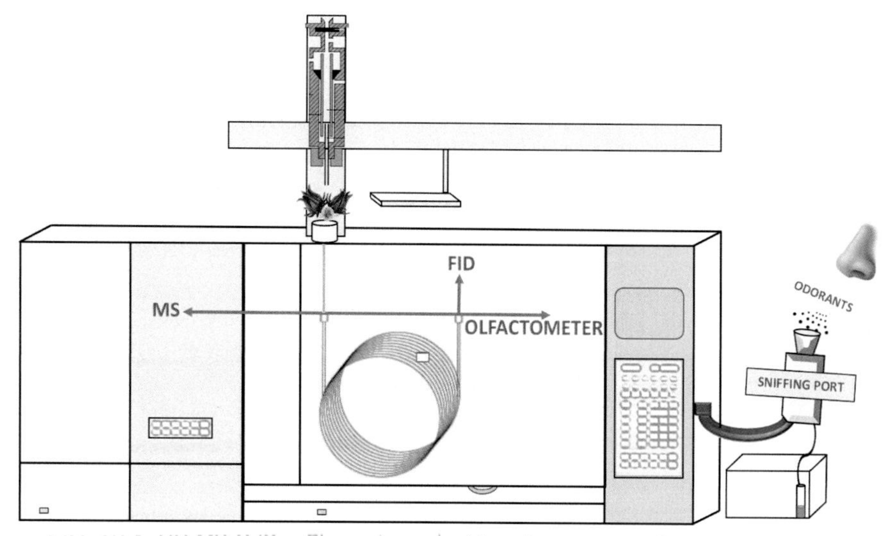

FIG. 10.1 GC MS-O (Gas Chromatography-Mass Spectrometry-Olfactometry) system.

prominent, as their FD values were the highest. Apart from these furans, hexanal (green, herbal), nonanal (herbal, grassy), and benzaldehyde (almond) were determined as aroma-active aldehydes with an FD value changing between 16 and 128. The aldehydes are of high importance, as their odor threshold values are prevalently lower than those of aroma compounds; hence, it is safe to conclude that they have considerably powerful effects on foods' characteristic aromas. Also, acetic acid was detected in Beninese roselle teas, having the highest FD value of 512 in cold infusions, following furan compounds. This acid was reported to be one of the aroma compounds found in Mexican roselle teas (Ramirez-Rodrigues et al., 2011), but in this study, it was the first time these compounds were attributed to be key odorants (Table 10.2). As a result of this study, the odor of Beninese roselle tea was reported to be characterized by caramel, bready, fruity, and herbal odor notes. This outcome was supported with applied representativeness tests. These tests include similarity and intensity tests to compare the obtained volatile extracts with the original sample and to check the efficiency of the extraction method. According to these tests, volatile extracts obtained from the LLE technique had 87% intensity and 62% similarity to the original sample, which can prove the applicability of LLE in olfactometric analysis.

The TIM procedure was applied to roselle teas during a study conducted by Juhari et al. (2018). The researchers collected roselle samples from eight different countries including Australia, China, Chad, Malaysia, Mexico, Nigeria, Sudan, and Thailand and investigated

TABLE 10.2
Key Odorants Reported in Roselle Teas From Different Origin.

	ODOR DESCRIPTION			
Key Odorants	**Mexican Roselle Tea[a]**	**Beninese Roselle Tea[b]**	**Roselle Tea From Different Countries[c]**	**Olfactometric Techniques**
3-Methyl-3-buten-2-one	nd	Fruity	nd	AEDA
α-Terpinolene	nd	nd	Woody, nutty, floral, medicinal, grassy, waxy, coconut, insectlike, plasticlike	DFM
2-Methyl-3-buten-2-ol	nd	Sweet, fruity	nd	AEDA
Butanal	nd	nd	Green, grassy, green apple, floral	DFM
Hexanal	Green, grass, nutty	Green, herbal	Grassy, pungent, chemical, gluelike, earthy, mushroom	AEDA, TIM, DFM
3-Octanone	Butter, cookie, baked	nd	nd	TIM
2-Hexenal	nd	nd	Grassy, fresh tree, sweet, pungent	DFM
3-Penten-2-ol	nd	Oily, herbal	nd	AEDA
Octanal	nd	Lemon, citrus	Leaf, grassy, mint, lemon, fruity, spicy, musty, cheeselike	TIM, DFM
DL-Limonene	nd	Citrusy, lemon	Fresh, orange, citrus, fruity	AEDA, DFM
1-Octen-3-one	Mushroom, dirt, green		Earthy, dried fruit, sweet, oily, grassy, forestlike	DFM
3-Hydroxy-3-methyl-2-butanone	nd	Fatty	nd	AEDA
(*E*)-2-Octenal	Rancid nuts	nd	Rancid, medicinal, cheeselike, sweet	TIM, DFM

2-Methylfuran	nd	nd	Sweet, vanilla, cheese, grassy, fresh, floral, leafy	DFM
1-Hydroxy-2-propanone	nd	Caramel	nd	AEDA
Nonanal	Fruity, green	Herbal, grassy	Nutty, dried leaf, mild cheese, grassy, herblike	AEDA, TIM, DFM
2-Methyl-1-penten-3-one	nd	nd	Citrusy, orange, dried flower, rose	DFM
Acetic acid	nd	Vinegary	nd	AEDA
Furfural	Sweet, baked bread	Caramel, bready	Sweet, fruity, floral, cheeselike	AEDA, TIM, DFM
1,8-Cineole	nd	nd	Sweet, fruity, lemongrass, grassy, insectlike, nutty	DFM
Benzaldehyde	nd	Almond	nd	AEDA
1-Octen-3-ol	Mushroom, dirt, metallic	nd	Garlic, mushroom, spice, berry, peppermint	TIM, DFM
5-Methyl-2-furfural	nd	Caramel, creamy	nd	AEDA
Decanal	nd	Sweet, nutty	nd	TIM
2-Ethylfuran	nd	nd	Sweet, fruity, woody, green	DFM
2-Methylbutanoic acid	nd	Cheesy, acidic	nd	AEDA
(*E*)-2-Nonenal	Cucumber, green, floral	nd	nd	TIM
Cumene	nd	nd	Minty, fresh, sweet, fruity, nutty, insectlike, citrus	DFM
Benzyl alcohol	nd	Floral	nd	AEDA
Linalool	Floral, woody, citrus		nd	TIM
2-Formylpyrrole	nd	Tea, herbal	nd	AEDA
1-Octanol	Fresh leather, chemical		nd	TIM
Pantolactone	nd	Burnt, spicy	nd	AEDA
Pentanal	nd	nd	Oily, pungent, sweet, grassy, musty	DFM
1-Nonanol	Chemical, painty	nd	nd	TIM
Methyl-2-furoate	nd	Caramel	nd	AEDA
Eugenol	Sweet spices	Spicy	nd	AEDA, TIM
(*E,E*)-2,4-Nonadienal	Rancid nuts, citrus, green	nd	nd	TIM
β-Pinene	nd	nd	Floral, sweet, grassy, gluelike, fruity	DFM
2-Undecenal	Green, grass	nd	nd	TIM
Geranylacetone	Fruitlike, apple sauce	nd	nd	TIM

AEDA, aroma extract dilution analysis; *DFM*, detection frequency method; *nd*, not determined; *TIM*, time-intensity method.

[a] Ramírez-Rodrigues et al., 2011.

[b] Zannou et al., 2020.

[c] Juhari et al., 2018.

the effects of drying processes, as well as cultivar changes, on key odorants of teas. After performing TIM following DHS by nine panelists, different groups of compounds, such as terpenes, aldehydes, ketones, esters, furans, and alcohols, were attributed to be key odorants in the aroma of roselle teas. The most abundant group was reported to be terpenes, with six aroma-active compounds that include *β*-pinene (sweet, green), cumene (fresh, green), limonene (citrus, fruity, minty), 1,8-cineole (sweet, minty, grassy), *p*-cymene (solventlike, fruity odor), and *α*-terpinolene (woody, pine odor) detected by GC-O (Table 10.2). The compounds detected by the highest number of panelists were *α*-terpinolene, butanal (green, grassy, floral), 2-hexenal (green, leaf), 1-octen-3-ol (garlic, mushroom, peppermint), and hexanal (grassy, pungent). These compounds can be counted among the most important odorants, as they were detected consistently in GC-O analysis. Also, it was reported that both drying and the cultivar differences had important effects on key odorants in roselle teas.

Ramirez-Rodrigues et al. (2011) have used Osme technique to perform olfactometric analysis of Mexican roselle teas. The Osme technique requires the assistance of computer tools, as well as trained participants, to sniff and report the intensities and duration of the odor of aroma-active compounds coming from the GC device (Miranda-Lopez et al., 1992). Four participants are generally required to perform the GC-O analysis with this technique. The panelists assessing the TIM procedure revealed a total of 22 aroma-active compounds in the samples. Among these, the most intense odorants were reported to be 1-octen-3-one (mushroom) and nonanal (green) followed by geranylacetone (fruitlike, apple sauce), eugenol (sweet spices), and (*E*)-2-nonenal (cucumber, green, floral). It was also reported that 1-octen-3-one, geranylacetone, and (*E*)-2-nonenal are compounds identified for the first time in roselle. Additionally, 1-octen-3-one had the highest intensity, and although this compound was reported to be one of as light-induced off-flavors in apple juice, it contributed to the aroma of roselle teas with favorable fruitlike odor notes.

4 ROSELLE PHENOLICS

H. sabdariffa L., namely, roselle or red sorrel, is an important member of the Malvaceae family, known to contain a great antioxidant potential, high phenolic content, and bright red color. In phenolic compounds, the most promoting part of the plant is declared to be its calyxes and petals (Tsai et al., 2002). These plant parts have been used since ancestral times, mainly in the preparation of different types of hot and cold beverages, with strong antioxidant characteristics. These qualities help treat chronic diseases, because of their antihypertensive, antihyperlipidemic, antiinflammatory, antimicrobial, and anticancer effects, among others. These positive effects have been related to the presence of anthocyanins, flavonoids, phenolic acids, and other organic acids specific to roselle samples (Borrás-Linares et al., 2015). It was previously elucidated that anthocyanins and other phenolic compounds are the principal sources of antioxidants in roselle (*H. sabdariffa*) samples. These components can scavenge free radicals, participate in the regeneration of other antioxidants, and protect cell constituents against oxidative damage. One of the main reasons for the high antioxidative effects against lipid peroxidation by the roselle plant is declared to be related to inhibiting the malondialdehyde formation (Jabeur et al., 2017). Acetaldehyde and malondialdehyde are two main derivatives of lipid peroxidation responsible for the formation of malondialdehyde-acetaldehyde adducts that are causing atherosclerotic and cardiac diseases (Zimmerman et al., 2017).

In some parts of the world, such as Mexico, roselle calyxes are traditionally used for preparing refreshing drinks (mostly as tea infusions), jams, jellies, liquors, flour for biscuits, wine, and food colorants (Cid-Ortega & Guerrero-Beltrán, 2015; Sirag et al., 2014). Tea infusions of *H. sabdariffa* L. are also commonly consumed as a traditional and healthy refresher in Benin, namely, "Sinko" (Zannou et al., 2020). Additionally, roselle leaves are also used for making pickles, cooking, and preparing soup because of their substantial folic acid and iron content.

Owing to their rich phytochemical content and health-promoting benefits, mainly derived from anthocyanins, other flavonoids, organic acids and their derivatives, vitamins (mainly C, B, and E), and *β*-carotene, *H. sabdariffa* L. and its products are considered to be a high-potency, functional food. Over time, an extensive literature has been developed concerning the potential benefits of roselle calyxes, because of their rich content of natural antioxidants, providing even higher levels than well-known antioxidant-containing fruits, such as blueberries and raspberries (Juliani et al., 2009).

The bioactive ingredients of roselle are quite unstable and may vary depending on the cultivar, drying conditions, extraction procedure, and infusion processes (Aurelio et al., 2008). Several studies focused on the quality loss during the drying process of leafy vegetables and medicinal and aromatic plants (MAPs), including *H. sabdariffa* L., *Moringa oleifera*, and *Melissa officinalis* (Branisa et al., 2017; Kumar et al., 2015; Saini et al.,

2014). The main objective of the drying process of plants, herbs, and flowers is to prevent microbial growth in and enzymatic deterioration of the plant material by way of reducing the internal humidity (water activity) with different methods, such as natural sunlight drying, convective oven drying, microwave drying, infrared drying, and freeze-drying. However, traditional unconscious drying processes, such as direct sun exposure, may lead to a significant loss of phenolics, especially anthocyanins. Kumar et al. (2015) reported that the processes of drying at room temperature and freeze-drying are the most suitable drying methods for the preservation of chlorophyll, ascorbic acid, and antioxidant compounds, whereas direct sun exposure, microwave, oven, cross-flow, and infrared drying processes resulted in a significant loss in total phenolic content, antioxidant capacity, and vitamin content. Therefore the protection of phenolics that exist in calyxes has direct influence on the health-promoting qualities and the economic value of the roselle plant.

Studies have reported that roselle contains 57% anthocyanin, 38% phenolic acid, and 5% flavor (Piovesana et al., 2019) (Fig. 10.1). Profiles of these compounds are given in Fig. 10.2. Deceptive data on the phenolic compounds obtained from detailed analysis by liquid chromatography (LC)-MS/MS are given in Table 10.3.

Roselle possesses a wide range of remarkable phenolic compounds, and in this chapter, the flavonoids, anthocyanins, and antioxidant properties of roselle will be summarized.

4.1 Roselle Anthocyanins

Anthocyanins are water-soluble, natural pigments that form the color of fruits, vegetables, flowers, and other plants ranging from pink to purple. They are the glycosylated polyhydroxy and polymethoxy products of flavylium salts (2-phenylbenzopyrylium) and are composed of some sugars and nonsugar (aglycone anthocyanidins) constituents present in glycoside that form in the cell cytoplasm. Generally, a sugar molecule is linked to anthocyanidins from the third carbon atom in a $C_6C_3C_6$ skeleton. The main differences between anthocyanins arise from the number and position of hydroxyl and methoxyl groups, the position of bound sugars, and the type of compounds that bind to sugars in the molecule.

About 700 anthocyanins, derived from 27 aglycons known as "anthocyanidins," have been identified in nature. Distinct from other flavonoids, anthocyanins are known for their high ability to absorb visible light. However, this delicate color feature of anthocyanins is highly dependent on the stability of the aqueous and pH levels of the internal plant media. The antioxidant activity of anthocyanins is fundamentally based on their chelating capacity by metal ions and protein-binding properties. The increase in the number of −OH groups in the structure, the *o*-dihydroxy structure in the β-ring, and the −OH bonding to the third and fourth carbon increases antioxidant activity (Castañeda-Ovando et al., 2009; Wallace et al., 2016). Anthocyanidins can be sorted in terms of their in vitro antioxidant activity, from the strongest to the weakest, as cyanidin, delphinidin, malvidin, peonidin, and petunidin. Anthocyanins have long been used in several industries as natural alternatives for synthetic colorants, as well as for pharmaceutical purposes as a rich source of precious phytochemicals.

Anthocyanins are the most abundant phytochemicals in *H. sabdariffa*, which provide its bright red color (Carvajal-Zarrabal et al., 2005). Delphinidin-3-sambubioside (namely, hibiscin), cyanidin-3-sambubioside, delphinidin-3-glucoside, cyanidin-3-glucoside, cyanin, and malvin are the well-known anthocyanins in roselle calyx extracts, reported in the literature (Figs. 10.3 and 10.4) (Alarcón-Alonso et al., 2012; Carvajal-Zarrabal et al., 2005; Pinela et al., 2019).

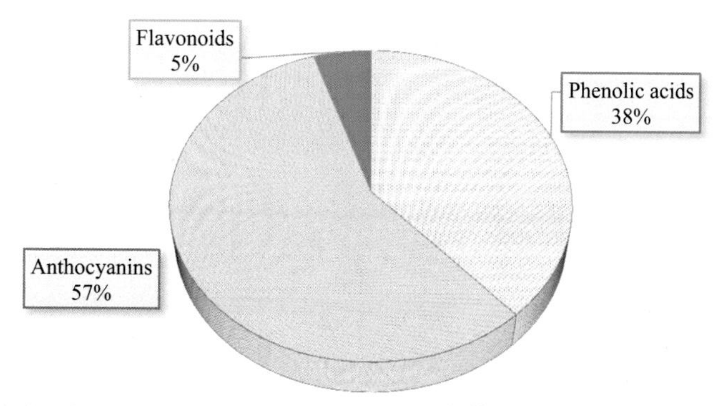

FIG. 10.2 Phenolic profile of *Hibiscus sabdariffa* (Piovesana et al., 2019).

TABLE 10.3
Chromatographic, Ultraviolet-Visible, and Mass Spectroscopic Characteristics of the Phenolic Compounds of *Hibiscus sabdariffa* L.

No	PHENOLIC COMPOUND	MOLECULAR FORMULA	λ_{MAX} (NM)	$[M+H]^+$	MS^2 (+)	$[M-H]^-$	MS^2 (−)	References
1	Hydroxycitric acid	$C_6H_7O_8$	261	–	–	207	115, 127, 189	Pérez-Ramírez et al. (2015), Piovesana et al. (2019), Borrás-Linares et al. (2015)
2	Hibiscus acid	$C_6H_5O_7$	261	–	–	189	127	Pérez-Ramírez et al. (2015), Piovesana et al. (2019), Borrás-Linares et al. (2015)
3	3-Caffeoylquinic acid	$C_{16}H_{18}O_9$	325	–	–	353	191, 179, 135	Piovesana et al. (2019), Borrás-Linares et al. (2015), Wang et al. (2014)
4	Delphinidin-3-sambubioside	$C_{26}H_{27}O_{16}$	280, 520,525	597	303	–	–	Pérez-Ramírez et al. (2015), Piovesana et al. (2019), Zhang et al. (2014)
5	3-*p*-Coumaroylquinic acid	$C_{16}H_{18}O_8$	310	339	147, 119	337	191, 163, 119	Piovesana et al. (2019)
6	Cyanidin-3-sambubioside	$C_{26}H_{27}O_{15}$	280, 518	581	287	–	–	Pérez-Ramírez et al. (2015), Piovesana et al. (2019), Zhang et al. (2014)
7	Cyanidin-3-*O*-glucoside	$C_{21}H_{21}O_{11}$	515, 269	449	287	–	–	Zhang et al. (2014)
8	Neochlorogenic acid	$C_{16}H_{18}O_9$	326	355	163	353	191, 179, 135	Pérez-Ramírez et al. (2015), Piovesana et al. (2019), Wang et al. (2014)
9	Cryptochlorogenic acid	$C_{16}H_{18}O_9$	323	355	163	353	191, 179, 173, 135	Pérez-Ramírez et al. (2015), Piovesana et al. (2019), Wang et al. (2014)
10	Myricetin-3-sambubioside	$C_{26}H_{28}O_{17}$	349	613	319	611	317	Pérez-Ramírez et al. (2015), Piovesana et al. (2019)
11	5-*p*-Coumaroylquinic acid	$C_{16}H_{18}O_8$	311	339	147	337	191, 173, 163, 119	Piovesana et al. (2019)
12	Myricetin-3-arabinogalactoside	$C_{26}H_{28}O1_6$	352	–	–	611	316, 317	Pérez-Ramírez et al. (2015)
13	Quercetin-3-sambubioside	$C_{26}H_{28}O_{16}$	354	597	303	595	301	Piovesana et al. (2019)

14	5-*O*-Caffeoylshikimic acid	$C_{16}H_{15}O_8$	292, 329	–	–	335	161, 135	Piovesana et al. (2019), Borrás-Linares et al. (2015)
15	Rutin	$C_{27}H_{30}O_{16}$	354	611	303	609	301	Pérez-Ramírez et al. (2015), Piovesana et al. (2019), Wang et al. (2014)
16	Quercetin-3-glucoside	$C_{21}H_{20}O_{12}$	353	–	–	463	301	Pérez-Ramírez et al. (2015), Piovesana et al. (2019), Borrás-Linares et al. (2015)
17	Kaempferol-3-*O*-rutinoside	$C_{27}H_{30}O_{15}$	345	–	–	593	285	Pérez-Ramírez et al. (2015), Piovesana et al. (2019)
18	*n*-Feruloyltyramine	$C_{18}H_{19}NO_4$	286	–	–	312	178, 135	Pérez-Ramírez et al. (2015)
19	Peonidin-3-sambubioside	$C_{27}H_{31}O_{15}$	528, 279	595	301	–	–	Zhang et al. (2014)

Data obtained from earlier studies Pérez-Ramírez, I. F., Castaño-Tostado, E., Ramírez-de León, J. A., Rocha-Guzmán, N. E., & Reynoso-Camacho, R. (2015). Effect of stevia and citric acid on the stability of phenolic compounds and in vitro antioxidant and antidiabetic capacity of a roselle (*Hibiscus sabdariffa* L.) beverage. *Food Chemistry, 172*, 885–892; Piovesana, A., Rodrigues, E., & Noreña C. P. Z. (2019). Composition analysis of carotenoids and phenolic compounds and antioxidant activity from hibiscus calyxes (*Hibiscus sabdariffa* L.) by HPLC-DAD-MS/MS. *Phytochemical Analysis, 30*(2), 208–217; Zhang, B., Mao, G., Zheng, D., Zhao, T., Zou, Y., Qu, H., Li, F., Zhu, B., Yang, L., & Wu, X. (2014). Separation, identification, antioxidant, and anti-tumor activities of *Hibiscus sabdariffa* L. extracts. *Separation Science and Technology, 49*(9), 1379–1388; Borrás-Linares, I., Fernández-Arroyo, S., Arráez-Roman, D., Palmeros-Suárez, P.A., Del Val-Díaz, R., Andrade-Gonzáles, I., Fernández-Gutiérrez, A., Gómez-Leyva, J.F., & Segura-Carretero, A. (2015). Characterization of phenolic compounds, anthocyanidin, antioxidant and antimicrobial activity of 25 varieties of Mexican Roselle (*Hibiscus sabdariffa*). *Industrial Crops and Products, 69*, 385–394; Wang, J., Cao, X., Jiang, H., Qi, Y., Chin, K. L., & Yue Y. (2014). Antioxidant activity of leaf extracts from different *Hibiscus sabdariffa* accessions and simultaneous determination five major antioxidant compounds by LC-Q-TOF-MS. *Molecules, 19*(12), 21226–21238.

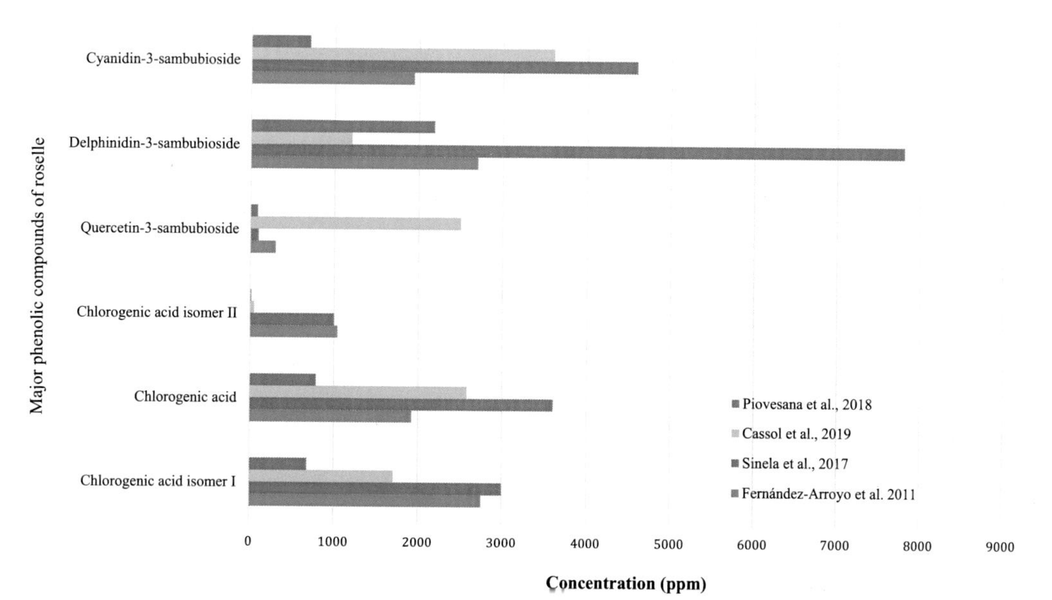

FIG. 10.[illegible] Distribution of six major phenolic compounds of *Hibiscus sabdariffa*.

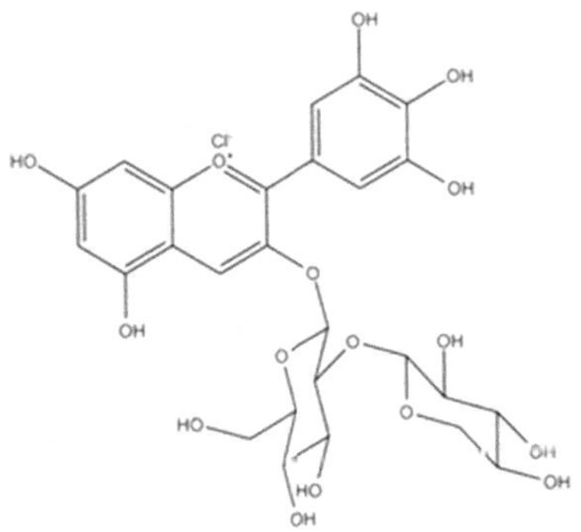

Delphinidin-3-O-sambubioside

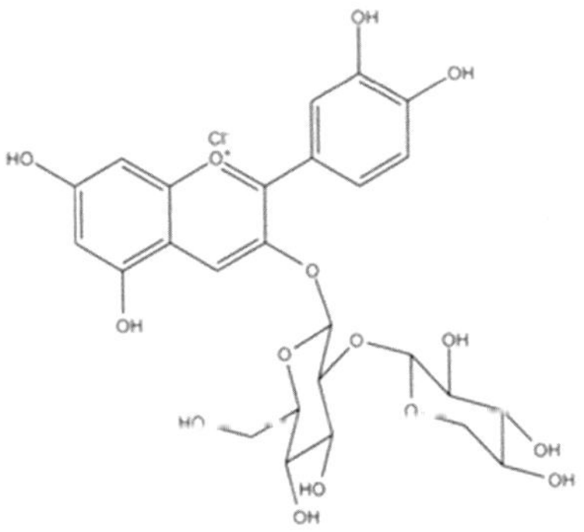

Cyanidin-3-sambubioside

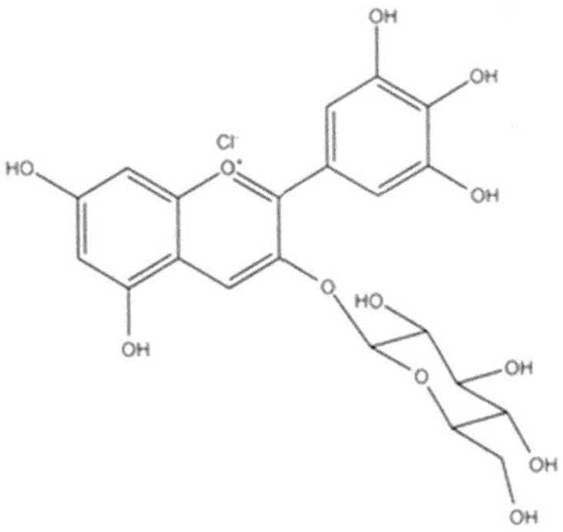

Delphinidin-3-glucoside

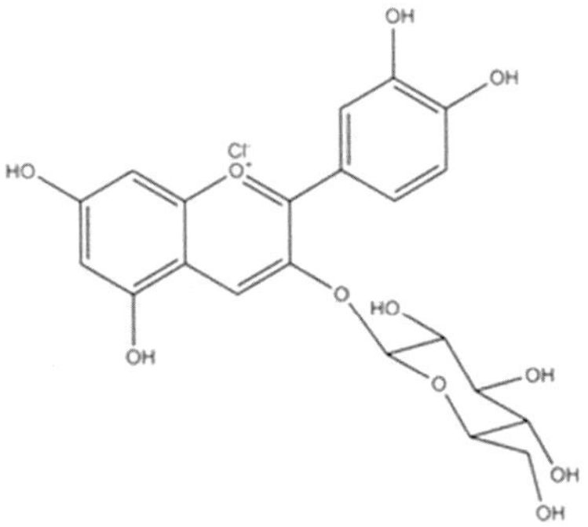

Cyanidin-3-glucoside

FIG. 10.4 Main anthocyanins found in *Hibiscus sabdariffa* L.

H. sabdariffa L. calyxes are generally harvested at high moisture content, and therefore they require an effective drying process. However, it is quite important to protect the sensitive bioactive compounds of the roselle plant during the drying process. Otherwise, the delicate roselle calyxes may lose their health-promoting effects. Previous studies have emphasized that the different drying processes result in alterations to the antioxidant capacity of MAPs.

For instance, in an earlier study, Branisa et al. (2017) performed a detailed comparison of six drying methods (shaded air-drying, freeze-drying, food dehydration, convection oven, microwave, and sundrying) and solvent mixtures to monitor the changes of carotenoids, chlorophyll, anthocyanins, and the antioxidant capacity of two different MAPs grown in Slovakia, *Urtica dioica* and *M. officinalis*. The researchers declared that among all the methods, the freeze-drying process is the most suitable method to preserve chlorophyll and carotenoid content. Oven and microwave drying methods negatively influenced the anthocyanin concentrations because of the applied high temperature. An exciting finding revealed by the study was that the ambient air-dried herbs possessed an even higher content of anthocyanins than the freeze-dried samples. Similarly, Komes et al. (2011) examined the drying methods to observe the changes in pigments of *M. officinalis* L., *Calendula officinalis*, and *Borago officinalis*, such as total phenols, anthocyanins, carotenoids, and chlorophylls *a* and *b*. According to the revealed results, fresh (non-dried) samples possessed the highest amount of polyphenols and carotenoids, while air-dried samples exhibited the highest anthocyanin content. As determined in the other studies, the freeze-drying method was found to be the most suitable drying technique to preserve the chlorophyll content of a plant. In line with the previously mentioned works, Kumar et al. (2015) constructed their work to observe the effects of different drying methods (infrared, cross-flow, microwave, air-, freeze-, oven, and sundrying) on the bioactive compounds of *H. sabdariffa* L. Researchers declared that the oven and microwave drying processes lead to a significant loss of bioactive compounds (including anthocyanins), while freeze- and air-drying were found to be the best drying methods for roselle calyxes. Additionally, researchers concluded that the cold-water extracts obtained by room-dried leaves exhibited a higher total phenolic content than the extracts that were obtained by using organic solvents or hot water.

The number of bioactive compounds present in the roselle sample depends on not only the drying process but also the variety. In this regard, Borrás-Linares et al. (2015) examined 25 different Mexican roselle varieties within the scope of comparing their phytochemical profiles, including phenolics, anthocyanins, flavonoids, and antioxidant and antibacterial characteristics. Among all the varieties, the "Tempranilla," "Talpa," and "Violenta" samples exhibited the highest anthocyanidin contents (4408, 4085, and 4246 mg/100 g dry calyx, respectively), whereas the anthocyanidin contents of the varieties "JB 01 SM," "Blanca 1" (from Guerrero region), and "Blanca 2" (from Jalisco region) were found to be 0 mg/100 g dry calyx (Fig. 10.5).

4.2 Roselle Flavonoids

Flavonoids, the most common phenolic group in nature, provide the organoleptic characteristics of plant-based materials, such as color, aroma, and taste (Peterson & Totlani, 2005). Furthermore, the regular consumption of flavonoids has several proven health benefits, such as anticancer effects, defense against neurologic diseases such as Alzheimer disease, and the

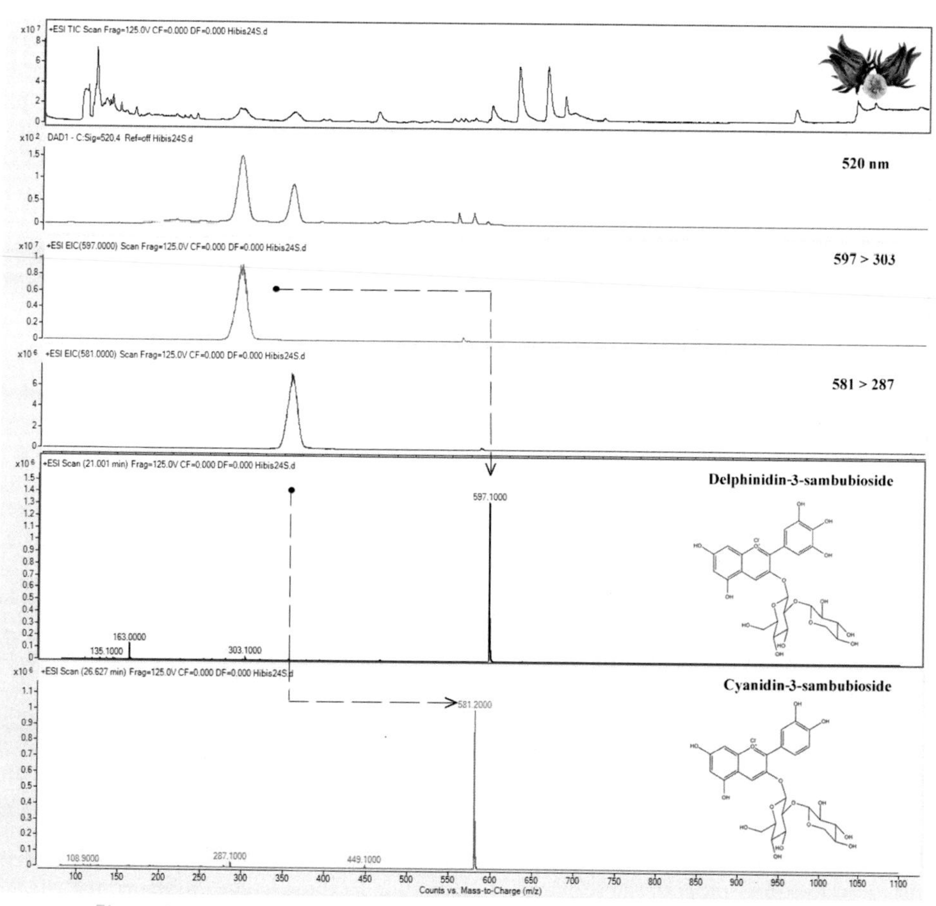

FIG. 10.5 Chromatograms of roselle anthocyanins (delphinidin-3-sambubioside and cyanidin-3-sambubioside).

protection of the circulatory system against atherosclerosis, cardiovascular diseases, etc. (Cid-Ortega & Guerrero-Beltrán, 2015). As there is a great potential for benefits among both plants and humans, phenolic compounds attract significant interest in many sectors and research (Panche et al., 2016).

H. sabdariffa is known for its rich flavonol and phenolic acid content. Among phenolics, roselle mainly possesses hibiscin, gossypetin, gossypetrin, quercetin, luteolin, myricetin, sabdaritrin, sabdaretin, hibiscetin, hibiscetrin, protocatechuic acid (PCA; 3,4-dihydroxybenzoic acid), and their derivatives. One of the most health-promoting bioactive compounds in roselle, hibiscetin, was identified for the first time in *H. sabdariffa* in the study by Perkin (1909), who examined all parts of the roselle (calyxes, corolla, epicalyx, and leaves) by means of an etheric extraction. In an earlier study, Rodríguez-Medina et al. (2009) identified 16 phenolic compounds, namely, hydroxycitric acid, hibiscus acid, chlorogenic acid and two isomers, myricetin 3-arabinogalactoside, quercetin-3-sambubioside, 5-*O*-caffeoylshikimic acid, quercetin-3-rutinoside, quercetin-3-glucoside, kaempferol 3-*O*-rutinoside, *N*-feruloyltyramine, kaempferol-3-(*p*-coumarylglucoside), quercetin, delphinidin-3-sambubioside, and cyanidin-3-sambubinoside, present in roselle aqueous extracts by using LC-DAD-ESI-TOF/IT (diode-array detector/electrospray ionization/time-of-flight/ion trap). Similarly, Piovesana et al. (2019) detected 20 phenolics in hibiscus calyxes by HPLC-DAD-MS/MS. According to revealed data, delphinidin-3-sambubioside was found to be the most abundant phenolic compound detected in hibiscus calyxes (218.2 mg/100 g fresh weight), followed by 3-caffeoylquinic acid (79.2 mg/100 g fresh weight) and cyanidin-3-sambubioside (70.4 mg/100 g fresh weight). Another point to be emphasized in the work of Piovesana et al. (2019) was the total flavonoid content was quantified to be about 26 mg/100 g fresh weight, mainly composed of quercetin, kaempferol, and myricetin. In another study, Sawabe et al. (2005) identified kaempferol-3-*O*-rutinoside, quercetin-3-*O*-rutinoside, and kaempferol-3-*O*-glucopyranoside as the major phenolics obtained from the roselle leaves. Cassol et al. (2019) investigated the difference in extraction yields of phenolics obtained by two different techniques: exhaustive extraction with acidified methanol and microwave-assisted extraction. Results showed that the samples obtained from the acidified methanolic extraction process possessed much higher phenolic contents than those obtained from the microwave-assisted extraction process. Researchers identified a total of 13 phenolic compounds, and among these compounds, 3-caffeoylquinic acid was found to be the most abundant one in both the extraction techniques.

4.3 Phenolic Acids in *Hibiscus sabdariffa* L.

Phenolic acids are known to be one of the major plant phenolics, which are mostly present in bound form. These bonds mainly occur between carboxylic groups and carbohydrates, amino acids, and alcohols, composing glycosides, amides, and phenol esters. The members of this phenolic group can be classified into two distinctive frameworks: hydroxycinnamic and hydroxybenzoic acids. Even though their basic structure is quite similar, they differ in positions and the number of hydroxyl groups (Stalikas, 2007). These compounds are considered well-known antioxidants, especially those with three hydroxyl groups, such as gallic acid. Gallic acid and its derivatives (gallic acid lauryl ester, gallic acid methyl ester, etc.) are commonly used as food additives owing to their high antioxidant potential (Aruoma et al., 1993). Another remarkable natural phenolic acid is PCA (3,4-dihydroxybenzoic acid), which is widely found in herbal teas of MAPs. PCA was declared to be a much more effective antioxidant than trolox by means of scavenging activity against 2,2-diphenyl-1-picryl-hydrazyl-hydrate (DPPH), 2,2′-azino-bis(3-ethylbenzothiazoline-6-sulfonic acid) (ABTS), and other radicals (Li et al., 2011).

PCA is also found in roselle calyx extracts, bringing a potent antioxidant property and making the roselle extracts a valuable health-promoting agent. The chemopreventive and anticarcinogenic properties of roselle PCA have been discussed by a great number of authors in the existing literature. Basically, PCA reduces lipid oxidation, which can be considered the main reason for cell destruction, and relatedly it possesses anticarcinogenic, particularly antiproliferative and apoptotic, effects against several cancers (breast cancer, gastric cancer, etc.) (Carvajal-Zarrabal et al., 2005; Kampa et al., 2004). In fact, PCA is declared to have a remarkable antileukemic effect when it is present in high concentrations (Abubakar et al., 2012; Lin et al., 2007).

Another important member of phenolic acids present in roselle is chlorogenic acid and its derivatives. Chlorogenic acids are water-soluble phenolic compounds, basically derived by the esterification reactions of some specific *trans*-cinnamic acids (Rodrigues & Bragagnolo, 2013). There exist several studies in the extant literature about the antiinflammatory, antibacterial, and several other health benefits of chlorogenic acids, including cardiovascular problems, diabetes, and Alzheimer disease. Coffee is one of the well-known chlorogenic acid intake sources in the human diet. Piovesana et al. (2019) detected chlorogenic acid and its derivatives in *H. sabdariffa* L. calyxes by means of HPLC-DAD-MS/MS. The phenolic profile and antioxidant capacity of tea infusions obtained from several plants, such as roselle tea leaves and MAPs, have always been discussed based upon LC. Kelebek et al. (2019) detected six chlorogenic acids in the same way as seen in St. Johns wort tea infusions (*Hypericum perforatum*) by using LC-DAD-ESI-MS/MS. Similarly, Kelebek (2016) carried out a detailed investigation on black tea phenolics (LC-DAD-ESI-MS/MS) and their antioxidant capacity (by means of DPPH and ABTS). Sonmezdag et al. (2018) characterized the phenolic structure of tea infusions of thyme (*Thymus vulgaris* L.) and identified the substantial types of phenolics, particularly luteolin-diglucuronide-glucuronide and PCA-hexoside.

In an earlier study, the total concentrations of roselle chlorogenic acid and its isomers were found to be in the range of 4.80–10.30 mg/g in dried leaves, representing the majority of the average composition. Researchers declared that neochlorogenic acid was found to be the most abundant chlorogenic acid derivative, with a concentration between 2.87 and 7.16 mg/g, representing 40% of the total chlorogenic acids determined to be present in the roselle samples. In the same study, the second most abundant phenolic acid was found to be cryptochlorogenic acid (ranging from 1.14 to 2.72 mg/g) followed by chlorogenic acid (0.42–1.49 mg/g) (Zhen et al., 2016).

4.4 Antioxidant Properties of Roselle

Antioxidants are key ingredients that are considered effective inhibitors of carcinogenesis, among other conditions, and are pathogenically associated with oxidative mechanisms. In the recent years, a large amount of information has emerged concerning the role of oxidative stress in the development of different diseases, such as the formation of certain types of cancer, cardiovascular diseases, and age-related degenerative chronic diseases in humans, and the possible therapeutic effects of antioxidants against these diseases (Ellnain-Wojtaszek et al., 2003). Although there already exist several natural metabolisms against free radicals, the human body still needs some exogenous intake of dietary antioxidants to balance the reactive oxygen. In this sense, recent studies focused on discovering new and cheap natural antioxidant sources and preserving their beneficial effects for an extended period.

There exist a great variety of natural antioxidants sourced from MAPs to fruits, vegetables, mushrooms, and algae. Mostly phenolics (especially flavonoids and anthocyanins), carotenoids, and vitamins possess antioxidant activity. However, the most abundant antioxidant source in nature is polyphenols, which can be present in different parts of plants, such as their roots, bark, leaves, fruit, flowers, stems, and seeds. These polyphenols mainly exist as conjugated to mono- and polysaccharides. Although fruits are the most well-known antioxidant source, tea infusions of some herbs and MAPs contain much more effective antioxidants owing to their high bioavailability in the human body.

Tsai et al. (2002) performed a comprehensive study on the antioxidant capacity of roselle petals by using oxygen radical absorbance capacity (ORAC), the ferric reducing ability of plasma (FRAP), and total antioxidant status assays. Researchers concluded that the antioxidant capacity increased by increasing the weight of the petals used and by extending the extraction time. In addition, researchers declared that the thermal applications during the processing of the roselle petals resulted in a slight decrease in their antioxidant activity. In another earlier study, Zhen et al. (2016) investigated the antioxidant activity of the leaves of 25 different roselle varieties and determined that the antioxidant capacity was between 101.5 and 152.5 μmol trolox/g by using the ABTS assay. Another previous research study focused on the antioxidant capacity of the extracts of two red and white roselle flowers, obtained by three different solvents (hexane, methanol, and ethyl acetate). According to the results, expectedly, red varieties possessed higher antioxidant capacities than the white varieties. At the same time, methanol was the most effective solvent in the case of antioxidant extraction in all samples. Researchers declared that the highest inhibition of peroxidation was found in the red roselle variety, as 80%, 65%, and 74% for methanol, ethyl acetate, and hexane extracts, respectively (Christian et al., 2006). In another study, Sáyago-Ayerdi et al. (2007) investigated the antioxidant capacity of roselle flowers by means of FRAP, ABTS, and ORAC assays and determined the antioxidant capacity (μmol trolox equivalents/g) to be 66.3, 90.8, and 303.5, respectively. Researchers explained the higher value of the ORAC assay with the polarity of the solvent used in the extraction procedure. Similarly, Sindi et al. (2014) performed an optimization study to examine the alterations in antioxidant capacity by means of DPPH, FRAP, and TEAC (total equivalent antioxidant capacity) assaying of roselle samples regarding the effect of using different solvents (water, methanol, ethyl acetate, and hexane) and different extraction times and temperatures. According to the results from the study, the most potent antioxidant activity obtained in the extracts was observed in those that were obtained by water with and without formic acid (72% and 70%, respectively, regarding other solvents) in the DPPH assay. In line with the study done by Sindi et al. (2014), Ramakrishna et al. (2008) examined the effects of using different solvents (methanol, acetone, and ethyl acetate) on the antioxidant capacity of roselle calyxes and fruits by means of the DPPH assay. Researchers concluded that methanol was the most effective solvent to obtain antioxidant agents of *H. sabdariffa* L. when compared with acetone and ethyl acetate.

On the other hand, roselle also possesses a yellowish green oil in its seeds with a 20% (in dry basis) oil yield approximately. This special seed oil is a good source of lipid-soluble natural antioxidants, especially γ-tocopherol and β-sitosterol. Moreover, it contains a significant amount of free fatty acids, including mainly linoleic (C18:2), palmitic (C16:0), and oleic (C18:1) acids (Mohamed et al., 2007).

5 CONCLUSIONS

It is evident that the consumption of roselle has a great potential for increase in the coming years, around the world. The aroma, key odorants, phenolic composition, and antioxidant characteristics of different roselle varieties and roselle teas were elucidated in this chapter. Roselle aroma is shown to be composed of a wide variety of aroma groups with different chemical properties, and within these groups, aldehydes, alcohols, and furans are the highest contributors to the overall aroma. These findings are supported with the GC-O methods that are used to elucidate the key odorants in roselle samples. According to the applied studies, key odorants including furfural and 2-methyl-furfural as furans with caramel odor notes, hexanal and nonanal with green notes, and α-terpinolene and linalool as terpenes with floral and citrus notes appear to contribute to the characteristic aroma of roselle teas. Additionally, roselle is rich in flavonoids and anthocyanins responsible for its attractive appearance. The specific phenolics present in abundance in roselle are delphinidin-3-sambubioside (hibiscin), cyanidin-3-sambubioside, delphinidin-3-glucoside, and cyanidin-3-glucoside. The presence of anthocyanins, flavonoids, phenolic acids, and other organic acids specific to roselle samples is also the reason for the high antioxidant capacity of the plant. The outcomes from the present work might be considered as supplying useful information concerning the aroma and phenolic properties of roselle and its products. Additionally, the extant literature highlights the potential of MAPs and species to be utilized as a source of bioactive compounds of functional herbal beverages or natural coloring agents in food colorants or as an important source of health-promoting content for pharmaceutical industries. In light of this information, it can be concluded that the remarkable bioactive compounds of roselle are suggested to be protected, evaluated, controlled, and optimized, especially when they are subjected to processing.

REFERENCES

Abubakar, M. B., Abdullah, W. Z., Sulaiman, S. A., & Suen, A. B. (2012). A review of molecular mechanisms of the anti-leukemic effects of phenolic compounds in honey. *International Journal of Molecular Sciences, 13*(11), 15054–15073.

Acree, T. E., Barnard, J., & Cunningham, D. (1984). A procedure for the sensory analysis of gas chromatographic effluents. *Food Chemistry, 14*(4), 273–286.

Acree, T. E., Shallenberger, R., Ebeling, Ludwig, S. P., Gogoris, A. C., & Arn, H. (July 1997). Food flavors: Formation, analysis and packaging influences. In E. T. Contis, C. T. Ho, C. J. Mussinan, T. H. Parliment, F. Shahidi, & A. M. Spanier (Eds.), *Proceedings of the 9th international flavor conference the george charalambous memorial symposium* (Vol. 1, p. 4). Elsevier Science.

Agoreyo, F. O., Agoreyo, B. O., & Onuorah, M. N. (2008). Effect of aqueous extracts of *Hibiscus sabdariffa* and *Zingiber officinale* on blood cholesterol and glucose levels of rats. *African Journal of Biotechnology, 7*(21), 3949–3951.

Alarcón-Alonso, J., Zamilpa, A., Aguilar, F. A., Herrera-Ruiz, M., Tortoriello, J., & Jimenez-Ferrera, E. (2012). Pharmacological characterization of the diuretic effect of *Hibiscus sabdariffa* Linn (Malvaceae) extract. *Journal of Ethnopharmacology, 139*(3), 751–756.

Ali, B. H., Wabel, N. A., & Blunden, G. (2005). Phytochemical, pharmacological and toxicological aspects of *Hibiscus sabdariffa* L.: A review. *Phytotherapy Research: An International Journal Devoted to Pharmacological and Toxicological Evaluation of Natural Product Derivatives, 19*(5), 369–375.

Aruoma, O. I., Murcia, A., Butler, J., & Halliwell, B. (1993). Evaluation of the antioxidant and prooxidant actions of gallic acid and its derivatives. *Journal of Agricultural and Food Chemistry, 41*(11), 1880–1885.

Aurelio, D. L., Edgardo, R. G., & Navarro-Galindo, S. (2008). Thermal kinetic degradation of anthocyanins in a roselle (*Hibiscus sabdariffa* L. cv.'Criollo') infusion. *International Journal of Food Science and Technology, 43*(2), 322–325.

Borrás-Linares, I., Fernández-Arroyo, S., Arráez-Roman, D., Palmeros-Suárez, P. A., Del Val-Díaz, R., Andrade-Gonzáles, I., Fernández-Gutiérrez, A., Gómez-Leyva, J. F., & Segura-Carretero, A. (2015). Characterization of phenolic compounds, anthocyanidin, antioxidant and antimicrobial activity of 25 varieties of Mexican roselle (*Hibiscus sabdariffa*). *Industrial Crops and Products, 69*, 385–394.

Branisa, J., Jomova, K., Porubska, M., Kollar, V., Simunkova, M., & Valko, M. (2017). Effect of drying methods on the content of natural pigments and antioxidant capacity in extracts from medicinal plants: A spectroscopic study. *Chemical Papers, 71*(10), 1993–2002.

Camelo-Méndez, G. A., Ragazzo-Sánchez, J. A., Jimenez-Aparicio, A. R., Vanegas-Espinoza, P. E., Paredes-López, O., & Del Villar-Martinez, A. A. (2013). Comparative study of anthocyanin and volatile compounds content of four varieties of Mexican roselle (Hibiscus sabdariffa L.) by multivariable analysis. *Plant Foods for Human Nutrition, 68*(3), 229–234.

Carvajal-Zarrabal, O., Waliszewski, S. M., Barradas-Dermitz, D., Orta-Flores, Z., Hayward-Jones, P., Nolasco-Hipólito, C., Angulo-Guerrero, O., Sànchez-Ricaño, R., & Trujillo, P. (2005). The consumption of *Hibiscus sabdariffa* dried calyx ethanolic extract reduced lipid profile in rats. *Plant Foods for Human Nutrition, 60*(4), 153.

Cassol, L., Rodrigues, E., & Noreña, C. P. Z. (2019). Extracting phenolic compounds from *Hibiscus sabdariffa* L. calyx using microwave assisted extraction. *Industrial Crops and Products, 133*, 168–177.

Castañeda-Ovando, A., Pacheco-Hernández de Lourdes, Páez-Hernández, E., Rodríguez, J. A., & Galán-Vidal, C. A. (2009). Chemical studies of anthocyanins: A review. *Food Chemistry, 113*(4), 859–871.

Chen, C.-C., Chou, F.-P., Ho, Y.-C., Lin, W.-L., Wang, C.-P., Kao, E.-S., Huang, A.-C., & Wang, C.-J. (2004). Inhibitory effects of *Hibiscus sabdariffa* L extract on low-density lipoprotein oxidation and anti-hyperlipidemia in fructose-fed and cholesterol-fed rats. *Journal of the Science of Food and Agriculture, 84*(15), 1989–1996.

Chen, S.-H., Huang, T. C., Ho, C. T., & Tsai, P. J. (1998). Extraction, analysis, and study on the volatiles in roselle tea. *Journal of Agricultural and Food Chemistry, 46*(3), 1101–1105.

Christian, K., Nair, M., & Jackson, J. C. (2006). Antioxidant and cyclooxygenase inhibitory activity of sorrel (*Hibiscus sabdariffa*). *Journal of Food Composition and Analysis, 19*(8), 778–783.

Cid-Ortega, S., & Guerrero-Beltrán, J. (2015). Roselle calyces (*Hibiscus sabdariffa*), an alternative to the food and beverages industries: A review. *Journal of Food Science & Technology, 52*(11), 6859–6869.

Contis, E. T., Ho, C.-T., Mussinan, C., Parliment, T., Shahidi, F., & Spanier, A. (1998). *Food flavors: Formation, analysis and packaging influences*. Elsevier.

Delahunty, C. M., Graham, E., & Dufour, J.-P. (2006). Gas chromatography-olfactometry. *Journal of Separation Science, 29*(14), 2107–2125.

Dravnieks, A., & O'Donnell, A. (1971). Principles and some techniques of high-resolution headspace analysis. *Journal of Agricultural and Food Chemistry, 19*(6), 1049–1056.

Ellnain-Wojtaszek, M., Kruczyński, Z., & Kasprzak, J. (2003). Investigation of the free radical scavenging activity of *Ginkgo biloba* L. leaves. *Fitoterapia, 74*(1–2), 1–6.

Farag, M. A., Rasheed, D. M., & Kamal, I. M. (2015). Volatiles and primary metabolites profiling in two *Hibiscus sabdariffa* (roselle) cultivars via headspace SPME-GC-MS and chemometrics. *Food Research International, 78*, 327–335.

Fernandes, L., Casal, S., Pereira, J. A., Saraiva, J. A., & Ramalhosa, E. (2017). Edible flowers: A review of the nutritional, antioxidant, antimicrobial properties and effects on human health. *Journal of Food Composition and Analysis, 60*, 38–50.

Fuller, G. H., Steltenkamp, R., & Tisserand, G. A. (1964). The gas chromatograph with human sensor: Perfumer model. *Annals of the New York Academy of Sciences, 116*(2), 711–724.

Gaya, I. B., Mohammad, O. M. A., Suleiman, A. M., Maje, M. I., & Adekunle, A. B. (2009). Toxicological and lactogenic studies on the seeds of *Hibiscus sabdariffa* Linn (Malvaceae) extract on serum prolactin levels of albino wistar rats. *The Internet Journal of Endocrinology, 5*(2), 1–6.

Gonzalez-Palomares, S., Estarrón-Espinosa, M., Gómez-Leyva, J. F., & Andrade-González, I. (2009). Effect of the temperature on the spray drying of roselle extracts (*Hibiscus sabdariffa* L.). *Plant Foods for Human Nutrition, 64*, 62–67.

Grosch, W. (1994). Determination of potent odourants in foods by aroma extract dilution analysis (AEDA) and calculation of odour activity values (OAVs). *Flavour and Fragrance Journal, 9*, 147–158.

Hainida, E., Ismail, A., Hashim, N., Mohd.-Esa, N., & Zakiah, A. (2008). Effects of defatted dried roselle (*Hibiscus sabdariffa* L.) seed powder on lipid profiles of hypercholesterolemia rats. *Journal of the Science of Food and Agriculture, 88*(6), 1043–1050.

Hidalgo-Villatoro, S. G., León Cifuentes-Reyes, W. A. de, Ruano-Solís, H. H., & Cano-Castillo, L. E. (2009). Caracterización de trece genotipos de rosa de Jamaica *Hibiscus sabdariffa* en Guatemala. *Agronomía Mesoamericana, 20*, 101–109.

Hopkins, A. L., Lamm, M. G., Funk, J. L., & Ritenbaugh, C. (2013). *Hibiscus sabdariffa* L. In the treatment of hypertension and hyperlipidemia: A comprehensive review of animal and human studies. *Fitoterapia, 85*, 84–94.

Ismail, A., Ikram, E. H. K., & Nazri, H. S. M. (2008). Roselle (*Hibiscus sabdariffa* L.) seeds-nutritional composition, protein quality and health benefits. *Food, 2*(1), 1–16.

Jabeur, I., Pereira, E., Barros, L., Calhelha, R. C., Soković, M. M., Oliveira, B. P. P., & Ferreira, I. C. F. R. (2017). *Hibiscus sabdariffa* L. as a source of nutrients, bioactive compounds and colouring agents. *Food Research International, 100*, 717–723.

Juhari, N. H., Bredie, W. L. P., Toldam-Andersen, T. B., & Petersen, M. A. (2018). Characterization of roselle calyx from different geographical origins. *Food Research International, 112*, 378–389.

Juhari, N. H., & Petersen, M. A. (September 2017). Aroma profile and proximate composition of roselle seeds: Effects of different origins and different sample preparation methods. In *Proceedings of the 15th weurman flavour research symposium, Graz, Austria* (pp. 18–22).

Juliani, R., Welch, C., Wu, Q., Diouf, B., Malainy, D., & Simon, J. (2009). Chemistry and quality of hibiscus (*Hibiscus sabdariffa*) for developing the natural-product industry in Senegal. *Journal of Food Science, 74*(2), 113–S121.

Junsathian, P., Yordtong, K., Corpuz, H. M., Katayama, S., Nakamura, S., & Rawdkuen, S. (2018). Biological and neuroprotective activity of Thai edible plant extracts. *Industrial Crops and Products, 124*, 548–554.

Kaisoon, O., Siriamornpun, S., Weerapreeyakul, N., & Meeso, N. (2011). Phenolic compounds and antioxidant activities of edible flowers from Thailand. *Journal of Functional Foods, 3*, 88–99.

Kampa, M., Vassilia-Ismini, A., George, N., Artemissia-Phoebe, N., Anastassia, N., Anastassia, H., Efstathia, B., Elena, K., George, B., Dimitrios, B., Achille, G., & Elias, C. (2004). Antiproliferative and apoptotic effects of selective phenolic acids on T47D human breast cancer cells. Potential mechanisms of action. *Breast Cancer Research, 6*(2), R63.

Kelebek, H. (2016). LC-DAD–ESI-MS/MS characterization of phenolic constituents in Turkish black tea: Effect of infusion time and temperature. *Food Chemistry, 204*, 227–238.

Kelebek, H., Selli, S., & Sevindik, O. (2020). Screening of phenolic content and antioxidant capacity of Okitsu Mandarin (*Citrus unshui* Marc.) fruits extracted with various solvents. *Journal of Raw Materials and Processed Foods, 1*, 7–12.

Kelebek, H., Sevindik, O., & Selli, S. (2019). LC-DAD-ESI-MS/MS-based phenolic profiling of St John's Wort Teas and their antioxidant activity: Eliciting infusion induced changes. *Journal of Liquid Chromatography & Related Technologies, 42*(1–2), 9–15.

Kesen, S. (2020). Characterization of aroma and aroma-active compounds of Turkish turmeric (*Curcuma longa*) extract. *Journal of Raw Materials to Processed Foods, 1*, 13–21.

Kesen, S., Kelebek, H., Sen, K., Ulaş, M., & Selli, S. (2013). GC-MS-olfactometric characterization of the key aroma compounds in Turkish olive oils by application of the aroma extract dilution analysis. *Food Research International, 54*, 1987–1994.

Komes, D., Belščak-Cvitanović, A., Horžić, D., Marković, K., & Kovačević Ganič, K. (2011). Characterisation of pigments and antioxidant properties of three medicinal plants dried under different drying conditions. In *Proceedings of the 11th international congress on engineering and food*.

Kumar, S. S., Manoj, P., Shetty, N. P., & Giridhar, P. (2015). Effect of different drying methods on chlorophyll, ascorbic acid and antioxidant compounds retention of leaves of *Hibiscus sabdariffa* L. *Journal of the Science of Food and Agriculture, 95*(9), 1812–1820.

Lin, H.-H., Chen, J.-H., Kuo, W.-H., & Wang, C.-J. (2007). Chemopreventive properties of *Hibiscus sabdariffa* L. on human gastric carcinoma cells through apoptosis induction and JNK/p38 MAPK signaling activation. *Chemico-Biological Interactions, 165*(1), 59–75.

Linssen, J. P. H., Janssens, J., Roozen, J., & Posthumus, M. (1993). Combined gas chromatography and sniffing port analysis of volatile compounds of mineral water packed in polyethylene laminated packages. *Food Chemistry, 46*(4), 367–371.

Li, X., Wang, X., Chen, D., & Chen, S. (2011). Antioxidant activity and mechanism of protocatechuic acid in vitro. *Functional Foods in Health and Disease, 1*(7), 232–244.

Maganha, E. G., da Costa Halmenschlager, R., Rosa, R. M., Henriques, J. A. P., de Paula Ramos, A. L. L., & Saffi, J. (2010). Pharmacological evidences for the extracts and secondary metabolites from plants of the genus *Hibiscus*. *Food Chemistry, 118*(1), 1–10.

McKay, D. L., Chen, O., Saltzman, E., & Blumberg, J. B. (2010). *Hibiscus sabdariffa* L. tea (tisane) lowers blood pressure in prehypertensive and mildly hypertensive adults. *The Journal of Nutrition, 140*(2), 298–303.

Miranda-Lopez, R., Libbey, L. M., Watson, B. T., & McDaniel, M. R. (1992). Odor analysis of pinot noir wines from grapes of different maturities by a gas chromatography-olfactometry technique (Osme). *Journal of Food Science, 57*, 985–993.

Mohamed, R., Fernandez, J., Pineda, M., & Aguilar, M. (2007). Roselle (*Hibiscus sabdariffa*) seed oil is a rich source of γ-tocopherol. *Journal of Food Science, 72*(3), 207–211.

Neuwinger, H. D. (2000). African traditional medicine: A dictionary of plant use and applications. With supplement: Search system for diseases. Stuttgart, Germany *Medical Pharmacology*, 29–30.

Ningrum, A., Schreiner, M., Luna, P., Khoerunnisa, F., & Tienkink, E. (2019). Free volatile compounds in red and purple roselle (*Hibiscus sabdariffa*) pomace from Indonesia. *Journal of Food Science*, 749–754.

Nzikou, J.-M., Bouanga-Kalou, G., Matos, L., Ganongo-Po, F. B., Mboungou-Mboussi, P. S., Moutoula, F. E., & Desobry, S. (2011). Characteristics and nutritional evaluation of seed oil from roselle (*Hibiscus sabdariffa* L.) in Congo-Brazzaville. *Current Research Journal of Biological Sciences, 3*(2), 141–146.

Panche, A., Diwan, A. D., & Chandra, S. R. (2016). Flavonoids: An overview. *Journal of Nutritional Science, 5*, 1–15.

Pandey, K. B., & Rizvi, S. I. (2009). Current understanding of dietary polyphenols and their role in health and disease. *Current Nutrition & Food Science, 5*(4), 249–263.

Pérez-Ramírez, I. F., Castaño-Tostado, E., Ramírez-de León, J. A., Rocha-Guzmán, N. E., & Reynoso-Camacho, R. (2015). Effect of stevia and citric acid on the stability of phenolic compounds and in vitro antioxidant and antidiabetic capacity of a roselle (*Hibiscus sabdariffa* L.) beverage. *Food Chemistry, 172*, 885–892.

Perkin, A. G. (1909). CCV.—the colouring matters of the flowers of *Hibiscus sabdariffa* and *Thespasia lampas*. *Journal of the Chemical Society Transactions, 95*, 1855–1860.

Peterson, D. G., & Totlani, V. M. (2005). Influence of flavonoids on the thermal generation of aroma compounds. In *Phenolic compounds in foods and natural health products* (pp. 143–160). ACS Publications.

Pinela, J., Prieto, M. A., Pereira, E., Jabeur, I., Barreiro, M. F., Barros, L., & Ferreira Isabel, C. F. R. (2019). Optimization of heat-and ultrasound-assisted extraction of anthocyanins from *Hibiscus sabdariffa* calyces for natural food colorants. *Food Chemistry, 275*, 309–321.

Pino, J. A., Eliosbel, M., & Rolando, M. (2006). Volatile constituents from tea of roselle (*Hibiscus sabdariffa* L.). *Revista CENIC Ciencias Quimicas, 37*(3), 127–129.

Piovesana, A., Eliseu, R., & Noreña, C. P. Z. (2019). Composition analysis of carotenoids and phenolic compounds and antioxidant activity from hibiscus calyces (*Hibiscus sabdariffa* L.) by HPLC-DAD-MS/MS. *Phytochemical Analysis, 30*(2), 208–217.

Plotto, A., Mazaud, F., Röttger, A., & Steffel, K. (2004). Hibiscus: Post-production management for improved market access. *Food and Agriculture Organization of the UN (FAO)*, 1–18.

Pollien, P., Ott, A., Montigon, F., Baumgartner, M., Muñoz-Box, R., & Chaintreau, A. (1997). Hyphenated headspace-gas chromatography-sniffing technique: Screening of impact odorants and quantitative aromagram comparisons. *Journal of Agricultural and Food Chemistry, 45*(7), 2630–2637.

Prenesti, E., Berto, S., Daniele, P. G., & Toso, S. (2007). Antioxidant power quantification of decoction and cold infusions of *Hibiscus sabdariffa* flowers. *Food Chemistry, 100*(2), 433–438.

Ramakrishna, B. V., Jayaprakasha, G. K., Jena, B. S., & Singh, R. P. (2008). Antioxidant activities of roselle (*Hibiscus sabdariffa*) calyces and fruit extracts. *Journal of Food Science and Technology-Mysore, 45*(3), 223–227.

Ramírez-Rodrigues, M., Balaban, M. O., Marshall, M., & Rouseff, R. L. (2011). Hot and cold water infusion aroma profiles of *Hibiscus sabdariffa*: Fresh compared with dried. *Journal of Food Science, 76*(2), 212–217.

Rodrigues, N. P., & Bragagnolo, N. (2013). Identification and quantification of bioactive compounds in coffee brews by HPLC–DAD–MSn. *Journal of Food Composition and Analysis, 32*(2), 105–115.

Rodríguez-Medina, I. C., Beltrán-Debón, R., Micol Molina, V., Alonso-Villaverde, C., Joven, J., Menéndez, J. A., Segura-Carretero, A., & Fernández-Gutiérrez, A. (2009). Direct characterization of aqueous extract of *Hibiscus sabdariffa* using HPLC with diode array detection coupled to ESI and ion trap MS. *Journal of Separation Science, 32*(20), 3441–3448.

Ross, I. A. (2003). *Hibiscus sabdariffa*. In *Medicinal plants of the world* (pp. 267–275). Totowa, NJ: Humana Press.

Saini, R. K., Shetty, N., Prakash, M., & Giridhar, P. (2014). Effect of dehydration methods on retention of carotenoids, tocopherols, ascorbic acid and antioxidant activity in Moringa oleifera leaves and preparation of a RTE product. *Journal of Food Science & Technology, 51*(9), 2176–2182.

Sawabe, A., Nesumi, C., Morita, M., Matsumoto, S., Matsubara, Y., & Komemushi, S. (2005). Glycosides in African dietary leaves, *Hibiscus sabdariffa*. *Journal of Oleo Science, 54*(3), 185–191.

Sáyago-Ayerdi, S., Arranz, S., Serrano, J., & Goñi, I. (2007). Dietary fiber content and associated antioxidant compounds in roselle flower (*Hibiscus sabdariffa* L.) beverage. *Journal of Agricultural and Food Chemistry, 55*(19), 7886–7890.

Schlotzhauer, William, Pair, Sam D., & Horvat, Robert J. (1996). Volatile constituents from the flowers of Japanese honeysuckle (*Lonicera japonica*). *Journal of Agricultural and Food Chemistry, 44*, 206–209.

Selli, S. (2007). Volatile constituents of orange wine obtained from moro oranges (*Citrus sinensis* [L.] Osbeck). *Journal of Food Quality, 30*, 330–341.

Selli, S., Canbas, A., Varlet, V., Kelebek, H., Prost, C., & Serot, T. (2008). Characterization of the most odor-active volatiles of orange wine made from a Turkish cv. Kozan (*Citrus sinensis* L. Osbeck). *Journal of Agricultural and Food Chemistry, 56*, 227–234.

Selli, S., Rannou, C., Prost, C., Robin, J., & Serot, T. (2006). Characterization of aroma-active compounds in rainbow trout (*Oncorhynchus mykiss*) eliciting an off-odor. *Journal of Agricultural and Food Chemistry, 54*, 9496–9502.

Sen, K., & Sonmezdag, A. S. (2020). Elucidation of phenolic profiling of cv. Antep karasi grapes using LC-DAD-ESI-MS/MS. *Journal of Raw Materials to Processed Foods, 1*, 1–6.

Serot, T., Prost, C., Visan, L., & Burcea, M. (2001). Identification of the main odor-active compounds in musts from French and Romanian hybrids by three olfactometric methods. *Journal of Agricultural and Food Chemistry, 49*(4), 1909–1914.

Sindi, H. A., Marshall, L. J., & Morgan, M. R. A. (2014). Comparative chemical and biochemical analysis of extracts of *Hibiscus sabdariffa*. *Food Chemistry, 164*, 23–29.

Sirag, N., Elhadi, M. M., Algaili, A. M., Mohamed, H. H., & Mohamed, O. (2014). Determination of total phenolic content and antioxidant activity of roselle (*Hibiscus sabdariffa* L.) calyx ethanolic extract. *Standard Research Journal of Pharmacy and Pharmacology, 1*(2), 034–039.

Sonmezdag, A. S., Kelebek, H., & Selli, S. (2018). Characterization of bioactive and volatile profiles of thyme (*Thymus vulgaris* L.) teas as affected by infusion times. *Journal of Food Measurement and Characterization, 12*(4), 2570–2580.

Stalikas, C. D. (2007). Extraction, separation, and detection methods for phenolic acids and flavonoids. *Journal of Separation Science, 30*(18), 3268–3295.

Tahir, H. E., Xiaobo, Z., Mariod, A. A., Mahunu, G. K., Abdualrahman, M. A. Y., & Tchabo, W. (2017). Assessment of antioxidant properties, instrumental and sensory aroma profile of red and white karkade/roselle (*Hibiscus sabdariffa* L.). *Journal of Food Measurement and Characterization, 11*(4), 1559–1568.

Takahashi, J. A., Rezende, F. A. G. G., Moura, M. A. F., Dominguete, L. C. B., & Sande, D. (2020). Edible flowers: Bioactive profile and its potential to be used in food development. *Food Research International, 129*, 108868.

Topi, D. (2020). Volatile and chemical compositions of freshly squeezed sweet lime (*Citrus limetta*) juices. *Journal of Raw Materials to Processed Foods, 1*, 22–27.

Tsai, P. J., McIntosh, J., Pearce, P., Blake, C., & Jordan, B. R. (2002). Anthocyanin and antioxidant capacity in roselle (*Hibiscus sabdariffa* L.) extract. *Food Research International, 35*(4), 351–356.

Van Ruth, S. (2001). Methods for gas chromatography-olfactometry: A review. *Biomolecular Engineering, 17*(4–5), 121–128.

Wallace, T. C., Slavin, M., & Frankenfeld, C. L. (2016). Systematic review of anthocyanins and markers of cardiovascular disease. *Nutrients, 8*(1), 32.

Wang, J., Cao, X., Jiang, H., Qi, Y., Chin, K. L., & Yue, Y. (2014). Antioxidant activity of leaf extracts from different *Hibiscus sabdariffa* accessions and simultaneous determination five major antioxidant compounds by LC-Q-TOF-MS. *Molecules, 19*(12), 21226–21238.

Yagoub, A. E.-G., Mohamed, B. E., Ahmed, A. H. R., & El Tinay, A. H. (2004). Study on furundu, a traditional Sudanese fermented roselle (*Hibiscus sabdariffa* L.) seed: Effect on in vitro protein digestibility, chemical composition, and functional properties of the total proteins. *Journal of Agricultural and Food Chemistry, 52*(20), 6143–6150.

Yang, Z., Baldermann, S., & Watanabe, N. (2013). Recent studies of the volatile compounds in tea. *Food Research International, 53*, 585–599.

Zannou, O., Kelebek, H., & Selli, S. (2020). Elucidation of key odorants in Beninese roselle (*Hibiscus sabdariffa* L.) infusions prepared by hot and cold brewing. *Food Research International*, 109133.

Zhang, B., Mao, G., Zheng, D., Zhao, T., Zou, Y., Qu, H., Li, F., Zhu, B., Yang, L., & Wu, X. (2014). Separation, identification, antioxidant, and anti-tumor activities of *Hibiscus sabdariffa* L. extracts. *Separation Science and Technology, 49*(9), 1379–1388.

Zhen, J., Villani, T. S., Guo, Y., Qi, Y., Chin, K., Pan, M.-H., Ho, C.-T., Simon, J. E., & Wu, Q. (2016). Phytochemistry, antioxidant capacity, total phenolic content and anti-inflammatory activity of *Hibiscus sabdariffa* leaves. *Food Chemistry, 190*, 673–680.

Zimmerman, M. C., Clemens, D. L., Duryee, M. J., Sarmiento, C., Chiou, A., Hunter, C. D., Tian, J., Klassen, L. W., O'Dell, J. R., Thiele, G. M., Mikuls, T. R., & Anderson Daniel, R. (2017). Direct antioxidant properties of methotrexate: Inhibition of malondialdehyde-acetaldehyde-protein adduct formation and superoxide scavenging. *Redox Biology, 13*, 588–593.

FURTHER READING

Fujioka, K., & Shibamoto, T. (2008). Chlorogenic acid and caffeine contents in various commercial brewed coffees. *Food Chemistry, 106*(1), 217–221.

Yang, L., Gou, Y., Zhao, T., Zhao, J., Li, F., Zhang, B., & Wu, X. (2012). Antioxidant capacity of extracts from calyx fruits of roselle (*Hibiscus sabdariffa* L.). *African Journal of Biotechnology, 11*(17), 4063–4068.

CHAPTER 11

Performance and Emission Characteristics of a Compression Ignition Engine Fueled With Roselle and Karanja Biodiesel

PANKAJ SHRIVASTAVA • UPENDRA RAJAK • PRERANA NASHINE • TIKENDRA NATH VERMA

1 INTRODUCTION

Recently, the world fossil fuel is found only in limited reserves, so it is imperative to investigate for economically, technically, and environmentally accepted alternative renewable fuels for a diesel engine, which can be produced from different feedstock that are effectively accessible in a country (Agarwal, 2007; Sivaramakrishnan, 2018). The other problems that arise due to the ignition of fossil fuel are being the prime source of ozone consumption, acid rain, environmental changes, and global warming, and it also has harmful effects on human health. Some sources are even predicting that the world fossil fuel reserve will be depleted by the year 2045 (Karthikeyan & Mahalakshmi, 2007; Panwar et al., 2010).

Diesel motors are broadly utilized in various applications because of their lean operation, higher thermal efficiency, and lower brake specific fuel consumption (BSFC), and they tend to emit lower greenhouse gases than a spark ignition engine. Different toxins released from a compression ignition (CI) engine rely upon numerous variables that incorporate design parameters, operating conditions, type of fuel used, and engine exhaust emission after treatment utilized (Abdel-Rahman, 1998). To solve these critical issues, extensive research work was carried out over the past few decades. Several researchers have proposed that the utilization of liquid biofuels in a small amount with diesel fuel or as a sole fuel after some engine modification will positively help find some solution for this critical issue (Atmanlı et al., 2014; Gautam & Agrawal, 2013). Biodiesel fuel has a remarkable potency as a substitute fuel for CI engine. Straight vegetable oil when directly used in a diesel engine for long term can cause several issues in engine deposits, sticking piston ring, choking of injector due to high viscosity, and lower fuel volatility (Dhinesh et al., 2016). However, to overcome this problem, the transesterification process is used to bring the properties of oil close to diesel oil and also to reduce the harmful effects (Annamalai et al., 2016; Demirbas, 2007). The most extensive transesterification process is used to derive biodiesel from different feedstock, which alludes to a chemical reaction with edible and nonedible oil in the presence of a catalyst (alkali, acid, and enzyme), with ethanol and methanol as an alcoholic agent. Various researchers are studying the effects of compression ratio (CR) as an input parameter, and the effects are analyzed in terms of engine performance and emission characteristics by using different biodiesel fuels (Sahoo & Das, 2009; Sharma et al., 2008).

Raheman and Ghadge (2008) found that when utilizing Mahua oil biodiesel and its blends as fuel, increasing the CR from 18 to 20 results in 33% higher brake thermal efficiency (BTE). Jindal et al. (2010) investigated the influence of CR in CI engine fueled with Jatropha methyl ester. They have found that the BTE is increased by 8.2% when combining the CR with an injection pressure to 18 and 250 bar compared with standard CR with an injection pressure of 17.5 and 210 bar at full load conditions. Muralidharan and Vasudevan (2011) investigated the use of waste cooking oil biodiesel and its blends as an alternative fuel in diesel engine, and their effect is analyzed by varying the CR. The results showed that better engine performance is observed for B40 blends than mineral diesel fuel at CR 21. EL_Kassaby and

Roselle. https://doi.org/10.1016/B978-0-323-85213-5.00015-9

Nemit_allah (2013) observed that the BTE is increased by 28% for B20 blend with an increase in CR from 14 to 18 while using waste cooking oil and its blends as a fuel. Hirkude and Padalkar (2014) investigated the effect on diesel engine characteristics by using waste fried oil and observed that BSFC is reduced while the exhaust gas temperature (EGT) is increased, with increase in CR (Bora & Ujjwal, 2016). Mishra (2017) observed that at full load conditions, BTE increased by 38.45% for B40 blend, which is 4% higher than diesel fuel in CR 21, by using waste cooking oil biodiesel fuel in CI engine. Sharma and Murgan (2015) experimented by using karanja methyl ester and its blend with diesel and have found that by varying the CR (17.5 and 18.5), the BSFC increases by increasing the percentage of karanja biodiesel in the blend.

Based on the gaps identified from the literature survey, the objective of this research work is to explore the technical feasibility by utilizing roselle and karanja biodiesel as an alternative fuel in diesel engine and analyze their effect by varying the CR (16.5, 17.5, and 18.5) and engine load (50% and 100%). However, as per the authors' knowledge, there is no relevant research found related to roselle biodiesel, and therefore this biodiesel that is derived from the roselle seeds is new. The raw material for the production of this biodiesel is readily available in the region of Raipur, Chhattisgarh, India (21.2514 degrees N, 81.6296 degrees E, and 298 m above sea level). To reduce the viscosity of the biodiesel, the transesterification technique is used, which eliminates the various operational difficulties associated with roselle oil and also improves the other biodiesel properties.

2 METHODOLOGY

2.1 Material

The biodiesel feedstock, that is, roselle and karanja seeds, is readily available in the region of Raipur (21.2514 degrees N, 81.6296 degrees E, and 298.16 m above mean sea level), Chhattisgarh, India. The next step for the production of biodiesel is to extract oil from the roselle and karanja seeds and then pass the oil through filter paper to remove impurities. The transesterification process is then carried out by using methanol (CH_3OH) as an alcoholic agent, and sodium hydroxide (NaOH) and potassium hydroxide (KOH) are used as catalysts.

2.2 Roselle and Karanja Biodiesel Production

A single-step transesterification technique is used for the production of biodiesel fuel from roselle oil. In a separate flask, a fixed amount of methanol and catalyst is mixed until they were fully dissolved and form a methoxide (Singh & Verma, 2019a, b). Roselle oil was then mixed with the methoxide and then the sample was heated up to a temperature of about 60–75°C. The mixture is then stirred by using a magnetic stirrer at around 360–380 rotations per minute (rpm) for at least 1 h. The mixture is then transferred to a separating funnel and is allowed to settle overnight at a room temperature for phase separation by gravity. Roselle biodiesel is formed at the top layer while glycerin floats at the bottom of the funnel. Roselle biodiesel is then washed with normal water until the washing water became neutral. Finally, the washed biodiesel is kept in an airtight container. The same procedure is followed for the production of karanja biodiesel from karanja seeds.

2.3 Properties of Diesel, Roselle Biodiesel, and Karanja Biodiesel

The important physicochemical properties of diesel, roselle biodiesel, and karanja biodiesel, along with the free fatty acid composition, are shown in Tables 11.1 and 11.2. For better engine performance, biodiesel with lower fatty acid composition is collected for conducting the experiment.

TABLE 11.1
Physicochemical Properties of Diesel, Roselle Biodiesel, and Karanja Biodiesel.

Fuel	B0	LA100	KB100	Unit
Density	830–840	878	880–913	kg/m^3
Viscosity at 40°C	2.5–3.11	5.64	3.99–5.71	mm^2/s
Heating value	42.2	38.72	38.91–42.13	MJ/kg
Cetane number	48	52	52	—

TABLE 11.2
Fatty Acid Composition of Roselle and Karanja Oil.

Fatty Acid Composition (%)	Palmitic Acid	Oleic Acid	Stearic Acid	Linolenic Acid	Linoleic Acid
Roselle oil (*Hibiscus sabdariffa*)	18.0	33.6	4.06	2.02	38.05
Karanja oil (*Pongamia pinnata*)	11.55	51.62	7.40	2.62	16.54

3 EXPERIMENTAL PROCEDURE

3.1 Experimental Setup

A four-stroke, single-cylinder, water-cooled CI engine was used for performing experiments in the Internal Combustion Engine Laboratory, Department of Mechanical Engineering, National Institute of Technology Manipur. In this investigation, diesel, roselle biodiesel, and karanja biodiesel (B0, LA100, and KB100) were used by varying the engine load (50% and 100%) and CR (16.5, 17.5, and 18.5), with a fixed injection timing of 23.5 degrees before top dead center at a constant engine speed of 1500 rpm. Fig. 11.1 shows the schematic layout of the engine setup. Table 11.3 shows the detailed specification of the experimental test engine.

3.2 Error Analysis and Uncertainties

In general, all experimental tests are subject to errors and uncertainties during the test. The uncertainty in the experimental test result rises due to selection of sensors, operating conditions, calibration of the experimental setup, experimental procedure, and observation reading. The summary of the apparatus used in the present investigation, including the range and precision of the instruments, is given in Table 11.4. To perform the uncertainty analysis, the following method is discussed by Rajak et al. (2018), Shrivastava et al. (2019a), Shrivastava, Salam, et al. (2020), Shrivastava, Verma et al. (2020), and Heywood (1988). The overall uncertainty analysis of the experiment was found out by using Eq. (11.1):

$$\text{Overall uncertainty }(\%) = \text{square root of}\left[(0.15)^2 + (1.0)^2 + (0.2)^2 + (0.5)^2 + (0.2)^2 + (1.0)^2 + (0.15)^2 + (1.0)^2 + (1.0)^2 + (1.0)^2 + (0.5)^2\right] \tag{11.1}$$

$$\text{Total percentage of uncertainty} = \pm 2.3\%$$

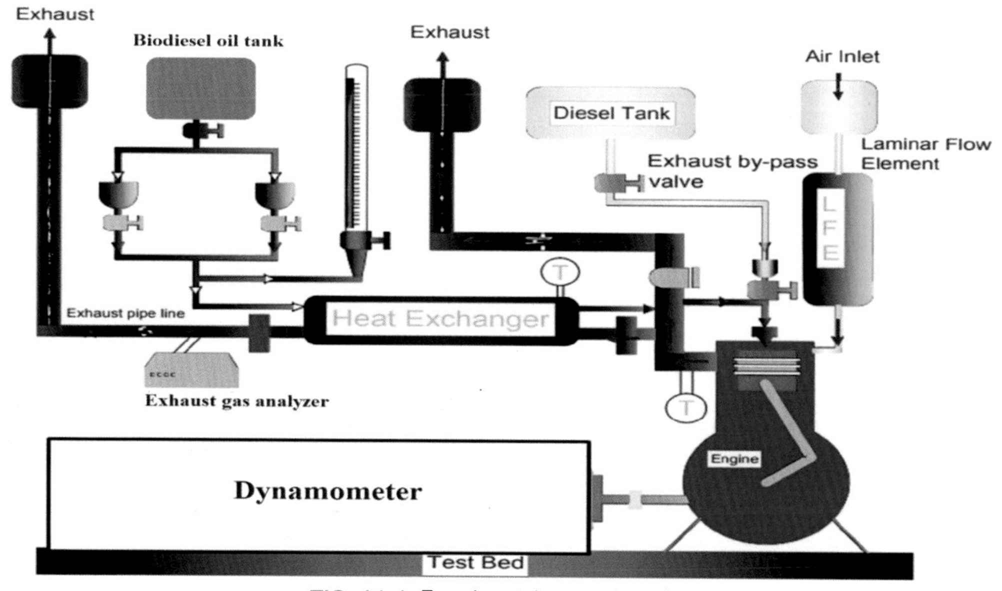

FIG. 11.1 Experimental research engine.

TABLE 11.3
Engine Specification.

Particulars	Specifications
Engine make	Legion brothers
Type	Water-cooled, four-stroke
Number of cylinders	1
Bore × stroke	80 × 110 mm
Connecting rod length	235 mm
Rated speed	1500 rpm
Compression ratio	17:1
Nozzle injection pressure	220 bar
Dynamometer	Eddy current type
Start of injection at	23.5 degrees bTDC
Nozzle hole	3
Valve orientation	Overhead
Orientation	Vertical

bTDC, before top dead center.

4 RESULTS AND DISCUSSION

4.1 Performance Characteristics

4.1.1 Brake thermal efficiency

The variation in BTE with different CRs and loads for various tested fuels is illustrated in Fig. 11.2. It is the ratio between the power output from the engine and the product of the lower heating value and fuel mass flow rate of the fuel. It is clear from the graph that BTE increases gradually with an increase in engine load because at higher engine load, there is reduction in heat loss and also increase in the engine power (Rajak & Verma, 2018a, b). Generally, an increase in CR results in improving the thermal efficiency of the engine. This may be due to the higher compressed air temperature at higher CR, which indicates better combustion of fuel (Shrivastava, Salam, et al., 2020, Shrivastava, Verma et al., 2020). At full load condition, the BTE was found to be about 31.6%, 32.4%, and 33.3% for karanja biodiesel and 31.2%, 32.1%, and 32.8% for roselle biodiesel at a CR of 16.5, 17.5, and 18.5, respectively. The diesel fuel has a higher BTE of 32.6%, 33.1%, and 33.7% at 16.5, 17.5, and 18.5 CR. The possible reason for this increase in BTE is the proper mixing of fuel and air and this leads to improvement in the quality of fuel spray phenomenon at a higher CR of 18.5 (Shrivastava, Salam, et al., 2020, Shrivastava, Verma et al., 2020).

4.1.2 Brake specific fuel consumption

The variation in BSFC at different CRs and loads for various tested fuels is shown in Fig. 11.3. It is seen from the graph that BSFC for all tested fuel decreases with an increase in the CR, whereas the BSFC for roselle and karanja biodiesel was noticed to be higher at all CRs. At CR 16.5, 17.5, and 18.5, the BSFC was found to be about 0.2, 0.283, and 0.278 kg/kW-h, respectively, for karanja biodiesel (KB100) and0.298, 0.289, and 0.283 kg/kW-h, respectively, for roselle biodiesel (LA100), compared to 0.263, 0.256, and 0.248 kg/kW-h, respectively, for diesel fuel. Therefore increasing the CR was more beneficial for biodiesel fuel than for

TABLE 11.4
Uncertainty of Experimental Test Engine.

Sr. No	Instrument	Accuracy	Percentage Uncertainty
1	Temperature sensor	±1°C	±0.15%
2	Speed sensor	±10 rpm	±1.0%
3	Load indicator	±0.1 kg	±0.2%
4	Pressure sensor	±1 bar	±1%
5	Crank angle encoder	±1 degree	±0.5%
6	Smoke meter	0.1%	±1.0%
7	Fuel measuring	±1 mL	±1.0%
8	Heat value measured	—	±0.5%
9	Test 350 gas analyzer		
	CO_2	±0.3%	±0.2%
	NO_x	±5%	±0.5%

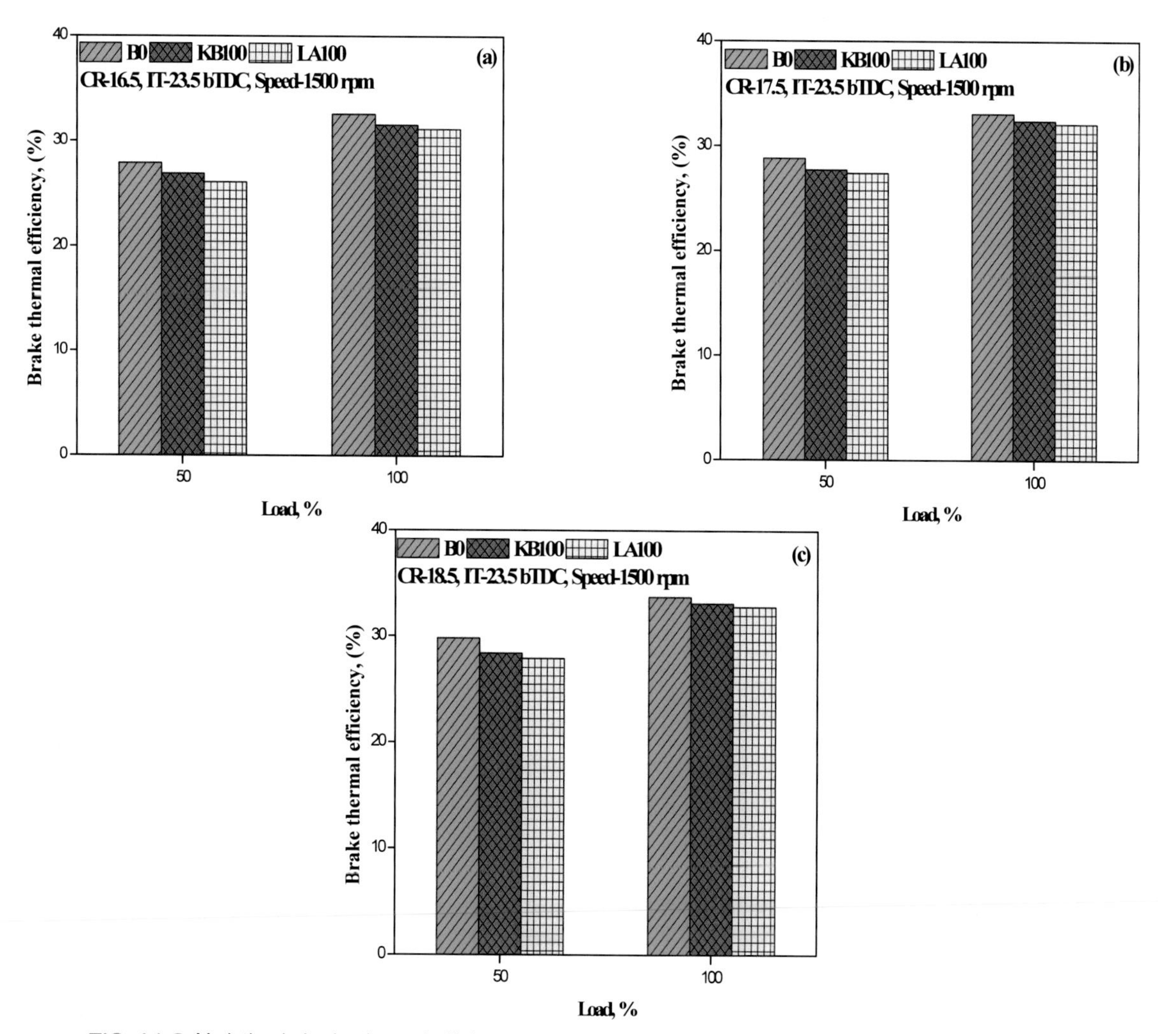

FIG. 11.2 Variation in brake thermal efficiency with load at different compression ratios (CRs). *bTDC*, before top dead center.

diesel fuel. This behavior is due to the higher viscosity and cetane number value for biodiesel fuel, which results in improving the combustion process at higher CRs compared to diesel fuel (Rajak & Verma, 2018a, b; Singh & Verma, 2019a, b).

4.1.3 Exhaust gas temperature

The variation in EGT at different CRs and loads for various tested fuels is shown in Fig. 11.4. It is an important parameter that gives qualitative information about the combustion of fuel. It is observed from the graph that with an increase in engine load, the EGT for all tested fuel increases, as this may be because of higher combustion temperature that is generated inside the combustion chamber when more fuel is burning with an increase in engine load (Rajak et al., 2019). At full load condition and at 16.5, 17.5, and 18.5 CR, the EGT is found to be about 300.72, 309.6, and 317.22°C, respectively, for KB100, while for LA100, it was observed to be about 294.17, 304.4, and 313.37°C, respectively, compared to 319.16, 324.59, and 338.75°C, respectively, for diesel fuel.

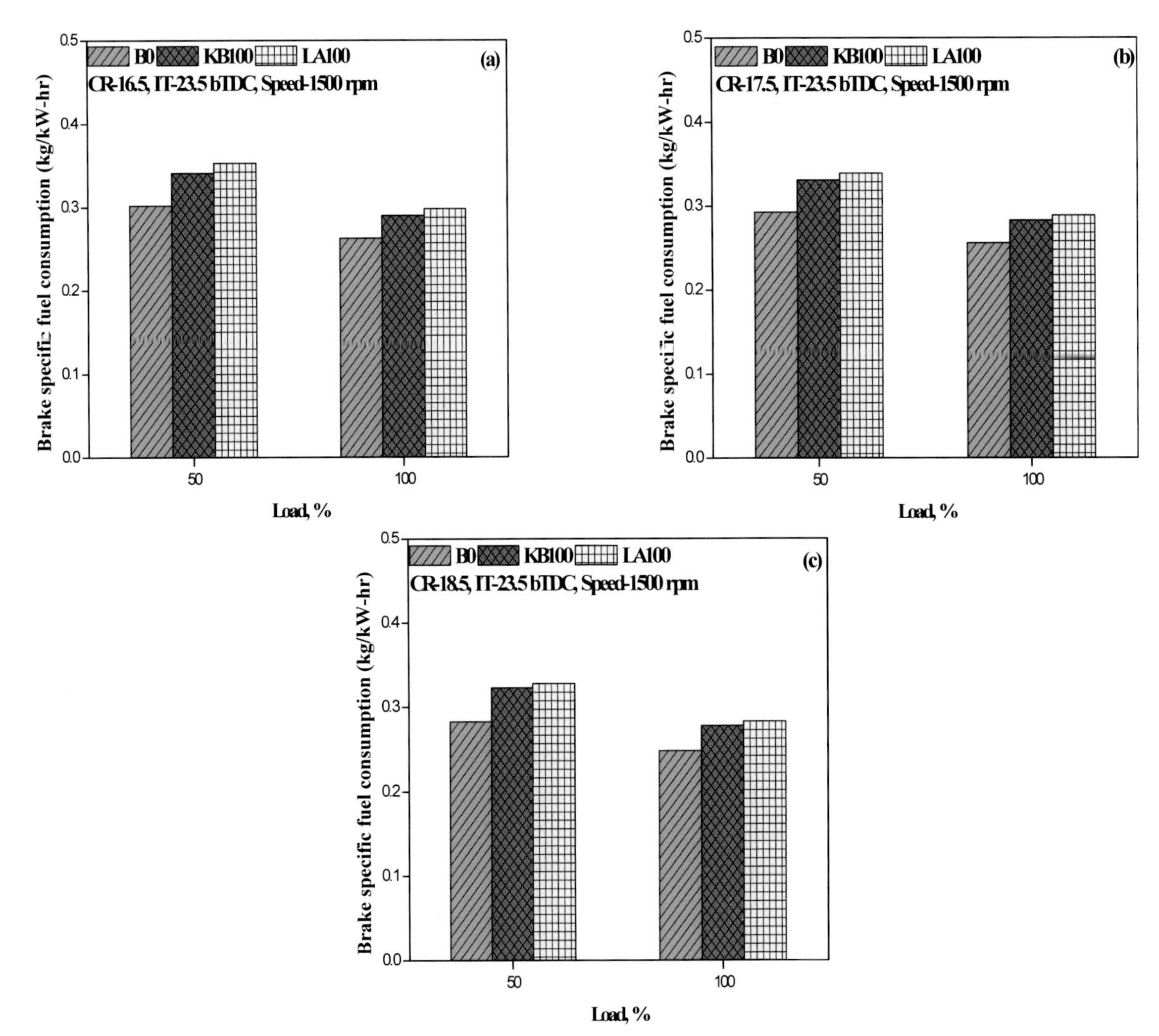

FIG. 11.3 Variation in brake specific fuel consumption with load at different compression ratios (CRs). *bTDC*, before top dead center.

4.2 Emission Parameters

4.2.1 CO_2 emission

The variation in carbon dioxide (CO_2) emission at different CRs and engine loads for various tested fuels is illustrated in Fig. 11.5. It is clear from the graph that reduction in smoke emission, with an increase in CR, is from 16.5 to 18.5. This is one possible reason for the reduction in smoke emission and is because of the presence of more oxygen in the biodiesel which results in complete combustion of the fuel (Shrivastava, Salam, et al., 2020, Shrivastava, Verma et al., 2020). At full load condition and at 16.5, 17.5, and 18.5 CR, the CO_2 emission is found to be about 843.96, 822.8, and 812.3 g/kWh, respectively, for KB100 and for LA100 it was recorded to be approximately 858.6, 838.6, and 824.5, respectively, compared to 828.9, 812.28, and 800.93 g/kWh, respectively, for diesel fuel.

4.2.2 NO_x emission

The variation in oxide of nitrogen (NO_x) emission for different tested fuel at different CRs and loads is presented in Fig. 11.6. It is the combination of nitrogen

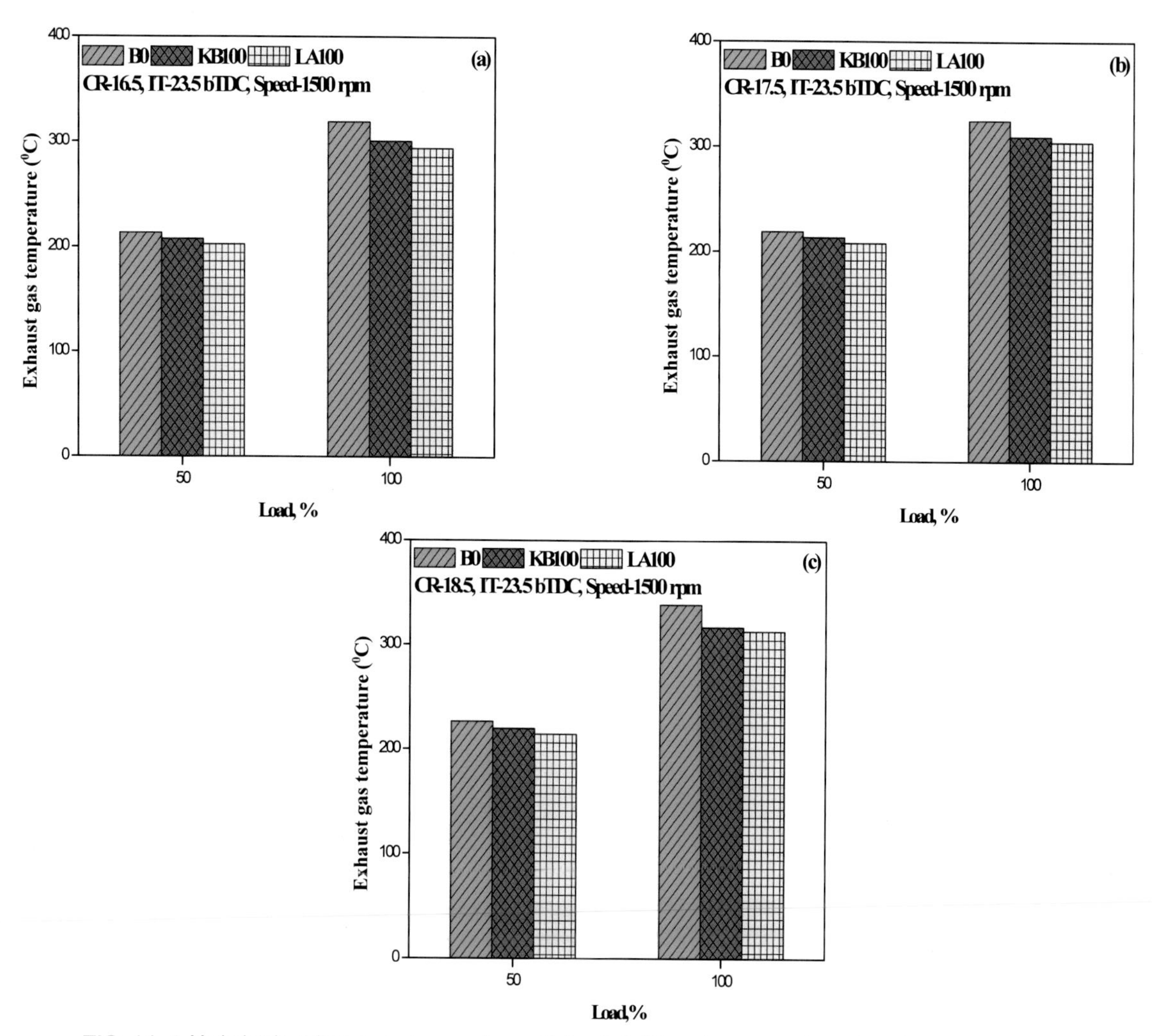

FIG. 11.4 Variation in exhaust gas temperature with load at different compression ratios (CRs). *bTDC*, before top dead center.

dioxide (NO_2) and nitric oxide (NO), which are formed from the reaction between nitrogen and oxygen gases in the air during the combustion process. The critical factors for the formation of NO_x emission are the flame temperature of combustion, the percentage of oxygen present in the fuel, and the combustion duration time (Holman, 1996; Shrivastava et al., 2019b; Singh et al., 2018). It is clear from the figure that at full load condition, NO_x emission followed an increasing trend with an increase in engine load as well as with CR from 16.5 to 18.5. Among all the tested fuel, it was observed that at higher engine load roselle and karanja biodiesel indicates a reduction in NO_x emission compared to standard diesel fuel, but NO_x emission increases with increase in CR. At full load condition and at 16.5, 17.5, and 18.5 CR, the NO_x emission was found to be about 2819.8, 2930.7, and 3014.5 ppm, respectively, for KB100 and for LA100 it was noticed to be about 3001.7, 3089.3, and 3152.9 ppm, respectively, compared to 3016, 3318.5, and 3420.5 ppm for diesel

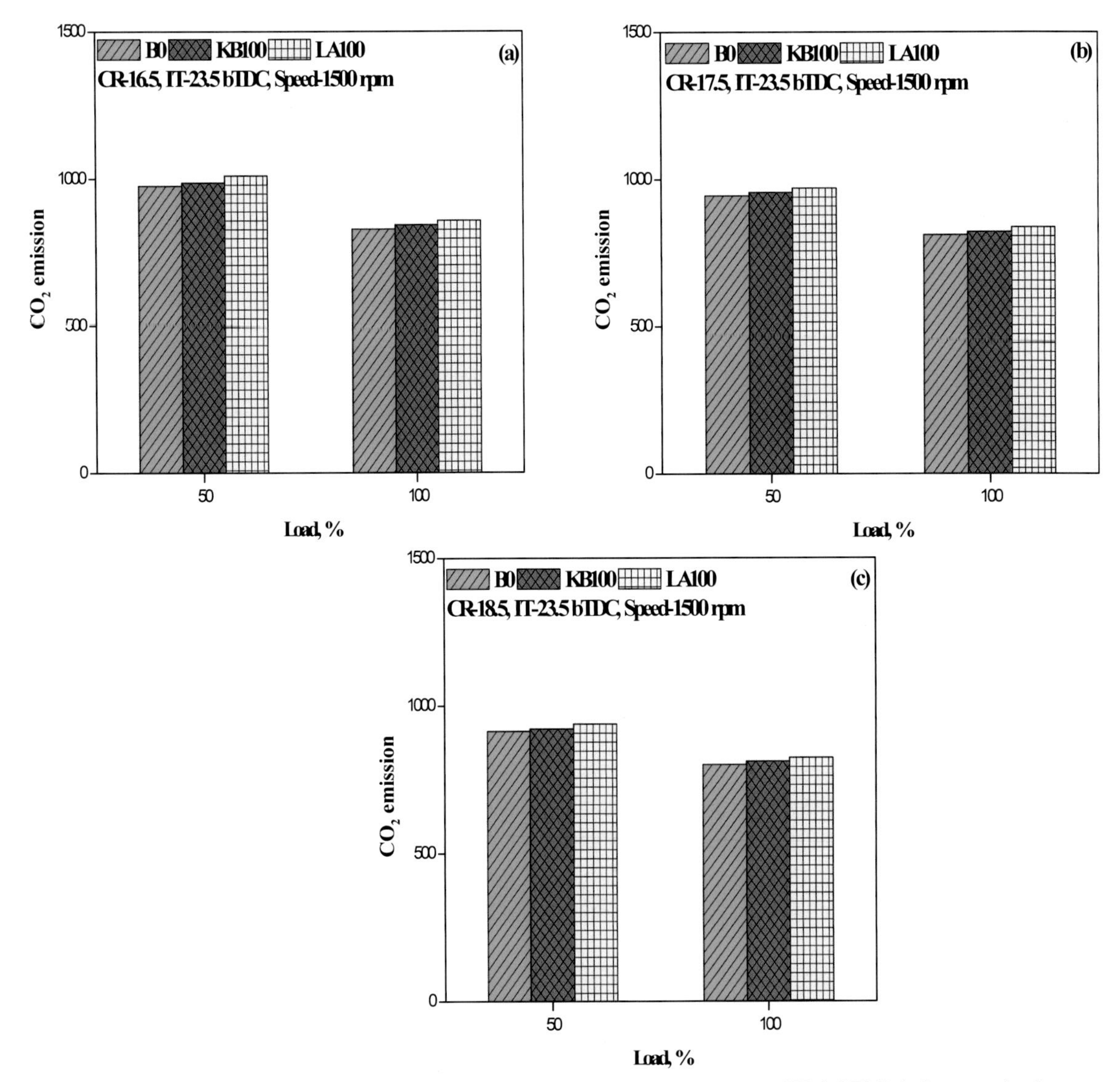

FIG. 11.5 Variation in CO_2 emission with load at different compression ratios (CRs). *bTDC*, before top dead center.

fuel, respectively. This may be because at higher CRs the air that enters inside the cylinder is appropriately mixed with the compressed air and results in a higher cylinder temperature, and this also indicates the complete combustion of the fuel (Shrivastava, Salam, et al., 2020, Shrivastava, Verma et al., 2020).

4.2.3 Smoke emission

The variation in smoke emission for various tested fuels at different CRs and engine loads is shown in Fig. 11.7. It is clear from the figure that smoke emission decreases gradually with an increase in CR. This may be due to the presence of more oxygen percentage

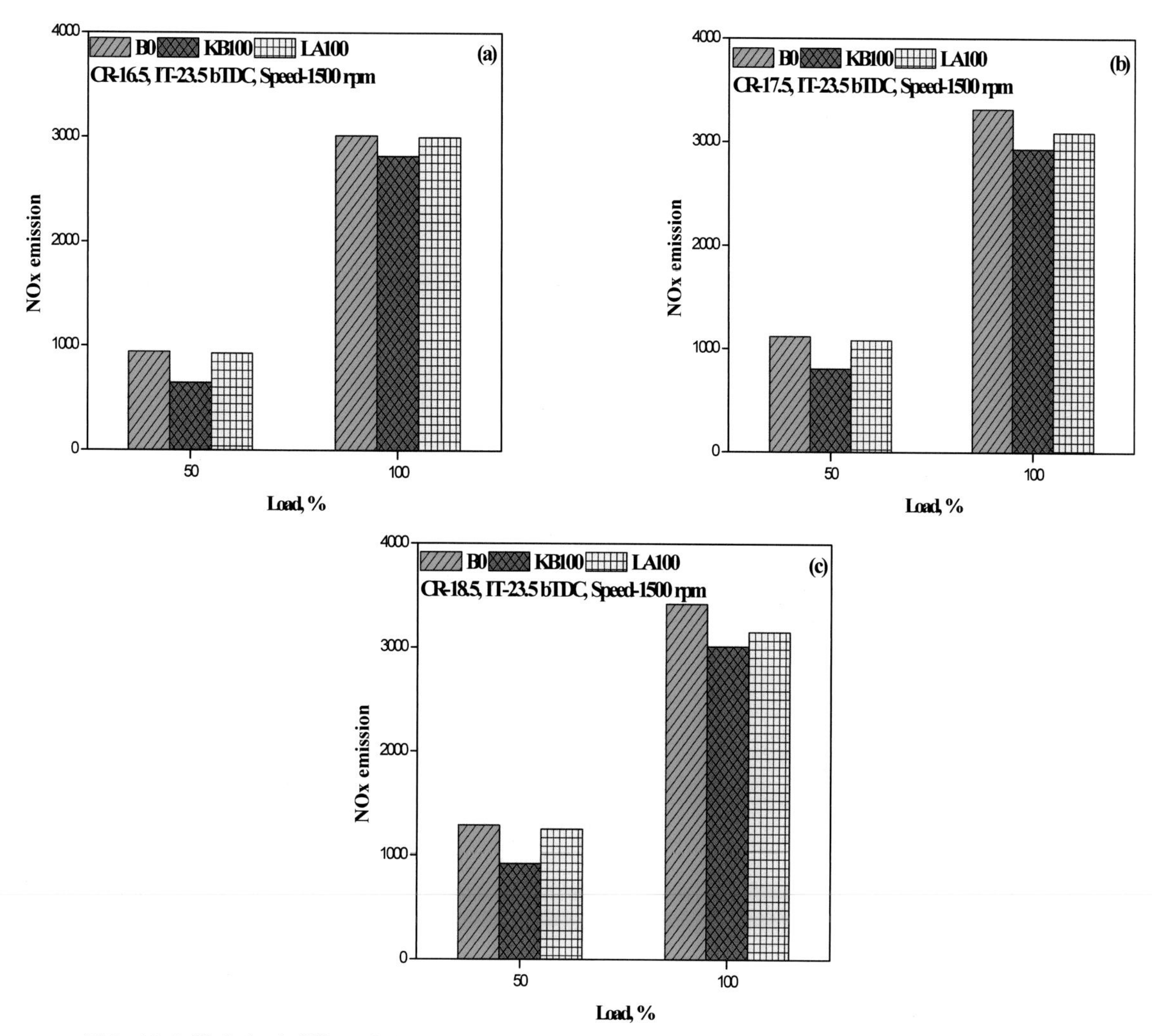

FIG. 11.6 Variation in NO_x emission with load at different compression ratios (CRs). *bTDC*, before top dead center.

in the biodiesel fuel, and oxygen is strongly connected with the carbon atom and so it is tough to break the bond between carbon and oxygen (Salam & Verma, 2019; Shrivastava & Verma, 2020a, b). At higher engine load and at 16.5, 17.5, and 18.5 CR, the smoke emission is found to be about 1.17, 1.16, and 1.14 Bosch smoke number (BSN), respectively, for KB100 and 1.07, 1.04, and 1.04 BSN, respectively, for LA100 compared to 1.55, 1.52, and 1.48 BSN, respectively, for diesel fuel.

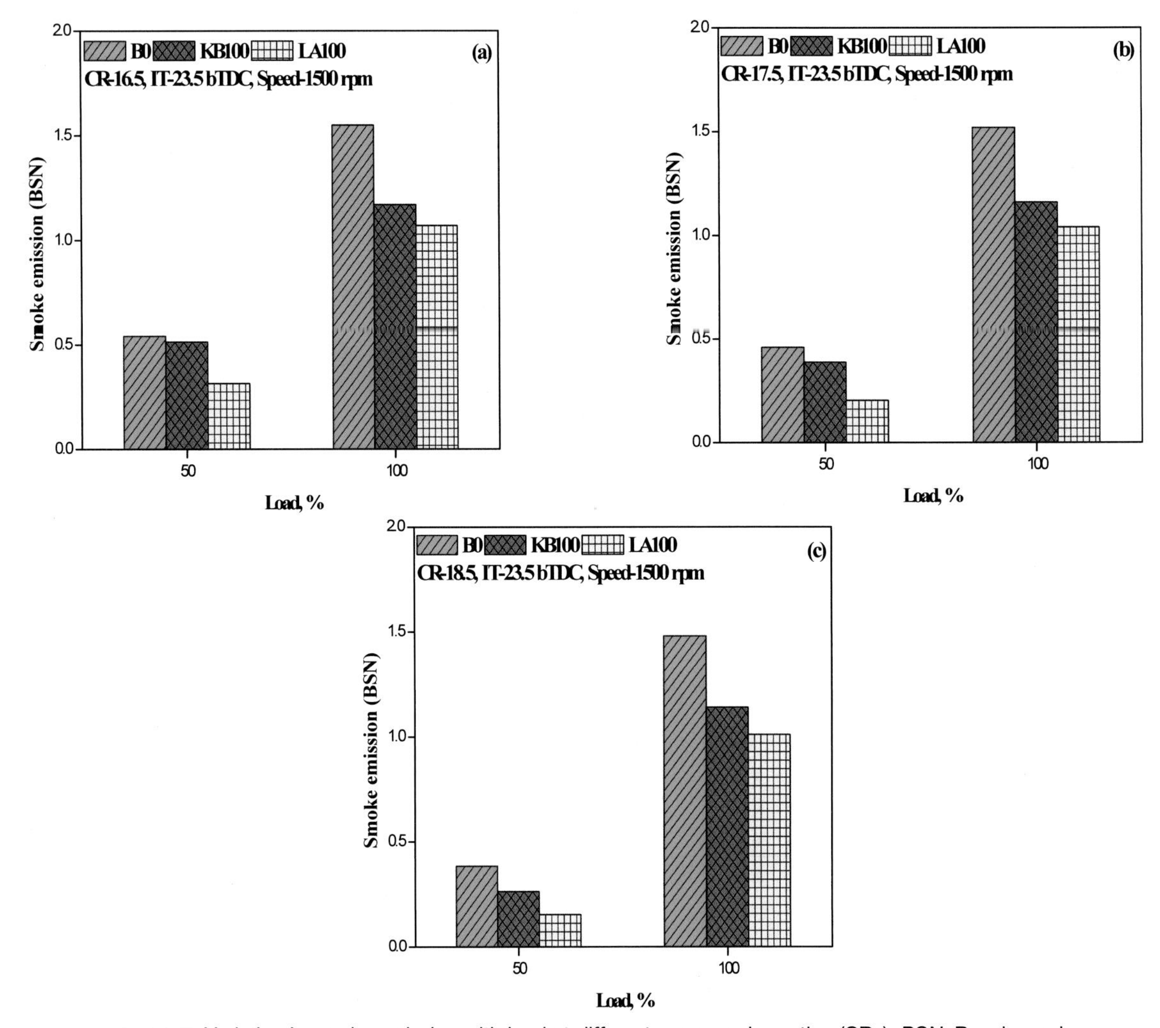

FIG. 11.7 Variation in smoke emission with load at different compression ratios (CRs). *BSN*, Bosch smoke number; *bTDC*, before top dead center.

5 CONCLUSIONS

In this research work, performance and emission characteristics of roselle and karanja biodiesel are investigated by varying the CRs (16.5, 17.5, and 18.5) and engine load (50% and 100% load) and comparing the results with those of standard diesel fuel. The following specific conclusions were made based on the experimental results:

- The BTE of roselle and karanja biodiesel increased by 4.53% and 4.87%, respectively, with increase in CR from 16.5 to 18.5.
- The BSFC decreased by 4.31% for KB100, 5.30% for LA100, and 5.70% for diesel fuel as CR increased from 16.5 to 18.5 and at full load condition.
- The reduction in smoke emission by 2.63% for KB100 and 5.90% for LA100 is recorded at CR, i.e., 18.5, higher than that of the original CR of 16.5.
- The EGT and NO_x emission increased while carbon dioxide emission decreased by raising the CR from 16.5 to 18.5.
- Higher CR results in better engine performance and emission characteristics for both roselle and karanja biodiesel fueled with a CI engine.

NOMENCLATURE

B0	Pure diesel
BSFC	Brake specific fuel consumption
BSN	Bosch smoke number
bTDC	Before top dead center
BTE	Brake thermal efficiency
CI	Compression ignition
CO_2	Carbon dioxide
CR	Compression ratio
EGT	Exhaust gas temperature
FFA	Free fatty acid
KB100	Karanja biodiesel
LA100	Roselle biodiesel
NO_x	Oxide of nitrogen
rpm	Rotation per minute

REFERENCES

Abdel-Rahman, A. A. (1998). On the emissions from internal-combustion engines: A review. *International Journal of Energy Research, 22*(6), 483–513. https://doi.org/10.1002/(SICI)1099-114X(199805)22:6%3C483::AID ER377%3E3.0.CO;2-Z

Agarwal, A. K. (2007). Biofuels (alcohols and biodiesel) applications as fuels for internal combustion engines. *Progress in Energy and Combustion Science, 33*(3), 233–271. https://doi.org/10.1016/j.pecs.2006.08.003

Annamalai, M., Dhinesh, B., Nanthagopal, K., Sivarama Krishnan, P., Lalvani, J. I. J. R., Parthasarathy, M., & Annamalai, K. (2016). An assessment on performance, combustion and emission behavior of a diesel engine powered by ceria nanoparticle blended emulsified biofuel. *Energy Conversion and Management, 123*, 372–380. https://doi.org/10.1016/j.enconman.2016.06.062

Atmanlı, A., Ileri, E., & Yüksel, B. (2014). Experimental investigation of engine performance and exhaust emissions of a diesel engine fueled with diesel–n-butanol–vegetable oil blends. *Energy Conversion and Management, 81*, 312–321. https://doi.org/10.1016/j.enconman.2014.02.049

Bora, B. J., & Saha, U. K. (2016). Experimental evaluation of a rice bran biodiesel–biogas run dual fuel diesel engine at varying compression ratios. *Renewable Energy, 87*, 782–790. https://doi.org/10.1016/j.renene.2015.11.002

Demirbas, A. (2007). Progress and recent trends in biofuels. *Progress in Energy and Combustion Science, 33*(1), 1–18. https://doi.org/10.1016/j.enconman.2008.09.001

Dhinesh, B., Lalvani, J. I. J. R., Parthasarathy, M., & Annamalai, K. (2016). An assessment on performance, emission and combustion characteristics of single cylinder diesel engine powered by *Cymbopogon flexuosus* biofuel. *Energy Conversion and Management, 117*, 466–474. https://doi.org/10.1016/j.enconman.2016.03.049

EL_Kassaby, M., & Nemit_allah, M. A. (2013). Studying the effect of compression ratio on an engine fueled with waste oil produced biodiesel/diesel fuel. *Alexandria Engineering Journal, 52*(1), 1–11. https://doi.org/10.1016/j.aej.2012.11.007

Gautam, A., & Agarwal, A. K. (2013). Experimental investigations of comparative performance, emission and combustion characteristics of a cottonseed biodiesel-fueled four-stroke locomotive diesel engine. *International Journal of Engine Research, 14*(4), 354–372. https://doi.org/10.1177/1468087412458215

Heywood, J. B. (1988). *Internal combustion engine fundamentals.* McGraw-Hill Publishers. ISBN 0-07-028637-X.

Hirkude, J., & Padalkar, A. S. (2014). Experimental investigation of the effect of compression ratio on performance and emissions of CI engine operated with waste fried oil methyl ester blend. *Fuel Processing Technology, 128*, 367–375. https://doi.org/10.1016/j.fuproc.2014.07.026

Holman, J. P. (1966). *Experimental methods for engineers* (p. 564).

Jindal, S., Nandwana, B. P., Rathore, N. S., & Vashistha, V. (2010). Experimental investigation of the effect of compression ratio and injection pressure in a direct injection diesel engine running on Jatropha methyl ester. *Applied Thermal Engineering, 30*(5), 442–448. https://doi.org/10.1016/j.applthermaleng.2009.10.004

Karthikeyan, R., & Mahalakshmi, N. V. (2007). Performance and emission characteristics of a turpentine–diesel dual fuel engine. *Energy, 32*(7), 1202–1209. https://doi.org/10.1016/j.energy.2006.07.021

Mishra, R. K. (2017). Effect of variable compression ratio on performance of a diesel engine fueled with karanja biodiesel and its blends. In *Materials science and engineering conference series* (Vol. 225 (1), p. 012064). https://iopscience.iop.org/article/10.1088/1757-899X/225/1/012064/meta.

Muralidharan, K., & Vasudevan, D. (2011). Performance, emission and combustion characteristics of a variable compression ratio engine using methyl esters of waste cooking oil and diesel blends. *Applied Energy, 88*(11), 3959–3968. https://doi.org/10.1016/j.apenergy.2011.04.014

Panwar, N. L., Shrirame, H. Y., Rathore, N. S., Jindal, S., & Kurchania, A. K. (2010). Performance evaluation of a diesel engine fueled with methyl ester of castor seed oil. *Applied Thermal Engineering, 30*(2–3), 245–249. https://doi.org/10.1016/j.applthermaleng.2009.07.007

Raheman, H., & Ghadge, S. V. (2008). Performance of diesel engine with biodiesel at varying compression ratio and ignition timing. *Fuel, 87*(12), 2659–2666. https://doi.org/10.1016/j.fuel.2008.03.006

Rajak, U., Nashine, P., Singh, T. S., & Verma, T. N. (2018). Numerical investigation of performance, combustion and emission characteristics of various biofuels. *Energy Conversion and Management, 156*, 235–252. https://doi.org/10.1016/j.enconman.2017.11.017

Rajak, U., Nashine, P., & Verma, T. N. (2019). Assessment of diesel engine performance using spirulina microalgae biodiesel. *Energy, 166*, 1025–1036. https://doi.org/10.1016/j.energy.2018.10.098

Rajak, U., & Verma, T. N. (2018a). Spirulina microalgae biodiesel–A novel renewable alternative energy source for compression ignition engine. *Journal of Cleaner Production, 201*, 343–357. https://doi.org/10.1016/j.jclepro.2018.08.057

Rajak, U., & Verma, T. N. (2018b). Effect of emission from ethylic biodiesel of edible and non-edible vegetable oil, animal fats, waste oil and alcohol in CI engine. *Energy Conversion and Management, 166*, 704–718. https://doi.org/10.1016/j.enconman.2018.04.070

Sahoo, P. K., & Das, L. M. (2009). Process optimization for biodiesel production from Jatropha, Karanja and Polanga oils. *Fuel, 88*(9), 1588–1594. https://doi.org/10.1016/j.fuel.2009.02.016

Salam, S., & Verma, T. N. (2019). Appending empirical modelling to numerical solution for behaviour characterisation of microalgae biodiesel. *Energy Conversion and Management, 180*, 496–510. https://doi.org/10.1016/j.enconman.2018.11.014

Sharma, A., & Murugan, S. (2015). Potential for using a tyre pyrolysis oil-biodiesel blend in a diesel engine at different compression ratios. *Energy Conversion and Management, 93*, 289–297. https://doi.org/10.1016/j.enconman.2015.01.023

Sharma, Y. C., Singh, B., & Upadhyay, S. N. (2008). Advancements in development and characterization of biodiesel: a review. *Fuel, 87*(12), 2355–2373. https://doi.org/10.1016/j.fuel.2008.01.014

Shrivastava, P., Salam, S., Verma, T. N., & Samuel, O. D. (2020). Experimental and empirical analysis of an IC engine operating with ternary blends of diesel, karanja and roselle biodiesel. *Fuel, 262*, 116608.

Shrivastava, P., & Verma, T. N. (2020a). An experimental investigation into engine characteristics fueled with lal ambari biodiesel and its blends. *Thermal Science and Engineering Progress, 17*, 100356.

Shrivastava, P., & Verma, T. N. (2020b). Effect of fuel injection pressure on the characteristics of CI engine fuelled with biodiesel from Roselle oil. *Fuel, 265*, 117005.

Shrivastava, P., Verma, T. N., & Pugazhendhi, A. (2019a). An experimental evaluation of engine performance and emission characteristics of CI engine operated with Roselle and Karanja biodiesel. *Fuel, 254*, 115652.

Shrivastava, P., Verma, T. N., & Pugazhendhi, A. (2019b). An experimental evaluation of engine performance and emission characteristics of CI engine operated with Roselle and Karanja biodiesel. *Fuel, 254*(115652), 1–12. https://doi.org/10.1016/j.fuel.2019.115652

Shrivastava, P., Verma, T. N., Samuel, O. D., & Pugazhendhi, A. (2020). An experimental investigation on engine characteristics, cost and energy analysis of CI engine fuelled with Roselle, Karanja biodiesel and its blends. *Fuel, 275*, 117891. https://doi.org/10.1016/j.fuel.2020.117891

Singh, T. S., & Verma, T. N. (2019a). Impact of tri-fuel on compression ignition engine emissions: Blends of waste frying oil–alcohol–diesel. In *Methanol and the alternate fuel economy* (pp. 135–156). Singapore: Springer. https://doi.org/10.1007/978-981-13-3287-6_7.

Singh, T. S., & Verma, T. N. (2019b). Taguchi design approach for extraction of methyl ester from waste cooking oil using synthesized CaO as heterogeneous catalyst: Response surface methodology optimization. *Energy Conversion and Management, 182*, 383–397. https://doi.org/10.1016/j.enconman.2018.12.077

Singh, T. S., Verma, T. N., Nashine, P., & Shijagurumayum, C. (2018). BS-III diesel vehicles in Imphal, India: An emission perspective. In *Air pollution and control* (pp. 73–86). Singapore: Springer.

Sivaramakrishnan, K. (2018). Investigation on performance and emission characteristics of a variable compression multi fuel engine fuelled with Karanja biodiesel–diesel blend. *Egyptian Journal of Petroleum, 27*(2), 177–186. https://doi.org/10.1016/j.ejpe.2017.03.001

CHAPTER 12

Application of Design for Sustainability to Develop Smartphone Holder Using Roselle Fiber-Reinforced Polymer Composites

S.M. SAPUAN • R.A. ILYAS • M.R.M. ASYRAF • A. SUHRISMAN • T.M.N. AFIQ • M.S.N. ATIKAH • R. IBRAHIM

1 INTRODUCTION

Natural fibers, such as corn (Ibrahim, Sapuan, et al., 2020; Sari et al., 2020), cassava (Ibrahim, Edhirej, et al., 2020), water hyacinth (Syafri, Sudirman, et al., 2019b), cogon fiber (Jumaidin et al., 2020; Jumaidin, Saidi, et al., 2019b), sugarcane (Asrofi, Sapuan, et al., 2020; Asrofi, Sujito, et al., 2020b; Jumaidin, Ilyas, et al., 2019), ramie (Syafri, Kasim, et al., 2019), ginger (Abral, Ariksa, et al., 2019a, 2020a), flax (Akonda et al., 2020), kenaf (Aisyah et al., 2019; Mazani et al., 2019), sisal (Yorseng et al., 2020), hemp (Arulmurugan et al., 2019), jute (Islam et al., 2019), oil palm fiber (Ayu et al., 2020), and sugar palm (Atiqah et al., 2019; Halimatul et al., 2019b; Hazrol et al., 2020; Maisara et al., 2019; Norizan et al., 2020; Nurazzi et al, 2019, 2020; Sapuan et al., 2020), have gained a considerable interest owing to their advantages, e.g., biodegradability, renewability, plenty availability, and good mechanical properties, compared with synthetic fibers (Ilyas & Sapuan, 2020a, 2020b; Nadlene et al., 2018; Nazrin et al., 2020). Natural plant fibers are cellular hierarchic biocomposites produced by nature and are basically semicrystalline cellulose microfibril-reinforced amorphous matrices made of hemicellulose, lignin, waxes, extractives, and trace elements (Ilyas et al., 2017; Ilyas, Sapuan, & Ishak, 2018; Ilyas, Sapuan, Ishak, Zainudin, et al., 2018; Ilyas, Sapuan, Ibrahim, et al., 2019; Ilyas, Sapuan, Ishak, & Zainudin, 2019, Ilyas, Sapuan, Atikah, et al., 2020). Lignocellulosic fibers consist, therefore, of a cemented microfibril aggregate. As a consequence, the structure of plants spans many length scales to provide maximum strength with a minimum of material. These advantages have pulled in material engineers to use natural fibers to be reinforced with either synthetic or biobased polymer materials as polymer composite reinforcement to reduce timber usage and lessen underutilization of natural fibers (Abral, Atmajaya, et al., 2020; Abral, Basri, et al., 2019; Halimatul et al., 2019a; Nadlene et al., 2016a). Besides that, these natural fibers have many potential applications from plastic packaging to scaffolds for tissue regeneration (Atikah et al., 2019; Ilyas, Sapuan, Atiqah, et al., 2020; Ilyas, Sapuan, Ibrahim, et al., 2019b, 2020b; Ilyas, Sapuan, Ishak, & Zainudin, 2018a, 2018b, 2018c; Ilyas, Sapuan, Sanyang, et al., 2018f; Sanyang et al., 2018). Various types of natural fibers can be found in tropical countries like Malaysia and Indonesia (Ishak et al., 2013). Roselle plant (*Hibiscus sabdariffa* L.), being a source of natural fiber, is abundantly available in nature and is grown in Borneo, Guyana, Malaysia, Sri Lanka, Togo, Indonesia, and Tanzania (Razali et al., 2015). Table 12.1 lists various types of natural fibers and their countries of origin. Therefore this chapter aims to design and fabricate a smartphone holder using product design specification (PDS) and hand layup process using roselle fiber-reinforced epoxy polymer.

1.1 Roselle

Roselle (*H. sabdariffa* L.) (family: Malvaceae) (Fig. 12.1) can be found in Asia and is widely grown in West and East Africa, India, China, and Southeast Asia, e.g., Malaysia, Thailand, Vietnam, Myanmar, and Indonesia, for various applications. Besides that, roselle plants are

Roselle. https://doi.org/10.1016/B978-0-323-85213-5.00006-8

TABLE 12.1
Natural Fibers and Origins (Nadlene et al., 2016a).

Natural Fiber	Origin
Flax	Borneo
Hemp	Yugoslavia, China
Sun hemp	Nigeria, Guyana, Sierra Leone, India
Ramie	Honduras, Mauritius
Jute	India, Egypt, Guyana, Jamaica, Ghana, Malawi, Sudan, Tanzania
Kenaf	Iraq, Tanzania, Jamaica, South Africa, Cuba, Togo
Roselle	Borneo, Guyana, Malaysia, Sri Lanka, Togo, Indonesia, Tanzania
Sisal	East Africa, Bahamas, Antigua, Kenya, Tanzania, India
Abaca	Malaysia, Uganda, Philippines, Bolivia
Coir	India, Sri Lanka, Philippines, Malaysia

FIG. 12.1 The roselle flower.

widely spread in the tropics and subtropics in the north and south hemispheres, as well as in many parts of Jamaica, Trinidad and Tobago, and Central America (Morton, 1974). The shrubbery is used in some countries for decorative purposes, while in others, seeds and petals are used for human consumption (Sánchez-Mendoza et al., 2008). The pharmacologic field has been centered in the recent years on the use of roselle calyces because of their potential advantages in alternative medicinal products. Researchers demonstrated the antioxidant, diuretic, fever relief, hypoglycemia, cholesterol, and hypertension reduction characteristics of roselle extracts (Cid-Ortega & Guerrero-Beltrán, 2015).

Besides applications for food and medical purposes, roselle plant is also used to make ropes, jute, and textile with its versatile fiber (Radzi et al., 2017). In Malaysia, roselle plants are cut off after a year of cultivation because of the fruit's deterioration quality beyond this period. As a result, the whole roselle plant becomes agriculture waste, and hence, efficient use of this plant's fiber is achievable through reinforcement in polymer composites. Before any industrial applications such as car manufacturing and building construction, it is crucial to understand the physical, thermal, mechanical, and chemical properties of roselle fiber (Razali et al., 2015). Roselle fiber is a type of natural fiber that attracts researchers' attention owing to its good tensile properties, toughness, water resistance, inflexibility, and coarseness (Nadlene et al., 2016b). These physical and mechanical properties were determined mainly by its chemical and physical compositions, including fiber structure, fiber-angle cellulose content, and cross-section angle, and by degree of polymerization (Idicula et al., 2005). Because its chemical properties are similar to the well-known jute fiber, roselle fiber is seen as a potential candidate for reinforcing polymer fiber application.

FIG. 12.2 **(A)** Roselle plant, **(B)** roselle stalks, **(C)** water retting process, and **(D)** roselle fiber (Nadlene et al., 2016a).

According to Nadlene et al. (2016a), roselle should be planted 4–5 months before the end of the rainy season. The period from planting to harvest is approximately 4–5 months, while the period from flowering to harvest is approximately 1.5 months. Therefore between December and January, a good planting period is established so that the plant may be harvested between May and June.

1.2 Fiber Production

Roselle is mainly grown for the bast fiber extracted from the stems that exist inside the phloem covered under the bark (Fig. 12.2A). The fiber networks provide mechanical strength to the plant by maintaining the phloem's conductive cells. Roselle stems are red, and the fibers were extracted using water retting process (Nadlene, 2016a). Roselle stalks were harvested in bud stage to obtain decent-quality fibers (Fig. 12.2B). They were bundled and retted for 3–4 days in water (Fig. 12.2C). The retted stem was washed in running water, which separated the fibers from the stem and cleaned them from impurities before sundrying (Kian et al., 2017). The final product is as illustrated in Fig. 12.2D.

1.3 Epoxy

Epoxy resins that were discovered in 1909 by Prileschajew are defined as prepolymers of low molecular weight containing more than one epoxy group. The epoxy group is also called the group of oxirane or ethoxyline and considered a representative class of the epoxy polymer (Yu et al., 2009). The bulk of the epoxy resins that is currently available is diglycidyl ether oligomers of bisphenol A (Roşu et al., 2002). These oligomers cure the epoxy resin and become a thermosetting polymer when they react with the hardener (Zhang et al., 2012).

Being one of the most common adhesive materials, epoxy is used in various applications. The best available adhesive and epoxy adhesives are used in the automotive and aerospace industries. Solvent-free epoxy adhesives provide water resistance, pressure resistance, and longevity, as well as chemical and thermal resistance. Epoxy coatings provide a solid, protective layer of excellent hardness that comes from epoxy resins that are excellent electric insulators, making them useful for the electronics industry. They are used in the manufacturing of generators, motors, transformers, and insulators.

Epoxy resin is a feasible polymer that has efficient power, strong endurance, and resilience and is resistant to moisture and chemicals. These features are providing epoxy resin with great electric insulating properties and contribute to its nonvolatility (Belaadi et al., 2014). Epoxy resin can be cured without pressure at ambient temperature via two approaches: using a curing agent or heat curing. Almost all materials such as wood, glass, natural fiber, and metal can be bonded using epoxy resin, which showed little or no shrinkage after healing (Feng & Guo, 2016). Table 12.2 shows the properties of epoxy resin, and Table 12.3 presents various natural fiber-reinforced composites from works of literature. Even though there are studies reported on using natural fiber-reinforced epoxy polymer composites, none has been found on the utilization of roselle fibers in epoxy resin matrix. Therefore the objective of this work is to design smartphone holders using PDS and fabricate them using the hand layup process. Roselle fiber-reinforced epoxy polymers were used to fabricate this product. It should be noted that the roselle fibers used in this work were not chemically treated or modified.

1.4 Polymer Composites

A composite consists of two or more materials mixed at the macroscopic stage but insoluble into each other. The embedded element is called the reinforcement, and the other element is called the matrix. Polymers are usually categorized into two different classes according to their origins: natural and synthetic. They are also commonly classified in thermosetting and thermoplastic polymers for most applications, especially in the field of polymer composites. Thermoplastic products are currently dominated by polypropylene, polyethylene, and polyvinyl chloride as matrices for biofibers. At the same time, phenol, epoxy, and polyester resins are the thermosets that are most commonly used (Sanjay et al., 2015). Over the past few decades, natural fibers have gained the attention of many researchers and scientists as alternative reinforcements for polymer composites because they are more advantageous than the conventional glass and carbon fibers (Saheb & Jog, 1999). These include flax, hemp, jute, sisal, kenaf, coir, kapok, banana, and many other natural fibers (Li et al., 2006). Natural fibers have superior mechanical characteristics in comparison with glass fibers in terms of flexibility and rigidity (Jarukumjorn & Suppakarn, 2009). The key benefits of polymers are easy processing and cost-saving. Incorporation of fibers with fillers modified the mechanical properties of the finished polymer composite to fulfill the strength and modulus specifications in most of their uses as well as widened their applications from computers to spacecraft (Saheb & Jog, 1999).

TABLE 12.2
Properties of Epoxy Resin.

Appearance	A Clear Pale Yellow Liquid
Specific gravity at 25°C (g/cm^3)	1.12
Solid content (%)	84
Tensile strength (MPa)	31
Flexural strength (MPa)	67
Impact strength (kg/m^2)	9

Data were taken from Mittal, V., Saini, R., & Sinha, S. (2016). Natural fiber-mediated epoxy composites - A review. *Composites Part B: Engineering*, *99*, 425–435. https://doi.org/10.1016/j.compositesb.2016.06.051.

TABLE 12.3
Reported Work on Natural Fiber-Reinforced Epoxy Composites.

Natural Fiber	Characteristics	References
Cellulose fiber	Mechanical properties	Low et al. (2007)
Bamboo fiber	Wear properties	Biswas and Satapathy (2010)
Sisal fiber	Tensile, wear, and water absorption characteristics	Mohan and Kanny (2011)
Hemp fiber	Impact and flexural properties	Wood et al. (2011)
Piassava fiber	Mechanical characteristics	Cristina et al. (2012)
Cellulose fiber	Toughness and impact strength	Alamri et al. (2012)
Recycled cellulose fiber	Flexural, fracture toughness, and water absorption	Alamri and Low (2012b)
Coir fiber	Tensile properties	Romli et al. (2012)
Jute fiber	Flexural and interlaminar shear strength	Mishra and Biswas (2013)
Flax fiber	Data related to Rosen model	Coroller et al. (2013)
Wood dust	Mechanical characteristics	Kumar and Duraibabu et al. (2014), Kumar, Kumar, et al. (2014)
Wood dust	Analysis tensile and flexural strength using Taguchi method	Suhvhqw et al. (2014)
Kenaf fiber	Tensile and flexural strength	Rq et al. (2014)
Ramie fiber	Mechanical strength	Gu et al. (2014)
Kenaf fiber	Validate the tensile strength between experimental and theoretic usage (rule of mixtures)	Mahjoub et al. (2014)
Cellulose fiber	Thermal properties	Alamri et al. (2012)
Sugar palm fiber	Fickian diffusivity	Leman et al. (2008)
Weave flax fiber	Water diffusivity	Newman (2009)
Flax fiber	Fracture and toughness	Liu and Hughes (2008)
Palm tree fiber	Dielectric properties	Amor et al. (2010)
Hemp yarn fiber	A study between experimental and numeric results	Vasconcellos and Touchard (2012)
Silk	Energy and failure response	Ataollahi et al. (2012)
Silk	Crashworthiness characteristics	Oshkovr et al. (2012)
Hemp fiber	Fatigue behavior	Vasconcellos, Touchard, and Chocinski-arnault (2014)
Flax fiber	Damping behavior	Guen et al. (2014)
Flax fiber	Fatigue behavior	Shaoxiong Liang et al. (2014)
Flax fiber	Crashworthiness characteristics	Yan et al. (2014a)
Flax fiber	Interfacial characteristics	Le et al. (2020)
Flax fiber	Fire reaction characteristics	Kandare et al. (2014)
Natural silk	Absorption energy	Eshkoor et al. (2014)
Flax fiber	Crushing property	Yan et al. (2014b)
Hemp fiber	Rheologic and thermal analysis	Landro and Janszen (2014)
Banana fiber	Tensile and flexural strength	Sapuan (2000)

TABLE 12.3
Reported Work on Natural Fiber-Reinforced Epoxy Composites.—cont'd

Natural Fiber	Characteristics	References
Tenax leaf fiber	Mechanical and thermal characteristics	Maria et al. (2010)
Bamboo fiber	Wear and frictional characteristics	Nirmal et al. (2012)
Flax fiber	Fatigue properties	Liang et al. (2012)
Jute fiber	Tensile characteristics	Hossain et al. (2013)
Flax fiber	Tensile and compressive characteristics	Muralidhar (2013)
Sisal fiber	Tensile and flexural properties	Irxqg et al. (2014)
Betelnut fiber	Wear and abrasion characteristics	Yousif et al. (2010)
Agave fiber	Mechanical characteristics	Mylsamy and Rajendran (2011a)
Flax fiber	Crashworthiness properties	Yan and Chouw (2013)
Phormium leaf fiber	Flexural and water absorption characteristics	Newman et al. (2007)
Sugar palm fiber	Tensile characteristics	Bachtiar et al. (2008)
Arenga pinnata fiber	Mechanical properties	Ali et al. (2010)
Agave fiber	Tensile and flexural characteristics	Mylsamy and Rajendran (2011b)
Kenaf fiber	Flexural characteristics	Yousif et al. (2012)
Fique fiber	Flexural properties	Hoyos and Vázquez (2012)
Recycled cellulose fiber	Fracture and flexural characteristics	Alamri and Low (2012a)
Flax fiber	Tensile properties	Scida et al. (2013)
Banana fiber	Viscoelastic behavior	Venkateshwaran et al., (2013)
Bamboo fiber	Tensile and elongation properties	Lu et al. (2013)
Natural silk fiber	Crashworthiness properties	Eshkoor et al. (2013)
Sisal fiber	Impact and flexural characteristics	Srisuwan et al. (2014)
Coconut fiber	Adhesion properties	Kumar and Duraibabu et al. (2014), Kumar, Kumar, et al. (2014)
Kenaf fiber	Mechanical characteristics	Fiore et al. (2015)
Luffa fiber	Mechanical and thermal behavior	Anbukarasi and Kalaiselvam (2014)
Kenaf fiber	Thermal properties	Azwa and Yousif (2013)
Coconut sheath fiber	Mechanical and thermal characteristics	Kumar, Duraibabu, et al. (2014), Kumar, Kumar, et al. (2014)
Abaca fiber	Thermal study	Liu et al. (2014)
Agave fiber	Water absorption study	Mylsamy and Rajendran (2011b)
Recycled cellulose fiber	Water absorption study	Alamri and Low (2012a)
Bamboo husk fiber	Mechanical and thermal characteristics	Shih (2007)
Jute	Wear properties	Ahmed et al. (2012)
Basalt fiber	Tensile and flexural characteristics	Ary Subagia et al. (2014)
Nanocellulose	Thermal and mechanical study	Gabr et al. (2014)

2 ROSELLE COMPOSITE SMARTPHONE HOLDER PRODUCT DEVELOPMENT USING THE DESIGN FOR SUSTAINABILITY APPROACH

The principal purpose of the task is to develop a new design of roselle fiber-reinforced polymer composite smartphone holder product via the design for sustainability (DfS) method. From this project, the expected output is to substitute the current products available in the market, which are mainly fabricated from rubber and wood materials. Moreover, the project aims to establish the applications of biocomposite products in the household accessories market. However, the design of the product is limited to the application of roselle fiber-reinforced polymer composite as a smartphone holder for desk use. As previously mentioned, the implementation of roselle fiber-reinforced polymer composite as substitutive materials was targeting to develop an environment-friendly and lightweight smartphone holder product while maintaining the mechanical strength to fulfill the intended functions (Asyraf et al., 2020b; Jaafar et al., 2018).

All in all, the product was developed via four main steps: conducting the market investigation, proposing conceptual designs, selecting final design, and creating the detailed design of the finished product. These include the smartphone holder mold shape and manufacturing of the smartphone holder mold and the smartphone holder. Each of the product development processes will be further elaborated in the subsequent subtopics.

2.1 Product Background and Market Investigation

From the statistics in 2019, around 18.4 million smartphone users in Malaysia use the gadget as their daily assistance device and to form communication networks with each other (Muller, 2019). Somehow the device has also been used to read e-books, watch videos, and have teleconferences with friends and colleagues, which require the users to hold it for a long period. Thus this brought ideas to designers to develop smartphone holders with various materials, shapes, and design to attract the consumers' attention. Usually, a smartphone holder is made from rubber sheet, metal, softwood, and plastic materials to execute the original function.

Pugh Total Design approach is one of the available concurrent engineering methods used by many designers in developing new products. The method started with market analysis to ensure all relevant facts about the intended product design are collected accordingly (Pugh, 1996). All information on the planned product design, e.g., existing product features, customer requirements, and product price, were acquired via informal interviews with users of the product and web browsing on current smartphone holder concepts (Asyraf et al., 2020c; Sapuan, 2014). One notable discovery from the performed market analysis was that up to date, similar smartphone holder products have not yet been produced using biocomposite materials, specifically roselle fiber-reinforced polymer composites. Additionally, there is a lack of smartphone holder designs available in the current market with engineering elements. Particularly, no similar smartphone holder of similar materials is being produced in the authors' institution (Universiti Putra Malaysia).

Thus the market analysis showed that the intended product design by the DfS approach would be a new market fragment to be reconnoitered using roselle fiber-reinforced polymer composites. Moreover, the implementation of roselle fiber-reinforced polymer composites would consume less raw material, reduce manufacturing costs, and reduce energy consumption. Lastly, the addition of engineering elements into the product features, such as good strength properties with the minimalist design, would draw consumers' recognition for the new roselle fiber-reinforced polymer composite smartphone holder in the market.

2.2 Product Design Specifications

Overall, a document, namely, PDS, would be identified and created according to the previous market analysis data to execute the product development process. In this stage, PDS aids to elaborate on the requirements of the product that should be fitted and well accepted by the customer's view. Generally, 32 elements have to be considered in the PDS document, as mentioned by Pugh (1991). From those elements, Fig. 12.3 specifies nine elements that were determined to be incorporated into the design of roselle composite smartphone holder.

From these identified PDS elements, durability was one of the vital aspects in developing roselle fiber-reinforced polymer composite smartphone holder. The durability performance of the newly invented product should be equivalent to or better than the existing conventional smartphone holders, which are commonly made of wood, plastic, and rubber. Besides, the weight and size of the elements from the PDS document are included as key factors for consumers when considering to buy a smartphone holder. Hence, the materials and manufacturing cost aspect has to be considered to maintain the affordable price to users, which consequently makes the product competitive in the market.

Another important element in PDS for a smartphone holder is safety. For roselle fiber-reinforced polymer composite smartphone holder, it must be free from sharp edges and small components to avoid from harming the users and their toddlers while handling the product. Furthermore, the product should be designed with simplicity and minimalist concepts as well as smooth surface finishing to maintain its marketability to the users. This dimension influences the design outcome and the manufacturing capacity. Descriptions of the roselle fiber-reinforced polymer composite smartphone holder were explained in the PDS listing, as summarized in Table 12.4.

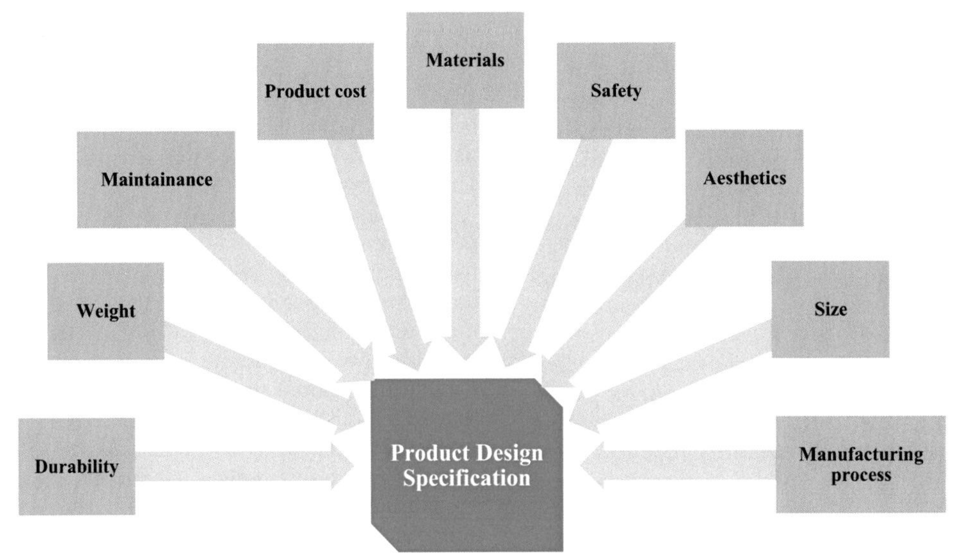

FIG. 12.3 Elements of product design specification considered for roselle fiber-reinforced polymer composite smartphone holder.

TABLE 12.4
Description of Roselle Fiber-Reinforced Polymer Composite Smartphone Holder PDS.

PDS Elements	Description
1. Durability	Must possess high strength and durability to hold a smartphone in either portrait or landscape view.
2. Weight	One product must weigh less than 200 g.
3. Maintenance	The product must be free from maintenance.
4. Product cost	The total cost of a single product must be less than RM30.
5. Materials	The product would be fabricated from roselle fiber-reinforced unsaturated polyester composites.
6. Safety	The product should have a smooth surface and round angle to avoid cuts and accidents to the user. Only a single component is used to maintain it from protruding.
7. Aesthetics	The product design should be simple and minimal with good quality and finishing.
8. Size	The width, length, and thickness of the holder must not exceed 100, 230, and 30 mm, respectively.
9. Manufacturing process	The product should be easily fabricated and manufactured either using the hand layup technique or an automated process.

PDS, product design specification.

TABLE 12.5
New Conceptual Designs of Roselle Fiber-Reinforced Polymer Composite Smartphone Holder and Its Descriptions.

	Conceptual Design	Description
1.		• Concept inspired by high-heeled shoes. • Lifts the smartphone higher than regular holder. • Good artistic structure.
2.		• Easy to manufacture and assemble. • Simple and minimalist design. • High stability owing to wider baseline.
3.		• Simple and minimalist design. • Easy to store. • Easy to fabricate and manufacture. • High stability owing to wider baseline.
4.		• The hollow triangular feature can be used to place stationeries (i.e., pen, pencil, key, etc.). • Three strips of lines for aesthetic purpose.
5.		• The hollow triangular feature can be used to place stationeries (i.e., pen, pencil, key, etc.). • Simple and minimalist design. • Isometric design, hence easy to fabricate.

2.3 Conceptual Design Development of Roselle Composite Hand Phone Holder

From the generated PDS elements, the subsequent stage in roselle fiber-reinforced polymer composite smartphone holder development is to make concept designs of the product. Later, the best concept design will be chosen among the proposed concept designs. The details for each activity are elaborated in the following subsections.

2.3.1 Generating conceptual design using the brainstorming method

Several generation technique ideas were used by designers to develop conceptual designs for products, e.g., Theory of Inventive Problem Solving (TRIZ); brainstorming; Strength, Weakness, Opportunity, and Threat (SWOT) analysis; gallery method; and Systematic Exploitation of Proven Ideas of Experience (Asyraf et al., 2019; Sapuan, 2005). For this project, the simplest way to generate the idea for conceptualizing the design concepts was using the brainstorming approach. A design focus group was formed and a discussion was held among the members of Advanced Engineering Materials and Composites (AEMC) Research Centre, Department Mechanical and Manufacturing Engineering, Universiti Putra Malaysia. Every concept design was outlined and listed based on the discussion outputs and the PDS documents of the previous design stage. In the end, around five conceptual designs of roselle fiber-reinforced polymer composite smartphone holder were developed. The details of each concept design are laid out in Table 12.5. Creative and innovative variations were made to add values to the ideas in addition to being manufactured using sugar palm composites. The solution and characteristics of each design were created using the DfS suggestion to enhance the application of biocomposite materials to gain the desired functionality.

2.3.2 Conceptual design using the Pugh concept selection method

At this point of stage, the best concept designs among the five alternatives of roselle fiber-reinforced polymer composite smartphone holder were selected via the Pugh concept selection method. There are other available methods to select best concepts, e.g., Analytic Hierarchy Process, Analytic Network Process, TOPSIS, and VIKOR (Asyraf et al., 2020a; Asyraf, Rafidah, et al., 2020d; Davoodi et al., 2008; Shaharuzaman et al., 2020). These methods implement the evaluation of PDS attributes as selection criteria among all concept designs being proposed. Material attributes in the PDS document were excluded from the evaluation process for conceptual design selection because all ideas used the same material. The scoring process for each concept design was executed by comparing the design's performance to the datum design (current product existing in the market with similar function to be used as a reference in the process). To be specific, the datum used in the project was the wooden smartphone holder.

Moreover, the scoring process among the concept designs was based on three criteria: "+" for performance better than datum, "−" for worse performance than datum, and "S" for equal performance as the datum. At the end of the process, the best concept design was selected depending on the highest value of nett score, which is the sum of "+" score subtracted by the sum of "−" score. The Pugh selection approach was chosen in this project because it was a quite straightforward technique to perform a simultaneous and systematic decision-making process. Table 12.6 displays the selection of best conceptual designs obtained using the Pugh design selection approach.

According to Table 12.6, the concept design 3 recorded the highest nett score value of +3 among all the other concept designs proposed. Concept design 3 has shown that it has better performance in terms of weight, production cost, and manufacturing process in comparison with the existing wooden smartphone holder. Thus concept design 3 was applied as the final design for the roselle fiber-reinforced polymer composite smartphone holder, as shown in Fig. 12.4.

Concept design 3 was selected as the final concept design for roselle fiber-reinforced polymer composite smartphone holder with regard to weight, cost, safety, and aesthetics. This concept design's features are symmetric and flat shaped, which marks the design as the simplest among all other designs. These would contribute to improving the fabricating process of the product, which would subsequently increase the product's manufacturability and productivity while reducing the total product cost by the simple mold design. Moreover, a long groove across the width was cut and removed from the product surface, which was the smartphone insertion slot. The product weight was reduced by removing the part that did not contribute to supporting the load, without compromising on the product strength and stiffness. All the edges have been

TABLE 12.6
Summary of Conceptual Design Selection Using the Pugh Selection Method.

	Conceptual Design of Roselle Fiber-Reinforced Polymer Composite Smartphone Holder					
	Existing	**1**	**2**	**3**	**4**	**5**
Selection Criteria						
Durability		+	−	S	+	S
Weight		−	−	+	−	−
Maintenance		−	−	S	−	S
Product cost		−	−	+	−	−
Safety		−	−	S	−	S
Aesthetics		+	+	S	+	+
Size		−	+	S	−	−
Manufacturing process (ease of manufacturing)		−	+	+	−	−
Total score	$\sum+$	2	3	3	2	1
	$\sum-$	7	5	0	6	4
	$\sum S$	0	0	5	4	3
Nett Score	$\sum+(-)\sum-$	−5	−2	3	−4	−3

FIG. 12.4 Best conceptual design selected for roselle composite smartphone holder.

made into a circular shape to eliminate sharp edges and improve product safety when in use.

As mentioned previously, these design features are termed as design for excellence (DfX), which promotes designers to integrate elements such as design for safety, design for manufacturing, design for ease of assembly, and ergonomics in their conceptual design solution. Thus good-quality end product and customer satisfaction, along with reduction in design errors as early as possible, are possible when the DfX is applied in the conceptual design stage of green product development (Jack, 2013).

2.4 Detailed Design of Roselle Composite Smartphone Holder

As a subsequent activity in the DfS approach, the detailed design was executed. At this level, the essential information to manufacture the selected concept design was well-defined. The information involved in this stage were product dimensions, bill of materials, and its tolerance. This information was illustrated using technical engineering drawing. In specific, this project used the computer-aided design software, CATIA V5, to generate technical drawings for selected roselle fiber-reinforced polymer composite smartphone holder designs. The same software was also being used to create the mold design for the fabrication of the product.

2.5 Roselle Composite Smartphone Holder Fabrication

After the detailed design stage, the last activity performed was to manufacture the roselle fiber-reinforced polymer composite smartphone holder. The fabrication process was divided into two stages: mold fabrication and end-product fabrication.

2.5.1 Mold fabrication

The mold used for this composite smartphone holder was made from rectangular 3D-printed material, with 200 mm in length and 150 mm in width, and it was coated with wax. Wax is a mold-releasing agent that was applied to the surface before the molding process so that the end product can be easily removed. The mold was designed in three parts for ease of molding operation. Part 1 was the left side and part 2 was the right side of the mold, which were made in a rectangular form of 200 mm in length and 75 mm in width, respectively, and coated with wax. Part 3 was the upper side of the mold, which was made in a rectangular shape that was 200 and 150 mm in width and length, respectively, and was coated with wax. The functions of the upper side were to cover the fiber after epoxy application and to avoid the debris from entering the composite during the curing process.

The 3D printing method was used to produce the mold because the size of the mold was small and the mold was compatible to be printed using the 3D printer (Ultimaker 3, Ultimaker, Netherlands) (Fig. 12.5A). Poly(lactic) acid (PLA) was used as the filament for the 3D printer because of its suitability for small production that required less energy and generated less waste. Besides, the PLA material can be easily disposed of at the end of its life cycle via the natural decomposition process. Fig. 12.5B,C shows the fabricated mold for the smartphone holder product.

2.5.2 Composite smartphone holder fabrication

For the production of the smartphone holder, epoxy resin that was cured by methyl ethyl ketone peroxide (MEKP) was used to be reinforced with roselle fiber. As shown in Table 12.7, epoxy, roselle fiber, and

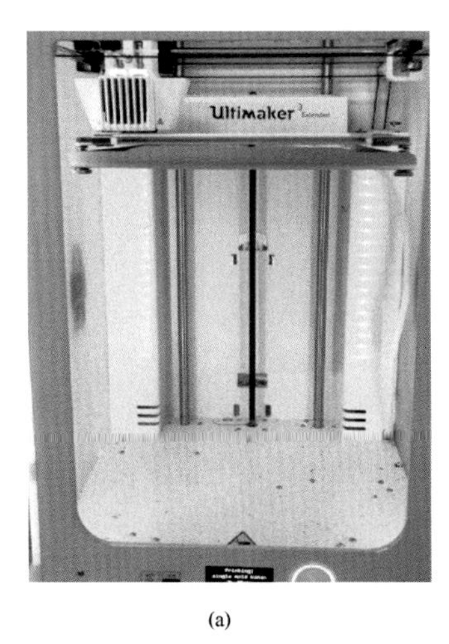

(a)

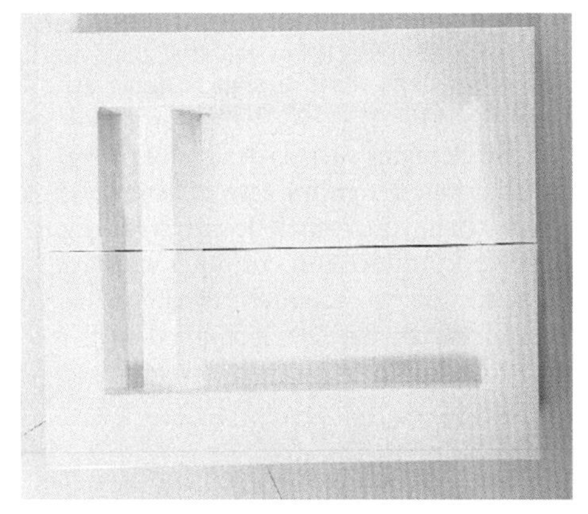

(b) (c)

FIG. 12.5 **(A)** 3D printer used. **(B,C)** Product mold.

TABLE 12.7
The Required Amount of Epoxy, MEKP Catalyst, and Roselle Fiber.

Item	Weight
Volume of composite mold	$= 411.752\ cm^3$
70% Of epoxy	$= 0.7\ cm^3 \times 411.752\ cm^3 \times 1.12\ (g/cm^3)$ $= 322.814\ g$
2% Of MEKP catalyst	$= 0.02\ cm^3 \times 411.752\ cm^3 \times 1.55\ (g/cm^3)$ $= 12.764\ g$
30% Of roselle powder	$= 0.30\ cm^3 \times 411.752\ cm^3 \times 1.4\ (g/cm^3)$ $= 172.936\ g$
Total weight	$= 322.814\ g + 12.764\ g + 172.936\ g$ $= 508.514\ g$

MEKP, methyl ethyl ketone peroxide.

MEKP were manually mixed according to their compositions. The mass of MEKP used was 2% of the epoxy resin's mass (Mazani et al., 2019). Both chemicals were measured in appropriate amounts and poured in a well-purified container. The solution was slightly agitated to stop the formation of air bubbles, which led to porosity in the structure of the final product.

Epoxy resin and MEKP were combined in a separate container. Before pouring the mixture into the mold, both the resin mixture and the powdered roselle fiber with a mass of 40% of the epoxy resin's mass were premixed. The mixture was then carefully poured into the combined part 1 and part 2 (clamped with G-clamp) of the mold, as shown in Fig. 12.6, by testing the bubble formation in the mixture. Then, part 3 of the mold was put on top of part 1 and part 2 of the mold and was manually pressed. The composite was subsequently placed under room temperature for 24 h to be healed. Next the composite was removed from the mold before finishing stage using sandpaper to remove irregular and excess edges followed by clear spray for appropriate surface finishing of the composite (Fig. 12.7).

FIG. 12.6 Part 1 and part 2 of the mold clamped using G-clamp.

FIG. 12.7 The final product.

3 CONCLUSIONS

As discussed in this chapter, the main direction for this work was to develop a new smartphone holder product from roselle fiber-reinforced composites. The roselle fiber-reinforced composite smartphone holder was intended to be a sustainable substitutive solution to the current products in the market. Throughout the development process of the product, several activities were carried out using the DfS approach, including market analysis, forming PDSs, creating smartphone holder concept designs, and creating detailed designs. For the detailed designs, the subactivities performed were holder mold production, manufacturing of plaque mold and plaque component, and final product assembly. The application of the DfS method to develop a smartphone holder is expected to nurture the interest of any manufacturer to develop any green product in the future.

ACKNOWLEDGMENTS

The authors would like to thank Dr. Mohd Radzi of the Linton University, Negeri Sembilan, Malaysia, and Dr. Nadlene Razali of the Universiti Teknikal Malaysia Melaka for their cooperation in the publication of this chapter of the book.

REFERENCES

Abral, H., Ariksa, J., Mahardika, M., Handayani, D., Aminah, I., Sandrawati, N., Pratama, A. B., Fajri, N., Sapuan, S. M., & Ilyas, R. A. (2020a). Transparent and antimicrobial cellulose film from ginger nanofiber. *Food Hydrocolloids, 98*, 105266. https://doi.org/10.1016/j.foodhyd.2019.105266

Abral, H., Ariksa, J., Mahardika, M., Handayani, D., Aminah, I., Sandrawati, N., Sapuan, S. M., & Ilyas, R. A. (2019a). Highly transparent and antimicrobial PVA based bionanocomposites reinforced by ginger nanofiber. *Polymer Testing,* 106186. https://doi.org/10.1016/j.polymertesting.2019.106186

Abral, H., Atmajaya, A., Mahardika, M., Hafizulhaq, F., Kadriadi, Handayani, D., Sapuan, S. M., & Ilyas, R. A. (2020b). Effect of ultrasonication duration of polyvinyl alcohol (PVA) gel on characterizations of PVA film. *Journal of Materials Research and Technology*, 1–10. https://doi.org/10.1016/j.jmrt.2019.12.078

Abral, H., Basri, A., Muhammad, F., Fernando, Y., Hafizulhaq, F., Mahardika, M., Sugiarti, E., Sapuan, S. M., Ilyas, R. A., & Stephane, I. (2019b). A simple method for improving the properties of the sago starch films prepared by using ultrasonication treatment. *Food Hydrocolloids, 93*, 276–283. https://doi.org/10.1016/j.foodhyd.2019.02.012

Ahmed, I., Zoranic, A., Javaid, S., & Iii, G. G. R. (2012). Mod-Checker : Kernel module integrity checking in the cloud environment. In *2012 41st International Conference on Parallel Processing Workshops*. IEEE. https://doi.org/10.1109/ICPPW.2012.46.

Aisyah, H. A., Paridah, M. T., Sapuan, S. M., Khalina, A., Berkalp, O. B., Lee, S. H., Lee, C. H., Nurazzi, N. M., Ramli, N., Wahab, M. S., & Ilyas, R. A. (2019). Thermal properties of woven kenaf/carbon fibre-reinforced epoxy hybrid composite panels. *International Journal of Polymer Science, 2019*, 1–8. https://doi.org/10.1155/2019/5258621

Akonda, M. H., Shah, D. U., & Gong, R. H. (2020). Natural fibre thermoplastic tapes to enhance reinforcing effects in composite structures. *Composites Part A: Applied Science and Manufacturing, 131*(November 2019), 105822. https://doi.org/10.1016/j.compositesa.2020.105822

Alamri, H., & Low, I. M. (2012a). Effect of water absorption on the mechanical properties of nano-filler reinforced epoxy nanocomposites. *Journal of Materials & Design, 42*, 214–222. https://doi.org/10.1016/j.matdes.2012.05.060

Alamri, H., & Low, I. M. (2012b). Mechanical properties and water absorption behaviour of recycled cellulose fi bre reinforced epoxy composites. *Polymer Testing, 31*(5), 620–628. https://doi.org/10.1016/j.polymertesting.2012.04.002

Alamri, H., Low, I. M., & Alothman, Z. (2012). Mechanical, thermal and microstructural characteristics of cellulose fibre reinforced epoxy/organoclay nanocomposites. *Composites Part B: Engineering, 43*(7), 2762–2771. https://doi.org/10.1016/j.compositesb.2012.04.037

Ali, A., Sanuddin, A. B., & Ezzeddin, S. (2010). The effect of aging on *Arenga pinnata* fiber-reinforced epoxy composite. *Materials & Design, 31*(7), 3550–3554. https://doi.org/10.1016/j.matdes.2010.01.043

Amor, I. B., Ghallabi, Z., Kaddami, H., Raihane, M., Arous, M., & Kallel, A. (2010). Experimental study of relaxation process in unidirectional (epoxy/palm tree fiber) composite. *Journal of Molecular Liquids, 154*(2–3), 61–68. https://doi.org/10.1016/j.molliq.2010.04.006

Anbukarasi, K., & Kalaiselvam, S. (2014). Study of effect of fibre volume and dimension on mechanical, thermal, and water absorption behaviour of luffa reinforced epoxy composites. *Journal of Materials & Design*. https://doi.org/10.1016/j.matdes.2014.10.078

Arulmurugan, M., Prabu, K., Rajamurugan, G., & Selvakumar, A. S. (2019). Impact of BaSO4 filler on woven Aloevera/Hemp hybrid composite: Dynamic mechanical analysis. *Materials Research Express, 6*(4), 045309. https://doi.org/10.1088/2053-1591/aafb88

Ary Subagia, I. D. G., Tijing, L. D., Kim, Y., Kim, C. S., Vista Iv, F. P., & Shon, H. K. (2014). Mechanical performance of multiscale basalt fiber-epoxy laminates containing tourmaline micro/nano particles. *Composites Part B: Engineering, 58*, 611–617. https://doi.org/10.1016/j.compositesb.2013.10.034

Asrofi, M., Sapuan, S. M., Ilyas, R. A., & Ramesh, M. (2020). Characteristic of composite bioplastics from tapioca starch and sugarcane bagasse fiber: Effect of time duration of ultrasonication (bath-type). *Materials Today: Proceedings*. https://doi.org/10.1016/j.matpr.2020.07.254

Asrofi, M., Sujito, Syafri, E., Sapuan, S. M., & Ilyas, R. A. (2020b). Improvement of biocomposite properties based tapioca starch and sugarcane bagasse cellulose nanofibers. *Key Engineering Materials, 849*, 96–101. https://doi.org/10.4028/www.scientific.net/KEM.849.96

Asyraf, M. R. M., Ishak, M. R., Sapuan, S. M., & Yidris, N. (2019). Conceptual design of creep testing rig for full-scale cross arm using TRIZ-morphological chart-analytic network process technique. *Journal of Materials Research and Technology, 8*(6), 5647–5658. https://doi.org/10.1016/j.jmrt.2019.09.033

Asyraf, M. R. M., Ishak, M. R., Sapuan, S. M., & Yidris, N. (2020a). Conceptual design of multi-operation outdoor flexural creep test rig using hybrid concurrent engineering approach. *Journal of Materials Research and Technology, 9*(2), 2357–2368. https://doi.org/10.1016/j.jmrt.2019.12.067

Asyraf, M. R. M., Ishak, M. R., Sapuan, S. M., Yidris, N., & Ilyas, R. A. (2020b). Woods and composites cantilever beam: A comprehensive review of experimental and numerical creep methodologies. *Journal of Materials Research and Technology*. https://doi.org/10.1016/j.jmrt.2020.01.013

Asyraf, M. R. M., Ishak, M. R., Sapuan, S. M., Yidris, N., Ilyas, R. A., Rafidah, M., & Razman, M. R. (2020c). Evaluation of design and simulation of creep test rig for full-scale cross arm structure. *Advances in Civil Engineering*. https://doi.org/10.1155/2019/6980918

Asyraf, M. R. M., Rafidah, M., Ishak, M. R., Sapuan, S. M., Ilyas, R. A., & Razman, M. R. (2020d). Integration of TRIZ, morphological chart and ANP method for development of FRP composite portable fire extinguisher. *Polymer Composites*, 1–6. https://doi.org/10.1002/pc.25587

Ataollahi, S., Taher, S. T., Eshkoor, R. A., Ariffin, A. K., & Azhari, C. H. (2012). Energy absorption and failure response of silk/epoxy composite square tubes : Experimental. *Composites Part B: Engineering, 43*(2), 542–548. https://doi.org/10.1016/j.compositesb.2011.08.019

Atikah, M. S. N., Ilyas, R. A., Sapuan, S. M., Ishak, M. R., Zainudin, E. S., Ibrahim, R., Atiqah, A., Ansari, M. N. M., & Jumaidin, R. (2019). Degradation and physical properties of sugar palm starch/sugar palm nanofibrillated cellulose bionanocomposite. *Polimery, 64*(10), 27–36. https://doi.org/10.14314/polimery.2019.10.5

Atiqah, A., Jawaid, M., Sapuan, S. M., Ishak, M. R., Ansari, M. N. M., & Ilyas, R. A. (2019). Physical and thermal properties of treated sugar palm/glass fibre reinforced thermoplastic polyurethane hybrid composites. *Journal of Materials Research and Technology, 8*(5), 3726–3732. https://doi.org/10.1016/j.jmrt.2019.06.032

Ayu, R. S., Khalina, A., Harmaen, A. S., Zaman, K., Isma, T., Liu, Q., Ilyas, R. A., & Lee, C. H. (2020). Characterization study of empty fruit bunch (EFB) fibers reinforcement in poly(butylene) succinate (PBS)/starch/glycerol composite sheet. *Polymers, 12*(7), 1571. https://doi.org/10.3390/polym12071571

Azwa, Z. N., & Yousif, B. F. (2013). Characteristics of kenaf fibre/epoxy composites subjected to thermal degradation. *Polymer Degradation and Stability, 98*(12), 2752–2759. https://doi.org/10.1016/j.polymdegradstab.2013.10.008

Bachtiar, D., Sapuan, S. M., & Hamdan, M. M. (2008). The effect of alkaline treatment on tensile properties of sugar palm fibre reinforced epoxy composites. *Materials & Design, 29*(7), 1285–1290. https://doi.org/10.1016/j.matdes.2007.09.006

Belaadi, A., Bezazi, A., Bourchak, M., Scarpa, F., & Zhu, C. (2014). Thermochemical and statistical mechanical properties of natural sisal fibres. *Composites Part B: Engineering, 67*, 481–489. https://doi.org/10.1016/j.compositesb.2014.07.029

Biswas, S., & Satapathy, A. (2010). A comparative study on erosion characteristics of red mud filled bamboo-epoxy and glass-epoxy composites. *Materials and Design, 31*(4), 1752–1767. https://doi.org/10.1016/j.matdes.2009.11.021

Cid-Ortega, S., & Guerrero-Beltrán, J. A. (2015). Roselle calyces (*Hibiscus sabdariffa*), an alternative to the food and beverages industries: A review. *Journal of Food Science & Technology, 52*(11), 6859–6869. https://doi.org/10.1007/s13197-015-1800-9

Coroller, G., Lefeuvre, A., Le, A., Bourmaud, A., Ausias, G., Gaudry, T., & Baley, C. (2013). Effect of flax fibres individualisation on tensile failure of flax/epoxy unidirectional composite. *Composites Part A: Engineering, 51*, 62–70. https://doi.org/10.1016/j.compositesa.2013.03.018

Cristina, D., Nascimento, O., Ferreira, A. S., Monteiro, S. N., Coeli, R., Aquino, M. P., & Kestur, S. G. (2012). Studies on the characterization of piassava fibers and their epoxy composites. *Composites Part A: Applied Science and Manufacturing, 43*(3), 353–362. https://doi.org/10.1016/j.compositesa.2011.12.004

Davoodi, M. M., Sapuan, S. M., & Yunus, R. (2008). Conceptual design of a polymer composite automotive bumper energy absorber. *Materials and Design, 29*(7), 1447–1452. https://doi.org/10.1016/j.matdes.2007.07.011

Eshkoor, R. A., Oshkovr, S. A., Sulong, A. B., Zulkifli, R., Ariffin, A. K., & Azhari, C. H. (2013). Comparative research on the crashworthiness characteristics of woven natural silk/epoxy composite tubes. *Materials and Design, 47*, 248–257. https://doi.org/10.1016/j.matdes.2012.11.030

Eshkoor, R. A., Ude, A. U., Oshkovr, S. A., Sulong, A. B., Zulki, R., Arif, A. K., & Azhari, C. H. (2014). Failure mechanism of woven natural silk/epoxy rectangular composite tubes under axial quasi-static crushing test using trigger mechanism. *International Journal of Impact Engineering, 64*, 53–61. https://doi.org/10.1016/j.ijimpeng.2013.09.004

Feng, J., & Guo, Z. (2016). Temperature-frequency-dependent mechanical properties model of epoxy resin and its composites. *Composites Part B: Engineering, 85*, 161–169. https://doi.org/10.1016/j.compositesb.2015.09.040

Fiore, V., Bella, G. Di, & Valenza, A. (2015). The effect of alkaline treatment on mechanical properties of kenaf fibers and their epoxy composites. *Composites Part B: Engineering, 68*, 14–21. https://doi.org/10.1016/j.compositesb.2014.08.025

Gabr, M. H., Phong, N. T., Okubo, K., Uzawa, K., Kimpara, I., & Fujii, T. (2014). Thermal and mechanical properties of electrospun nano-celullose reinforced epoxy nanocomposites. *Polymer Testing, 37*, 51–58. https://doi.org/10.1016/j.polymertesting.2014.04.010

Guen, M. Le, Newman, R. H., Fernyhough, A., & Staiger, M. P. (2014). Tailoring the vibration damping behaviour of flax fibre-reinforced epoxy composite laminates via polyol additions. *Composites Part A: Applied Science and Manufacturing, 67*, 37–43. https://doi.org/10.1016/j.compositesa.2014.08.018

Gu, Y., Tan, X., Yang, Z., Li, M., & Zhang, Z. (2014). Hot compaction and mechanical properties of ramie fabric/epoxy composite fabricated using vacuum assisted resin infusion molding. *Materials & Design (1980-2015), 56*, 852–861. https://doi.org/10.1016/j.matdes.2013.11.077

Halimatul, M. J., Sapuan, S. M., Jawaid, M., Ishak, M. R., & Ilyas, R. A. (2019a). Effect of sago starch and plasticizer content on the properties of thermoplastic films: Mechanical testing and cyclic soaking-drying. *Polimery, 64*(6), 32–41. https://doi.org/10.14314/polimery.2019.6.5

Halimatul, M. J., Sapuan, S. M., Jawaid, M., Ishak, M. R., & Ilyas, R. A. (2019b). Water absorption and water solubility properties of sago starch biopolymer composite films filled with sugar palm particles. *Polimery, 64*(9), 27–35. https://doi.org/10.14314/polimery.2019.9.4

Hazrol, M. D., Sapuan, S. M., Ilyas, R. A., Othman, M. L., & Sherwani, S. F. K. (2020). Electrical properties of sugar palm nanocrystalline cellulose reinforced sugar palm starch nanocomposites. *Polimery, 65*(05), 363–370. https://doi.org/10.14314/polimery.2020.5.4

Hossain, R., Islam, A., Van Vuurea, A., & Verpoest, I. (2013). Tensile behavior of environment friendly jute epoxy laminated composite. *Procedia Engineering, 56*, 782–788. https://doi.org/10.1016/j.proeng.2013.03.196

Hoyos, C. G., & Vázquez, A. (2012). Flexural properties loss of unidirectional epoxy/fique composites immersed in water and alkaline medium for construction application. *Composites : Part B: Engineering, 43*, 3120–3130. https://doi.org/10.1016/j.compositesb.2012.04.027

Ibrahim, M. I., Edhirej, A., Sapuan, S. M., Jawaid, M., Ismarrubie, N. Z., & Ilyas, R. A. (2020). Extraction and characterization of Malaysian cassava starch, peel, and bagasse, and selected properties of the composites. In R. Jumaidin, S. M. Sapuan, & H. Ismail (Eds.), *Biofiller-reinforced biodegradable polymer composites* (1st ed., pp. 267–283). CRC Press.

Ibrahim, M. I., Sapuan, S. M., Zainudin, E. S., Zuhri, M. Y., Edhirej, A., & Ilyas, R. A. (2020). Characterization of corn fiber-filled cornstarch biopolymer composites. In R. Jumaidin, S. M. Sapuan, & H. Ismail (Eds.), *Biofiller-reinforced biodegradable polymer composites* (1st ed., pp. 285–301). CRC Press.

Idicula, M., Neelakantan, N. R., Oommen, Z., Joseph, K., & Thomas, S. (2005). A study of the mechanical properties of randomly oriented short banana and sisal hybrid fiber reinforced polyester composites. *Journal of Applied Polymer Science, 96*(5), 1699–1709. https://doi.org/10.1002/app.21636

Ilyas, R. A., & Sapuan, S. M. (2020a). The preparation methods and processing of natural fibre bio-polymer composites. *Current Organic Synthesis, 16*(8), 1068–1070. https://doi.org/10.2174/157017941608200120105616

Ilyas, R. A., & Sapuan, S. M. (2020b). Biopolymers and biocomposites: Chemistry and technology. *Current Analytical Chemistry, 16*(5), 500–503. https://doi.org/10.2174/157341101605200603095311

Ilyas, R., Sapuan, S., Atikah, M., Asyraf, M., Rafiqah, S. A., Aisyah, H., Nurazzi, N. M., & Norrrahim, M. (2020c). Effect of hydrolysis time on the morphological, physical, chemical, and thermal behavior of sugar palm nanocrystalline cellulose (*Arenga pinnata* (*Wurmb.*) *Merr*). *Textile Research Journal*. https://doi.org/10.1177/0040517520932393

Ilyas, R. A., Sapuan, S. M., Atiqah, A., Ibrahim, R., Abral, H., Ishak, M. R., Zainudin, E. S., Nurazzi, N. M., Atikah, M. S. N., Ansari, M. N. M., Asyraf, M. R. M., Supian, A. B. M., & Ya, H. (2020a). Sugar palm (*Arenga pinnata* [*Wurmb.*] *Merr*) starch films containing sugar palm nanofibrillated cellulose as reinforcement: Water barrier properties. *Polymer Composites, 41*(2), 459–467. https://doi.org/10.1002/pc.25379

Ilyas, R. A., Sapuan, S. M., Ibrahim, R., Abral, H., Ishak, M. R., Zainudin, E. S., Atikah, M. S. N., Mohd Nurazzi, N., Atiqah, A., Ansari, M. N. M., Syafri, E., Asrofi, M., Sari, N. H., & Jumaidin, R. (2019a). Effect of sugar palm nanofibrillated cellulose concentrations on morphological, mechanical and physical properties of biodegradable films based on agro-waste sugar palm (*Arenga pinnata* (*Wurmb.*) *Merr*) starch. *Journal of Materials Research and Technology, 8*(5), 4819–4830. https://doi.org/10.1016/j.jmrt.2019.08.028

Ilyas, R. A., Sapuan, S. M., Ibrahim, R., Abral, H., Ishak, M. R., Zainudin, E. S., Atiqah, A., Atikah, M. S. N., Syafri, E., Asrofi, M., & Jumaidin, R. (2020b). Thermal, biodegradability and water barrier properties of bio-nanocomposites based on plasticised sugar palm starch and nanofibrillated celluloses from sugar palm fibres. *Journal of Biobased Materials and Bioenergy, 14*(2), 234–248. https://doi.org/10.1166/jbmb.2020.1951

Ilyas, R. A., Sapuan, S. M., Ishak, M. R., & Zainudin, E. S. (2017). Effect of delignification on the physical, thermal, chemical, and structural properties of sugar palm fibre. *BioResources, 12*(4), 8734–8754. https://doi.org/10.15376/biores.12.4.8734-8754

Ilyas, R. A., Sapuan, S. M., & Ishak, M. R. (2018a). Isolation and characterization of nanocrystalline cellulose from sugar palm fibres (*Arenga Pinnata*). *Carbohydrate Polymers, 181*, 1038–1051. https://doi.org/10.1016/j.carbpol.2017.11.045

Ilyas, R. A., Sapuan, S. M., Ishak, M. R., & Zainudin, E. S. (2018b). Water transport properties of bio-nanocomposites reinforced by sugar palm (*Arenga Pinnata*) nanofibrillated cellulose. *Journal of Advanced Research in Fluid Mechanics and Thermal Sciences Journal, 51*(2), 234–246.

Ilyas, R. A., Sapuan, S. M., Ishak, M. R., & Zainudin, E. S. (2018c). Sugar palm nanocrystalline cellulose reinforced sugar palm starch composite: Degradation and water-barrier properties. *IOP Conference Series: Materials Science and Engineering, 368*, 012006. https://doi.org/10.1088/1757-899X/368/1/012006

Ilyas, R. A., Sapuan, S. M., Ishak, M. R., & Zainudin, E. S. (2019). Sugar palm nanofibrillated cellulose (*Arenga pinnata* (*Wurmb.*) *Merr*): Effect of cycles on their yield, physic-chemical, morphological and thermal behavior. *International Journal of Biological Macromolecules, 123*. https://doi.org/10.1016/j.ijbiomac.2018.11.124

Ilyas, R. A., Sapuan, S. M., Ishak, M. R., Zainudin, E. S., & Atikah, M. S. N. (2018e). Characterization of sugar palm nanocellulose and its potential for reinforcement with a starch-based composite. In *Sugar palm biofibers, biopolymers, and biocomposites* (pp. 189–220). CRC Press. https://doi.org/10.1201/9780429443923-10.

Ilyas, R. A., Sapuan, S. M., Sanyang, M. L., Ishak, M. R., & Zainudin, E. S. (2018f). Nanocrystalline cellulose as reinforcement for polymeric matrix nanocomposites and its potential applications: A review. *Current Analytical Chemistry, 14*(3), 203–225. https://doi.org/10.2174/1573411013666171003155624

Irxqg, L. V., Wkh, W., Lq, F., Rulhqwdwlrq, X., Ileuhv, R. I., Ehwwhu, J., Dqg, W., Xudo, I. O. H., Lq, S., Wr, F., & Pdw, W. K. H. (2014). Tensile and flexural properties of sisal fibre reinforced epoxy composite: A comparison between unidirectional and mat form of fibres. *Procedia Materials Science, 5*, 2434–2439. https://doi.org/10.1016/j.mspro.2014.07.489

Ishak, M. R., Sapuan, S. M., Leman, Z., Rahman, M. Z. A. A., Anwar, U. M. K. K., & Siregar, J. P. (2013). Sugar palm (*Arenga pinnata*): Its fibres, polymers and composites. *Carbohydrate Polymers, 91*(2), 699–710. https://doi.org/10.1016/j.carbpol.2012.07.073

Islam, F., Islam, N., Shahida, S., Karmaker, N., Koly, F. A., Mahmud, J., Keya, K. N., & Khan, R. A. (2019). Mechanical and interfacial characterization of jute fabrics reinforced unsaturated polyester resin composites. *Nano Hybrids and Composites, 25*, 22–31. https://doi.org/10.4028/www.scientific.net/NHC.25.22

Jaafar, C. N. A., Rizal, M. A. M., & Zainol, I. (2018). Effect of kenaf alkalization treatment on morphological and mechanical properties of epoxy/silica/kenaf composite. *International Journal of Engineering and Technology, 7*, 258–263. https://doi.org/10.14419/ijet.v7i4.35.22743

Jack, H. (2013). Universal design topics. In *Engineering design, planning, and management* (1st ed.). Academic Press. pp. 323–290.

Jarukumjorn, K., & Suppakarn, N. (2009). Effect of glass fiber hybridization on properties of sisal fiber – polypropylene composites. *Composites Part B: Engineering, 40*(7), 623–627. https://doi.org/10.1016/j.compositesb.2009.04.007

Jumaidin, R., Ilyas, R. A., Saiful, M., Hussin, F., & Mastura, M. T. (2019). Water transport and physical properties of sugarcane bagasse fibre reinforced thermoplastic potato starch biocomposite. *Journal of Advanced Research in Fluid Mechanics and Thermal Sciences, 61*(2), 273–281.

Jumaidin, R., Khiruddin, M. A. A., Asyul Sutan Saidi, Z., Salit, M. S., & Ilyas, R. A. (2020). Effect of cogon grass fibre on the thermal, mechanical and biodegradation properties of thermoplastic cassava starch biocomposite. *International Journal of Biological Macromolecules, 146*(xxxx), 746–755. https://doi.org/10.1016/j.ijbiomac.2019.11.011

Jumaidin, R., Saidi, Z. A. S., Ilyas, R. A., Ahmad, M. N., Wahid, M. K., Yaakob, M. Y., Maidin, N. A., Rahman, M. H. A., & Osman, M. H. (2019b). Characteristics of cogon grass fibre reinforced thermoplastic cassava starch biocomposite: Water absorption and physical properties. *Journal of Advanced Research in Fluid Mechanics and Thermal Sciences, 62*(1), 43–52.

Kandare, E., Luangtriratana, P., & Kandola, B. K. (2014). Fire reaction properties of flax/epoxy laminates and their balsa-core sandwich composites with or without fire protection. *Composites Part B: Engineering, 56*, 602–610. https://doi.org/10.1016/j.compositesb.2013.08.090

Kian, L. K., Jawaid, M., Ariffin, H., & Alothman, O. Y. (2017). Isolation and characterization of microcrystalline cellulose from roselle fibers. *International Journal of Biological Macromolecules, 103*, 931–940. https://doi.org/10.1016/j.ijbiomac.2017.05.135

Kumar, S. M. S., Duraibabu, D., & Subramanian, K. (2014). Studies on mechanical , thermal and dynamic mechanical properties of untreated (raw) and treated coconut sheath fiber reinforced epoxy composites. *Journal of Materials & Design, 59*, 63–69. https://doi.org/10.1016/j.matdes.2014.02.013

Kumar, R., Kumar, K., Sahoo, P., & Bhowmik, S. (2014). Study of mechanical properties of wood dust reinforced epoxy composite. *MSPRO, 6*(Icmpc), 551–556. https://doi.org/10.1016/j.mspro.2014.07.070

Landro, L. Di, & Janszen, G. (2014). Composites with hemp reinforcement and bio-based epoxy matrix. *Composites Part B: Engineering, 67*, 220–226. https://doi.org/10.1016/j.compositesb.2014.07.021

Le, A., Kervoelen, A., Le, A., Nardin, M., & Baley, C. (2020). Interfacial properties of flax fibre – epoxy resin systems : Existence of a complex interphase. *Composites Science and Technology, 100*(2014), 152–157. https://doi.org/10.1016/j.compscitech.2014.06.009

Leman, Z., Sapuan, S. M., Saifol, A. M., Maleque, M. A., & Ahmad, M. M. H. M. (2008). Moisture absorption behavior of sugar palm fiber reinforced epoxy composites. *Materials & Design, 29*, 1666–1670. https://doi.org/10.1016/j.matdes.2007.11.004

Liang, S., Gning, P. B., & Guillaumat, L. (2012). A comparative study of fatigue behaviour of flax/epoxy and glass/epoxy composites. *Composites Science and Technology, 72*(5), 535–543. https://doi.org/10.1016/j.compscitech.2012.01.011

Liang, S., Gning, P., & Guillaumat, L. (2014). Properties evolution of flax/epoxy composites under fatigue loading. *International Journal of Fatigue, 63*, 36–45. https://doi.org/10.1016/j.ijfatigue.2014.01.003

Li, X., Tabil, L. G., Panigrahi, S., & Crerar, W. J. (2006). *The influence of fiber content on properties of injection molded flax fiber-HDPE biocomposites*.

Liu, Q., & Hughes, M. (2008). The fracture behaviour and toughness of woven flax fibre reinforced epoxy composites. *Composites Part A: Engineering, 39*, 1644–1652. https://doi.org/10.1016/j.compositesa.2008.07.008

Liu, K., Zhang, X., Takagi, H., Yang, Z., & Wang, D. (2014). Effect of chemical treatments on transverse thermal conductivity of unidirectional abaca fiber/epoxy composite. *Composites Part A: Applied Science and Manufacturing, 66*, 227–236. https://doi.org/10.1016/j.compositesa.2014.07.018

Low, I. M., McGrath, M., Lawrence, D., Schmidt, P., Lane, J., Latella, B. A., & Sim, K. S. (2007). Mechanical and fracture properties of cellulose-fibre-reinforced epoxy laminates. *Composites Part A: Applied Science and Manufacturing, 38*(3), 963–974. https://doi.org/10.1016/j.compositesa.2006.06.019

Lu, T., Jiang, M., Jiang, Z., Hui, D., Wang, Z., & Zhou, Z. (2013). Effect of surface modification of bamboo cellulose fibers on mechanical properties of cellulose/epoxy composites. *Composites Part B: Engineering*. https://doi.org/10.1016/j.compositesb.2013.02.031

Mahjoub, R., Yatim, J. M., Rahman, A., Sam, M., Mahjoub, R., Yatim, J. M., Rahman, A., & Sam, M. (2014). Characteristics of continuous unidirectional kenaf fiber reinforced epoxy composites. *Journal of Materials & Design*. https://doi.org/10.1016/j.matdes.2014.08.010

Maisara, A. M. N., Ilyas, R. A., Sapuan, S. M., Huzaifah, M. R. M., Nurazzi, N. M., & Saifulazry, S. O. A. (2019). Effect of fibre length and sea water treatment on mechanical properties of sugar palm fibre reinforced unsaturated polyester composites. *International Journal of Recent Technology and Engineering, 8*(2S4), 510–514. https://doi.org/10.35940/ijrte.b1100.0782s419

Maria, I., Rosa, D., Santulli, C., & Sarasini, F. (2010). Mechanical and thermal characterization of epoxy composites reinforced with random and quasi-unidirectional untreated *Phormium tenax* leaf fibers. *Materials and Design, 31*(5), 2397–2405. https://doi.org/10.1016/j.matdes.2009.11.059

Mazani, N., Sapuan, S. M., Sanyang, M. L., Atiqah, A., & Ilyas, R. A. (2019). Design and fabrication of a shoe shelf from kenaf fiber reinforced unsaturated polyester composites. Issue 2000. In *Lignocellulose for future bioeconomy* (pp. 315–332). Elsevier. https://doi.org/10.1016/B978-0-12-816354-2.00017-7.

Mishra, V., & Biswas, S. (2013). Physical and mechanical properties of bi-directional jute fiber epoxy composites. *Procedia Engineering, 51*(NUiCONE 2012), 561–566. https://doi.org/10.1016/j.proeng.2013.01.079

Mittal, V., Saini, R., & Sinha, S. (2016). Natural fiber-mediated epoxy composites - A review. *Composites Part B: Engineering, 99*, 425–435. https://doi.org/10.1016/j.compositesb.2016.06.051

Mohan, T. P., & Kanny, K. (2011). Water barrier properties of nanoclay filled sisal fibre reinforced epoxy composites. *Composites Part A: Applied Science and Manufacturing, 42*(4), 385–393. https://doi.org/10.1016/j.compositesa.2010.12.010

Morton, J. F. (1974). Renewed interest in roselle (*Hibiscus sabdariffa* L.), the long-forgotten "Florida Cranberry.". *Florida State Horticultural Society Proceeding*, 415–425.

Muller, J. (2019). *Smartphone users in Malaysia 2017-2023*. Statista.

Muralidhar, B. A. (2013). Tensile and compressive behaviour of multilayer flax-rib knitted preform reinforced epoxy composites. *Materials and Design, 49*, 400–405. https://doi.org/10.1016/j.matdes.2012.12.040

Mylsamy, K., & Rajendran, I. (2011a). Influence of alkali treatment and fibre length on mechanical properties of short Agave fibre reinforced epoxy composites. *Materials and Design, 32*(8–9), 4629–4640. https://doi.org/10.1016/j.matdes.2011.04.029

Mylsamy, K., & Rajendran, I. (2011b). The mechanical properties, deformation and thermomechanical properties of alkali treated and untreated Agave continuous fibre reinforced epoxy composites. *Materials and Design, 32*(5), 3076–3084. https://doi.org/10.1016/j.matdes.2010.12.051

Nadlene, R., Sapuan, S. M., Jawaid, M., Ishak, M. R., & Yusriah, L. (2016a). A review on roselle fiber and its composites. *Journal of Natural Fibers, 13*(1), 10–41. https://doi.org/10.1080/15440478.2014.984052

Nadlene, R., Sapuan, S. M., Jawaid, M., Ishak, M. R., & Yusriah, L. (2016b). The effects of chemical treatment on the structural and thermal, physical, and mechanical and morphological properties of roselle fiber-reinforced vinyl ester composites. *Polymer Composites, 39*(1), 274–287. https://doi.org/10.1002/pc.23927

Nadlene, R., Sapuan, S. M., Jawaid, M., Ishak, M. R., & Yusriah, L. (2018). The effects of chemical treatment on the structural and thermal, physical, and mechanical and morphological properties of roselle fiber-reinforced vinyl ester composites. *Polymer Composites, 39*(1), 274–287. https://doi.org/10.1002/pc.23927

Nazrin, A., Sapuan, S. M., Zuhri, M. Y. M., Ilyas, R. A., Syafiq, R., & Sherwani, S. F. K. (2020). Nanocellulose reinforced thermoplastic starch (TPS), polylactic acid (PLA), and polybutylene succinate (PBS) for food packaging applications. *Frontiers in Chemistry, 8*(213), 1–12. https://doi.org/10.3389/fchem.2020.00213

Newman, R. H. (2009). Auto-accelerative water damage in an epoxy composite reinforced with plain-weave flax fabric. *Composites Part A: Applied Science and Manufacturing, 40*(10), 1615–1620. https://doi.org/10.1016/j.compositesa.2009.07.010

Newman, R. H., Clauss, E. C., Carpenter, J. E. P., & Thumm, A. (2007). Epoxy composites reinforced with deacetylated *Phormium tenax* leaf fibres. *Composites Part A: Applied Science and Manufacturing, 38*, 2164–2170. https://doi.org/10.1016/j.compositesa.2007.06.007

Nirmal, U., Hashim, J., & Low, K. O. (2012). Adhesive wear and frictional performance of bamboo fibres reinforced epoxy composite. *Tribiology International, 47*, 122–133. https://doi.org/10.1016/j.triboint.2011.10.012

Norizan, M. N., Abdan, K., Ilyas, R. A., & Biofibers, S. P. (2020). Effect of fiber orientation and fiber loading on the mechanical and thermal properties of sugar palm yarn fiber reinforced unsaturated polyester resin composites. *Polimery, 65*(2), 34–43. https://doi.org/10.14314/polimery.2020.2.5

Nurazzi, N. M., Khalina, A., Sapuan, S. M., & Ilyas, R. A. (2019). Mechanical properties of sugar palm yarn/woven glass fiber reinforced unsaturated polyester composites : Effect of fiber loadings and alkaline treatment. *Polimery, 64*(10), 12–22. https://doi.org/10.14314/polimery.2019.10.3

Nurazzi, N. M., Khalina, A., Sapuan, S. M., Ilyas, R. A., Rafiqah, S. A., & Hanafee, Z. M. (2020). Thermal properties of treated sugar palm yarn/glass fiber reinforced unsaturated polyester hybrid composites. *Journal of Materials Research and Technology*, *9*(2), 1606–1618. https://doi.org/10.1016/j.jmrt.2019.11.086

Oshkovr, S. A., Eshkoor, R. A., Taher, S. T., Ariffin, A. K., & Azhari, C. H. (2012). Crashworthiness characteristics investigation of silk/epoxy composite square tubes. *Composite Structures*, *94*(8), 2337–2342. https://doi.org/10.1016/j.compstruct.2012.03.031

Pugh, S. (1991). Total design: Integrated methods for successful product engineering. *Quality and Reliability Engineering International*, *7*(2), 119. https://doi.org/10.1002/qre.4680070210

Pugh, S. (1996). Concept selection–A method that works. In D. Clausing, & R. Andrade (Eds.), *Creating innovative products using total design*. Addison-Wesley Publishing Company.

Radzi, A. M., Sapuan, S. M., Jawaid, M., & Mansor, M. R. (2017). Influence of fibre contents on mechanical and thermal properties of roselle fibre reinforced polyurethane composites. *Fibers and Polymers*, *18*(7), 1353–1358. https://doi.org/10.1007/s12221-017-7311-8

Razali, N., Salit, M. S., Jawaid, M., Ishak, M. R., & Lazim, Y. (2015). A study on chemical composition, physical, tensile, morphological, and thermal properties of roselle fibre: Effect of fibre maturity. *BioResources*, *10*(1). https://doi.org/10.15376/biores.10.1.1803-1824

Romli, F. I., Nizam, A., Shakrine, A., Rafie, M., Laila, D., & Abdul, A. (2012). Factorial study on the tensile strength of a coir fiber- reinforced epoxy composite. *Procedia - Social and Behavioral Sciences*, *3*, 242–247. https://doi.org/10.1016/j.aasri.2012.11.040

Roşu, D., Caşcaval, C. N., Musta, F., & Ciobanu, C. (2002). Cure kinetics of epoxy resins studied by non-isothermal DSC data. *Thermochimica Acta*, *383*(1–2), 119–127. https://doi.org/10.1016/S0040-6031(01)00672-4

Rq, R., Lq, G., & Dqg, D. (2014). Mechanical and microstructure characterization of coconut spathe fibers and kenaf bast fibers reinforced epoxy polymer matrix composites. *Procedia Materials Science*, *5*, 2330–2337. https://doi.org/10.1016/j.mspro.2014.07.476

Saheb, D. N., & Jog, J. P. (1999). Natural fiber polymer composites: A review. *Advances in Polymer Technology*, *18*(4), 351–363. https://doi.org/10.1002/(SICI)1098-2329(199924)18:4<351::AID-ADV6>3.0.CO;2-X

Sánchez-Mendoza, J., Domínguez-López, A., Navarro-Galindo, S., & López-Sandoval, J. A. (2008). Some physical properties of Roselle (*Hibiscus sabdariffa* L.) seeds as a function of moisture content. *Journal of Food Engineering*, *87*(3), 391–397. https://doi.org/10.1016/j.jfoodeng.2007.12.023

Sanjay, M. R., Arpitha, G. R., & Yogesha, B. (2015). Study on mechanical properties of natural - glass fibre reinforced polymer hybrid composites : A review. *Materials Today: Proceedings*, *2*(4–5), 2959–2967. https://doi.org/10.1016/j.matpr.2015.07.264

Sanyang, M. L., Ilyas, R. A., Sapuan, S. M., & Jumaidin, R. (2018). Sugar palm starch-based composites for packaging applications. In *Bionanocomposites for packaging applications* (pp. 125–147). Springer International Publishing. https://doi.org/10.1007/978-3-319-67319-6_7

Sapuan, S. M. (2005). A conceptual design of the concurrent engineering design system for polymeric-based composite automotive pedals. *American Journal of Applied Sciences*, *2*(2), 514–525.

Sapuan, S. M. (2006). Mechanical properties of woven banana fibre reinforced epoxy composites. *Materials & Design*, *27*, 689–693. https://doi.org/10.1016/j.matdes.2004.12.016

Sapuan, S. M. (2014). *Tropical natural fibre composites*. Singapore: Springer. https://doi.org/10.1007/978-981-287-155-8

Sapuan, S. M., Aulia, H. S., Ilyas, R. A., Atiqah, A., Dele-Afolabi, T. T., Nurazzi, M. N., Supian, A. B. M., & Atikah, M. S. N. (2020). Mechanical properties of longitudinal basalt/woven-glass-fiber-reinforced unsaturated polyester-resin hybrid composites. *Polymers*, *12*(10), 2211. https://doi.org/10.3390/polym12102211

Sari, N. H., Pruncu, C. I., Sapuan, S. M., Ilyas, R. A., Catur, A. D., Suteja, S., Sutaryono, Y. A., & Pullen, G. (2020). The effect of water immersion and fibre content on properties of corn husk fibres reinforced thermoset polyester composite. *Polymer Testing*, *91*, 106751. https://doi.org/10.1016/j.polymertesting.2020.106751

Scida, D., Assarar, M., Poilâne, C., & Ayad, R. (2013). Influence of hygrothermal ageing on the damage mechanisms of flax-fibre reinforced epoxy composite. *Composites Part B: Engineering*, *48*, 51–58. https://doi.org/10.1016/j.compositesb.2012.12.010

Shaharuzaman, M. A., Sapuan, S. M., Mansor, M. R., & Zuhri, M. Y. M. (2020). Conceptual design of natural fiber composites as a side-door impact beam using hybrid approach. *Journal of Renewable Materials*, *8*(5), 549–563. https://doi.org/10.32604/jrm.2020.08769

Shih, Y. (2007). Mechanical and thermal properties of waste water bamboo husk fiber reinforced epoxy composites. *Materials Science and Engineering: A*, *446*, 289–295. https://doi.org/10.1016/j.msea.2006.09.032

Srisuwan, S., Prasoetsopha, N., & Suppakarn, N. (2014). The effects of alkalized and silanized woven sisal fibers on mechanical properties of natural rubber modified epoxy resin. *Energy Procedia*, *56*, 19–25. https://doi.org/10.1016/j.egypro.2014.07.127

Suhvhqw, Q. W. K. H., Ri, H. U. D., Ghyhorsphqw, S., Duh, F., Xvhg, E., Ri, E., Hdvh, W. K. H., Pdqxidfwxulqj, L. Q., Orz, D. Q. G., & Wr, Z. (2014). Optimization of mechanical properties of epoxy based wood dust reinforced green composite using Taguchi method. *Procedia Materials Science*, *5*, 688–696. https://doi.org/10.1016/j.mspro.2014.07.316

Syafri, E., Kasim, A., Abral, H., & Asben, A. (2019). Cellulose nanofibers isolation and characterization from ramie using a chemical-ultrasonic treatment. *Journal of Natural Fibers*, *16*(8), 1145–1155. https://doi.org/10.1080/15440478.2018.1455073

Syafri, E., Sudirman, M., Yulianti, E., Deswita, Asrofi, M., Abral, H., Sapuan, S. M., Ilyas, R. A., & Fudholi, A. (2019b). Effect of sonication time on the thermal stability, moisture absorption, and biodegradation of water hyacinth (*Eichhornia crassipes*) nanocellulose-filled bengkuang (*Pachyrhizus erosus*) starch biocomposites. *Journal of Materials Research and Technology*, *8*(6), 6223–6231. https://doi.org/10.1016/j.jmrt.2019.10.016

de Vasconcellos, D. S., Sarasini, F., Touchard, F., Pucci, M., Santulli, C., Tirillò, J., & Sorrentino, L. (2014). Reinforced epoxy composites Influence of low velocity impact on fatigue behaviour of woven hemp fibre reinforced epoxy composites. *Composites Part B: Engineering*. https://doi.org/10.1016/j.compositesb.2014.04.025

Vasconcellos, D., & Touchard, F. (2012). Experimental and numerical investigation of the interface between epoxy matrix and hemp yarn. *Composites Part A: Applied Science and Manufacturing, 43*(11), 2046–2058. https://doi.org/10.1016/j.compositesa.2012.07.015

de Vasconcellos, D. S., Touchard, F., & Chocinski-arnault, L. (2014). Tension – tension fatigue behaviour of woven hemp fibre reinforced epoxy composite : A multi-instrumented damage analysis. *International Journal of Fatigue, 59*, 159–169. https://doi.org/10.1016/j.ijfatigue.2013.08.029

Venkateshwaran, N., Perumal, A. E., & Arunsundaranayagam, D. (2013). Fiber surface treatment and its effect on mechanical and visco-elastic behaviour of banana/epoxy composite. *Materials and Design, 47*, 151–159. https://doi.org/10.1016/j.matdes.2012.12.001

Wood, B. M., Coles, S. R., Maggs, S., Meredith, J., & Kirwan, K. (2011). Use of lignin as a compatibiliser in hemp/epoxy composites. *Composites Science and Technology, 71*(16), 1804–1810. https://doi.org/10.1016/j.compscitech.2011.06.005

Yan, L., & Chouw, N. (2013). Crashworthiness characteristics of flax fibre reinforced epoxy tubes for energy absorption application. *Materials and Design, 51*, 629–640. https://doi.org/10.1016/j.matdes.2013.04.014

Yan, L., Chouw, N., & Jayaraman, K. (2014a). Lateral crushing of empty and polyurethane-foam filled natural flax fabric reinforced epoxy composite tubes. *Composites Part B: Engineering, 63*, 15–26. https://doi.org/10.1016/j.compositesb.2014.03.013

Yan, L., Chouw, N., & Jayaraman, K. (2014b). Effect of triggering and polyurethane foam-filler on axial crushing of natural flax/epoxy composite tubes. *Journal of Materials & Design, 56*, 528–541. https://doi.org/10.1016/j.matdes.2013.11.068

Yorseng, K., Rangappa, S. M., Pulikkalparambil, H., Siengchin, S., & Parameswaranpillai, J. (2020). Accelerated weathering studies of kenaf/sisal fiber fabric reinforced fully biobased hybrid bioepoxy composites for semi-structural applications: Morphology, thermo-mechanical, water absorption behavior and surface hydrophobicity. *Construction and Building Materials, 235*, 117464. https://doi.org/10.1016/j.conbuildmat.2019.117464

Yousif, B. F., Nirmal, U., & Wong, K. J. (2010). Three-body abrasion on wear and frictional performance of treated betelnut fibre reinforced epoxy (T-BFRE) composite. *Materials and Design, 31*(9), 4514–4521. https://doi.org/10.1016/j.matdes.2010.04.008

Yousif, B. F., Shalwan, A., Chin, C. W., & Ming, K. C. (2012). Flexural properties of treated and untreated kenaf/epoxy composites. *Materials & Design, 40*, 378–385. https://doi.org/10.1016/j.matdes.2012.04.017

Yu, S., Yang, S., & Cho, M. (2009). Multi-scale modeling of cross-linked epoxy nanocomposites. *Polymer, 50*(3), 945–952. https://doi.org/10.1016/j.polymer.2008.11.054

Zhang, J., Dong, H., Tong, L., Meng, L., Chen, Y., & Yue, G. (2012). Investigation of curing kinetics of sodium carboxymethyl cellulose/epoxy resin system by differential scanning calorimetry. *Thermochimica Acta, 549*, 63–68. https://doi.org/10.1016/j.tca.2012.09.015

FURTHER READING

Ilyas, R. A., Sapuan, S. M., Ishak, M. R., & Zainudin, E. S. (2018d). Development and characterization of sugar palm nanocrystalline cellulose reinforced sugar palm starch bionanocomposites. *Carbohydrate Polymers, 202*, 186–202. https://doi.org/10.1016/j.carbpol.2018.09.002

CHAPTER 13

Development of Roselle Fiber-Reinforced Polymer Biocomposite Mug Pad Using the Hybrid Design for Sustainability and Pugh Method

R.A. ILYAS • M.R.M. ASYRAF • S.M. SAPUAN • T.M.N. AFIQ • A. SUHRISMAN • M.S.N. ATIKAH • R. IBRAHIM

1 INTRODUCTION

Natural fiber is a fiber that is produced by geologic, animal, and plant processes. It has historically been used since ancient times to manufacture textiles, papers, clothes, and garments (Ilyas et al., 2021). It can be used as a component of composite materials, where the orientation of the fiber impacts the properties. Natural fibers, such as sugar palm (Atiqah et al., 2019; Halimatul et al., 2019b; Hazrol et al., 2020; Maisara et al., 2019; Norizan et al., 2020; Nurazzi et al., 2019; Nurazzi, Khalina, Sapuan, et al., 2020; Sapuan et al., 2020), oil palm fiber (Ayu et al., 2020), jute (Islam, Islam, Karmaker et al., 2019, Islam, Islam, Shahida et al., 2019), hemp (Arulmurugan et al., 2019), sisal (Yorseng et al., 2020), kenaf (Aisyah et al., 2019; Mazani et al., 2019), flax (Akonda et al., 2020), ginger (Abral, Ariksa, et al., 2019, Abral, Ariksa, et al., 2020), ramie (Syafri, Kasim, et al., 2019), sugarcane (Asrofi, Sapuan, et al., 2020; Asrofi, Sujito, et al., 2020; Jumaidin, Ilyas, et al., 2019), cogon fiber (Jumaidin, Saidi, et al., 2019; Jumaidin et al., 2020), water hyacinth (Syafri, Sudirman, et al., 2019), cassava (Ibrahim, Edhirej, et al., 2020), and corn (Ibrahim, Sapuan, et al., 2020; Sari et al., 2020), have gained a huge attention because of their advantages, e.g., ease of availability, renewability, biodegradability, and good mechanical properties compared with synthetic fibers (Ilyas & Sapuan, 2020a, 2020b; Nadlene et al., 2018; Nazrin et al., 2020). Therefore by encouraging research and development, natural fibers could be better than synthetic fibers in terms of cost-effectiveness and lightweight feature. Natural fibers derived from plants mainly consist of cellulose, hemicellulose, lignin, pectin, and other waxy substances (Ilyas, Sapuan, Asyraf, et al., 2020, Ilyas, Sapuan, Atikah, et al., 2020, Ilyas, Sapuan, Atiqah, et al., 2020, Ilyas, Sapuan, Ibrahim, et al., 2020; Ilyas et al., 2017; Ilyas, Sapuan, & Ishak, 2018; Ilyas, Sapuan, Ishak, Zainudin, Atikah, 2018; Ilyas, Sapuan, Ibrahim, et al., 2019; Ilyas, Sapuan, Ishak, & Zainudin, 2019). Cellulose fibers are composed of a cemented microfibril aggregate. As a result, the structure of plants spans many length scales to provide maximum strength with a minimum of material. These advantages have allowed engineers to use natural fibers to reinforce polymer composites to reduce the utilization of forest sources as well as minimize the surplus of natural fibers (Abral, Atmajaya, et al., 2020; Abral, Basri, et al., 2019; Halimatul et al., 2019a; Nadlene et al., 2016a). Besides that, these natural fibers have huge potential applications from plastic packaging to scaffolds for tissue regeneration (Atikah et al., 2019; Ilyas, Sapuan, Atiqah, et al., 2020; Ilyas, Sapuan, Ibrahim, et al., 2019, 2020; Ilyas, Sapuan, Ishak, & Zainudin, 2018b, 2018c, 2018d; Ilyas, Sapuan, Sanyang, et al., 2018f; Sanyang et al., 2018).

1.1 Roselle

Roselle (*Hibiscus sabdariffa* L.) (family: Malvaceae) is a species of hibiscus probably native to West and East Africa and Southeast Asia including Northeastern India. It is broadly cultivated in India and Malaysia. This plant can be found in the tropics and subtropics in the north and south hemisphere and in many parts of Jamaica, Trinidad, Tobago, and Central America (Morton, 1974). People worldwide use the shrubbery for decorative

Roselle. https://doi.org/10.1016/B978-0-323-85213-5.00002-0

purposes; meanwhile, the seeds and petals are used for human consumption (Sánchez-Mendoza et al., 2008). In the recent years, the pharmacologic field is accentuated because of roselle calyces' potential in alternative medicinal products. Researchers revealed that roselle extracts possess antioxidant, diuretic, fever relief, hypoglycemia, cholesterol reduction, and hypertension characteristics (Cid-Ortega & Guerrero-Beltrán, 2015).

This versatile plant's applications are utilized not only for food and medicinal purposes but also for making ropes, jute, and textile with its fiber (Radzi et al., 2017). In tropical countries like Malaysia, roselle fruit's quality is best within the first year of cultivation. Beyond this period, the plant is cut off and becomes agricultural waste; hence, the idea of efficient utilization of roselle fiber came in to convert this waste to reinforcement in polymer composites. Before considering the industrial applications of this composite, it is crucial to understand the physical, thermal, and mechanical properties of roselle fiber (Razali et al., 2015). Researchers show interest in roselle fiber because of its advantages; it is a natural fiber with good tensile properties, toughness, water resistance, inflexibility, and coarseness (Nadlene et al., 2016b). Natural fibers' physical characteristics are determined by their chemical and physical composition, including fiber structure, fiber-angle cellulose content, cross-section angle, and polymerization degree (Idicula et al., 2005).

Having similar chemical properties to well-known jute fibers, roselle fiber (shown in Fig. 13.1) has high potential as a polymer reinforcing material regarding its mechanical properties to manufacture composite goods.

1.2 Fiber Production

Several researchers explored roselle bast fiber's potential as reinforcement materials because of its similar characteristics to other known natural fibers, such as jute. The fiber is located under roselle stalk's bark and extracted via a series of water retting steps, as shown in Fig. 13.2. High-quality fiber is obtainable by harvesting during the bud stage. The stalks were packed and water-treated for 3–4 days. The retted stalks were washed under flowing water and separated from the stalks, washed, and sundried (Nadlene et al., 2016a). The final product is illustrated in Fig. 13.2E.

1.3 Unsaturated Polyester

Unsaturated polyester (UP) resin is one of the highest potential and widespread used resins as well as one of the most flexible synthetic copolymers in composite advancement (Nurazzi & Laila, 2017). UP is a high-performance engineering polymer that has been used in various engineering applications. UP is a thermosetting polymer that functions to copolymerize and cure the resin when dissolved in the styrene monomer. The backbone's unsaturated nature enables creating a 3D network to react with double bonds in the styrene monomer, through peroxide initiators (Malik et al., 2000). The ester group's positions and reactive sites in the molecular structure of UP are illustrated in Fig. 13.3.

FIG. 13.1 **(A)** Roselle stem. **(B)** Matured roselle.

FIG. 13.2 Extraction of roselle fiber: **(A)** roselle plant, **(B)** stalks in bundle form, **(C)** water retting process, **(D)** removing the fibers from the stalks, and **(E)** the final form of roselle fibers (Athijayamani et al., 2009; Nadlene et al., 2016a).

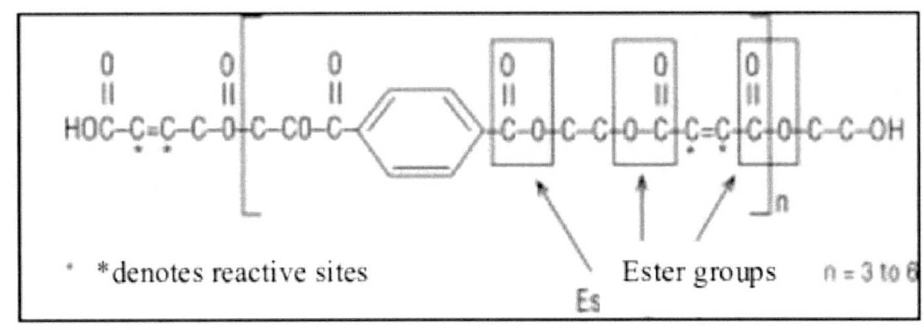

FIG. 13.3 Ester group position (C—O—O—C) and reactive sites (C*—C*) located inside the unsaturated polyester molecular structure (Nurazzi & Laila, 2017). * Denotes reactive sites.

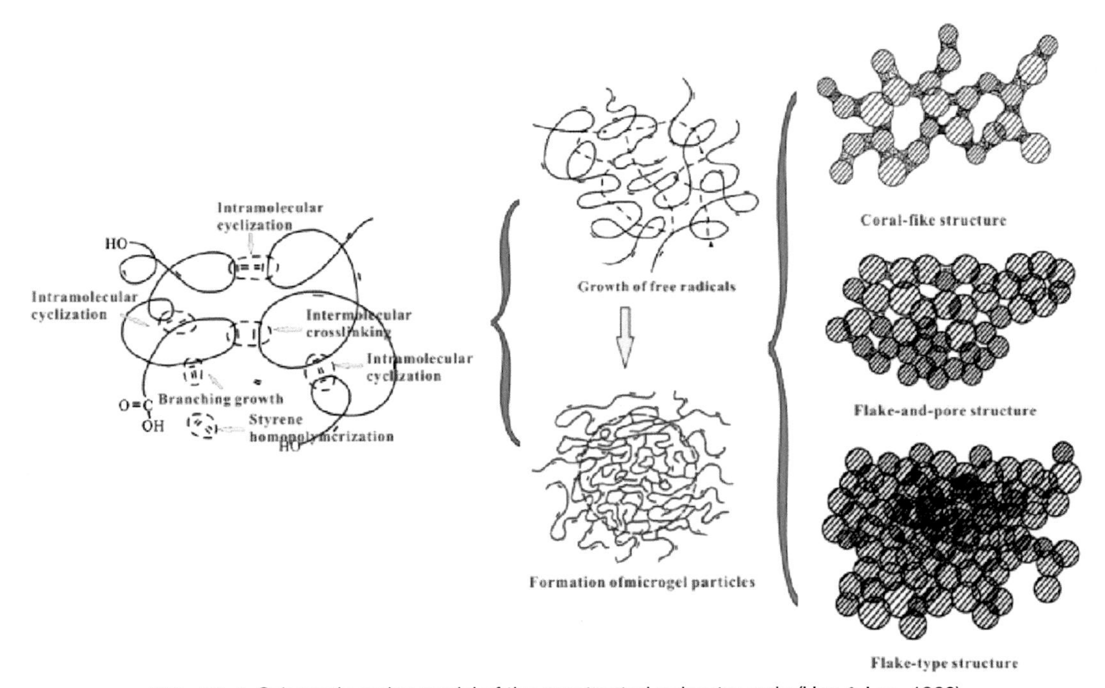

FIG. 13.4 Schematic curing model of the unsaturated polyester resin (Hsu & Lee, 1989).

Radical curing of the UP resin involves various phases of chain initiation, chain development, and chain termination, as well as chain transfers (Ardhyananta et al., 2017). Free radical initiators are primarily used in practical applications, and there are many methods to initiate free radicals, such as heat, light, electron beams, ultrasonic waves, etc. (Gao et al., 2019). Fig. 13.4 illustrates the curing process of the UP resin.

The benefits of UP are its dimensional flexibility; reasonable price; simplicity of handling, packaging, and manufacturing; and good mechanical, electric, and chemical balances (Nurazzi & Laila, 2017). Specific particular formulations provide high corrosion and thermal resistance. According to Mishra et al. (2003), UP resin is possibly the best option to match performance with structural capability.

The use of natural fibers to reinforce UP resin in composite materials has more benefits than plastics, glass, carbon, and aramid fibers, such as weight reduction, high durability, high specific rigidity and toughness, biodegradability, and dynamic properties (Azeez et al., 2020). There are various studies on the characteristics of natural fiber-reinforced UP composites, as shown in Table 13.1.

TABLE 13.1
Reported Works on Natural Fiber-Reinforced Unsaturated Polyester Composites.

Natural Fiber	Characteristics	Reference
Banana	Tensile strength, tensile modulus, elongation at break, bending strength, and bending modulus	Hossain et al. (2020)
Coconut inflorescence	Tensile and flexural behavior	Karthik and Arunachalam (2020)
Sugar palm yarn/glass	Thermal properties	Nurazzi, Khalina, Sapuan, et al. (2020)
Jute and E-Glass	Mechanical and interfacial properties	Keya, Kona, Razzak and Khan (2019)
Jute	Effect of weave Structure and yarn density	Keya, Kona and Khan (2019)
Sugar palm yarn	Mechanical and thermal properties	Nurazzi, Khalina, Chandrasekar, et al. (2020)
Sisal	Tensile testing, flexural test, Izod impact testing, thermal insulation measurements, room temperature water absorption, and boiling water temperature absorption	Biswas et al. (2019)
Jute	Mechanical and interfacial characterization	Islam, Islam, Shahida, et al. (2019)
Cane	Physical properties of fiber, mechanical properties of fiber, and mechanical properties of composite	Begum and Tanvir (2019)
E-glass	Tensile, bending properties, and impact	Islam, Islam, Karmaker, et al. (2019)
Alkali-treated *Ensete* stem	Tensile, flexural, surface morphology, and dynamic mechanical properties	Negawo et al. (2018)
Jute	Tensile strength, tensile modulus, elongation at break, and impact strength	Keya et al. (2020)
Nanocellulose	Mechanical properties of composites, dynamic mechanical properties of composites, microscopic analysis, water sorption studies, and rheologic analysis	Chirayil et al. (2019)
Kenaf	Mechanical properties and thermal characteristics	Rosamah et al. (2018)
Cellulose	Flexural properties	Sood and Dwivedi (2018)
Jute	Flexural properties and vibration behavior	Murdani et al. (2017)
Coir and cotton	Tensile, flexural, and impact strength	Balaji and Senthil Vadivu (2017)

2 APPLICATION OF DESIGN FOR SUSTAINABILITY IN ROSELLE BIOCOMPOSITE MUG PAD PRODUCT DEVELOPMENT

Design for sustainability (DfS) approach is a concurrent engineering method used to design and develop new roselle fiber-reinforced polymer biocomposite mug pad product. This product was proposed to replace the current mug pad in the market, which is usually made from ceramics, glass, stainless steel, and synthetic rubber. By implementing this roselle fiber as a key material in product fabrication, this research could add up the current number of biocomposite products in the global kitchenware market. Specifically, the proposed mug pad design's raw material was restricted only to roselle biocomposite. As mentioned earlier, the roselle fiber was chosen to reinforce UP as a newly designed biocomposite mug pad. The fiber itself is considered environmentally friendly and obtained from renewable resources. Moreover, this plant-based fiber possesses lightweight material with good mechanical strength to serve the intended functions (Ilyas, Sapuan, Asyraf, et al., 2020, Ilyas, Sapuan, Atikah, et al., 2020, Ilyas, Sapuan, Atiqah, et al., 2020, Ilyas, Sapuan, Ibrahim, et al., 2020; Ilyas et al., 2021; Ilyas, Sapuan, Atiqah, et al., 2020).

Before conducting prototype fabrication, the evaluation process was executed, including market investigation, development of concept design, final design selection, and detailed design, as previously done by other researchers (Asyraf, Ishak, Sapuan, & Yidris, 2019; Asyraf, Rafidah et al., 2020, Asyraf, Ishak, Sapuan, Yidris, 2020a). The detailed design was composed of product molding, manufacturing process, and the final product, which is further discussed in the following sections.

3 PRODUCT BACKGROUND AND MARKET INVESTIGATION

Mug pad is an everyday kitchenware used to place a hot mug or cup on the table to avoid damage to the table surface and cup slipping. It also functions to keep the temperature of the liquid inside the mug or cup for an extended period. Mug pads are commonly made from insulation materials, e.g., rubber, glass, synthetic plastic, and compact wood dust. These insulation materials also aid to reduce the risk of skin burn while handling the hot mug or cup. Hence, this issue has brought many ideas to designers to fabricate mug pad from various materials, shapes, and designs to get the customer's attention for marketing strategy.

Pugh Total Design is a concurrent engineering approach applied to select the best concept designs based on the attribute scoring value. This method implements market analysis to attain all appropriate information about the intended product design and is collected accordingly (Pugh, 1996). This information, including the existing product features, customer requirements, and price of the product, will be processed in the planning of the expected product design using the Pugh Total Design method (Asyraf, Ishak, Sapuan, Yidris, Ilyas et al., 2020; Sapuan, 2014). Informal interviews with users of the product and web browsing on current mug pad concepts were performed to acquire this information. The market analysis currently shows no mug pad made from biocomposite material, especially roselle fiber composites. Besides, no development has been carried out to develop a mug pad using the engineering elements and perspective. In specific, the authors' institution (Universiti Putra Malaysia) also did not produce any mug pad product from roselle fiber biocomposite.

From the market analysis, it was found that the intended product design by DfS and Pugh Total Design approaches would be a new market sector to be explored using natural fiber biocomposites, especially roselle biocomposites. On top of that, the application of roselle biocomposites could reduce the raw material cost, as the fiber itself is obtained from agricultural waste (Asyraf, Ishak, Razman, & Chandrasekar, 2019; Asyraf, Ishak, Sapuan, Yidris, & Ilyas, 2020b; Asyraf, Ishak, Sapuan, Yidris, Shahroze, et al., 2020; Ilyas, Sapuan, Asyraf, et al., 2020; Johari et al., 2020b, 2020a, 2019). Moreover, roselle fiber biocomposite mug pad shows low overall manufacturing cost and energy consumption. Lastly, incorporating engineering elements and perspective into the new product could draw consumers' recognition for the new roselle fiber biocomposite mug pad.

4 PRODUCT DESIGN SPECIFICATIONS

Based on previous market analysis, product design specification (PDS) was created to develop a product that is well-suited to the customer's views and demands. According to Pugh (1996), there are 32 elements that have to be considered in PDS elements. Specifically, for this project, only nine elements were implemented and used to develop a design for the roselle biocomposite mug pad (Fig. 13.5).

Within these selected PDS elements, durability is recognized as the main aspect of developing the roselle fiber biocomposite mug pad to at least achieve the same performance from its predecessor made from wood, glass, synthetic plastic, and rubber. On top of that, weight and size attributes were also established as crucial components in consumers' consideration

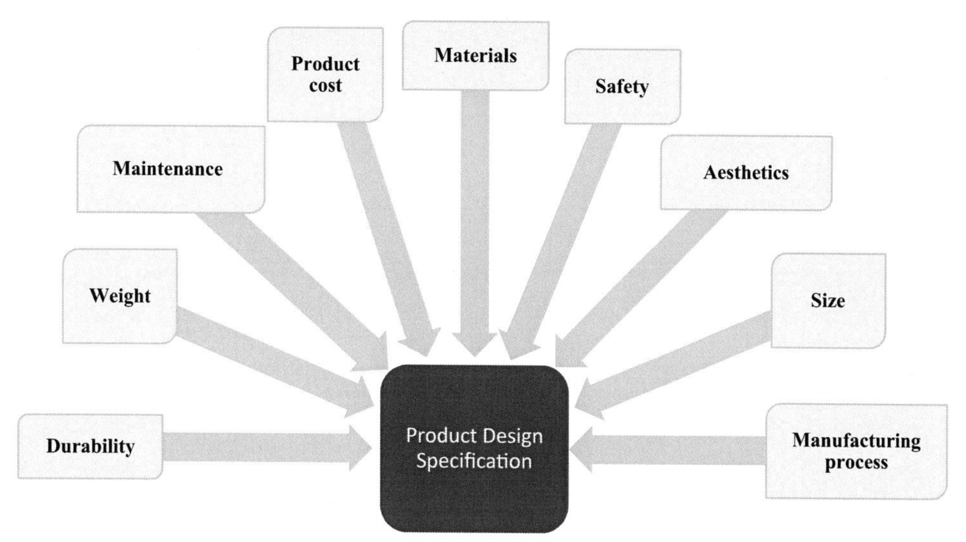

FIG. 13.5 Product design specification elements considered for roselle biocomposite mug pad product.

when buying mug pads. Apart from that, the cost element (raw material and manufacturing costs) is an essential attribute needed to embed and consider to attain low price value to ensure the product remains competitive in the current market. Another vital element in the PDS document for the mug pad is safety. In this case, the roselle biocomposite mug pad has to be designed without any sharp edges and small components to avoid any accidents to the users and their toddlers.

Moreover, the product design is proposed to have high aesthetic value with smooth surface finishing and unique design to attract customers' attention. The fabrication process of the biocomposite mug pad product has to be conducted using the hand layup method because it is the simplest way to produce it. In this case, precise and accurate dimensions for the final product could be created with optimum manufacturing capacity. Table 13.2 displays the overall description of the roselle biocomposite mug pad based on the PDS element listing.

TABLE 13.2
Description of PDS Elements for the Roselle Biocomposite Mug Pad Product.

PDS Elements	Description
1. Durability	The mug pad has to be durable, strong, and heat resistant.
2. Weight	The total weight should less than 50 g.
3. Maintenance	It has to be free from maintenance.
4. Product cost	The total cost of the mug pad must be less than RM10.
5. Material	The mug pad could be fabricated from roselle fiber-reinforced unsaturated polyester resin biocomposite.
6. Safety	The product should have a smooth surface and round angle to avoid cuts and accidents to the user. Only a single component is used to maintain it from protruding.
7. Aesthetics	The product design should be simple and have a good aesthetic shape with good quality and finishing.
8. Size	The maximum width and length of the plaque must be 100 mm, and thickness must not exceed 10 mm.
9. Manufacturing process	The product must be easy to fabricate and manufactured using either the hand layup technique or an automated process.

PDS, product design specification.

TABLE 13.3
Proposed Conceptual Designs of Roselle Fiber Biocomposite Mug Pad.

	Conceptual Design	Description
1.		• Concept inspired by the bottle cap. • Composed of the main pad and two stand legs. • Has a good artistic design.
2.		• Composed of bottom rectangular plate and a circular pad on the top. • Simple and minimalist design. • High stability owing to a wider baseline.
3.		• Concept inspired by the square coffee table. • Composed of four legs, one ring to hold the mug, and one square pad. • Complex shape with artistic value.
4.		• Simple and minimalist design. • Easy for storage. • Easy to fabricate and manufacture. • High stability. • High strength owing to wider baseline. • Concept inspired by a leaf.
5.		• Simple and minimalist design • Circular and isometric design, and hence easy to fabricate. • Easy for storage. • Easy to fabricate and manufacture. • High stability.

5 CONCEPTUAL DESIGN OF ROSELLE BIOCOMPOSITE MUG PAD PRODUCT

According to the listed PDS elements in Table 13.2, the following stage of the development process of biocomposite mug pad is conceptual design development. At this point, the most promising design will be selected among the listed concept designs. The techniques, steps, and other specifications for each of the conceptual design activities are explained in the subsequent subtopics.

6 IDEA GENERATION: BRAINSTORMING METHOD

The idea generation tools are required to develop conceptual design, such as brainstorming, SWOT (strength, weakness, opportunity, and threat) analysis, mind mapping, Theory of Inventive Problem Solving (TRIZ), and morphologic chart (Sapuan, 2005, 2015). For this project, a brainstorming method was implemented to generate ideas for the product's concept designs. A focus group was formed to comprehensively discuss and produce ideas on the conceptual designs of the biocomposite mug pad among members of the Advanced Engineering Materials and Biocomposites (AEMC) Research Centre, Department of Mechanical and Manufacturing Engineering, Universiti Putra Malaysia. From the brainstorming outputs, every concept design was listed based on the previous PDS document. Specifically, for this research activity, five design concepts of roselle fiber biocomposite mug pad with their details were produced, and they are tabulated in Table 13.3. Creative and innovative variations were made to add value to the ideas, in addition to manufacturing using roselle biocomposites. Along the idea generation process, the DfS suggestions were used to create solutions and features for each design to improve the application of biocomposite products without losing their functionality.

7 SELECTION OF CONCEPTUAL DESIGN USING THE PUGH METHOD

All five conceptual designs of the roselle fiber biocomposite mug pad were developed based on the

TABLE 13.4
Summary of Conceptual Design Selection Using the Pugh Selection Method.

Selection Criteria	Conceptual Design of Roselle Biocomposite Smartphone Holder					
	Existing	1	2	3	4	5
Durability		–	–	–	S	S
Weight		+	–	–	+	–
Maintenance		–	+	–	+	+
Product cost		–	+	–	+	+
Safety		–	–	–	+	+
Aesthetics		+	–	+	S	–
Size		+	S	–	S	S
Manufacturing process (ease of manufacturing)		–	–	–	+	+
Total score	$\sum+$	3	2	1	5	4
	$\sum-$	5	5	7	0	2
	$\sum S$	0	1	0	3	2
Nett score	$\sum+ (-) \sum-$	−2	−3	−6	5	2

discussion conducted from the brainstorming method. These conceptual designs were then followed by the concept design selection process using the Pugh concept selection method. For this step, there are various methods that can be used to identify the best concept design to be fabricated, such as Analytic Network Process (Asyraf, Rafidah, Ishak, Sapuan, et al., 2020), TOPSIS (Zheng & Lin, 2017), Analytic Hierarchy Process (Mansor et al., 2014), and VIKOR (Shaharuzaman et al., 2020). Based on the Pugh concept selection method, it was implemented based on the aforementioned PDS elements as the selection criteria for concept design selection. Overall, only the material elements in the PDS document are excluded from the evaluation activity. All concept designs in this project used similar material (roselle fiber-reinforced UP biocomposite). To evaluate each concept design, a scoring process was conducted to compare the proposed designs with datum (reference during comparison process) based on the prospective elements earlier. The datum selected was glass mug pad. The Pugh selection concept method implemented three selection criteria: "−" for worse performance than datum, "+" for performance better than datum, and "S" for equal performance as datum. At the end, the best design was chosen based on the highest scoring value of nett score by summing the "+" score values and subtracting it by the sum of "−" score values. The Pugh method was used specifically for this project because of conducting the simultaneous and systematic design selection process. Table 13.4 depicts the Pugh design selection of roselle fiber biocomposite mug pad.

Based on Table 13.4, the finding shows that concept design no. 4 has the best nett score value (+6) compared with the other concept designs proposed. At this point, concept design no. 4 displays six better attributes than the datum design, which was a glass mug pad. Hence, concept design no. 4 was chosen as the final design for the roselle fiber biocomposite mug pad, which is shown in Fig. 13.6.

Design no. 4 was chosen due to the improvement in attributes, such as weight, performance, product cost, safety, price, and manufacturing process, from the datum design. In general, the concept design was considered symmetric, flat shaped, and inspired by a leaf, representing nature-based products. In conjunction with this matter, this design's manufacturing process

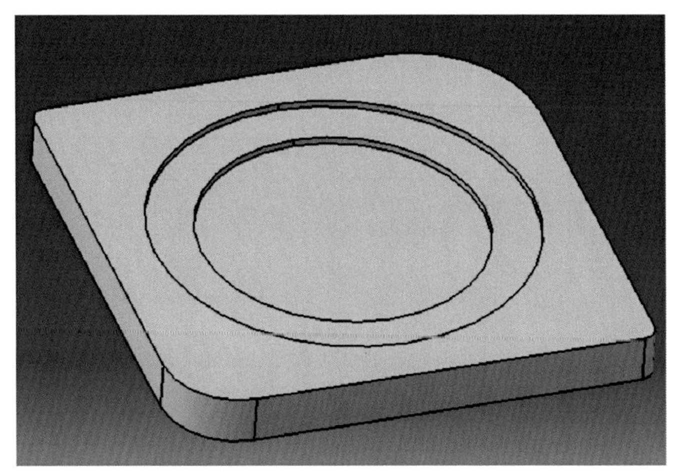

FIG. 13.6 Best conceptual design selected for roselle biocomposite mug pad.

was deemed simple and easy to perform by hand layup. Subsequently, it would increase the product's manufacturability and productivity while reducing the total product cost. It exhibits lightweight in terms of weight and performance because the roselle fiber has a low density and good tensile strength, which is further discussed in the following section.

On top of that, the chosen design has two distinct circular slots; hence, it can be used with two different mug sizes. These slots' function is to avoid slipping of the mug. These slots also aid in decreasing the total weight of the product by removing some areas. All the edges were made into a circular shape to eliminate any sharp edges and improve product safety when in use.

8 DETAILED DESIGN OF ROSELLE BIOCOMPOSITE MUG PAD

The detailed design was performed for the following activities in the DfS approach. The critical details to fabricate the selected conceptual design were well-defined at this phase. The product dimensions, tolerance, and material bill were included in the description. These details were represented by engineering drawing techniques using the computer-aided design software, CATIA V5, to produce technical drawings for the selected roselle biocomposite mug pad. Similar software was also used to design the mold for the fabrication of the product.

9 ROSELLE BIOCOMPOSITE MUG PAD FABRICATION

The final process after the detailed design phase was the fabrication of the roselle composite mug pad. The fabrication process was split into two main steps: the fabrication of molds and the end product.

9.1 Mold Fabrication

The mold was designed in two halves for ease of molding operation. Part 1, the female part was the lower fixed mold half; and part 2, also a female part is the upper removable mold half, as illustrated in Fig. 13.7. A mold-releasing agent was applied to the mold surface using Miracle Gloss prior to the molding process so that the end product can be easily removed.

The 3D printing fabrication method was used to produce the small-sized mold that was compatible to be printed using the 3D printer Anycubic 13 Mega, as illustrated in Fig. 13.8A. Poly(lactic) acid (PLA) was used as the 3D printer's filament owing to its suitability for a low production process that requires less energy and generates little waste. Furthermore, PLA material can be easily disposed through the natural decomposition process at the end of its life cycle stage. Both features strongly adhere to the DfS approach to develop a green product. Fig. 13.8B shows the fabricated mold for the mug pad product.

9.2 Biocomposite Mug Pad Fabrication

As for the composite's mug pad fabrication, neat UP resin polymer and methyl ethyl ketone peroxide (MEKP) as catalyst were used. UP, roselle fiber, and MEKP were manually mixed on a composition basis, as shown in Table 13.5. Based on the work of Mazani et al. (2019), the mass of the MEKP catalyst used was 2% of the resin's mass. Both chemicals were measured in a sufficient volume and poured into a well-cleaned container. The mixture was then gently agitated to avoid

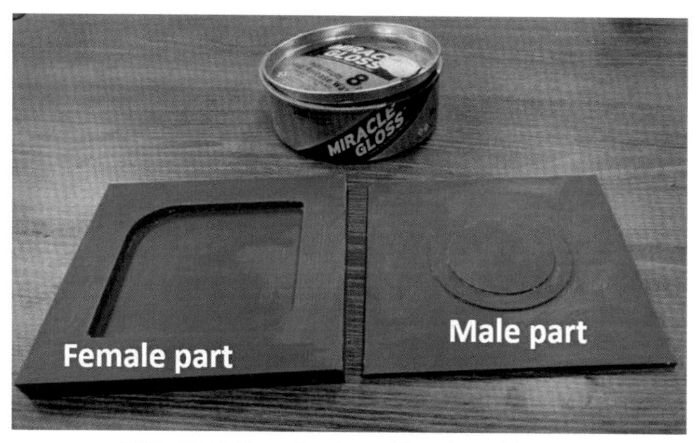

FIG. 13.7 Waxed male and female mold parts.

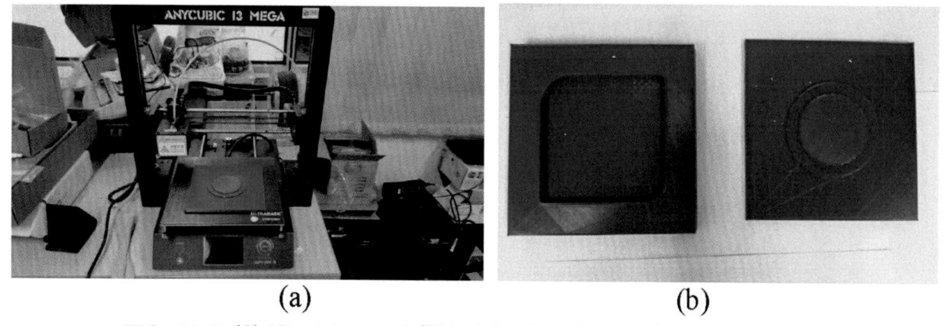

FIG. 13.8 **(A)** 3D printer used. **(B)** Poly(lactic) acid-based mug pad product.

TABLE 13.5
The Required Amount of UP, MEKP Catalyst, and Roselle Fiber.

Item	Weight
Volume of composite mold	$= 86.36\ cm^3$
70% Of neat UP	$= 0.7 \times 86.36\ cm^3 \times 1.12\ (g/cm^3)$ $= 67.71\ g$
2% Of MEKP catalyst	$= 0.02 \times 86.36\ cm^3 \times 1.55\ (g/cm^3)$ $= 2.68\ g$
30% Of roselle fiber	$= 0.3 \times 86.36\ cm^3 \times 1.4\ (g/cm^3)$ $= 36.27\ g$
Total weight	$= 67.71\ g + 2.68\ g + 36.27\ g$ $= 106.66\ g$

MEKP, methyl ethyl ketone peroxide; *UP*, unsaturated polyester.

the formation of air bubbles to prevent porosity in the structure of the final product.

Then around 30% of powdered roselle fiber was prepared in a small container. Both the resin mixture and the powdered roselle fiber were adequately mixed and carefully poured into part 1 of the mold by monitoring the formation of bubbles in the mixture, as shown in Fig. 13.9. Then the upper removable mold half (part 2) was placed on top of the lower fixed mold half (part 1) and was manually pressed. Following

FIG. 13.9 Mixture poured into mold.

FIG. 13.10 Final product.

that, the composite was placed in room temperature to be cured for 24 h. After that, the cured composite was removed from the mold. Sandpaper was used to remove irregular and excess edges, followed by clear spray for appropriate surface finishing of the composite end product, as shown in Fig. 13.10.

10 CONCLUSIONS

This chapter's key goal is to develop a new mug pad product using roselle composites using the DfS approach. The roselle mug pad composite is believed to be an alternative green solution to the existing product in the market. The tasks carried out during the product development based on the DfS approach were market investigation, PDS, mug pad conceptual design, detailed design of mug pad mold and mug pad, fabrication of mug pad mold and mug pad components, and product assembly. In this initiative, the success of the DfS approach implementation in this project is hoped to spur greater interest from manufactures in developing more sustainable products in the future.

REFERENCES

Abral, H., Ariksa, J., Mahardika, M., Handayani, D., Aminah, I., Sandrawati, N., Pratama, A. B., Fajri, N., Sapuan, S. M., & Ilyas, R. A. (2020). Transparent and antimicrobial cellulose film from ginger nanofiber. *Food Hydrocolloids, 98*, 105266. https://doi.org/10.1016/j.foodhyd.2019.105266

Abral, H., Ariksa, J., Mahardika, M., Handayani, D., Aminah, I., Sandrawati, N., Sapuan, S. M., & Ilyas, R. A. (2019). Highly transparent and antimicrobial PVA based bionanocomposites reinforced by ginger nanofiber. *Polymer Testing*, 106186. https://doi.org/10.1016/j.polymertesting.2019.106186

Abral, H., Atmajaya, A., Mahardika, M., Hafizulhaq, F., Kadriadi, Handayani, D., Sapuan, S. M., & Ilyas, R. A. (2020). Effect of ultrasonication duration of polyvinyl alcohol (PVA) gel on characterizations of PVA film. *Journal of Materials Research and Technology*, 1–10. https://doi.org/10.1016/j.jmrt.2019.12.078

Abral, H., Basri, A., Muhammad, F., Fernando, Y., Hafizulhaq, F., Mahardika, M., Sugiarti, E., Sapuan, S. M., Ilyas, R. A., & Stephane, I. (2019). A simple method for improving the properties of the sago starch films prepared by using ultrasonication treatment. *Food Hydrocolloids, 93*, 276–283. https://doi.org/10.1016/j.foodhyd.2019.02.012

Aisyah, H. A., Paridah, M. T., Sapuan, S. M., Khalina, A., Berkalp, O. B., Lee, S. H., Lee, C. H., Nurazzi, N. M., Ramli, N., Wahab, M. S., & Ilyas, R. A. (2019). Thermal properties of woven kenaf/carbon fibre-reinforced epoxy hybrid composite panels. *International Journal of Polymer Science, 2019*, 1–8. https://doi.org/10.1155/2019/5258621

Akonda, M. H., Shah, D. U., & Gong, R. H. (2020). Natural fibre thermoplastic tapes to enhance reinforcing effects in composite structures. *Composites Part A: Applied Science and Manufacturing, 131*(November 2019), 105822. https://doi.org/10.1016/j.compositesa.2020.105822

Ardhyananta, H., Puspadewa, F. D., Wicaksono, S. T., Widyastuti, Wibisono, A. T., Kurniawan, B. A., Ismail, H., & Salsac, A. V. (2017). Mechanical and thermal properties of unsaturated polyester/vinyl ester blends cured at room temperature. *IOP Conference Series: Materials Science and Engineering, 202*(1). https://doi.org/10.1088/1757-899X/202/1/012088

Arulmurugan, M., Prabu, K., Rajamurugan, G., & Selvakumar, A. S. (2019). Impact of BaSO4 filler on woven aloevera/Hemp hybrid composite: Dynamic mechanical analysis. *Materials Research Express, 6*(4), 045309. https://doi.org/10.1088/2053-1591/aafb88

Asrofi, M., Sapuan, S. M., Ilyas, R. A., & Ramesh, M. (2020). Characteristic of composite bioplastics from tapioca starch and sugarcane bagasse fiber: Effect of time duration of ultrasonication (Bath-Type). *Materials Today: Proceedings*. https://doi.org/10.1016/j.matpr.2020.07.254

Asrofi, M., Sujito, Syafri, E., Sapuan, S. M., & Ilyas, R. A. (2020). Improvement of biocomposite properties based tapioca starch and sugarcane bagasse cellulose nanofibers. *Key Engineering Materials, 849*, 96–101. https://doi.org/10.4028/www.scientific.net/KEM.849.96

Asyraf, M. R. M., Ishak, M. R., Razman, M. R., & Chandrasekar, M. (2019). Fundamentals of creep, testing methods and development of test rig for the full-scale crossarm: A review. *Jurnal Teknologi, 81*(4), 155–164.

Asyraf, M. R. M., Ishak, M. R., Sapuan, S. M., & Yidris, N. (2019). Conceptual design of creep testing rig for full-scale cross arm using TRIZ-Morphological chart-analytic network process technique. *Journal of Materials Research and Technology, 8*(6), 5647–5658. https://doi.org/10.1016/j.jmrt.2019.09.033

Asyraf, M. R. M., Ishak, M. R., Sapuan, S. M., & Yidris, N. (2020). Conceptual design of multi-operation outdoor flexural creep test rig using hybrid concurrent engineering approach. *Journal of Materials Research and Technology, 9*(2), 2357–2368. https://doi.org/10.1016/j.jmrt.2019.12.067

Asyraf, M. R. M., Ishak, M. R., Sapuan, S. M., Yidris, N., & Ilyas, R. A. (2020). Woods and composites cantilever beam: A comprehensive review of experimental and numerical creep methodologies. *Journal of Materials Research and Technology*. https://doi.org/10.1016/j.jmrt.2020.01.013

Asyraf, M. R. M., Ishak, M. R., Sapuan, S. M., Yidris, N., Ilyas, R. A., Rafidah, M., & Razman, M. R. (2020). Evaluation of design and simulation of creep test rig for full-scale cross arm structure. *Advances in Civil Engineering*. https://doi.org/10.1155/2019/6980918

Asyraf, M. R. M., Ishak, M. R., Sapuan, S. M., Yidris, N., Shahroze, R. M., Johari, A. N., Rafidah, M., & Ilyas, R. A. (2020). Creep test rig for cantilever beam: Fundamentals, prospects and present views. *Journal of Mechanical Engineering and Sciences, 14*.

Asyraf, M. R. M., Rafidah, M., Ishak, M. R., Sapuan, S. M., Ilyas, R. A., & Razman, M. R. (2020). Integration of TRIZ, morphological chart and ANP method for development of FRP composite portable fire extinguisher. *Polymer Composites*, 1–6. https://doi.org/10.1002/pc.25587

Athijayamani, A., Thiruchitrambalam, M., Natarajan, U., & Pazhanivel, B. (2009). Effect of moisture absorption on the mechanical properties of randomly oriented natural fibers/polyester hybrid composite. *Materials Science and Engineering A, 517*, 344–353. https://doi.org/10.1016/j.msea.2009.04.027

Atikah, M. S. N., Ilyas, R. A., Sapuan, S. M., Ishak, M. R., Zainudin, E. S., Ibrahim, R., Atiqah, A., Ansari, M. N. M., & Jumaidin, R. (2019). Degradation and physical properties of sugar palm starch/sugar palm nanofibrillated cellulose bionanocomposite. *Polimery, 64*(10), 27–36. https://doi.org/10.14314/polimery.2019.10.5

Atiqah, A., Jawaid, M., Sapuan, S. M., Ishak, M. R., Ansari, M. N. M., & Ilyas, R. A. (2019). Physical and thermal properties of treated sugar palm/glass fibre reinforced thermoplastic polyurethane hybrid composites. *Journal of Materials Research and Technology, 8*(5), 3726–3732. https://doi.org/10.1016/j.jmrt.2019.06.032

Ayu, R. S., Khalina, A., Harmaen, A. S., Zaman, K., Isma, T., Liu, Q., Ilyas, R. A., & Lee, C. H. (2020). Characterization study of empty fruit bunch (EFB) fibers reinforcement in poly(butylene) succinate (PBS)/starch/glycerol composite sheet. *Polymers, 12*(7), 1571. https://doi.org/10.3390/polym12071571

Azeez, T. O., Onukwuli, D. O., Nwabanne, J. T., & Banigo, A. T. (2020). Cissus populnea fiber - unsaturated polyester composites: Mechanical properties and interfacial adhesion. *Journal of Natural Fibers, 17*(9), 1281–1294. https://doi.org/10.1080/15440478.2018.1558159

Balaji, V., & Senthil Vadivu, K. (2017). Mechanical characterization of coir fiber and cotton fiber reinforced unsaturated polyester composites for packaging applications mechanical characterization of coir fiber and cotton fiber reinforced. *Journal of Applied Packaging Research, 9*(2), 12–19.

Begum, M. H. A., & Tanvir, N. I. (2019). Fabrication and characterization of cane fiber reinforced unsaturated polyester resin composites. *Bangladesh Journal of Scientific and Industrial Research, 54*(3), 247–256. https://doi.org/10.3329/bjsir.v54i3.42677

Biswas, B., Hazra, B., Sarkar, A., Bandyopadhyay, N. R., & Mitra, B. C. (2019). Influence of ZrO_2 incorporation on sisal fiber reinforced unsaturated polyester composites. *Polymer Composites*. https://doi.org/10.1002/pc.25087

Chirayil, C. J., George, C., Hosur, M., & Thomas, S. (2019). Chapter 12. Nanocellulose-reinforced unsaturated polyester composites. In *Unsaturated polyester resins*. Elsevier Inc.. https://doi.org/10.1016/B978-0-12-816129-6.00012-0

Cid-Ortega, S., & Guerrero-Beltrán, J. A. (2015). Roselle calyces (*Hibiscus sabdariffa*), an alternative to the food and beverages industries: A review. *Journal of Food Science and Technology, 52*(11), 6859–6869. https://doi.org/10.1007/s13197-015-1800-9

Gao, Y., romero, P., Zhang, H., Huang, M., & Lai, F. (2019). Unsaturated polyester resin concrete: A review. *Construction and Building Materials, 228*, 116709. https://doi.org/10.1016/j.conbuildmat.2019.116709

Halimatul, M. J., Sapuan, S. M., Jawaid, M., Ishak, M. R., & Ilyas, R. A. (2019a). Effect of sago starch and plasticizer content on the properties of thermoplastic films: Mechanical testing and cyclic soaking-drying. *Polimery, 64*(6), 32–41. https://doi.org/10.14314/polimery.2019.6.5

Halimatul, M. J., Sapuan, S. M., Jawaid, M., Ishak, M. R., & Ilyas, R. A. (2019b). Water absorption and water solubility properties of sago starch biopolymer composite films filled with sugar palm particles. *Polimery, 64*(9), 27–35. https://doi.org/10.14314/polimery.2019.9.4

Hazrol, M. D., Sapuan, S. M., Ilyas, R. A., Othman, M. L., & Sherwani, S. F. K. (2020). Electrical properties of sugar palm nanocrystalline cellulose reinforced sugar palm starch nanocomposites. *Polimery, 65*(05), 363–370. https://doi.org/10.14314/polimery.2020.5.4

Hossain, M., Mobarak, M. Bin, Rony, F. K., Sultana, S., Mahmud, M., & Ahmed, S. (2020). Fabrication and characterization of banana fiber reinforced unsaturated polyester resin based composites. *Nano Hybrids and Composites, 29*, 84–92. https://doi.org/10.4028/www.scientific.net/NHC.29.84

Hsu, C. P., & Lee, L. J. (1989). Microstructure formation in the cure of low shrinkage unsaturated polyester resin. *Annual Technical Conference of the Society of Plastics Engineers, 29*, 598–603.

Ibrahim, M. I., Edhirej, A., Sapuan, S. M., Jawaid, M., Ismarrubie, N. Z., & Ilyas, R. A. (2020). Extraction and characterization of Malaysian cassava starch, peel, and bagasse, and selected properties of the composites. In R. Jumaidin, S. M. Sapuan, & H. Ismail (Eds.), *Biofiller-reinforced biodegradable polymer composites* (1st ed., pp. 267–283). CRC Press.

Ibrahim, M. I., Sapuan, S. M., Zainudin, E. S., Zuhri, M. Y., Edhirej, A., & Ilyas, R. A. (2020). Characterization of corn fiber-filled cornstarch biopolymer composites. In R. Jumaidin, S. M. Sapuan, & H. Ismail (Eds.), *Biofiller-reinforced biodegradable polymer composites* (1st ed., pp. 285–301). CRC Press.

Idicula, M., Neelakantan, N. R., Oommen, Z., Joseph, K., & Thomas, S. (2005). A study of the mechanical properties of randomly oriented short banana and sisal hybrid fiber reinforced polyester composites. *Journal of Applied Polymer Science, 96*(5), 1699–1709. https://doi.org/10.1002/app.21636

Ilyas, R. A., & Sapuan, S. M. (2020a). The preparation methods and processing of natural fibre bio-polymer composites. *Current Organic Synthesis, 16*(8), 1068–1070. https://doi.org/10.2174/157017941608200120105616

Ilyas, R. A., & Sapuan, S. M. (2020b). Biopolymers and biocomposites: Chemistry and technology. *Current Analytical Chemistry, 16*(5), 500–503. https://doi.org/10.2174/157341101605200603095311

Ilyas, R. A., Sapuan, S. M., Asyraf, M. R. M., Atikah, M. S. N., Ibrahim, R., & Dele-Afolabia, T. T. (2020). Introduction to biofiller reinforced degradable polymer composites. In S. M. Sapuan, R. Jumaidin, & I. Hanafi (Eds.), *Biofiller reinforced biodegradable polymer composites*. CRC press.

Ilyas, R., Sapuan, S., Atikah, M., Asyraf, M., Rafiqah, S. A., Aisyah, H., Nurazzi, N. M., & Norrrahim, M. (2020). Effect of hydrolysis time on the morphological, physical, chemical, and thermal behavior of sugar palm nanocrystalline cellulose (*Arenga pinnata* (Wurmb.) Merr). *Textile Research Journal*. https://doi.org/10.1177/0040517520932393

Ilyas, R. A., Sapuan, S. M., Atiqah, A., Ibrahim, R., Abral, H., Ishak, M. R., Zainudin, E. S., Nurazzi, N. M.,

Atikah, M. S. N., Ansari, M. N. M., Asyraf, M. R. M., Supian, A. B. M., & Ya, H. (2020). Sugar palm (*Arenga pinnata* [Wurmb.] Merr) starch films containing sugar palm nanofibrillated cellulose as reinforcement: Water barrier properties. *Polymer Composites, 41*(2), 459–467. https://doi.org/10.1002/pc.25379

Ilyas, R. A., Sapuan, S. M., Ibrahim, R., Abral, H., Ishak, M. R., Zainudin, E. S., Atikah, M. S. N., Mohd Nurazzi, N., Atiqah, A., Ansari, M. N. M., Syafri, E., Asrofi, M., Sari, N. H., & Jumaidin, R. (2019). Effect of sugar palm nanofibrillated cellulose concentrations on morphological, mechanical and physical properties of biodegradable films based on agro-waste sugar palm (*Arenga pinnata* (Wurmb.) Merr) starch. *Journal of Materials Research and Technology, 8*(5), 4819–4830. https://doi.org/10.1016/j.jmrt.2019.08.028

Ilyas, R. A., Sapuan, S. M., Ibrahim, R., Abral, H., Ishak, M. R., Zainudin, E. S., Atiqah, A., Atikah, M. S. N., Syafri, E., Asrofi, M., & Jumaidin, R. (2020). Thermal, biodegradability and water barrier properties of bio-nanocomposites based on plasticised sugar palm starch and nanofibrillated celluloses from sugar palm fibres. *Journal of Biobased Materials and Bioenergy, 14*(2), 234–248. https://doi.org/10.1166/jbmb.2020.1951

Ilyas, R. A., Sapuan, S. M., & Ishak, M. R. (2018). Isolation and characterization of nanocrystalline cellulose from sugar palm fibres (*Arenga pinnata*). *Carbohydrate Polymers, 181*, 1038–1051. https://doi.org/10.1016/j.carbpol.2017.11.045

Ilyas, R. A., Sapuan, S. M., Ishak, M. R., & Zainudin, E. S. (2017). Effect of delignification on the physical, thermal, chemical, and structural properties of sugar palm fibre. *BioResources, 12*(4), 8734–8754. https://doi.org/10.15376/biores.12.4.8734-8754

Ilyas, R. A., Sapuan, S. M., Ishak, M. R., & Zainudin, E. S. (2018b). Water transport properties of bio-nanocomposites reinforced by sugar palm (*Arenga pinnata*) nanofibrillated cellulose. *Journal of Advanced Research in Fluid Mechanics and Thermal Sciences Journal, 51*(2), 234–246.

Ilyas, R. A., Sapuan, S. M., Ishak, M. R., & Zainudin, E. S. (2018c). Sugar palm nanocrystalline cellulose reinforced sugar palm starch composite: Degradation and water-barrier properties. *IOP Conference Series: Materials Science and Engineering, 368*, 012006. https://doi.org/10.1088/1757-899X/368/1/012006

Ilyas, R. A., Sapuan, S. M., Ishak, M. R., & Zainudin, E. S. (2018d). Development and characterization of sugar palm nanocrystalline cellulose reinforced sugar palm starch bionanocomposites. *Carbohydrate Polymers, 202*, 186–202. https://doi.org/10.1016/j.carbpol.2018.09.002

Ilyas, R. A., Sapuan, S. M., Ishak, M. R., & Zainudin, E. S. (2019). Sugar palm nanofibrillated cellulose (*Arenga pinnata* (Wurmb.) Merr): Effect of cycles on their yield, physic-chemical, morphological and thermal behavior. *International Journal of Biological Macromolecules, 123*. https://doi.org/10.1016/j.ijbiomac.2018.11.124

Ilyas, R. A., Sapuan, S. M., Ishak, M. R., Zainudin, E. S., & Atikah, M. S. N. (2018e). Characterization of sugar palm nanocellulose and its potential for reinforcement with a starch-based composite. In *Sugar palm biofibers, biopolymers, and biocomposites* (pp. 189–220). CRC Press. https://doi.org/10.1201/9780429443923-10.

Ilyas, R. A., Sapuan, S. M., Norrrahim, M. N. F., Yasim-Anuar, T. A. T., Kadier, A., Kalil, M. S., Atikah, M. S. N., Ibrahim, R., Asrofi, M., Abral, H., Nazrin, A., Syafiq, R., Aisyah, H. A., & Asyraf, M. R. M. (2021). Nanocellulose/starch biopolymer nanocomposites: Processing, manufacturing, and applications. In F. M. Al-Oqla (Ed.), *Advanced processing, properties, and application of strach and other bio-based polymer* (1st ed.). Elsevier Inc.. https://doi.org/10.1016/B978-0-12-189661-8.00006-8

Ilyas, R. A., Sapuan, S. M., Sanyang, M. L., Ishak, M. R., & Zainudin, E. S. (2018). Nanocrystalline cellulose as reinforcement for polymeric matrix nanocomposites and its potential applications: A review. *Current Analytical Chemistry, 14*(3), 203–225. https://doi.org/10.2174/1573411013666171003155624

Islam, N., Islam, F., Karmaker, N., Koly, F. A., Mahmud, J., Keya, K. N., & Khan, R. A. (2019). Fabrication and characterization of E-glass fiber reinforced unsaturated polyester resin based composite materials. *Nano Hybrids and Composite, 24*, 1–7. https://doi.org/10.4028/www.scientific.net/NHC.24.1

Islam, F., Islam, N., Shahida, S., Karmaker, N., Koly, F. A., Mahmud, J., Keya, K. N., & Khan, R. A. (2019). Mechanical and interfacial characterization of jute fabrics reinforced unsaturated polyester resin composites. *Nano Hybrids and Composite, 25*, 22–31. https://doi.org/10.4028/www.scientific.net/NHC.25.22

Johari, A. N., Ishak, M. R., Leman, Z., Yusoff, M. Z. M., & Asyraf, M. R. M. (2020a). Creep behaviour monitoring of short-term duration for fiber-glass reinforced composite cross-arms with unsaturated polyester resin samples using conventional analysis. *Journal of Mechanical Engineering and Sciences, 14*(3).

Johari, A. N., Ishak, M. R., Leman, Z., Yusoff, M. Z. M., & Asyraf, M. R. M. (2020b). Influence of $CaCO_3$ in pultruded glass fibre/unsaturated polyester composite on flexural creep behaviour using conventional and TTSP methods. *Polimery, 65*(10), 46–54. https://doi.org/10.14314/polimery.2020.11.6

Johari, A. N., Ishak, M. R., Leman, Z., Yusoff, M. Z. M., Asyraf, M. R. M., Ashraf, W., & Sharaf, H. K. (2019). Fabrication and cut-in speed enhancement of savonius vertical axis wind turbine (SVAWT) with hinged blade using fiberglass composites. *Seminar Enau Kebangsaan, 1997*, 978–983.

Jumaidin, R., Ilyas, R. A., Saiful, M., Hussin, F., & Mastura, M. T. (2019). Water transport and physical properties of sugarcane bagasse fibre reinforced thermoplastic potato starch biocomposite. *Journal of Advanced Research in Fluid Mechanics and Thermal Sciences, 61*(2), 273–281.

Jumaidin, R., Khiruddin, M. A. A., Asyul Sutan Saidi, Z., Salit, M. S., & Ilyas, R. A. (2020). Effect of cogon grass fibre on the thermal, mechanical and biodegradation properties of thermoplastic cassava starch biocomposite. *International*

Journal of Biological Macromolecules, 146, 746–755. https://doi.org/10.1016/j.ijbiomac.2019.11.011

Jumaidin, R., Saidi, Z. A. S., Ilyas, R. A., Ahmad, M. N., Wahid, M. K., Yaakob, M. Y., Maidin, N. A., Rahman, M. H. A., & Osman, M. H. (2019b). Characteristics of cogon grass fibre reinforced thermoplastic cassava starch biocomposite: Water absorption and physical properties. *Journal of Advanced Research in Fluid Mechanics and Thermal Sciences, 62*(1), 43–52.

Karthik, S., & Arunachalam, V. P. (2020). Investigation on the tensile and flexural behavior of coconut inflorescence fiber reinforced unsaturated polyester resin composites. *Materials Research Express, 7*(1). https://doi.org/10.1088/2053-1591/ab6c9d

Keya, K. N., Kona, N. A., Hossain, S., Islam, N., & Khan, R. A. (2020). Preparation and characterization of polypropylene and unsaturated polyester resin filled with jute fiber. *Advanced Materials Research, 1156*, 60–68. https://doi.org/10.4028/www.scientific.net/AMR.1156.60

Keya, K. N., Kona, N. A., & Khan, R. A. (2019). Studies on the mechanical characterization of jute fiber reinforced unsaturated polyester resin based composites: Effect of weave structure and yarn density. *Advanced Materials Research, 1156*(2012), 69–78. https://doi.org/10.4028/www.scientific.net/amr.1156.69

Keya, K. N., Kona, N. A., Razzak, M., & Khan, R. A. (2019). The comparative studies of mechanical and interfacial properties between jute and E-glass fiber-reinforced unsaturated polyester resin based composites. *Materials Engineering Research, 2*(1), 98–105. https://doi.org/10.25082/mer.2020.01.001

Maisara, A. M. N., Ilyas, R. A., Sapuan, S. M., Huzaifah, M. R. M., Nurazzi, N. M., & Saifulazry, S. O. A. (2019). Effect of fibre length and sea water treatment on mechanical properties of sugar palm fibre reinforced unsaturated polyester composites. *International Journal of Recent Technology and Engineering, 8*(2S4), 510–514. https://doi.org/10.35940/ijrte.b1100.0782s419

Malik, M., Choudhary, V., & Varma, I. K. (2000). Current status of unsaturated polyester resins. *Journal of Macromolecular Science - Polymer Reviews, 40*(2–3), 139–165. https://doi.org/10.1081/MC-100100582

Mansor, M. R., Sapuan, S. M., Zainudin, E. S., Nuraini, A. A., & Hambali, A. (2014). Conceptual design of kenaf fiber polymer composite automotive parking brake lever using integrated TRIZ-morphological chart-analytic hierarchy process method. *Materials and Design, 54*, 473–482. https://doi.org/10.1016/j.matdes.2013.08.064

Mazani, N., Sapuan, S. M., Sanyang, M. L., Atiqah, A., & Ilyas, R. A. (2019). Design and fabrication of a shoe shelf from kenaf fiber reinforced unsaturated polyester composites. *Lignocellulose for Future Bioeconomy*, (2000), 315–332. https://doi.org/10.1016/B978-0-12-816354-2.00017-7

Mishra, S., Mohanty, A. K., Drzal, L. T., Misra, M., Parija, S., Nayak, S. K., & Tripathy, S. S. (2003). Studies on mechanical performance of biofibre/glass reinforced polyester hybrid composites. *Composites Science and Technology, 63*(10), 1377–1385. https://doi.org/10.1016/S0266-3538(03)00084-8

Morton, J. F. (1974). Renewed interest in Roselle (*Hibiscus sabdariffa* L.), the long-forgotten "Florida cranberry". *Florida State Horticultural Society*, 415–425.

Murdani, A., Hadi, S., & Amrullah, U. S. (2017). Flexural properties and vibration behavior of jute/glass/carbon fiber reinforced unsaturated polyester hybrid composites for wind turbine blade. *Key Engineering Materials, 748*, 62–68. https://doi.org/10.4028/www.scientific.net/KEM.748.62

Nadlene, R., Sapuan, S. M., Jawaid, M., Ishak, M. R., & Yusriah, L. (2016a). A review on roselle fiber and its composites. *Journal of Natural Fibers, 13*(1), 10–41. https://doi.org/10.1080/15440478.2014.984052

Nadlene, R., Sapuan, S. M., Jawaid, M., Ishak, M. R., & Yusriah, L. (2016b). The effects of chemical treatment on the structural and thermal, physical, and mechanical and morphological properties of roselle fiber-reinforced vinyl ester composites. *Polymer Composites, 39*(1), 274–287. https://doi.org/10.1002/pc.23927

Nadlene, R., Sapuan, S. M., Jawaid, M., Ishak, M. R., & Yusriah, L. (2018). The effects of chemical treatment on the structural and thermal, physical, and mechanical and morphological properties of roselle fiber-reinforced vinyl ester composites. *Polymer Composites, 39*(1), 274–287. https://doi.org/10.1002/pc.23927

Nazrin, A., Sapuan, S. M., Zuhri, M. Y. M., Ilyas, R. A., Syafiq, R., & Sherwani, S. F. K. (2020). Nanocellulose reinforced thermoplastic starch (TPS), polylactic acid (PLA), and polybutylene succinate (PBS) for food packaging applications. *Frontiers in Chemistry, 8*(213), 1–12. https://doi.org/10.3389/fchem.2020.00213

Negawo, T. A., Polat, Y., Buyuknalcaci, F. N., Kilic, A., Saba, N., & Jawaid, M. (2018). Mechanical, morphological, structural and dynamic mechanical properties of alkali treated ensete stem fibers reinforced unsaturated polyester composites. *Composite Structures*. https://doi.org/10.1016/j.compstruct.2018.09.043

Norizan, M. N., Abdan, K., Ilyas, R. A., & Biofibers, S. P. (2020). Effect of fiber orientation and fiber loading on the mechanical and thermal properties of sugar palm yarn fiber reinforced unsaturated polyester resin composites. *Polimery, 65*(2), 34–43. https://doi.org/10.14314/polimery.2020.2.5

Nurazzi, N. M., Khalina, A., Chandrasekar, M., Aisyah, H. A., Rafiqah, S. A., Ilyas, R. A., & Hanafee, Z. M. (2020). Effect of fiber orientation and fiber loading on the mechanical and thermal properties of sugar palm yarn fiber reinforced unsaturated polyester resin composites. *Polimery, 65*(02), 115–124. https://doi.org/10.14314/polimery.2020.2.5

Nurazzi, N. M., Khalina, A., Sapuan, S. M., & Ilyas, R. A. (2019). Mechanical properties of sugar palm yarn/woven glass fiber reinforced unsaturated polyester composites : Effect of fiber loadings and alkaline treatment. *Polimery, 64*(10), 12–22. https://doi.org/10.14314/polimery.2019.10.3

Nurazzi, N. M., Khalina, A., Sapuan, S. M., Ilyas, R. A., Rafiqah, S. A., & Hanafee, Z. M. (2020). Thermal properties of treated sugar palm yarn/glass fiber reinforced unsaturated polyester hybrid composites. *Journal of Materials*

Research and Technology, 9(2), 1606–1618. https://doi.org/10.1016/j.jmrt.2019.11.086

Nurazzi, M., & Laila, D. (2017). A review : Fibres, polymer matrices and composites. *Science and Technology, 25*(October), 1085–1102.

Pugh, S. (1996). Concept selection–a method that works. In D. Clausing, & R. Andrade (Eds.), *Creating innovative products using total design*. Addison-Wesley Publishing Company.

Radzi, A. M., Sapuan, S. M., Jawaid, M., & Mansor, M. R. (2017). Influence of fibre contents on mechanical and thermal properties of roselle fibre reinforced polyurethane composites. *Fibers and Polymers, 18*(7), 1353–1358. https://doi.org/10.1007/s12221-017-7311-8

Razali, N., Salit, M. S., Jawaid, M., Ishak, M. R., & Lazim, Y. (2015). A study on chemical composition, physical, tensile, morphological, and thermal properties of roselle fibre: Effect of fibre maturity. *BioResources, 10*(1), 1803–1823. https://doi.org/10.15376/biores.10.1.1803-1824

Rosamah, E., S, A. K. H. P., Yap, S. W., Saurabh, C. K., Tahir, P. M., Dungani, R., & Owolabi, A. F. (2018). The role of bamboo nanoparticles in kenaf fiber reinforced unsaturated polyester composites. *Journal of Renewable Materials, 6*(1), 75–86. https://doi.org/10.7569/JRM.2017.634152

Sánchez-Mendoza, J., Domínguez-López, A., Navarro-Galindo, S., & López-Sandoval, J. A. (2008). Some physical properties of Roselle (*Hibiscus sabdariffa* L.) seeds as a function of moisture content. *Journal of Food Engineering, 87*(3), 391–397. https://doi.org/10.1016/j.jfoodeng.2007.12.023

Sanyang, M. L., Ilyas, R. A., Sapuan, S. M., & Jumaidin, R. (2018). Sugar palm starch-based composites for packaging applications. In *Bionanocomposites for packaging applications* (pp. 125–147). Springer International Publishing. https://doi.org/10.1007/978-3-319-67319-6_7.

Sapuan, S. M. (2005). A conceptual design of the concurrent engineering design system for polymeric-based composite automotive pedals. *American Journal of Applied Sciences, 2*(2), 514–525.

Sapuan, S. M. (2014). *Tropical natural fibre composites*. Springer Singapore. https://doi.org/10.1007/978-981-287-155-8

Sapuan, S. M. (2015). Concurrent engineering in natural fibre composite product development. *Applied Mechanics and Materials, 761*, 59–62. https://doi.org/10.4028/www.scientific.net/amm.761.59

Sapuan, S. M., Aulia, H. S., Ilyas, R. A., Atiqah, A., Dele-Afolabi, T. T., Nurazzi, M. N., Supian, A. B. M., & Atikah, M. S. N. (2020). Mechanical properties of longitudinal basalt/woven-glass-fiber-reinforced unsaturated polyester-resin hybrid composites. *Polymers, 12*(10), 2211. https://doi.org/10.3390/polym12102211

Sari, N. H., Pruncu, C. I., Sapuan, S. M., Ilyas, R. A., Catur, A. D., Suteja, S., Sutaryono, Y. A., & Pullen, G. (2020). The effect of water immersion and fibre content on properties of corn husk fibres reinforced thermoset polyester composite. *Polymer Testing, 91*, 106751. https://doi.org/10.1016/j.polymertesting.2020.106751

Shaharuzaman, M. A., Sapuan, S. M., Mansor, M. R., & Zuhri, M. Y. M. (2020). Conceptual design of natural fiber composites as a side-door impact beam using hybrid approach. *Journal of Renewable Materials, 8*(5), 549–563. https://doi.org/10.32604/jrm.2020.08769

Sood, M., & Dwivedi, G. (2018). Effect of fiber treatment on flexural properties of natural fiber reinforced composites : A review. *Egyptian Journal of Petroleum, 27*(4), 775–783. https://doi.org/10.1016/j.ejpe.2017.11.005

Syafri, E., Kasim, A., Abral, H., & Asben, A. (2019). Cellulose nanofibers isolation and characterization from ramie using a chemical-ultrasonic treatment. *Journal of Natural Fibers, 16*(8), 1145–1155. https://doi.org/10.1080/15440478.2018.1455073

Syafri, E., Sudirman, M., Yulianti, E., Deswita, Asrofi, M., Abral, H., Sapuan, S. M., Ilyas, R. A., & Fudholi, A. (2019). Effect of sonication time on the thermal stability, moisture absorption, and biodegradation of water hyacinth (*Eichhornia crassipes*) nanocellulose-filled bengkuang (*Pachyrhizus erosus*) starch biocomposites. *Journal of Materials Research and Technology, 8*(6), 6223–6231. https://doi.org/10.1016/j.jmrt.2019.10.016

Yorseng, K., Rangappa, S. M., Pulikkalparambil, H., Siengchin, S., & Parameswaranpillai, J. (2020). Accelerated weathering studies of kenaf/sisal fiber fabric reinforced fully biobased hybrid bioepoxy composites for semi-structural applications: Morphology, thermo-mechanical, water absorption behavior and surface hydrophobicity. *Construction and Building Materials, 235*, 117464. https://doi.org/10.1016/j.conbuildmat.2019.117464

Zheng, F., & Lin, Y. C. (2017). A fuzzy TOPSIS expert system based on neural networks for new product design. In *Proceedings of the 2017 IEEE international conference on applied system innovation: Applied system innovation for modern Technology* (Vol. 2017, pp. 598–601). ICASI. https://doi.org/10.1109/ICASI.2017.7988494

CHAPTER 14

The Effect of Fiber Length on Mechanical and Thermal Properties of Roselle Fiber-Reinforced Polylactic Acid Composites via ANSYS Software Analysis

S.N. AIN • R. NADLENE • R.A. ILYAS • A.M. RADZI • D. SIVAKUMAR

1 INTRODUCTION

Nowadays, natural fibers such as sisal, hemp, roselle, and pineapple leaf fiber have been used widely in industries. Researchers are actively exploring the biodegradable composites in line with the green technology objective and to find a better polymer material such as polylactic acid (PLA) to be blended with natural fiber as an alternative source to minimize the usage of synthetic fiber. In order to evaluate the ability of the biodegradable composite product to be stable in market, there are some aspects that need to be considered based on the applications of composite, thermal properties, and mechanical properties. The main factor that can affect the mechanical properties of natural fiber-reinforced composites is the surface adhesion between fiber and polymer. Fibre pullout is the best term of fibre breakage which leads to the failure of composites samples.

Natural fibers such as roselle have a big ability to be used widely in composite materials and to be one of the main sources in the future to replace synthetic fiber (Nadlene et al., 2015). This is because roselle fiber has low density that makes the fiber superficial and the cost to produce or fabricate this product is very cheap, compared with other synthetic fiber production process. PLA is a plant-derived polyester that is known as a bioplastic because it is obtained from renewable resources such as organic resources. Polymer polylactic acid (PLA) is a thermoplastic material that can be treated in both hot and cold conditions several times because of its mechanical properties, and its properties can be maintained at a certain rate or level. The advantages of PLA are because of its properties such as diaphaneity, which is good in optical transparency, and it has high performance in terms of bonding strength between fiber-reinforced PLA when compared with other synthetic polymers. Nevertheless, PLA has disadvantages including low thermal stability and low crystallization ability.

The use of biodegradable polymer was rapidly expanding or developing in both the industrial sector and academia. Over the years the development of biodegradable polymer products as a substitute to petroleum usage has been the main focus for researchers and the industrial sector. The main factor in the industrial sector and for researchers to replace petroleum as the main product used in daily life is that biodegradable polymer products can be an eco-friendly product and their use can prevent plastic waste.

Besides, biodegradable polymers were popular in the industrial sector because they are of low cost, can be decomposed easily, and are reusable. Natural fibers have many advantages in terms of ecosystem and are more biodegradable. However, it is not flawless. One of the advantage of the natural fiber is variation of fiber length will affect less on the regular and uniform yarn that result from manmade fiber material. Previous research shows that increase in kenaf fiber length will increase the tensile strength (Ghadakpour et al., 2020). In the literature, there are no reported studies on evaluating the effect of roselle fiber size on the properties of composites by using software analysis.

Roselle. https://doi.org/10.1016/B978-0-323-85213-5.00012-5

In this research, the length of fibers varies from 2 to 5 mm. The properties of roselle fiber-reinforced PLA composites will be evaluated by using ANSYS in terms of mechanical and thermal properties.

2 THE VERSATILITY OF ROSELLE FIBER

Traditionally, roselle seeds are used as feedstocks for chicken. Over the past few decades, roselle fibers have been used for ropes, heavy-duty cables, and the navy and merchant marine due to their sustainability and lack of decay for a long duration in seawater or freshwater. Fishing nets made from roselle fiber are also extremely resistant and break only under great strain (Julian, 1949). It has been demonstrated that roselle is high in calcium, niacin, riboflavin, protein, carbohydrates, flavonoids, acids, minerals, and vitamins (Mohamad et al., 2002). Thus roselle is very useful in medical applications (Tori Husdcon, 2011; Mungole & Chaturvedi, 2011). The plant has been reported to have antihypertensive, hepatoprotective, antihyperlipidemic, anticancer, and antioxidant properties (Mahadevan & Kamboj, 2009). In addition, roselle calyces have also been used as natural food colorants (Selim et al., 1993). Finally, fresh roselle or dry calyces are used to produce juices (Wilson, 2009; Grace, 2008). Roselle fibers have been widely used as a reinforcement material for composites and textile engineering (Managooli, 2009; Das Gupta, 1959; Wester, 1907).

Roselle fiber is currently used as a new alternative material for natural fiber. Some the researchers have conducted studies on the capability of roselle fibers to be blended with polymers to form composites. In the literature, there are a few studies on roselle fiber-reinforced thermoplastic materials but publication for this fiber is limited.

Roselle fiber is considered a natural fiber with good mechanical properties. Several papers have reviewed the mechanical properties of roselle natural fibers. Table 14.1 compare the properties of natural fibers with roselle fibers (Chandramohan & Marimuthu, 2011). As observed in Table 14.1, roselle fibers exhibit good tensile strength relative to other natural fibers.

Investigated the mechanical properties of roselle fiber-reinforced isotactic polypropylene (iPP) composites. The authors investigated the effect of the type, size, and content of roselle fibers and the presence and content of a commercial compatibilizer on the resulting mechanical properties. The samples were divided into four groups: bast fiber, core fiber, whole stalk fiber (40% bast fiber and 60% core fiber), and core added whole stalk fiber (20% bast fiber and 80% core fiber). All the prepared samples underwent several mechanical tests such as tensile tests, flexural tests, and Izod impact tests. In addition, the thermal degradation behavior of the fibers was investigated, and the microstructures of the impact test samples were examined using scanning electron microscopy. The results indicated that the addition of pure roselle bast fiber improved the tensile, flexural, and impact properties, whereas the addition of the core fiber only improved the Young's and flexural moduli of the composites. Increasing the bast fiber content or increasing the lengths or aspect ratios of the fiber improved the behavior of the composites (Junkasem et al., 2006). A failure result in the impact test indicated poor interfacial bonding between the iPP molecule and the fiber, while the addition of maleic anhydride improved the mechanical properties and interfacial bonding of the composites and fiber.

The effect of water absorption in a natural fiber is critical in composites. Many studies have been

TABLE 14.1
Properties of Natural Fibers.

Type of Fiber	Density (kg/m^3)	Water Absorption %	Modulus of Elasticity, E (GPa)	Tensile Strength (MPa)
Sisal	800–700	56	15	268
Roselle	800–750	40–50	17	170–350
Banana	950–750	60	23	180–430
Date palm	463	60–65	70	125–200
Coconut	145–380	130–180	19–26	120–200
Reed	490	100	37	70–140

performed to understand the behavior of natural fibers in wet conditions. Athijayamani et al. (2009) conducted an experiment to study the effect of moisture absorption on the mechanical properties of randomly oriented natural fibers in polyester hybrid composites. Roselle and sisal hybrid natural fiber samples were prepared as reinforcement materials. The weight ratio between sisal and roselle was 1:1, and the samples were grouped into three different lengths of 10, 50, and 150 mm with fiber contents of 10%, 20%, and 30%, respectively. Under dry and wet conditions, an increase in the fiber content and length increased the mechanical properties of the hybrid composites in terms of the tensile and flexural strength; however, the dry condition results were better. This result might be due to the exposure of moisture, which causes degradation of the fiber and matrix bonding. The presence of hydrogen bonding between the water molecules and cellulose fiber decreased the mechanical properties (Athijayamani et al., 2009). For the impact test, scattered results were observed for both the dry and wet conditions of the natural fiber. A possible explanation is that the fiber was pulled out from the matrix. Therefore cavities existed that allowed water to flow along the fiber-matrix interfaces by capillary action after composite failure.

3 METHODOLOGY

Fig. 14.1 shows the flowchart of the analysis of the effect of length on the composite sample. Roselle fiber with different fibre length were analyzed, and PLA was used as a binder to analyze in the ANSYS software.

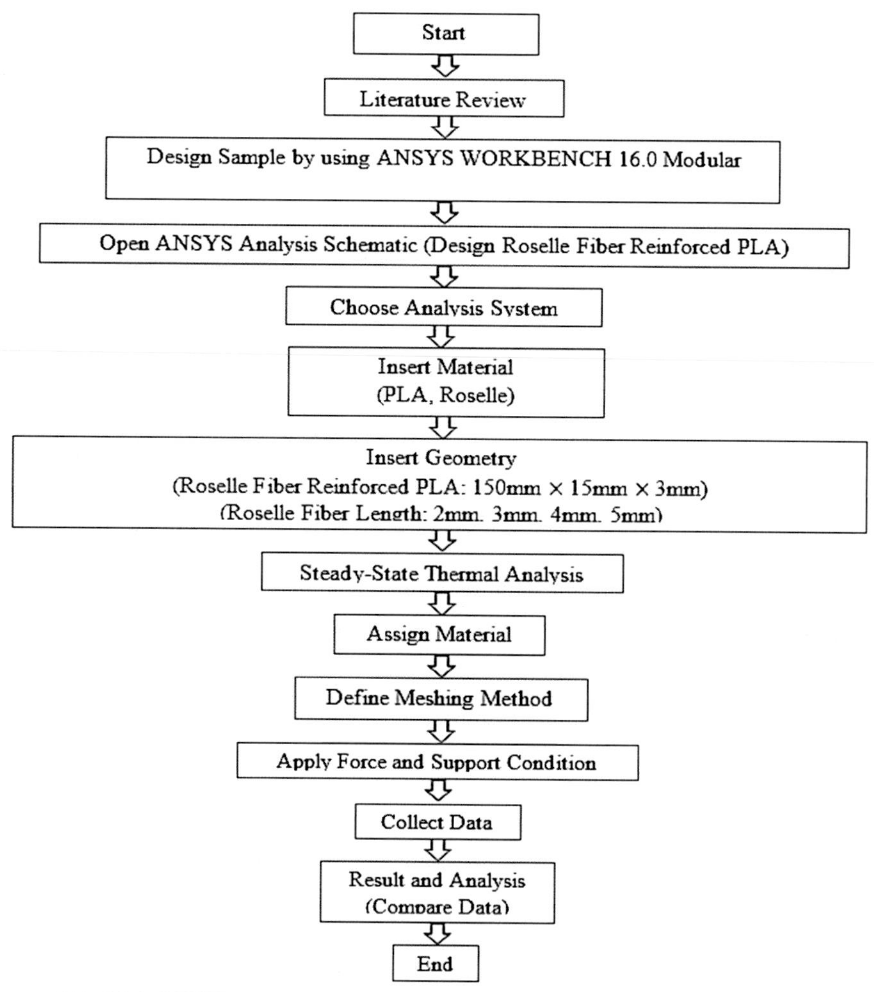

FIG. 14.1 Flowchart of the data setup in the ANSYS software. *PLA*, polylactic acid.

3.1 ANSYS Software

Finite element method (FEM) is the most common method for effective analysis of complex shapes, various materials, boundary conditions, and structures of loading. The finite element analysis process consists of three parts: modeling is the preprocessing step, analysis is the postprocessing step, and finally the result evaluation solution stage. Key preprocessing targets include 1D, 2D, or 3D modeling of the problem; assigning correct material parameters, components, meshing, and material properties; and applying proper structural or thermal boundary conditions with loads such as thermal, mechanical, electric, or magnetic loads depending on the application requirements. In the research program the drafted specimen model is opened and placed in Grafic file (IGS) format, and the ANSYS program is used for intent analysis. The layers of the composite materials are added after loading them onto ANSYS. Postprocessing is done to measure the reaction of a structure under any loading, such as static, impact, thermal, fatigue, and torque before the point of production of the component, and the effects of the finite element analysis are shown in terms of charts, graphs, maps, deflected structural patterns, and simulation. It also compares the experimental results, the theoretic model results, and the stress values obtained from the finite element analysis (Balasubramanian et al., 2020).

3.2 Analysis

Four samples were designed using ANSYS Workbench Geometry-Design Modular 16.0 with 3D model prototype. The ASTM D3039 was used in this finite element analysis and set as a modular schematic, as shown in Fig. 14.2. PLA with the dimension of 150 mm ×× 15 mm ×× 3 mm and the roselle fiber with different lengths of 2, 3, 4, and 5 mm were drawn, as illustrated in Fig. 14.4. The design was draw in an XY-axis plane. Fig. 14.3 shows the PLA design dimension. The type of body element for this design sample is the solid type.

3.2.1 Steady-state thermal

The first step is to open the ANSYS Workbench 16.0 and choose "Steady-State Thermal" in the project schematic in the toolbox. Fig. 14.5 shows the steady-state process that was chosen to analyze the three test samples for thermal degradation. In this analysis the degree of freedom for fiber reinforced PLA was not been used.

3.2.2 Define material

The second step is to define the material in engineering data for roselle fiber and PLA based on the properties of the material in the engineering data source and the filter engineering data. The engineering data sources tab is used to key in the material used; for example, in this analysis, two materials are used, roselle fiber and PLA

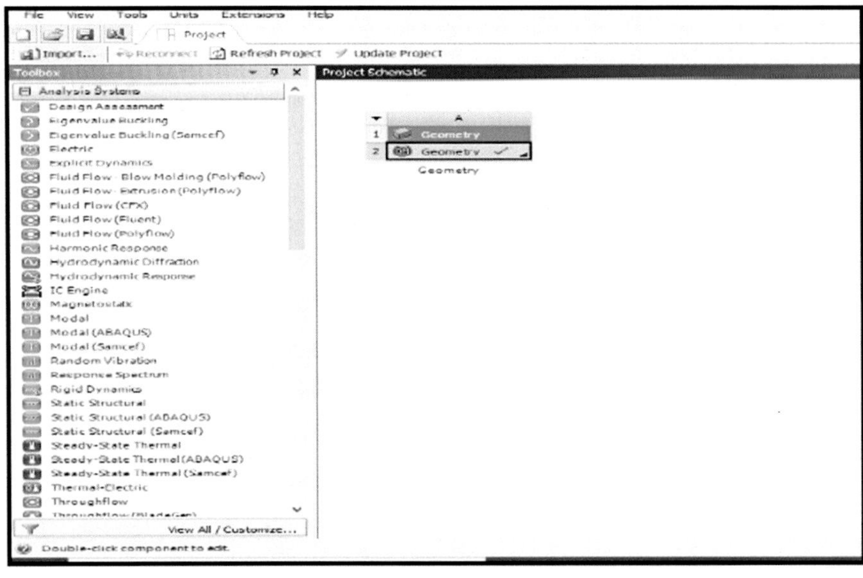

FIG. 14.2 The design modular schematic.

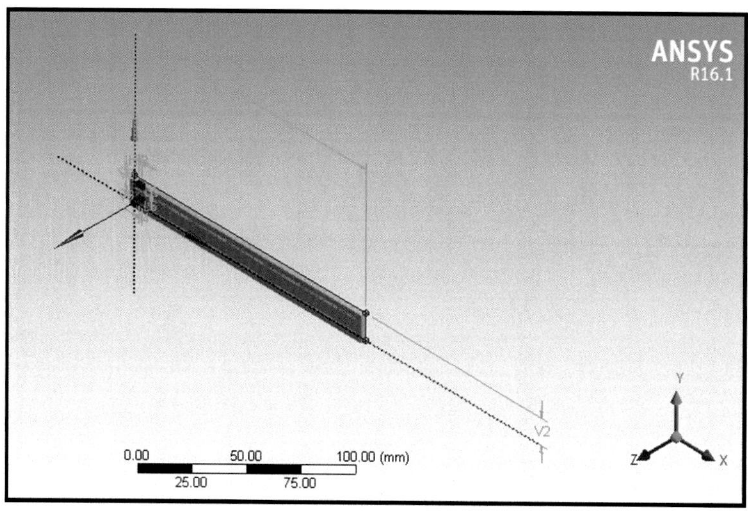

FIG. 14.3 Polylactic acid dimension.

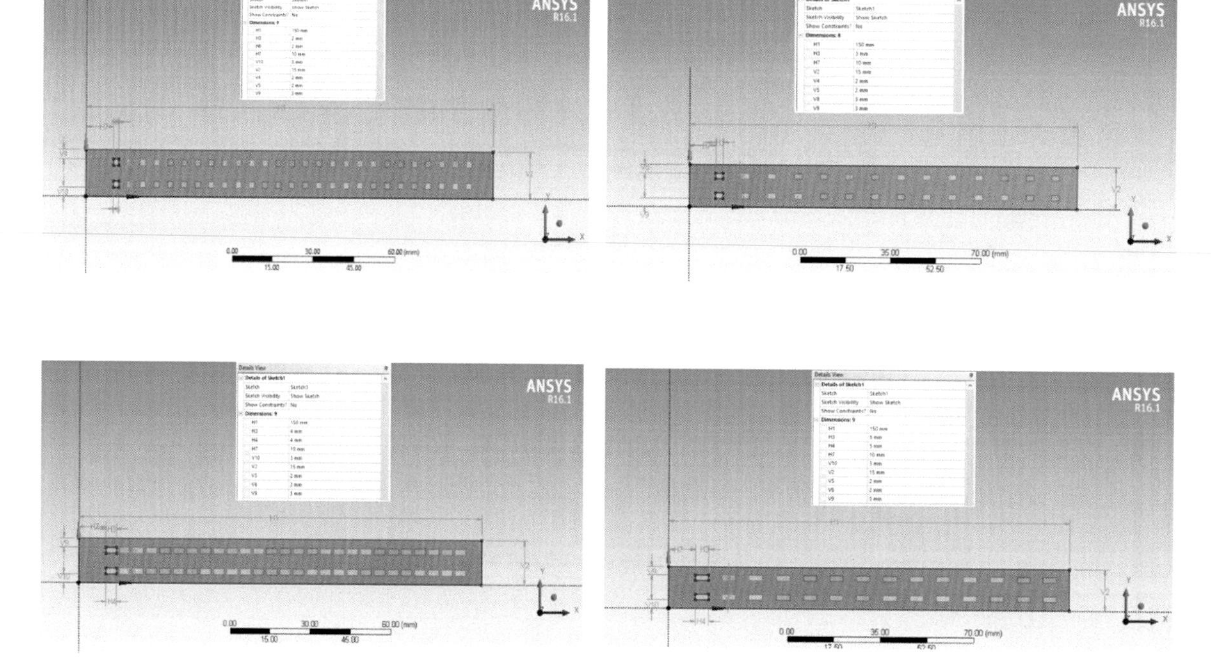

FIG. 14.4 The schematic design drawing for roselle fiber of 2–5 mm length with polylactic acid.

The filter engineering data is used to key in the properties of the roselle fiber and PLA based on the process and result to define. Fig. 14.6 shows the engineering data, engineering data source, and filter engineering data as the first step to assign the material. Figs. 14.7 and 14.8 show how the PLA and roselle fiber material properties were chosen in the filter engineering data.

Fig. 14.9 shows the material characteristics for PLA in ANSYS Workbench 16.0 that was applied in the analysis of steady-state thermal and static structural, where the

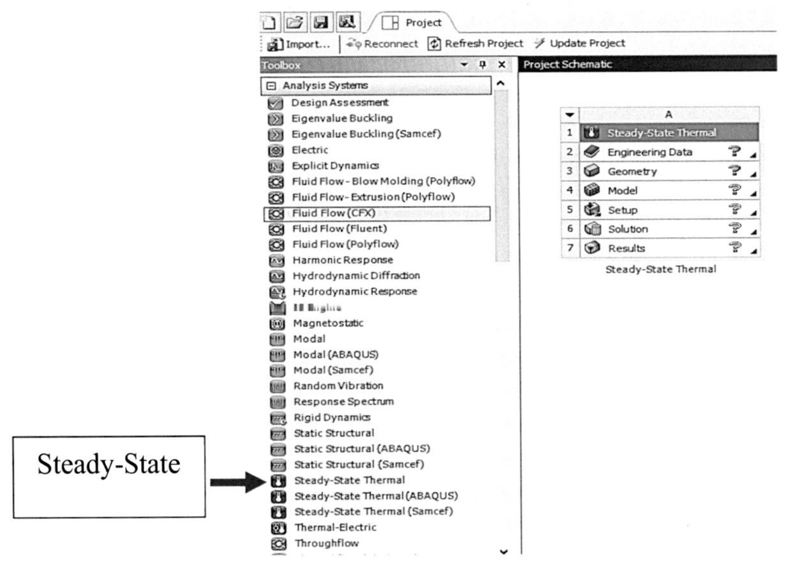

FIG. 14.5 Steady-state thermal.

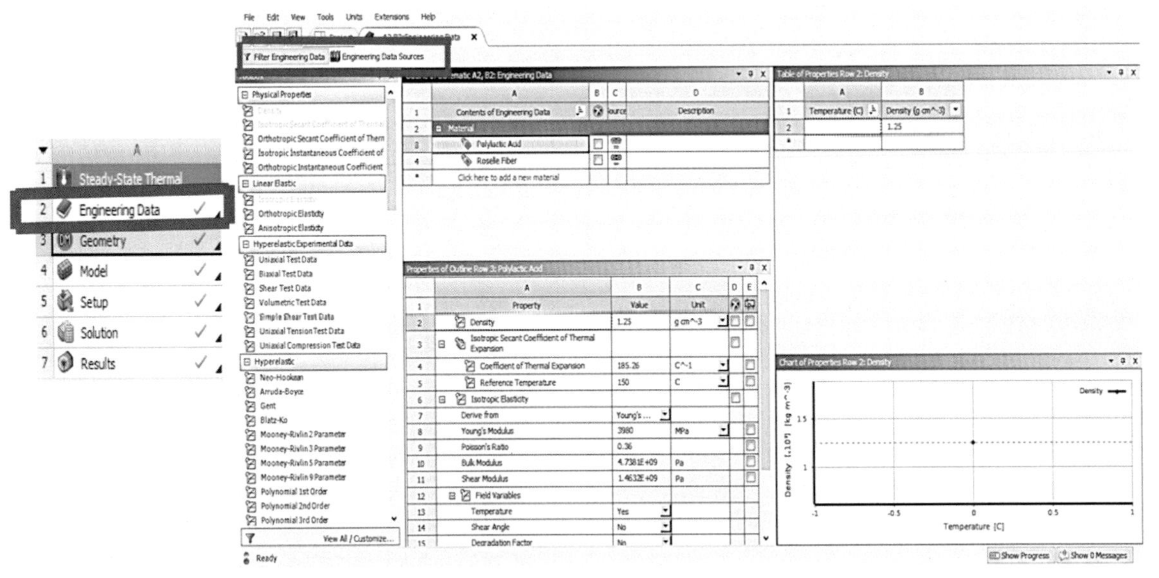

FIG. 14.6 The engineering data, engineering data source, and filter engineering data.

value of isotropic secant coefficient of thermal expansion is 150°C, the Young modulus is 3980 MPa, the Poisson ratio is 0.36, bulk modulus is 4738.1 MPa, shear modulus is 1463.2 MPa, yield strength is 52 MPa, compressive yield strength is 50 MPa, tensile ultimate strength is 73 MPa, and compressive ultimate strength is 70 MPa.

Fig. 14.10 shows the material characteristics for roselle fiber in ANSYS Workbench 16.0 that was applied in the analysis of steady-state thermal and static structural, where the value of the Young modulus is 53,000 MPa, the Poisson ratio is 0.3, bulk modulus is 44,167 MPa, shear modulus is 20,385 MPa, and tensile yield strength is 930 MPa.

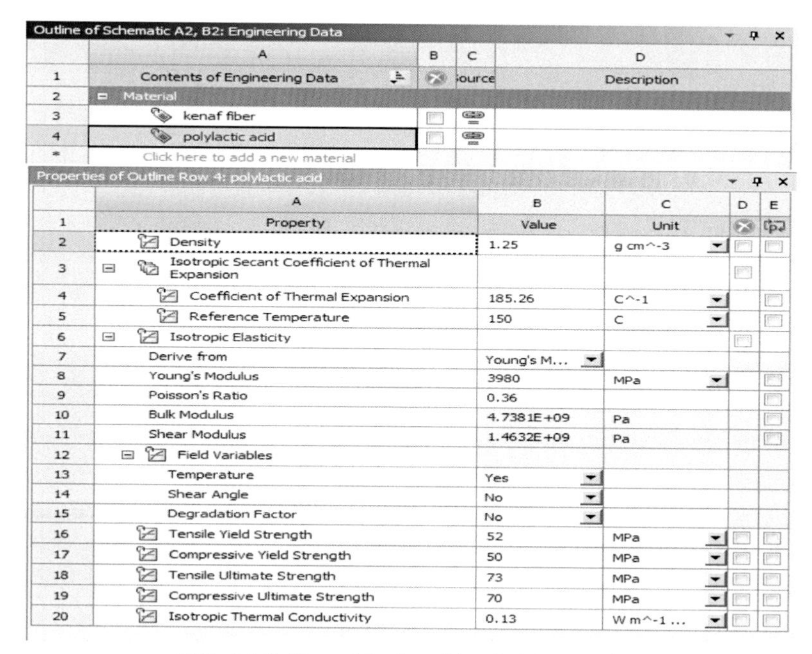

Outline of Schematic A2, B2: Engineering Data

	A	B	C	D
1	Contents of Engineering Data		Source	Description
2	Material			
3	kenaf fiber			
4	polylactic acid			
*	Click here to add a new material			

Properties of Outline Row 4: polylactic acid

	A	B	C	D	E
1	Property	Value	Unit		
2	Density	1.25	g cm^-3		
3	Isotropic Secant Coefficient of Thermal Expansion				
4	Coefficient of Thermal Expansion	185.26	C^-1		
5	Reference Temperature	150	C		
6	Isotropic Elasticity				
7	Derive from	Young's M...			
8	Young's Modulus	3980	MPa		
9	Poisson's Ratio	0.36			
10	Bulk Modulus	4.7381E+09	Pa		
11	Shear Modulus	1.4632E+09	Pa		
12	Field Variables				
13	Temperature	Yes			
14	Shear Angle	No			
15	Degradation Factor	No			
16	Tensile Yield Strength	52	MPa		
17	Compressive Yield Strength	50	MPa		
18	Tensile Ultimate Strength	73	MPa		
19	Compressive Ultimate Strength	70	MPa		
20	Isotropic Thermal Conductivity	0.13	W m^-1 ...		

FIG. 14.7 The polylactic acid material properties.

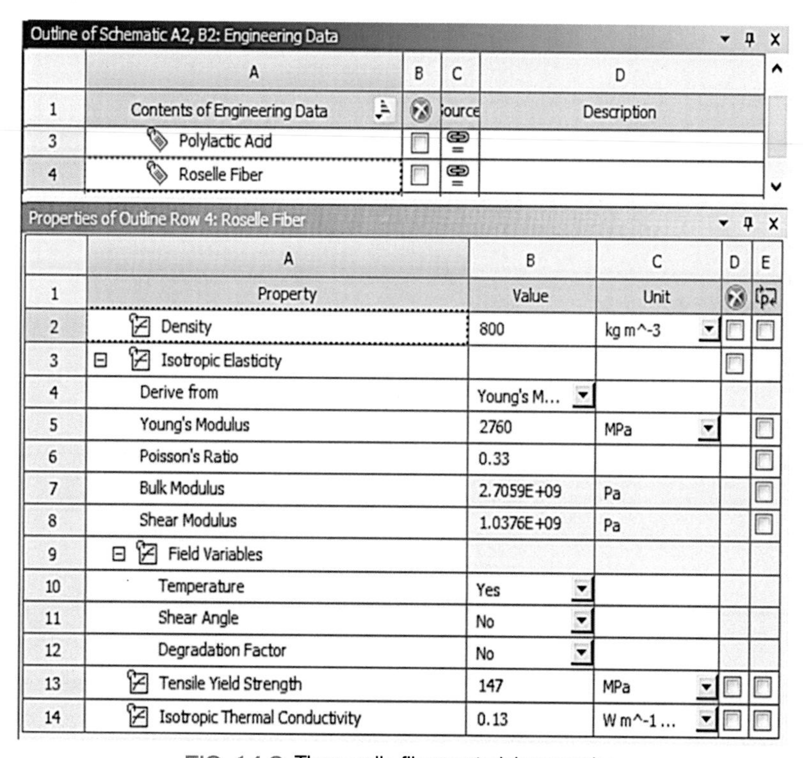

Outline of Schematic A2, B2: Engineering Data

	A	B	C	D
1	Contents of Engineering Data		Source	Description
3	Polylactic Acid			
4	Roselle Fiber			

Properties of Outline Row 4: Roselle Fiber

	A	B	C	D	E
1	Property	Value	Unit		
2	Density	800	kg m^-3		
3	Isotropic Elasticity				
4	Derive from	Young's M...			
5	Young's Modulus	2760	MPa		
6	Poisson's Ratio	0.33			
7	Bulk Modulus	2.7059E+09	Pa		
8	Shear Modulus	1.0376E+09	Pa		
9	Field Variables				
10	Temperature	Yes			
11	Shear Angle	No			
12	Degradation Factor	No			
13	Tensile Yield Strength	147	MPa		
14	Isotropic Thermal Conductivity	0.13	W m^-1 ...		

FIG. 14.8 The roselle fiber material properties.

polylactic acid

TABLE 40
polylactic acid > Constants

Density	1.25e-006 kg mm^-3
Coefficient of Thermal Expansion	185.26 C^-1
Thermal Conductivity	1.3e-004 W mm^-1 C^-1
Specific Heat	1.8e+006 mJ kg^-1 C^-1

TABLE 41
polylactic acid > Isotropic Secant Coefficient of Thermal Expansion

Reference Temperature C
150

TABLE 42
polylactic acid > Isotropic Elasticity

Temperature C	Young's Modulus MPa	Poisson's Ratio	Bulk Modulus MPa	Shear Modulus MPa
	3980	0.36	4738.1	1463.2

TABLE 43
polylactic acid > Tensile Yield Strength

Tensile Yield Strength MPa
52

TABLE 44
polylactic acid > Compressive Yield Strength

Compressive Yield Strength MPa
50

TABLE 45
polylactic acid > Tensile Ultimate Strength

Tensile Ultimate Strength MPa
73

TABLE 46
polylactic acid > Compressive Ultimate Strength

Compressive Ultimate Strength MPa
70

FIG. 14.9 Material characteristics of polylactic acid.

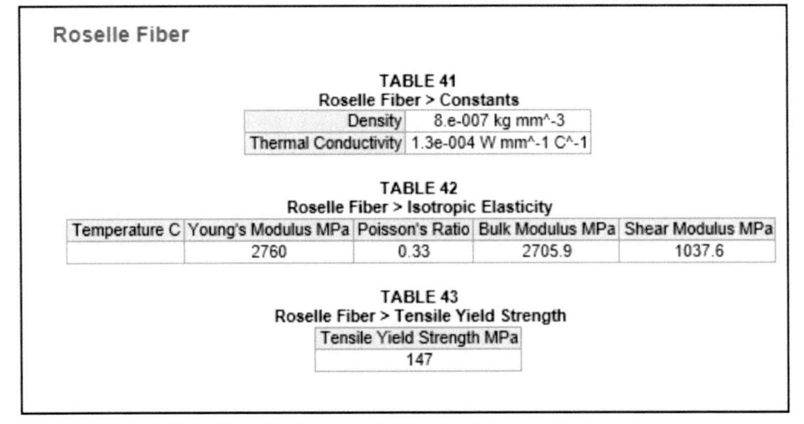

Roselle Fiber

TABLE 41
Roselle Fiber > Constants

Density	8.e-007 kg mm^-3
Thermal Conductivity	1.3e-004 W mm^-1 C^-1

TABLE 42
Roselle Fiber > Isotropic Elasticity

Temperature C	Young's Modulus MPa	Poisson's Ratio	Bulk Modulus MPa	Shear Modulus MPa
	2760	0.33	2705.9	1037.6

TABLE 43
Roselle Fiber > Tensile Yield Strength

Tensile Yield Strength MPa
147

FIG. 14.10 Material characteristics of roselle fiber.

3.2.3 Define design

The third step is the design modular to be inserted in the geometry. The design drawing for roselle fiber of 2–5 mm length was inserted into the geometry set. Fig. 14.11 shows the design sample for roselle fiber and PLA that has already been assigned in the geometry set.

Next, to insert the geometry, we need to double-click the model to perform a sample design model for the meshing process to take place. Fig. 14.12 shows the sample design model performed in steady-state thermal ANSYS Workbench 16.0.

Then assign the material in the outline project model based on the design modular and the material that have

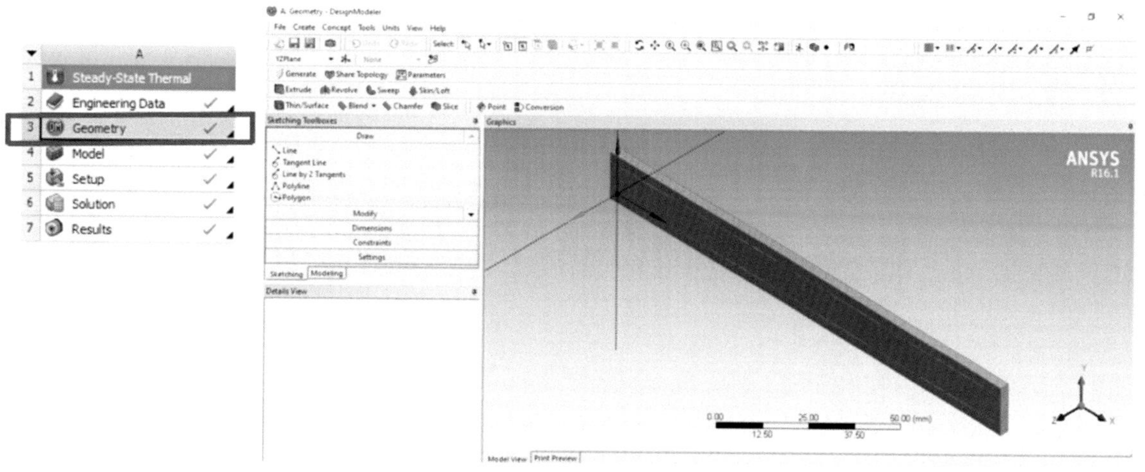

FIG. 14.11 Geometry setting.

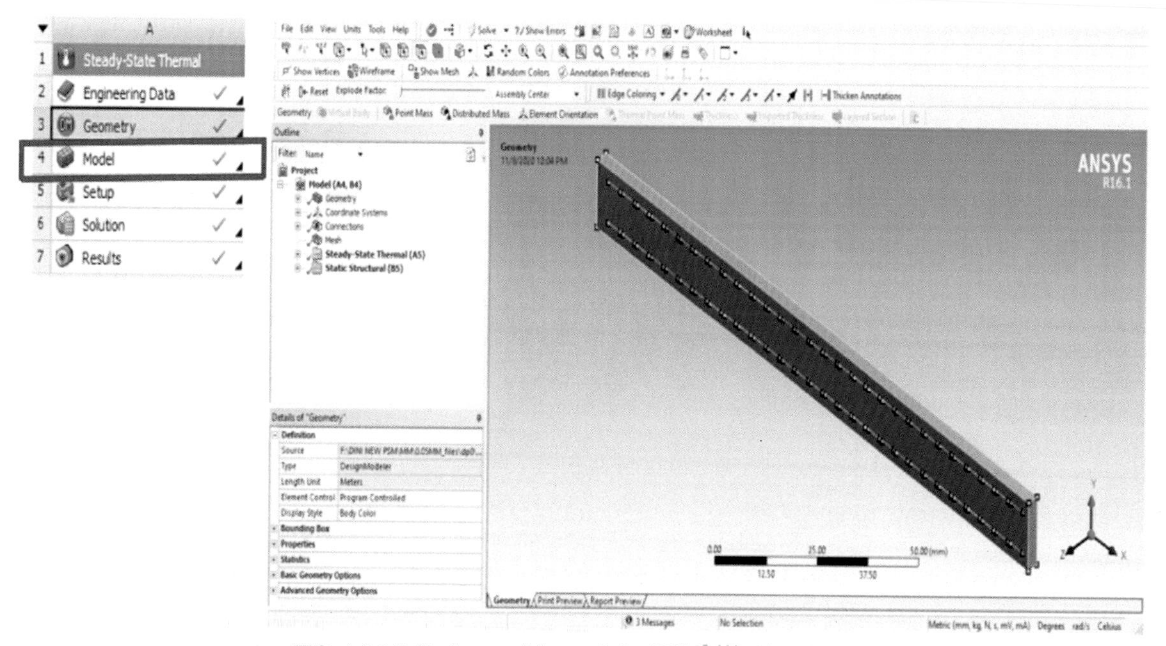

FIG. 14.12 Design model sample in ANSYS Workbench 16.0.

already been set in the engineering data. This is explained in more detail in Fig. 14.13.

3.2.4 Define meshing

Meshing is the importance part in the analysis. In order to define meshing, choose the generate mesh to proceed the mesh process. Fig. 14.14 shows the correct method to define the meshing process.

3.2.5 Apply temperature

The boundary condition for temperature has been applied by clicking Steady-State Thermal (AS) and then choosing the temperature and convection properties at the top to apply at the model surface. Fig. 14.15 shows the process for setting the temperature and convection.

Next, the first temperature was applied on the top of the sample model surface, which is PLA surface with a value of 150°C, as shown in Fig. 14.16.

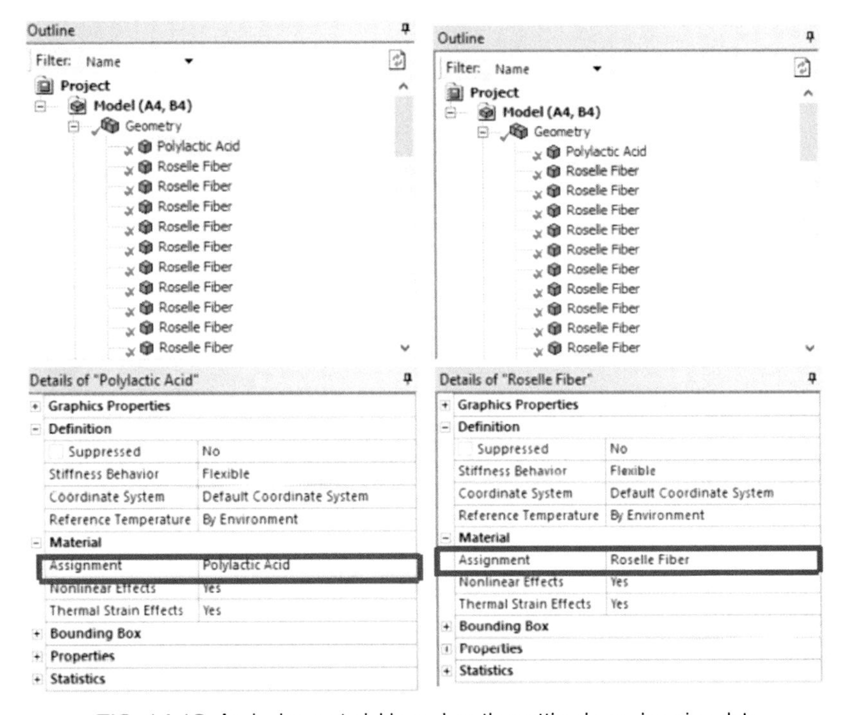

FIG. 14.13 Assigning material based on the setting in engineering data.

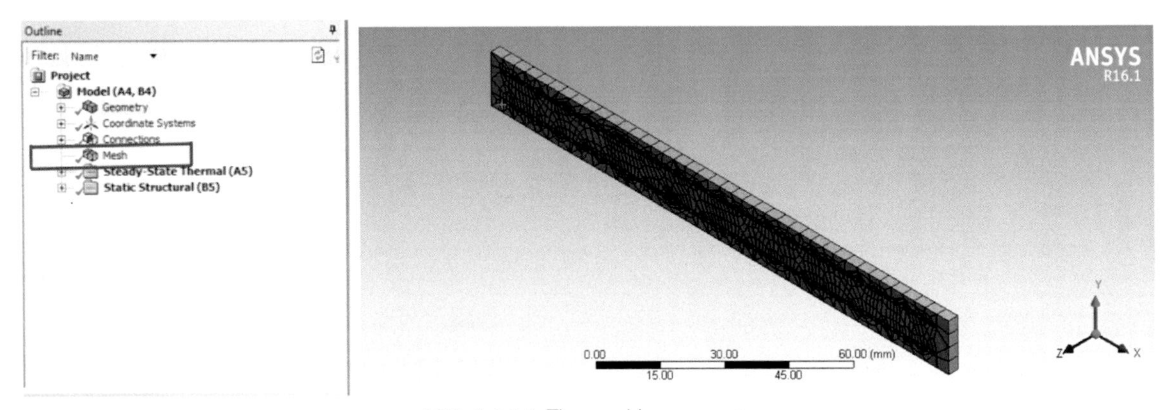

FIG. 14.14 The meshing process.

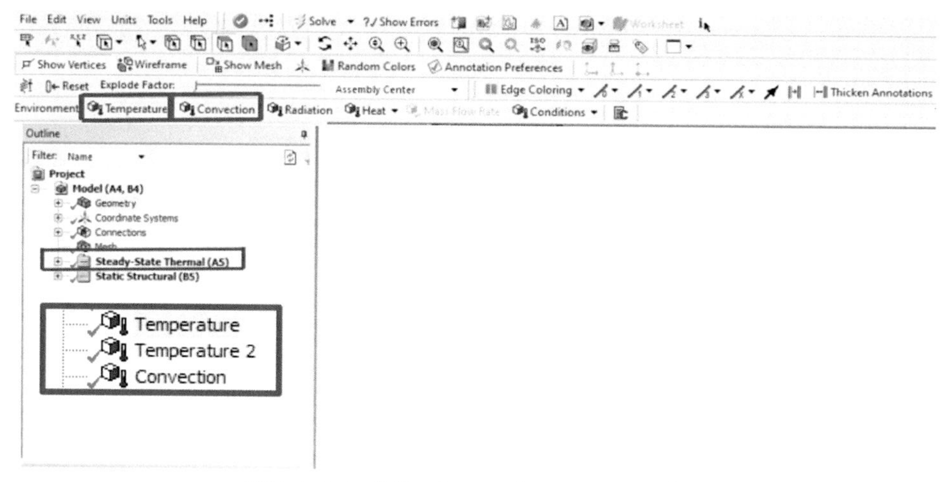

FIG. 14.15 Setting temperature and convection.

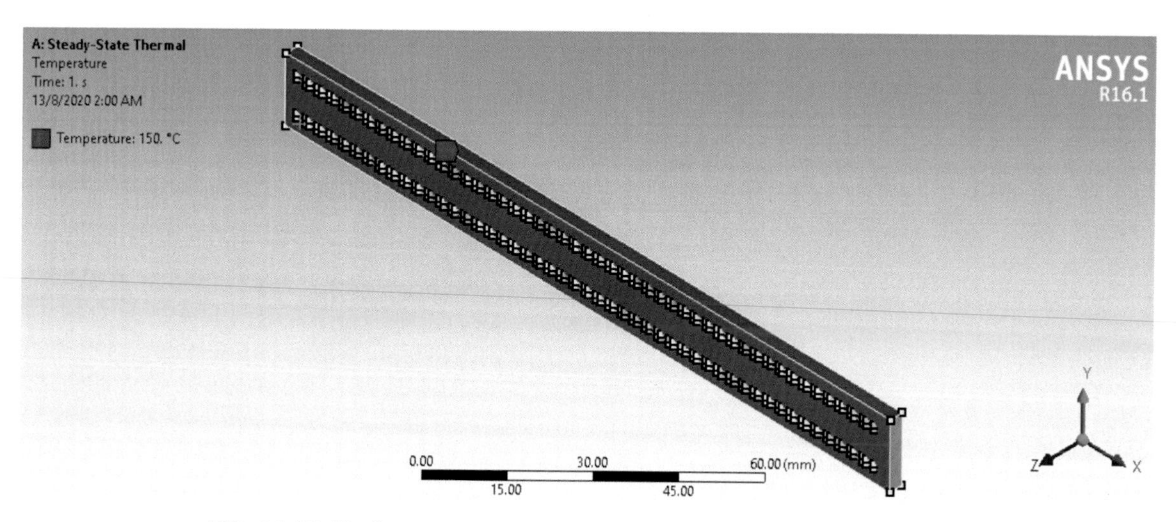

FIG. 14.16 The first temperature applied on the top surface of the sample model.

The second temperature was applied on the bottom of the sample model surface, which is PLA surface with a value of 75°C, as shown in Fig. 14.17.

In the next step, convection temperature was applied at the front and back sides of the sample model surface with a value of 150°C and film coefficient of 0.13 $W^2/mm°C$ for PLA material. Figs. 14.18 and 14.19 show the front side and back side of the sample model, respectively.

After setting up all the temperature and convection values in the sample model surface, the thermal properties were applied at Solution (A6) in the project outline. Right click the thermal properties and choose to evaluate all the results to get the thermal degradation data. Fig. 14.20 shows the detailed process to setup the thermal properties and evaluate the thermal degradation result.

3.2.6 Apply static structural

Next, return to the project schematic and choose the static structural to be linked with the steady-state thermal in ANSYS Workbench 16.0 design schematics to do some analysis after applying some force on the sample model design. Then double-click the setup features to open the model sample and define the meshing for the sample model. Fig. 14.21 shows the process in

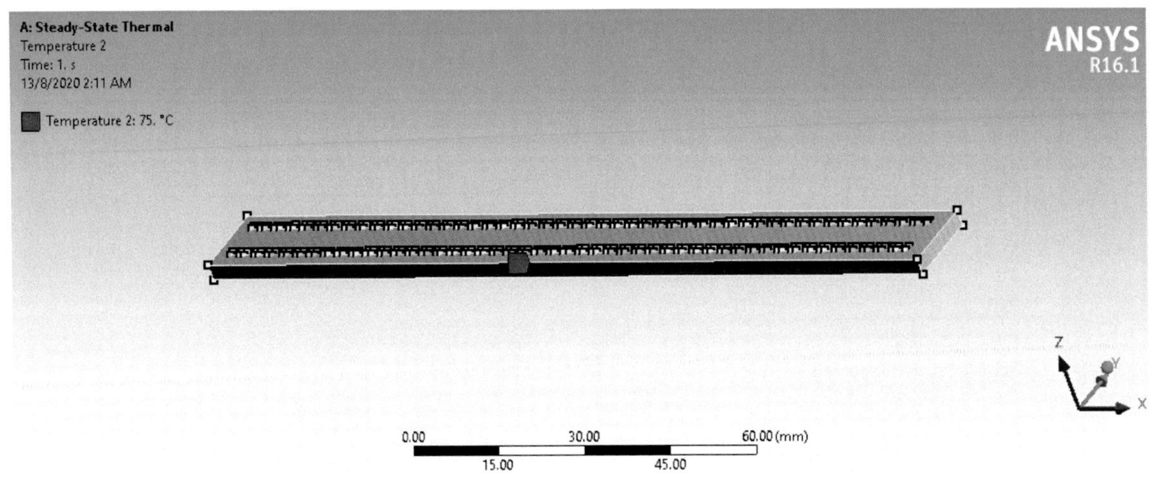

FIG. 14.17 The second temperature applied on the bottom surface of the sample model.

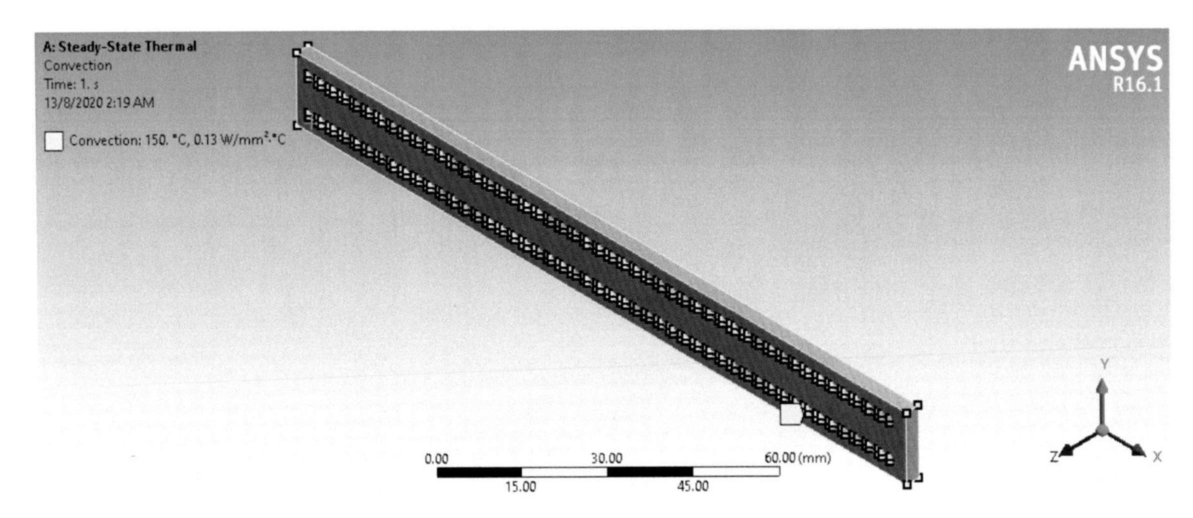

FIG. 14.18 Convection applied at the front side of the sample model.

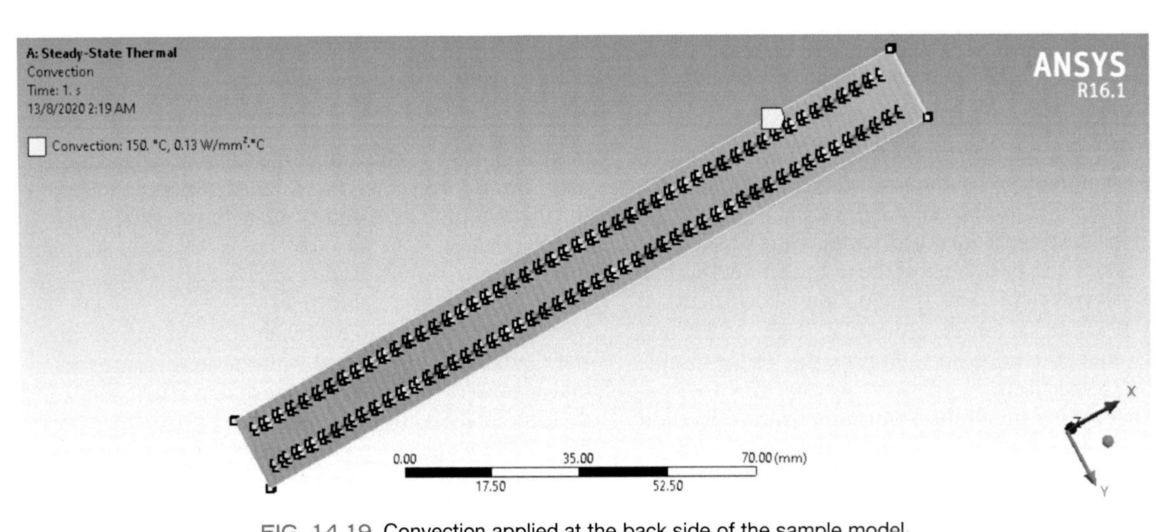

FIG. 14.19 Convection applied at the back side of the sample model.

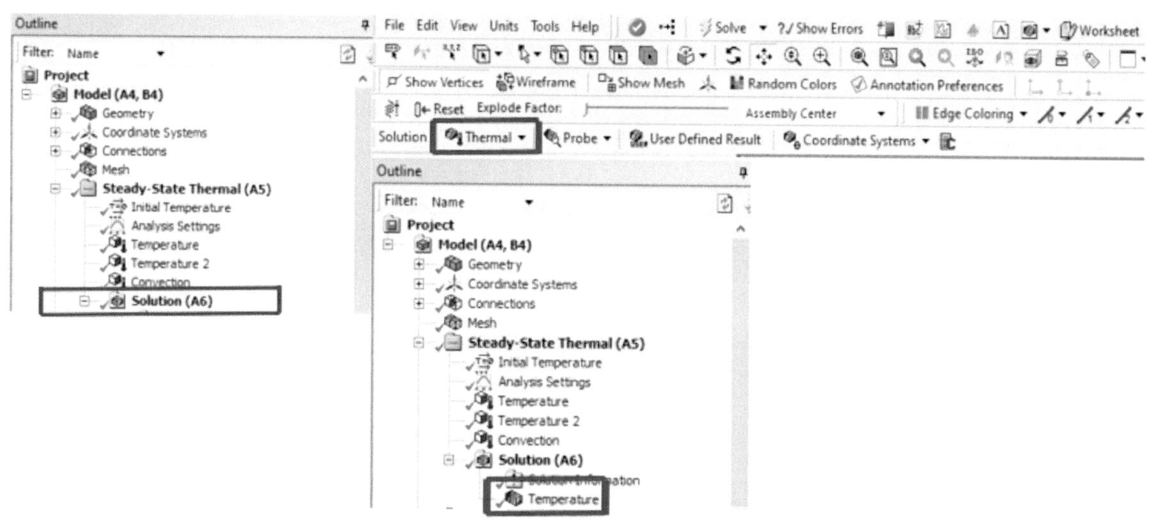

FIG. 14.20 Setting up the thermal properties and evaluating the result for thermal degradation.

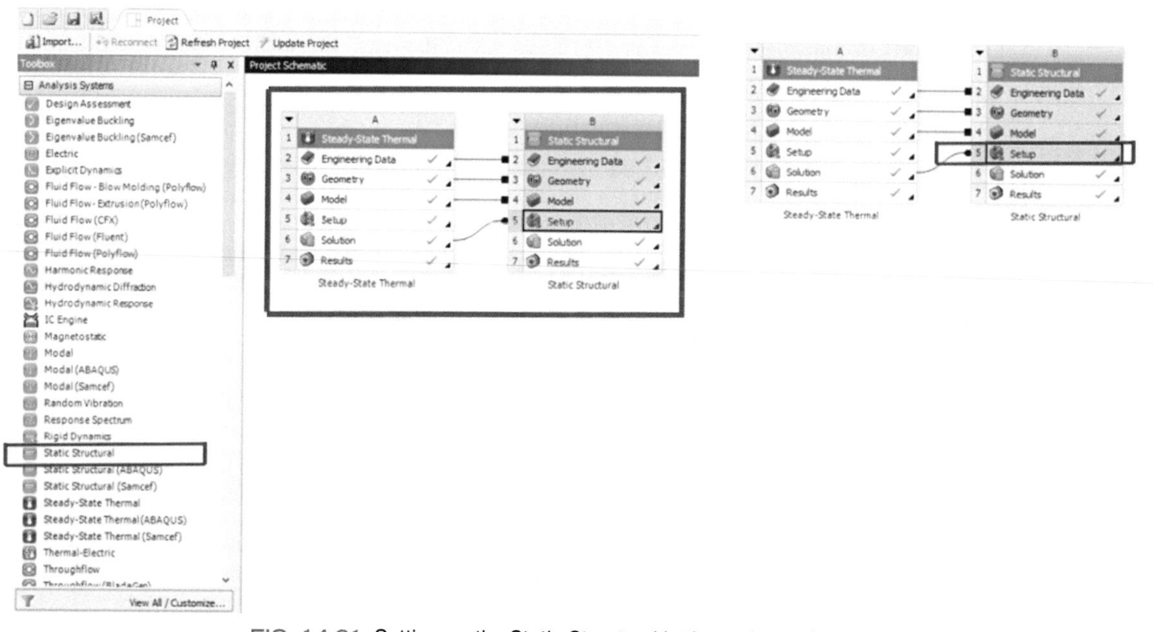

FIG. 14.21 Setting up the Static Structural in the project schematic.

setting up the static structural analysis in the project schematic. We use the global coordinate system for this static structural.

3.2.7 Applied force and support condition

After defining meshing as shown in Fig. 14.14, two forces were applied at the right side and left side of the design model. The amount of the force applied to the design model is 10,000 N and the boundary condition for the static structural was applied at the bottom surface of the roselle fiber-reinforced PLA, with the elongation or displacement for the sample model to be stretched to 1600 mm. Fig. 14.22 shows the correct way to set up the force and displacement in the sample design model.

After completing setting up the location of applied load, force, and displacement in the sample design

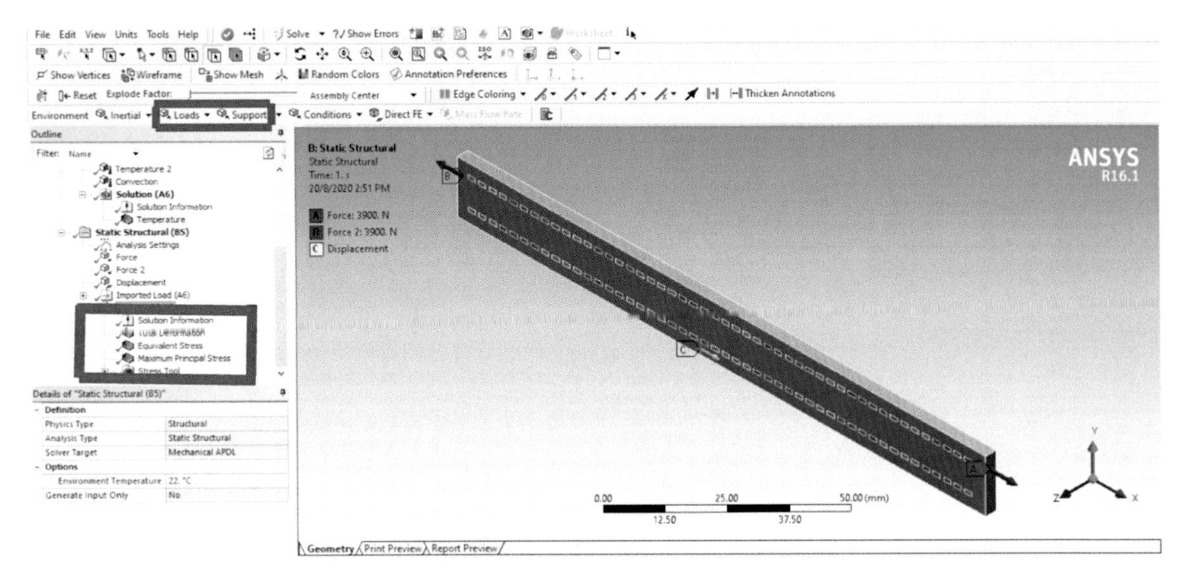

FIG. 14.22 The force and displacement were applied.

model click the Solution (B6) to apply the result for total deformation, equivalent stress, and maximum principal stress. All the results can be select at the deformation and stress properties in the top toolbar. Then evaluate the results of the properties. Fig. 14.23 shows the process to set up the total deformation, maximum stress, and equivalent stress (tensile strength) of roselle fiber-reinforced PLA composites.

4 RESULTS AND DISCUSSION

4.1 Meshing

Meshing is a crucial part in finite element analysis. The model is recommended to be less complicated design and should use the automated meshing process to make it easier to observe the model structure without detecting prestress in the object model. The automated meshing process is also used to observe the stress distributed over the model of the surface object. The meshing process, whether it is conformal or nonconformal meshing process, has the ability to control the interface type for solid model of roselle fiber reinforced with PLA. Meshing process also has been used to deal with large mesh sizes and for meshing different types of interface with different materials or physical properties (ANSYS Inc. 2007).

Fig. 14.24 shows that different lengths of roselle fiber will give different geometric shapes after the meshing process is done on the surface of PLA reinforced with roselle fiber model design. There are four sample designs that have an automated mesh in 1D element. The number of nodes and elements depends on the shape's function because in finite element analysis the model will be divided into small parts and all the parts will be known as the finite element for 1D element model. The element will be connecting all the nodes to become one object or solid body, and this will produce a certain shape in the meshing process and the number of nodes is usually more than the number of elements.

Fig. 14.26 shows the 2–5 mm length of roselle fiber-reinforced PLA. The fiber length of 2 mm gives 48,986 number of nodes and the number of elements that contribute to the meshing process is 8059. Other details of results can be found in Table 14.2, which shows the detailed number of nodes and elements for the sample testing of 2- to 5-mm-length roselle fiber reinforced with PLA.

The results show that the 2-mm roselle fiber reinforced with PLA gives the highest number of nodes (with the value of 48,986) and a high number of elements (with the value of 8059). This might be due to the 2-mm fiber length will produce more small parts and thus more nodes and elements will be required to connect the small parts to become one shape function or one solid body, when compared with other roselle fiber lengths.

4.2 Thermal Deformation

Table 14.3 shows the results for temperature deformation of four different fiber lengths with three different forces applied at the left side and right side of the roselle

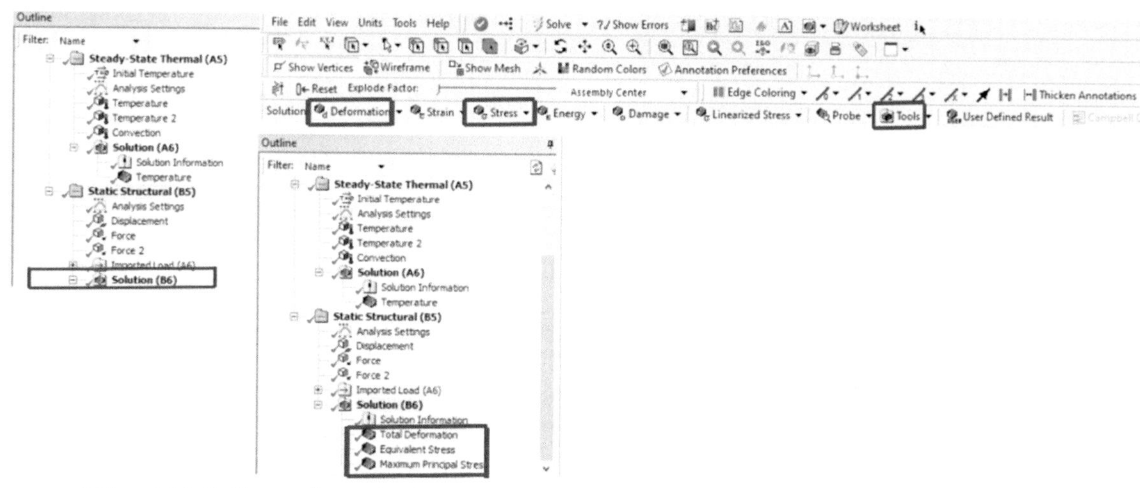

FIG. 14.23 Setting up the total deformation, maximum stress, and equivalent stress for the sample design model.

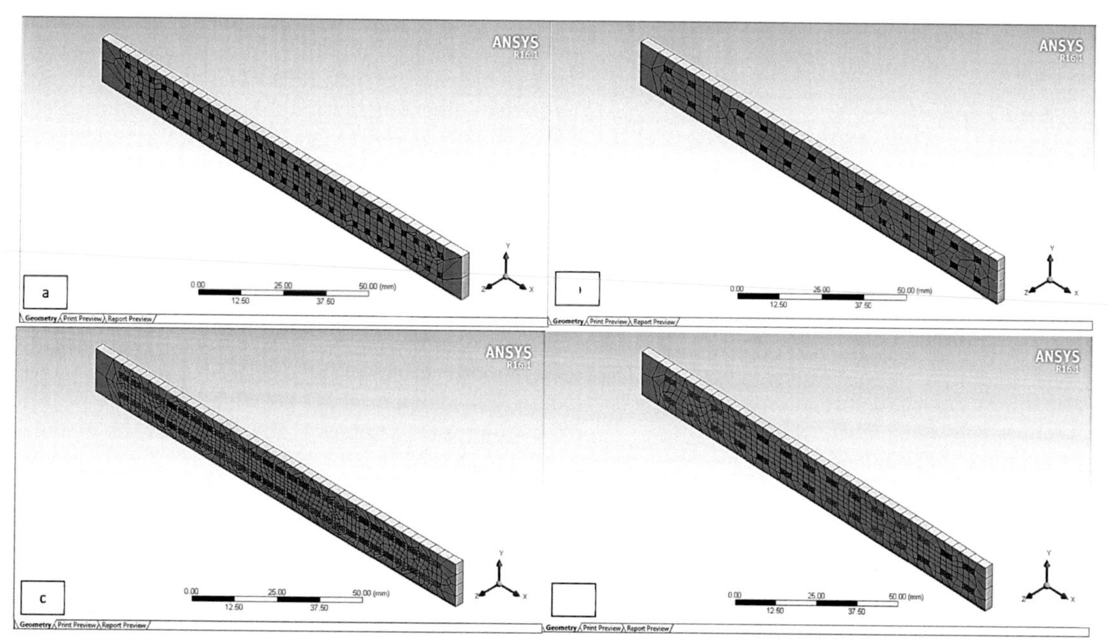

FIG. 14.24 The meshing for **(A)** 2, **(B)** 3, **(C)** 4, and **(D)** 5 mm lengths of roselle fiber reinforced with polylactic acid.

fiber-reinforced PLA bar design model. Fig. 14.25 illustrates the results of 10,000 N of force applied to four different lengths of roselle fiber-reinforced PLA composite bar, based on thermal deformation.

The obtained data shows the maximum thermal deformation for the roselle fiber with length 2 mm occurs at a higher temperature of 218.4°C compared with other roselle fiber lengths in this sample. This might be because the shortest fibers restrict the thermal deformation of composites, as their contact surface area is less than that of long fibers. The thermal distribution is limited in case of short fibers. The natural fiber act as a medium that can stored temperature as it is only can be fully degrade in high temperature (Ibrahim et al.,

TABLE 14.2
The Number of Nodes and Elements for 2- to 5-mm-Length Roselle Fibers Reinforced With Polylactic Acid.

	LENGTH			
Statistic	**2**	**3**	**4**	**5**
Nodes	48,986	36,601	36,587	18,130
Element	8059	5541	5472	2678

TABLE 14.3
The Thermal Deformation Result for the Different Lengths of Roselle Fiber and Different Load Applied on the Surface of the Composite Sample.

TEMPERATURE DEFORMATION (°C)		
	FORCE	
Roselle Fiber Sample Length (mm)	F = 10,000 N	
	Min	**Max**
2	75	218.4
3	62.341	185.51
4	75	180.22
5	49.54	150

2012). The lowest thermal deformation temperature is for the 5-mm-length roselle fiber, which is 150°C.

Fig. 14.25 shows the simulation model in ANSYS for thermal analysis. Thermal deformation of composites occurs under conditions of 10,000 N force applied. It can be seen from the figure that the shape of the samples becomes longer according to the temperature. For a quick example, at 150°C, the sample composites with 5 mm length of fiber have the most thermal deformation. As the temperature increases, the shape of the sample will change horizontally. The analysis of thermal deformation indicated that the samples show a change in color with different lengths of fiber.

The result was in good agreement with the previous researcher who states the difference in thermal expansion at different lengths occurs at high temperatures. According to the literature, composites with high fiber lengths have a good mechanical constraint to the deformation of polymer matrix caused by thermal stress distribution, which can lead to decrease in composite failure (Hao et al., 2020). Thermal deformation can lead to thermal expansion, which can change or alter the sample in terms of size, shape, area, or volume content, or the appearance of the sample result. In this case, when a temperature of 150°C is applied to the top and 75°C to the bottom of the surface, the PLA with a thermal expansion coefficient of 0.13 $W^2/mm°C$ can increase the volume of the material and the material will start to expand proportionally with the length of the roselle fiber-reinforced PLA.

4.3 Total Deformation

Table 14.4 and Fig. 14.26 show the result for the total deformation of four different lengths with force applied at the left side and right side of the roselle fiber-reinforced PLA bar design model and the modeling analysis, respectively.

The obtained data shows that maximum total deformation occurs in the fiber length of 2 mm, which is 1,213,403 mm, followed by the fiber length of 3 mm, which is 119,480 mm. The 5-mm-length roselle fiber gives a better result because it has lower total deformation than other lengths and the tendency of the roselle fiber-reinforced PLA to fail is less. It can be concluded that the increase in length affects the total deformation of roselle fiber-reinforced PLA composites. The fiber limits the flexibility of the composites and improves the rigidity. The result, as supported by previous research, shows that when the roselle fiber length increases, the rigidity will increase because of the low friction between the matrix and the fibers, which leads to the smooth surface of the roselle fiber to be expanded more (Ghadakpour et al., 2020). The displacement of the composites samples is the ability of the material to be elongated or expanded when force was applied, without the material being fractured.

Total deformation refers to the displacement of the composites due to the stresses applied on the surface area of the roselle fiber. When the force was applied at the end of the roselle fiber-reinforced PLA composites, they will be elongated or expanded to a certain size.

4.4 Equivalent Stress (von-Mises)

Table 14.5 and Fig. 14.27 show the results for the equivalent stress (von-Mises) of four different fiber lengths with force applied at the left side and right side of the roselle fiber-reinforced PLA bar design model.

Table 14.6 shows the results of maximum equivalent stress (von-Mises) for the four different lengths of roselle fiber-reinforced PLA. The result shows the 2-mm-length roselle fiber-reinforced PLA composites have

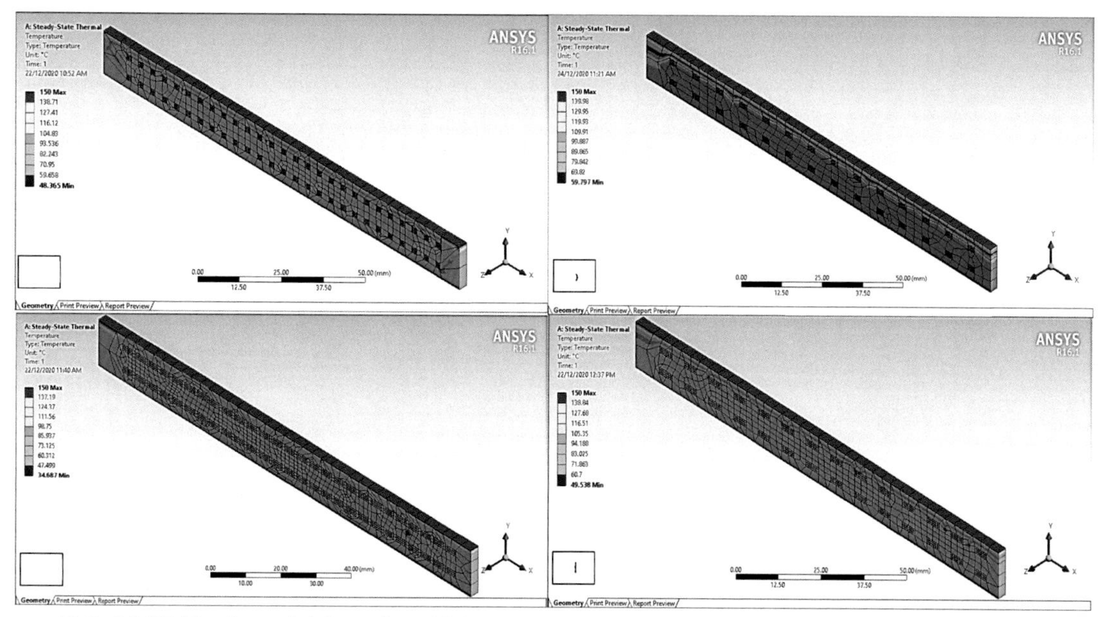

FIG. 14.25 The thermal deformation of **(A)** 2, **(B)** 3, **(C)** 4, and **(D)** 5 mm lengths of roselle fiber-reinforced polylactic acid.

TABLE 14.4
The Total Deformation Result for the Different Lengths of Roselle Fiber at 10,000 N on the Surface of the Composite Sample.

TOTAL DEFORMATION (MM)		
Roselle Fiber Sample Length (mm)	F = 10,000 N	
	Min	Max
2	3999.5	121,340
3	7532.5	119,480
4	9469.9	116,000
5	8009.8	118,300

the highest equivalent stress (von-Mises) at 1284 MPa and the 5-mm-length roselle fiber-reinforced PLA composites have the lowest equivalent stress (von-Mises) at 89.321 MPa. The equivalent stress (von-Mises) is also known as the tensile strength. Thus this result shows roselle fibers with a length of 5 mm have a good indication of tensile strength compare with the other fiber lengths. This might be due to the strong bonding or adhesion between the roselle fiber and PLA after force is applied at the end of the sample. In addition, short fibers have more surface interaction or contact area, which can limit the transformation of stress between the fiber and polymers. This can be supported by the previous researchers who stated that the PLA matrix also plays a major role in increasing the tensile strength by preventing the transmission of cracks from fiber to fiber, which could otherwise result in a catastrophic fracture (Kumar & Prakash, 2020). In addition, materials with high tensile strength have more ability to resist force.

Based on the result of data analysis, when the length of the roselle fiber increases, the equivalent stress (von-Mises) will decrease. This phenomenon is the common definition of the distortion energy that is used to evaluate the failure of the brittle material for polymer matrix material thermoplastics based on the equivalent stress (von-Mises). Thus the higher tensile strength, the lower the tendency of the structure of roselle fiber-reinforced PLA to fail. This can be supported by previous researchers who state that when the equivalent stress (von-Mises) is equal to or greater than the uniaxial failure force the material will be ductile for the polymer matrix composite material (Tu et al., 2019). The increase in tensile strength will give some advantages to the roselle fiber-reinforced PLA composites, including good adhesion bonding between the fiber

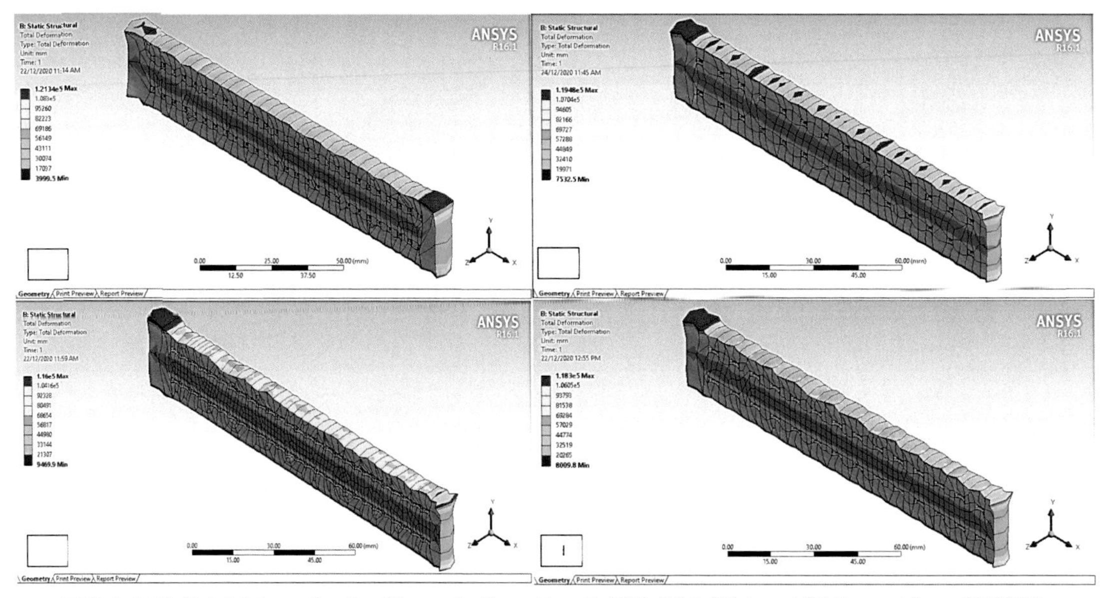

FIG. 14.26 Total deformation (mm) in roselle fiber of length **(A)** 2, **(B)** 3, **(C)** 4, and **(D)** 5 mm at force 10,000 N.

TABLE 14.5
The Equivalent Stress (von-Mises) Results for the Different Lengths of Roselle Fiber and Different Load Applied on the Surface of the Composite Sample.

EQUIVALENT STRESS (MPA)		
	FORCE	
	F = 10,000 N	
Roselle Fiber Sample Length (mm)	**Min**	**Max**
2	3.8366	128.420
3	0.5545	89.321
4	1.9218	87.053
5	2.1621	85.059

and matrix, less microcracking, less fiber pullout, and less fiber breakage at the end in composite material. The literature states that the maximum tensile strength of the composite is due to the strong adhesion between fibers and matrix, which enables a uniform transition of stress from the matrix to fiber (Ramasubbu & Madasamy, 2020).

4.5 Maximum Principal Stress

Table 14.6 and Fig. 14.28 show the results of maximum principal stress (MPa) of different lengths (2–5 mm), with force applied at the left and right sides of the roselle fiber-reinforced PLA bar design model.

The roselle fiber-reinforced PLA composites of 2 mm length show the highest maximum stress of 140 MPa due to the applied force and the composites of 5 mm length show the lowest maximum principal stress. The maximum principal stress is known as the highest stress that can be accommodated by the particles of things or materials when force was applied at the surface area of the sample test. Thus this data shows that the roselle fiber length of 2 mm gives a good result.

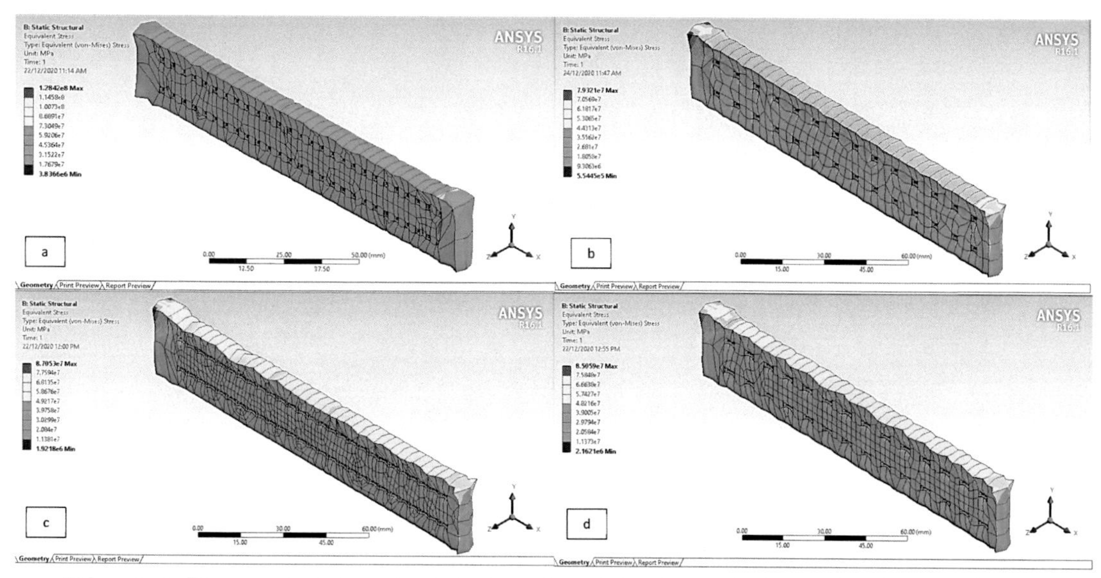

FIG. 14.27 The equivalent stress (von-Mises) of **(A)** 2, **(B)** 3, **(C)** 4, and **(D)** 5 mm lengths of roselle fiber-reinforced polylactic acid.

TABLE 14.6
The Maximum Principal Stress Results for the Different Lengths of Roselle Fiber and Different Load Applied on the Surface of the Composite Sample.

MAXIMUM PRINCIPAL STRESS (MPA)		
	FORCE	
	F = 10,000 N	
Roselle Fiber Sample Length (mm)	Min	Max
2	−31.0310	140.520
3	−9.5178	82.534
4	−32.9570	77.244
5	−19.773	76.173

According to the data analysis, the increase in roselle fiber length will decrease the stress. This might be because the amount of fiber is decreased when the fiber length is increased and the material will experience brittle phase when elongation and crack occur. This can be supported by a previous researcher who states that based on the maximum principal stress theory, a brittle material breaks down when the overall main stress in the specimen exceeds some limiting value for the sample test (Muhammad Rifai, 2006). From the stress-strain graph, the strength and the elasticity of the roselle fiber-reinforced PLA composite can be determined after reaching the maximum stress, which is when the tensile strength will decrease because fracture occurs after cracking.

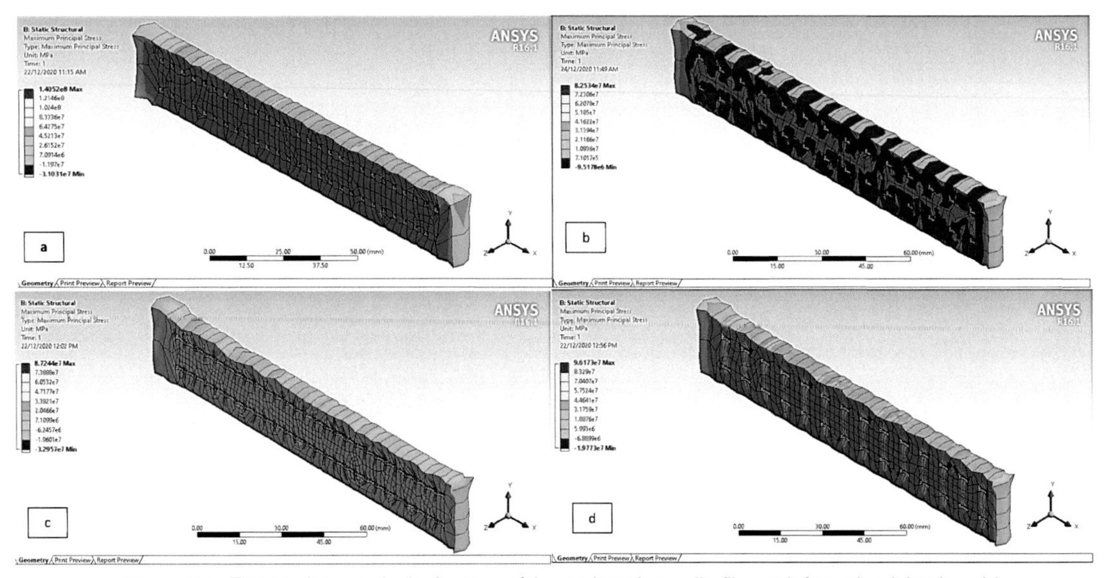

FIG. 14.28 The maximum principal stress of 2-mm-length roselle fiber-reinforced polylactic acid.

5 CONCLUSIONS AND RECOMMENDATIONS

The mechanical properties of roselle fiber-reinforced PLA were analyzed by using ANSYS Workbench 16.0. The sample designed in ANSYS Workbench 16.0 design modular with the dimension of 150 mm ×× 15 mm × 3 mm for PLA in 3D. The effect of fiber length has been analyzed by using the ANSYS Workbench simulation. The number of nodes and elements was determined through the automated meshing process.

The results of analysis from the meshing process in the steady-state thermal and static structural show that the roselle fiber with length 2 mm gives the optimum results in terms of thermal deformation and tensile strength compared with the fibers of other lengths. From the results, it can be seen that the 2-mm-length fiber gives the highest maximum principal stress and equivalent stress.

It is recommended to perform further research on the improvement in results analysis. By performing the experiment together with the analysis the results is more accurate. The other analysis parameters should also be included, such as diameter, shape, and density of the fiber, in order to obtain accurate results. Furthermore, the study of fiber orientation for the roselle fiber-reinforced PLA composites should be considered to evaluate the effect on the composite properties.

REFERENCES

ANSYS Inc. (2007). *ANSYS meshing* (Vol. 4).

Athijayamani, A., Thiruchitrambalam, M., Natarajan, U., & Pazhanivel, B. (2009). Effect of moisture absorption on the mechanical properties of randomly oriented natural fibers/polyester hybrid composite. *Materials Science and Engineering: A, 517*(1–2), 344–353.

Balasubramanian, K., Rajeswari, N., & Vaidheeswaran, K. (2020). Analysis of mechanical properties of natural fibre composites by experimental with FEA. xxxx *Materials Today: Proceedings,* 2–6. https://doi.org/10.1016/j.matpr.2020.01.098

Chandramohan, D., & Marimuthu, K. (2011). A review on natural fibers. *International Journal of Research and Reviews in Applied Sciences, 8*(2), 194–206.

Das Gupta, P. C. (1959). The hemicelluloses of roselle fiber (*Hibiscus sabdariffa*). *Textile Research Journal, 30*(3), 237.

Ghadakpour, M., Janalizadeh, A., & Soleimani, S. (2020). Transportation geotechnics investigation of the kenaf fiber hybrid length on the properties of the cement-treated sandy soil. *Transportation Geotechnics, 22*(August 2019), 100301. https://doi.org/10.1016/j.trgeo.2019.100301

Grace, F. (2008). *Investigation the ssuitability of* Hibiscus sabdariffa *calyx extract as colouring agent for paediatric syrups*. Kwame Nkrumah University of Science and Technology.

Hao, X., Zhou, H., Mu, B., Chen, L., Guo, Q., Yi, X., Sun, L., Wang, Q., & Ou, R. (2020). Effects of fiber geometry and orientation distribution on the anisotropy of mechanical properties, creep behavior, and thermal expansion of natural fiber/HDPE composites. *Composites Part B: Engineering, 185*, 107778. https://doi.org/10.1016/j.compositesb.2020.107778

Ibrahim, M. S., Sapuan, S. M., & Faieza, A. A. (2012). Mechanical and thermal properties of composites from unsaturated polyester filled with oil palm ash. *Journal of Mechanical Engineering and Sciences, 2*(June), 133–147.

Julian, C. C. (1949). Roselle-A potentially important plant fiber. *Economic Botany, 3*(1), 89–103.

Junkasem, J., Menges, J., & Supaphol, P. (2006). Mechanical properties of injection-molded isotactic polypropylene/roselle fiber composites. *Journal of Applied Polymer Science, 101*, 3291–3300.

Kumar, A. A. J., & Prakash, M. (2020). Mechanical and morphological characterization of basalt/cissus quadrangularis hybrid fiber reinforced polylactic acid composites. *Journal of Mechanical Engineering Science, 0*(0), 1–13. https://doi.org/10.1177/0954406220911072

Mahadevan, N., & Kamboj, P. (2009). *Hibiscus sabdariffa* Linn. – an overview. *Natural Product Radiance, 8*(1), 77–83.

Managooli, V. A. (2009). *Dyeing mesta (*Hibiscus sabdariffa*) fibre with natural colourant.*

Mohamad, O., Nazir, B. M., Rahman, M. A., & Herman, S. (2002). Roselle : A new crop in Malaysia. *Buletin Persatuan Genetik Malaysia, 37*, 12–13.

Muhammad, R. (2006). *Failure theories*. Academia.Edu. https://www.academia.edu/29913856/Failure_Theories_Teori_Kegagalan_.

Mungole, A., & Chaturvedi, A. (2011). *Hibiscus sabdariffa* L a rich source of secondary metabolisme. *International Journal of Pharmaceutical Sciences Review and Research, 6*(1), 83–87.

Nadlene, R., Sapuan, S. M., Jawaid, M., & Ishak, M. R. (2015). Material characterization of roselle fibre (*Hibiscus sabdariffa* L.) as potential reinforcement material for polymer composites. *Fibres and Textiles in Eastern Europe, 6*(114), 23–30. https://doi.org/10.5604/12303666.1167413

Ramasubbu, R., & Madasamy, S. (2020). Fabrication of automobile component using hybrid natural fiber reinforced polymer composite fabrication of automobile component using hybrid natural fiber. *Journal of Natural Fibers, 00*(00), 1–11. https://doi.org/10.1080/15440478.2020.1761927

Selim, K. A., Khalil, K. E., Abdel-Bary, M. S., & Abdel-Azeim, N. A. (1993). Extraction, encapsulation and utilization of red pigments from roselle (*Hibiscus sabdariffa* L.) as natural food colourants.

Tori Husdcon, N. D. (2011). A research review on the use of *Hibiscus sabdariffa. Better Medicine - National Network of Holistic Practitioner Communities*, 1–6. https://www.todaysdietitian.com/whitepapers/Hibiscus_Sabdariffa.pdf

Tu, S., Ren, X., He, J., & Zhang, Z. (2019). Stress – strain curves of metallic materials and post - necking strain hardening characterization : A review. *Fatigue and Fracture of Engineering Materials and Structures*, 1–7. https://doi.org/10.1111/ffe.13134

Wester, P. J. (1907). Roselle : Its culture and uses. *U.S. Department of Agriculture, 312*(October), 1–16.

Wilson, W. (2009). *Discover the many uses of the roselle plant.* NParks, 2009 http://mygreenspace.nparks.gov.sg/discover-the-many-uses-of-the-roselle-plant/.

FURTHER READING

Roselle. (2013). *Encyclopaedia Britannica*.

CHAPTER 15

The Influence of Fiber Size Toward Mechanical and Thermal Properties of Roselle Fiber-Reinforced Polylactide (PLA) Composites by Using Ansys Software

R.N.D. AQILAH • R. NADLENE • R.A. ILYAS • A.M. RADZI • N.B. RAZALI

1 INTRODUCTION

Natural fibers have become the main attraction for researchers. Natural fibers are used as an alternative resource to petroleum-based fibers due to environmental issues and depleting fossil fuels. Natural fibers have been able to benefit both the economy and environment. However, the selection of fiber depends on the fiber type and application (Tholibon et al., 2019). Natural fibers are suggested to replace synthetic fibers due to numerous benefits like biodegradability, renewability, low cost, and lightweight (El-Shekeil et al., 2012; Frone et al., 2013; Sanjay et al., 2019; El-Shekeil et al., 2014; Kargarzadeh et al., 2018). However, inconsistency is the main problem of natural fibers. In this area, the polymers can be affected by the hydrophobicity versus hydrophilicity of natural fibers (Balla et al., 2019; El-Shekeil et al., 2014; Venkateshwaran et al., 2013).

Besides that, the hydrophobic property of the polymer creates a problem in adhesion and wettability. To overcome these problems, compatibilizers are used as the interface. The common process usually used for this purpose for natural fibers is alkali treatment. In this process, fibers are immersed in an alkali solution for a certain time. Alkali treatment can eliminate waxes and impurities from the superficial of the fiber and create a rough surface for good compatibility with the matrix as biocomposites (Azammi et al., 2019; Ilyas et al., 2017; Sanjay et al., 2017). In this research, roselle fiber was chosen as reinforcement material because of its outstanding properties and environmental considerations based on previous research.

Roselle fiber has potential as an unconventional medium to substitute synthetic fibers as reinforcement in a composite. In subsidizing to a better environment, roselle fiber-reinforced composite has been used in the industrial sector and rural areas (Raman Bharath et al., 2015). In the composite industry, fiber size affects the mechanical and thermal performance of a material. The finer the fiber size, the higher the tensile strength of the material (Pickering et al., 2016).

1.1 Fibers

Fibers can be divided into two types which are natural and synthetic fibers. Synthetic fibers are normally more costly than natural fibers and are usually used in composite material manufacturing (Ilyas and Sapuan, 2020a, 2020b). Fibers are frequently used in applications such as sports goods, electrical insulators, boat hulls, and aerospace. Recently, engineering industries have replaced synthetic fibers with natural fibers, which are acquired from nature (Pandey, 2015, p. 64).

1.2 Natural Fibers

Natural fibers are renewable and ecological materials that are made up of natural fibers surrounded in a polymer matrix that could remain as an organic source like PLA. Natural fiber with thermoplastics is not easily decomposed but can simply be reused instead of thermosets. A natural fiber that has been used in a polymer

[illegible]

matrix has improved specific strength, degradability, and no itching complication as compared to synthetic fibers.

In general, natural fiber has been widely commercialized as a reinforcement material in product composites. They are also not harmful to the environment and have many advantageous properties. The use of natural fiber-reinforced composites as a replacement material in engineering such as structure, paperboard, and construction is highly recommended because of its superior properties and low cost has gained attention from industries. Ecological benefits of natural fibers instead of synthetic fibers have been demanded in the manufacturing and processing of reinforced composites. Researchers are now actively doing research on natural fiber in order to minimize the usage of artificial fibers in the future.

Natural fiber composites have obtained attention regarding their acceptability and uses. In structural applications, the mechanical properties of a material are the most significant for assessing the performance of the material under pressure (Drahansky et al., 2016). By using natural fiber to substitute the man-made material to strengthen cement wall sections, it will be able to reduce the effect on the environment concerning structural and insulation purposes. Natural fibers have been employed in structural applications such as in the automotive industry, aerospace industry, and civil engineering industry.

The research development in natural fiber was done on modified fiber, fiber behavior, and interfacial bonding in composites products as well as composite processing. Natural fibers have lower strength and stiffness compared to glass fibers. However, natural fibers have superior mechanical properties in terms of tensile and flexural strength. Mechanical interlocking happens when the surface of the fiber is rough and high in interfacial shear strength but has less impact on the transverse tensile strength (Pickering et al., 2016).

Eventhough natural fiber has more benefits, but it is not flawless. the main problem of natural fibers is the interfacial bonding between fibers n polymer. This is due to the different characteristics of these two materials; natural fiber is hydrophilic while the polymer is hydrophobic behavior. Therefore, the bonding between natural fibers and matrices is not completely strong. The reduction of mechanical properties is because natural fibers have high moisture absorption. Moreover, thermal degradation problems are correlated to natural fiber composites. Degradation happens when natural fibers go through overheating process.

1.2.1 Roselle Fiber

Roselle is cultivated primarily for the bast fiber obtained from its stems. Roselle fibers have attracted many researchers to explore their potential as reinforcement materials due to their similar properties to other established natural fibers such as jute (Wester, 1907). The cross section of roselle fiber is almost same like a bast fibre such as kenaf. On the most outer part, the bark provides protection from extreme temperature changes and excessive loss of moisture while hardening the stem. The fibers are hidden under the bark inside the phloem. The fibers keep the stem strong by supporting the conductive cells of the phloem. Xylem, also known as the woody core, is found in the inner part of the plant (Cellulose Fibers: Bio- and Nano-Polymer Composites, 2011).

1.3 Polymers

Polymers are usually made by merging macromolecules that are made up of atoms bonded together in a repeating form. The arrangement of a polymer can be simply seen as a chain. The chains are connected by links (Pandey, 2015, p. 64; Sari et al., 2020). Polymers can be either nonbiodegradable or biodegradable made of repeating monomer-like chains. In order to achieve a healthy and clean environment, engineering industries have developed environmental materials to substitute nonbiodegradable materials (Jayanth et al., 2018).

1.3.1 Roselle fiber-reinforced polymer

Currently, natural fiber composites have attracted interest from the automotive industry and general engineering applications due to their high mechanical properties and biodegradable characteristics. In addition, many researchers are now working and focusing their work on green materials (Chauhan & Chaudan, 2013). Roselle fiber is considered a natural fiber with good mechanical properties. Several papers have reviewed the mechanical properties of roselle natural fibers. Table 15.1 compares the properties of natural fibers with roselle fibers (Chandramohan & Marimuthu, 2011). Conducted several experiments to characterize roselle fiber-reinforced polymer composites (Ramu & Sakthivel, 2013). In this study, composites containing different ratios of roselle fibers with and without moisture were compared in terms of their mechanical properties such as tensile strength, flexural strength, impact strength, and Young's modulus. The aspect ratio of the fiber length was kept constant at 15 cm. The results indicated that the roselle fiber content affected the mechanical properties of the composites. An increase in fiber content resulted in an

TABLE 15.1
Properties of Natural Fibers.

Fibers Type	Density kg/m^3	Water Absorption %	Modulus of Elasticity E(GPa)	Tensile Strength (MPa)
Sisal	800–700	56	15	268
Roselle	800–750	40–50	17	170–350
Banana	950–750	60	23	180–430
Date palm	463	60–65	70	125–200
Coconut	145–380	130–180	19–26	120–200

increase flexural strength, elastic modulus, and tensile strength of the composites. However, for the impact test, the addition of roselle fiber to the polyester composites produced scattered results, where the highest impact strength was 10 wt.% of fibers for both types of samples. The presence of moisture for the impact test yielded better results. The authors concluded that the impact resistance of roselle fiber composites is highly affected by the interfacial bond strength.

Athijayamani et al. (2011) discussed the mechanical properties of several natural fiber composites. These authors studied short roselle and sisal hybrid fiber-reinforced polyester composites and analyzed the samples according to their weight ratio (wt%) while keeping the fiber length constant. Next, scanning electron microscopy (SEM) was used to investigate the fractured surfaces of the composites. It was observed that when the fiber content was increased, the tensile and flexural strengths also increased. However, in contrast to the impact test result, a negative hybrid effect was observed when fiber content was increased (Athijayamani et al., 2011). It was concluded that it is important to determine the optimum weight percentage of both fibers to obtain acceptable mechanical properties for this hybrid polyester composite.

Recently, several studies have been performed on roselle fiber composites for biomedical applications because they have high cellulose content, are lightweight and have high tensile strength (Thiruchitrambalam et al., 2010). Conducted a study on the development and evaluation of a novel roselle graft copolymer. An optimized reaction parameter of graft copolymerization of a methyl acrylate (MA) monomer onto roselle stem fibers was used as the principal monomer to determine the additive effect of a secondary monomer binary vinyl monomeric mixture on the percentage of grafting, properties, and behavior of the fiber. The graft copolymers were characterized and assessed for different physicochemical changes such as moisture absorption at various relative humidity levels, swelling behavior in various solvents, dye uptake, and chemical resistance against 1N NaOH and 1N HCl. It was observed that increasing the percentage of grafting and crystallinity reduced the crystallinity index and increased the physicochemico-thermal resistance, hydrophobicity, and affinity for organic solvents due to morphological transformations in these modified graft copolymers (Chauhan & Kaith, 2011). The authors further explored this topic by using the optimized parameter for the binary mixture with six different vinyl monomers to screen the changes in properties of the fiber. These graft copolymers were then characterized using advanced analytical techniques, and their mechanical stresses were assessed after being incorporated in phenol formaldehyde polymer matrix-based composites as reinforcements. The use of the graft copolymers in phenol formaldehyde as reinforcements was observed to increase the chemical resistance and mechanical strength of the materials (Chauhan & Kaith 2012a, 2012b).

1.3.2 Biodegradable plastic of natural fibers

In the past few years, various researchers have found that the use of naturally sourced materials is the solution for the waste issues due to their biodegradability. This solution has gained huge attention from industries and users. Biodegradable plastics have the potential for biodegradation only under certain circumstances which must be met to fully utilize the biodegradability of plastic materials (Rujnić-Sokele & Pilipović, 2017). Disposability or recyclability is important in processing composite materials as it brings environmental awareness (Drahansky et al., 2016).

PLA is a biodegradable composite that is usually used in food packaging, which is balanced in terms of mechanical properties, but degradation is necessary during composting. Additionally, the usage of biodegradable plastics is highly required in food packaging industries (Rujnić-Sokele & Pilipović, 2017; Ilyas et al. 2018, 2019). These plastics are disposed in the soil once they are completely dissolved into carbon dioxide and water by microorganisms. Besides that, biodegradable plastics are widely used in commercial goods like ballpoint pens, toothbrushes, garbage bags, fishing lines, tennis racket strings, and wrapping paper (Ochi, 2012).

1.3.3 PLA

PLA is usually used in composite manufacturing because it has a significant factor in the ecobiodegradable polymer industry and features one in all the top favorable candidates for future improvements (Castro-Aguirre et al., 2018; Murariu et al., 2012; Nazrin et al., 2020; Siakeng et al., 2019). PLA has good processability, outstanding electrical resistivity, and high dielectric breakdown intensity over room temperature (Hassan et al., 2019). PLA can be strengthened by natural fibers to produce fully biodegradable and ecological-friendly biocomposites (Dhannush et al., 2020).

According to the use of roselle fibers encourages faster degradation, which in effect enhances biocomposite biodegradation of PLA. Natural fiber-reinforced thermoplastic polymers enhance the mechanical properties of a material (Sanjay et al., 2017). PLA production is a common concept as it reflects the accomplishment of the vision of manufacturing cost-efficient, nonrenewable plastics. The tremendous advantage of PLA as a bioplastic is its flexibility and the fact that it degrades naturally when exposed to the environment.

1.4 Finite Element Analysis by Using ANSYS

Modeling analysis is a significant field of study as a hands-on use of finite element technique that is usually used to illustrate inconsistency to the test result by using ANSYS software. This method is normally used for numerous types of mechanical analysis. Even though it is practical, the difference between the expected result and the experimental result still happens. The use of FEA is to mend the geometry and mechanical properties of the composite structures tested in the finite element expectation (Sani et al., 2016).

ANSYS has become one of the popular design analytics resources available. The finite element method is one of the most complicated analysis solutions. Due to the complexity of the material properties, many engineering analytical solutions are insufficient for certain engineering issues because of the nature of the properties of materials, the boundary conditions, and the structure itself (Jeyasekaran et al., 2016). But, some types of elements and mesh refinements are changing and are commonly used for complex structure analysis (Barbero and Shahbazi 2017).

The fiber composite model can be generated numerically in a short period by using the software for the analysis of definite elements and several tests can be carried out electronically. Nevertheless, it is necessary to do a verification process on results obtained from analytical studies and simulations (Roslan et al., 2014).

Therefore, for this project, the effect of fiber size on the mechanical and thermal properties of roselle fiber-reinforced PLA composite was evaluated.

The improvement of superior engineering yields from natural resources is growing globally due to renewable and environmental problems. Among the numerous kinds of natural resources, roselle plants have been widely utilized over previous years (Raman Bharath et al., 2015). There are still restricted works on fully biodegradable composites also known as biocomposites (Ramli et al., 2017). The reasons for the increasing popularity of biocomposites or natural fiber composites (NFCs) are its abundancy, competitive quality, and environmental friendliness (Pandey, 2015, p. 64).

PLA is currently being studied and exploited as a polymer material which has received extensive devotion for typical consumption, like in packaging of materials, fabrication of fibers, and composites for a range of hands-on and mechanical uses. PLA is one of the eco biodegradable polymers and the characteristic shows it is the top favorable candidate for future improvements in green technology (Siakeng et al., 2019). PLA is categorized in terms of its biodegradable thermoplastic chemical compound nature, which is arranged from the monomers derivative from renewable agricultural sources like corn and sugarcane (Touri et al., 2019). However, despite these benefits, many application continues to be restricted because of its high cost, brittleness, and low thermal stability (Espinach et al., 2018).

The problems from previous studies are mostly about the weakness of interfacial adhesion between polymer and fiber. By introducing alkaline treatment of the fibers, it can enhance the performance of mechanical properties due to improved adhesion, thus affecting rough surface at the fiber matrix (Ramli et al., 2017). Roselle fiber is a natural fiber that can replace glass fibers and synthetic fibers owing to good mechanical properties. Instead, the effect of fiber size and dimension used in roselle fiber-reinforced PLA is partially discussed in previous research. In this project, the effect of fiber size on the mechanical and thermal properties was found to influence the performance of the composite formed.

This chapter aims to evaluate the effect of fiber size on the mechanical and thermal properties of roselle fiber-reinforced PLA composites by using ANSYS simulation software. To achieve the main goal of this chapter, the scope of this study was set according to the time given. The scopes of this project include the characterization of the composites done by using ANSYS software. Besides that, three parameters were subjected to different roselle fiber size to analyze the effect on the mechanical and thermal properties.

2 MATERIALS AND METHOD

In this chapter, the samples were designed as a model in ANSYS Workbench for design analysis. Three parameters were subjected to different roselle fiber size to analyze the effect on the mechanical and thermal properties. This technique involved many steps in order to find the optimal parameters on the roselle fiber size.

2.1 Engineering Determining Methods

2.1.1 Numerical method

This method was used to analyze the behavior of the physical properties of the model using a set of differential equations through digital computing. The numerical method can be categorized into three categories to understand meshing, which are finite differential method (FDM), finite element method (FEM), and finite volume method (FVM). In this project, FEM was used as shown in Fig. 15.1.

2.2 Finite Element Analysis Process

2.2.1 Flowchart

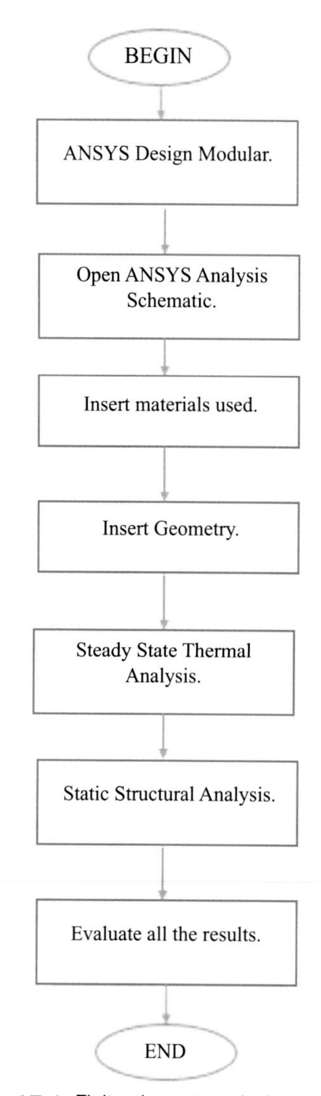

FIG. 15.1 Finite element analysis process.

2.2.2 Geometrical model

The samples were designed as a model in ANSYS Workbench for design analysis. Three samples with three parameters were subjected to different roselle fiber sizes. The roselle fiber-reinforced PLA was designed at 150 mm (length) × 15 mm (width) × 3 mm (thickness) according to the ASTM D3039 standard. The thickness of the roselle fibers were of indifferent sizes of 0.05, 0.08, and 0.1 mm. The different results were compared and discussed.

2.2.3 Material

2.2.3.1 Polylactic acid (Fig. 15.2)

2.2.3.1.1 Roselle Fiber (Fig. 15.3).

2.2.4 Element type

Meshing is an important procedure in finite element analysis process to make the model less complicated. An integrated meshing process is used to allow an easier analysis of the structural model without evaluating the prepressure on the object model. In this project, the meshing method was used to observe the stress distributed on the surface of the object model by using an automated meshing procedure. For this project, the type of element used is "all body in Solid condition." This procedure improves the analysis based on the content used, make data easier to access, and make the analysis more reliable in other terms such as time.

Polylactic acid

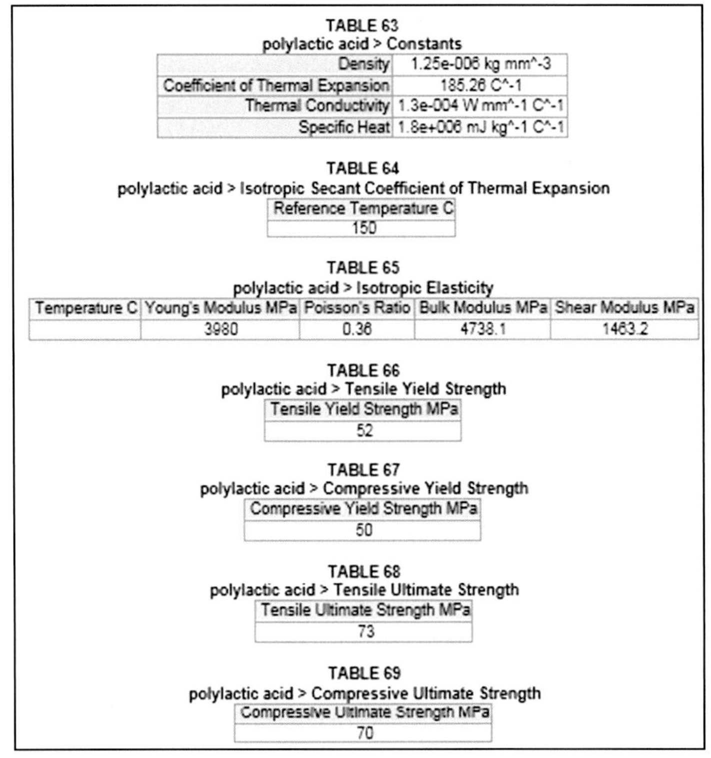

TABLE 63
polylactic acid > Constants

Density	1.25e-006 kg mm^-3
Coefficient of Thermal Expansion	185.26 C^-1
Thermal Conductivity	1.3e-004 W mm^-1 C^-1
Specific Heat	1.8e+006 mJ kg^-1 C^-1

TABLE 64
polylactic acid > Isotropic Secant Coefficient of Thermal Expansion

Reference Temperature C
150

TABLE 65
polylactic acid > Isotropic Elasticity

Temperature C	Young's Modulus MPa	Poisson's Ratio	Bulk Modulus MPa	Shear Modulus MPa
	3980	0.36	4738.1	1463.2

TABLE 66
polylactic acid > Tensile Yield Strength

Tensile Yield Strength MPa
52

TABLE 67
polylactic acid > Compressive Yield Strength

Compressive Yield Strength MPa
50

TABLE 68
polylactic acid > Tensile Ultimate Strength

Tensile Ultimate Strength MPa
73

TABLE 69
polylactic acid > Compressive Ultimate Strength

Compressive Ultimate Strength MPa
70

FIG. 15.2 Characteristics of polylactic acid applied to the model.

Roselle Fibre

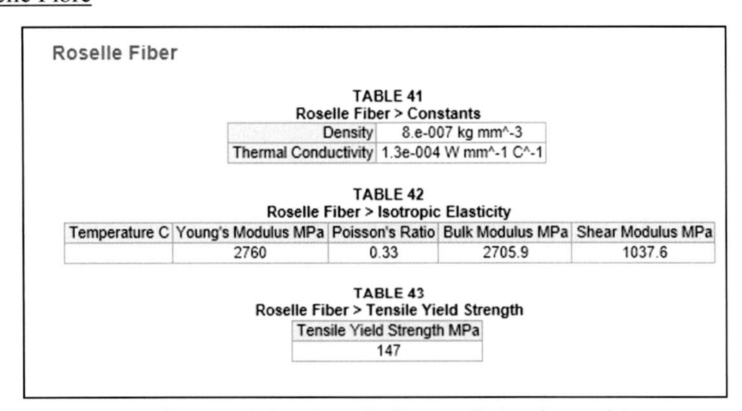

Roselle Fiber

TABLE 41
Roselle Fiber > Constants

Density	8.e-007 kg mm^-3
Thermal Conductivity	1.3e-004 W mm^-1 C^-1

TABLE 42
Roselle Fiber > Isotropic Elasticity

Temperature C	Young's Modulus MPa	Poisson's Ratio	Bulk Modulus MPa	Shear Modulus MPa
	2760	0.33	2705.9	1037.6

TABLE 43
Roselle Fiber > Tensile Yield Strength

Tensile Yield Strength MPa
147

FIG. 15.3 Characteristics of roselle fiber applied to the model.

2.2.5 Boundary condition and loading method

A boundary condition for the model was done by setting a known value for displacement or equivalent load. It can set either the load or the displacement for a given node but not both. Loads and constraints reflect the impact of the model on the surrounding environment. The interface bonding between the fiber and the matrix was defined as bonded conditions when boundary conditions were applied for uniaxial tensile conditions. All loads such as pressure, and thermal load were converted to related loads to allow the displacement equation to be solved.

2.2.6 ANSYS workbench setup

In this research, the model was designed in ANSYS Workbench as no import from other software was needed. Since the model dealt with thermal analysis, the "Steady-State Thermal" was used for the analysis system.

Procedure:

- Open the ANSYS Workbench 16.1.
- Choose the Analysis tool as "Steady-State Thermal" from the left tool selection bar.
- Numerous models for analysis can be used to run and the results can be extracted from the workbench (Fig. 15.4).

Step 1: Insert material in engineering data (Fig. 15.5).

Step 2: Insert material properties for RS and PLA (Fig. 15.6).

Step 3: Insert Geometry from the design modular.

The model design in the software was carried according to the dimension of the roselle-reinforced PLA dimension of 150 mm (length) × 15 mm (width) × 3 mm (thickness) and with different roselle fiber sizes of 0.05, 0.08, and 0.1 mm as illustrated in Fig. 15.7.

Step 4: Assign the materials (Fig. 15.8).

Step 5: Right click at "Mesh" and choose "to generate the mesh" (Fig. 15.9).

Step 6: Click "Steady-State Thermal."

Set the temperature surface place and convection surface place (Fig. 15.10).

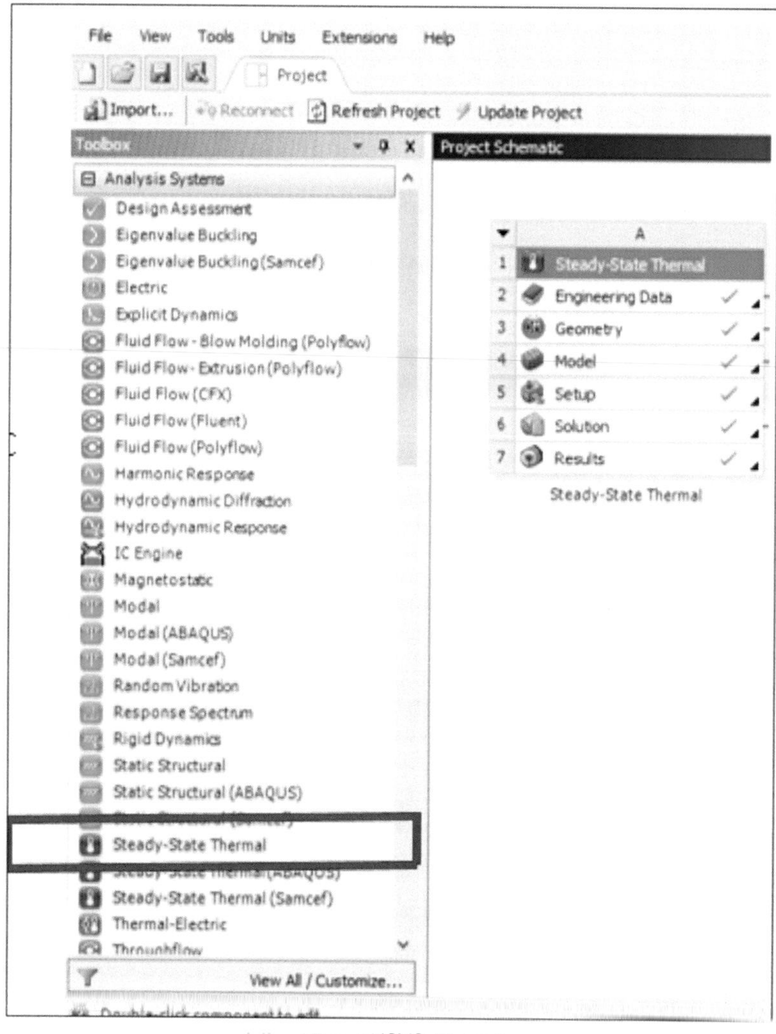

FIG. 15.4 ANSYS 16.1 setup.

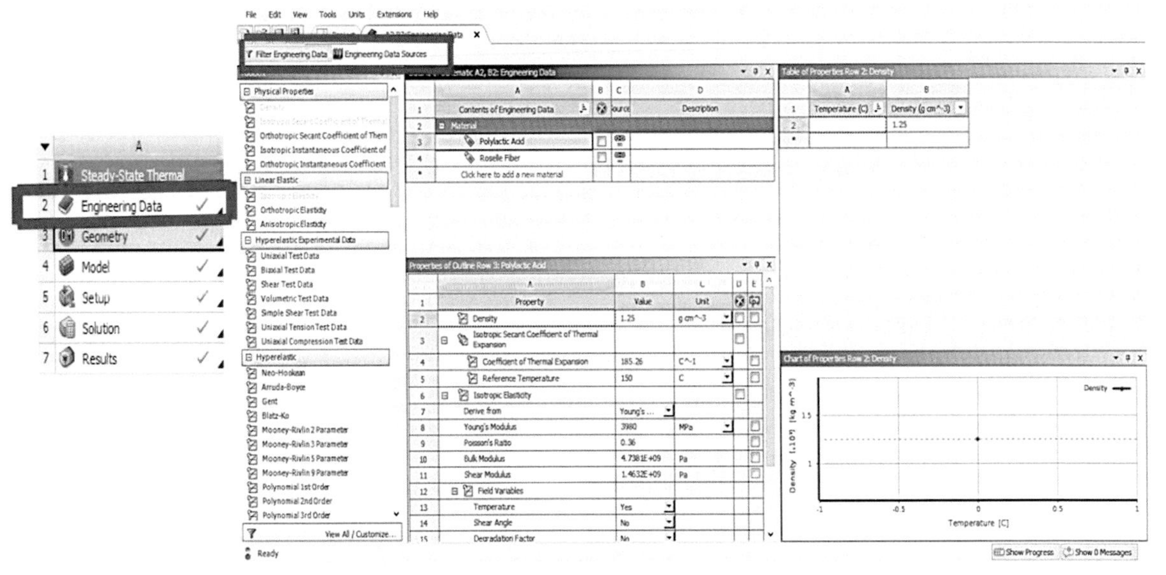

FIG. 15.5 Initial setup for ANSYS.

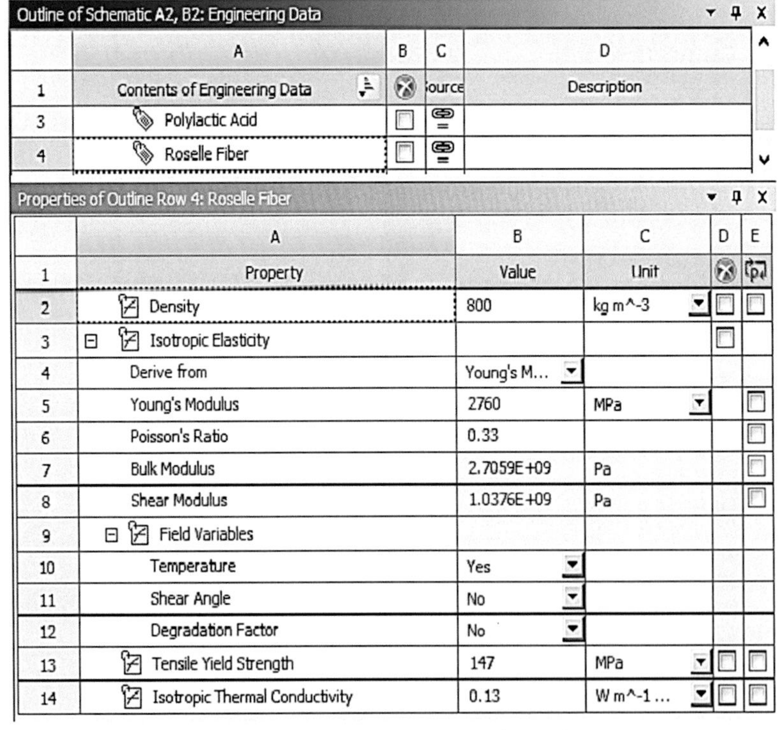

FIG. 15.6 Set for the materials used.

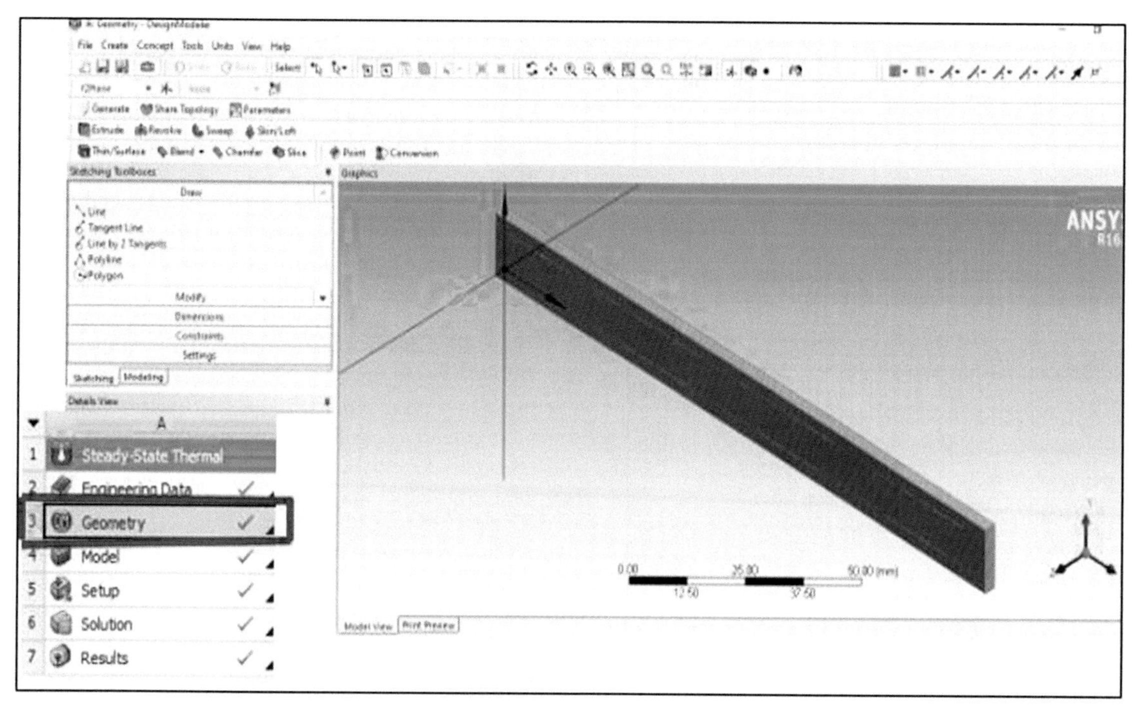

FIG. 15.7 Design the model in ANSYS.

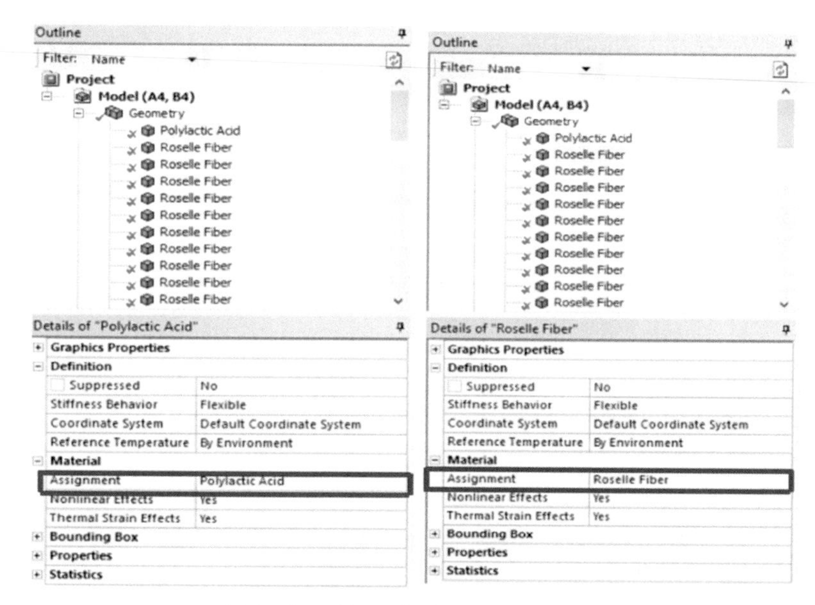

FIG. 15.8 Setup for the materials used.

Step 7: Choose temperature and evaluate all the results.

Boundary condition for thermal analysis (Figs. 15.11–15.14):

- Temperature: 150°C (at the top), 75°C (at the bottom).
- Convection coefficient: 0.13 $W/mm^2.°C$

Step 8: Choose the static structural and link with the steady-state thermal (double click setup) (Fig. 15.15–15.17).

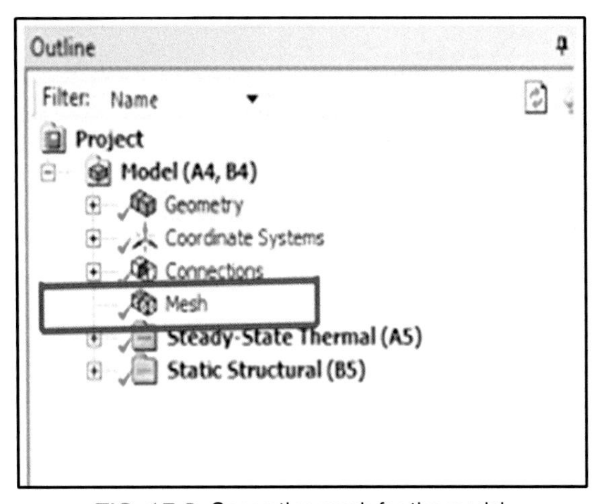

FIG. 15.9 Generating mesh for the model.

3 RESULTS AND DISCUSSIONS

The 3D model for analysis was done in ANSYS software to acquire a simulated image that indicated different mechanical and thermal properties that affected the model. All the results of the pictorial form are discussed in this chapter. The results shown from the constructed and plotted graphs were based on data obtained during the analysis. This chapter discusses the maximum stress, deformation, and thermal deformation expected from ANSYS 16.1.

3.1 Tensile Strength Analysis

3.1.1 FEA result analysis

The model was generated using ANSYS. Fig. 15.18 displays a model of a tensile test specimen for network geometry obtained with 3393 elements and 28,744 nodes for 0.05 mm fiber size, 3421 elements and 28,955 nodes for 0.08 mm fiber size, 3450 elements, and 29,183 nodes for 0.10 mm fiber size. The test process was simulated by running the simulation program ANSYS. Computer simulations for various samples were conducted by using different loads.

The FEA simulation design of the composites tensile test sample is shown in Figs. 15.19–15.21. The maximum stress distribution and deformation for the composites were obtained under conditions of tensile load. The loads of 1.3 and 10 kN force were used for the loading conditions in this simulation. For 0.05 mm fiber size, the maximum tensile stress that

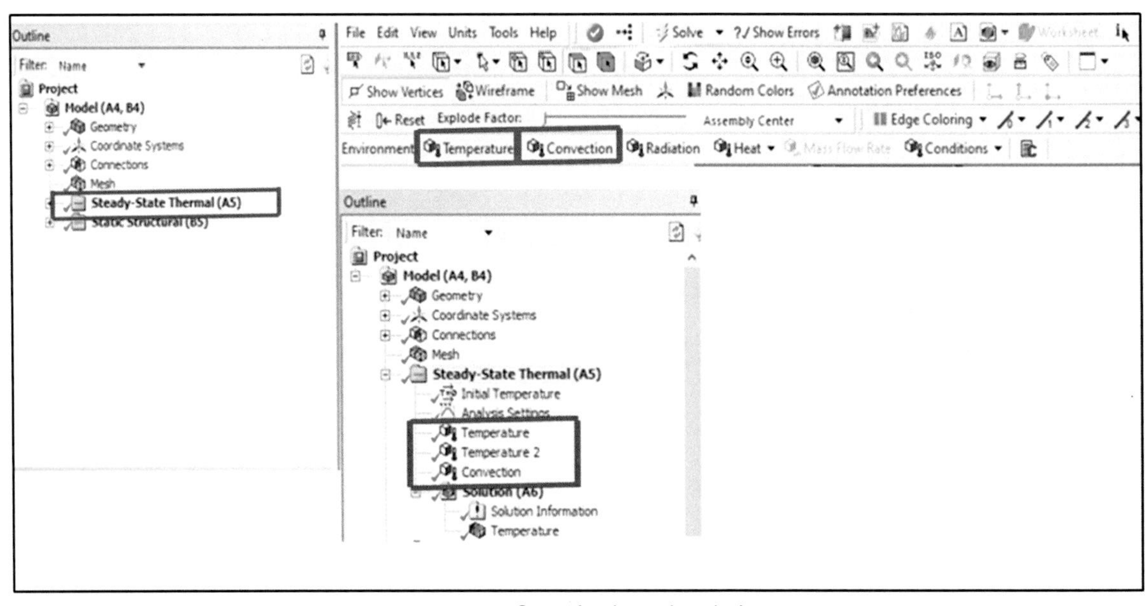

FIG. 15.10 Setup for thermal analysis.

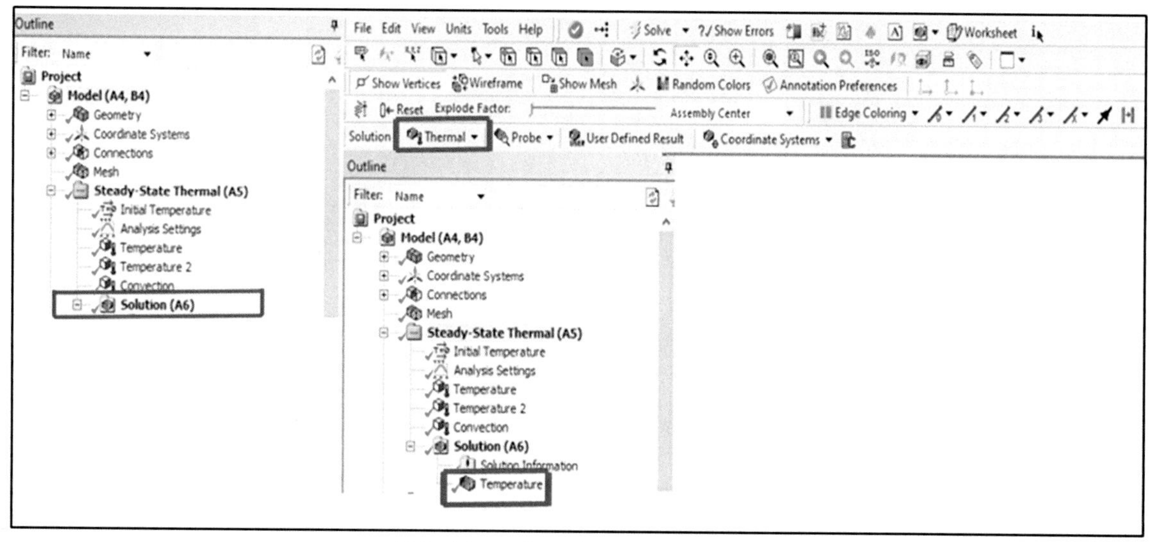

FIG. 15.11 Setup for thermal analysis.

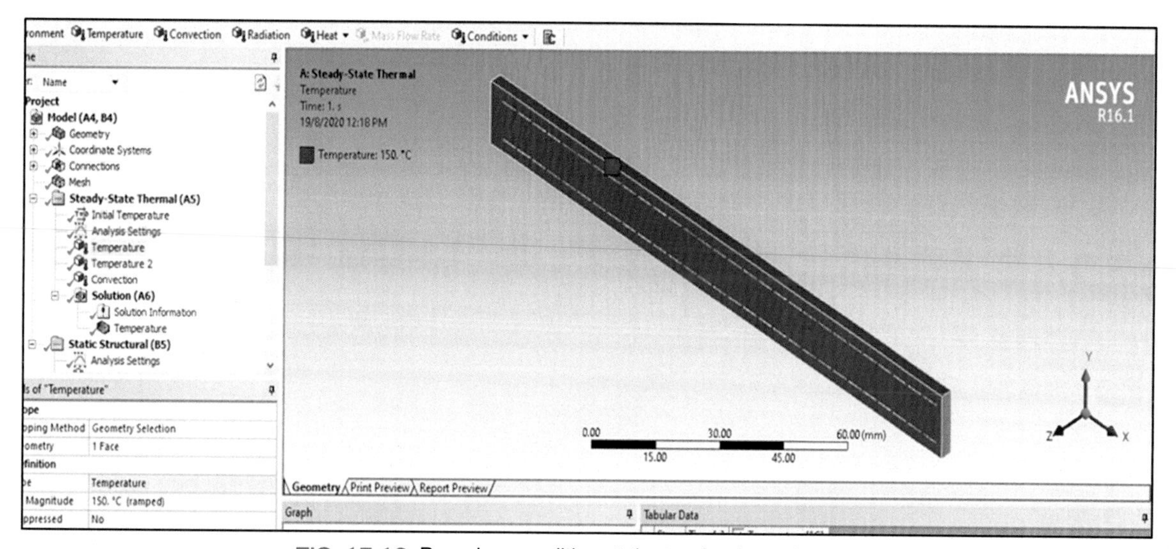

FIG. 15.12 Boundary condition at the top for thermal analysis.

acted on the sample was 39.52 MPa when 1.3 kN force was applied at the left and right sides of the sample. When 10 kN force was applied for 0.05 mm fiber size, the maximum tensile stress was 304.05 MPa. For 0.08 mm fiber size, the maximum tensile stress that acted on the sample was 1181.20 MPa when both loads were applied at the left and right sides of the sample. For 0.1 mm fiber size, a maximum tensile stress of 1090.40 MPa was recorded when both loads were applied. The analysis of the tensile test indicated that the composites in their elastic state were brittle but exhibited linear deformation.

For 0.05 mm fiber size, a slight deformation of 1600 mm was observed when both loads were applied. Therefore, the sample did not change much in shape compared to the other samples. The lower the deformation of the material; the better is the performance of mechanical and thermal properties. As for the 0.08 and 0.1 mm fiber sizes, the samples deformed easily because the contour was totally different from

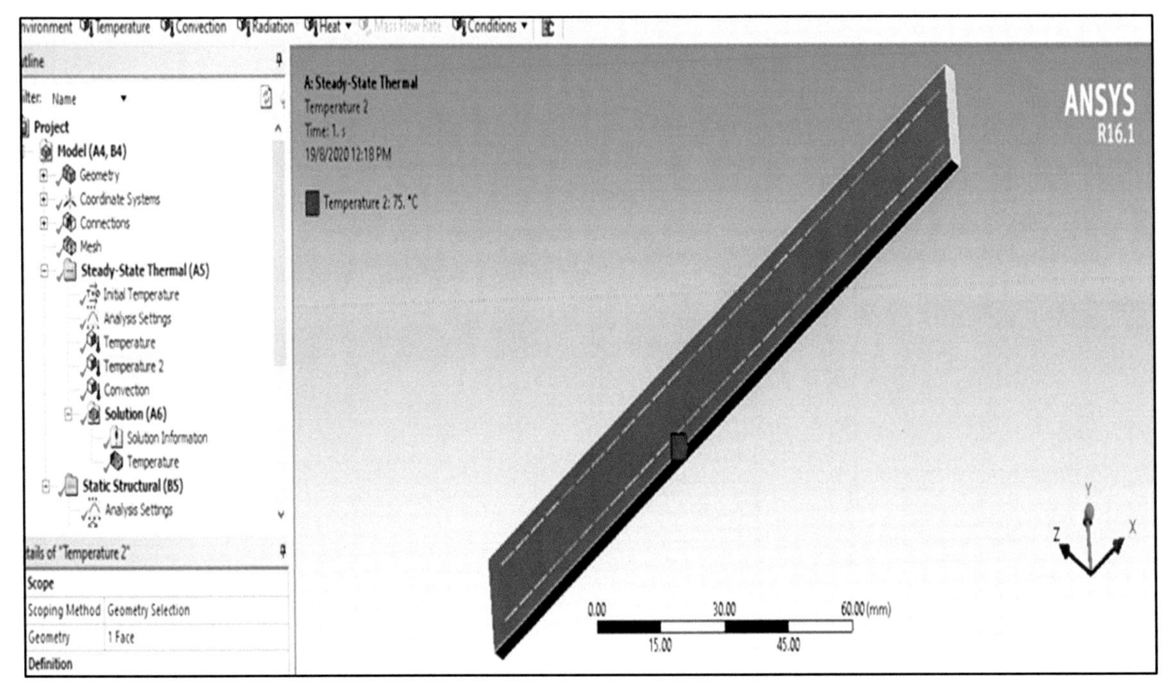

FIG. 15.13 Boundary condition at the bottom for thermal analysis.

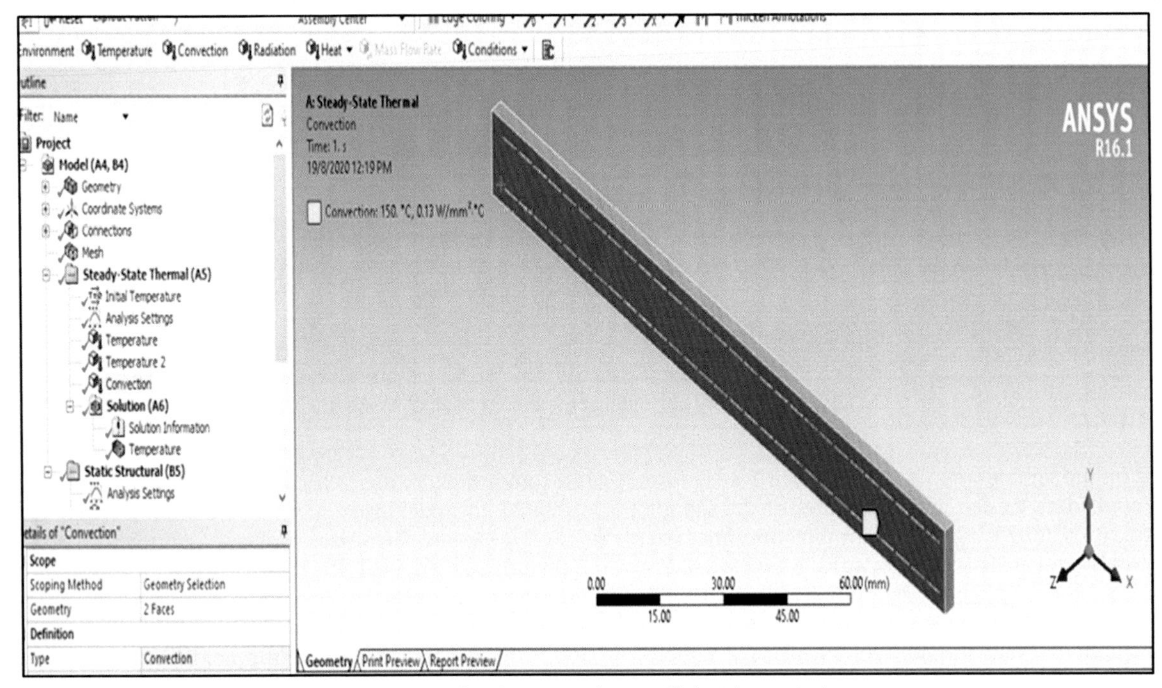

FIG. 15.14 Set the convection coefficient for analysis.

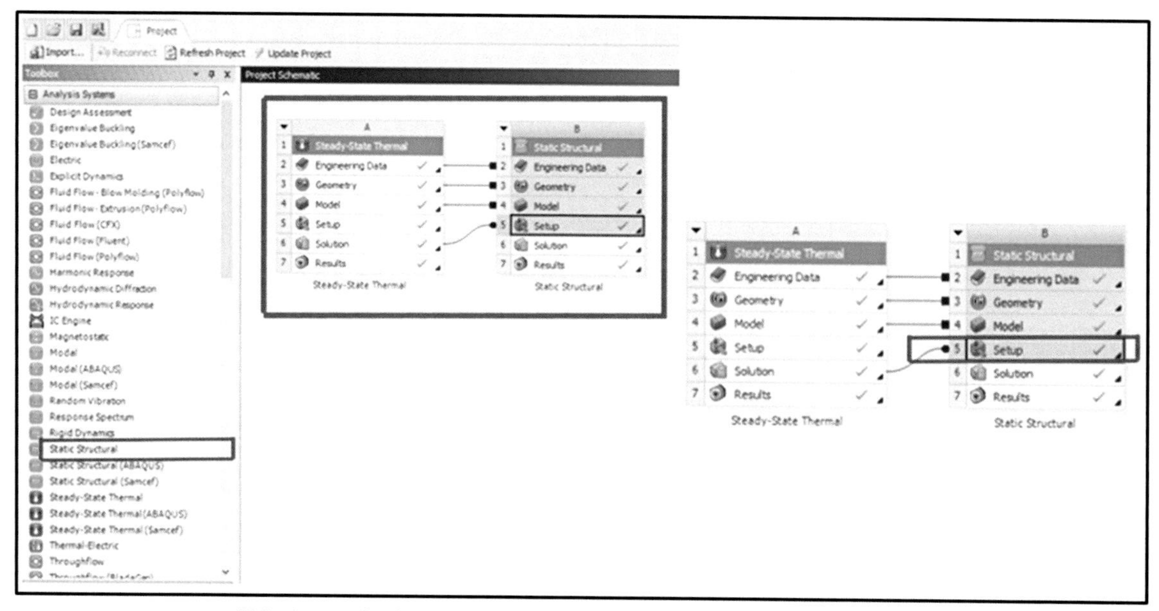

FIG. 15.15 Static structural and link with the steady state thermal system.

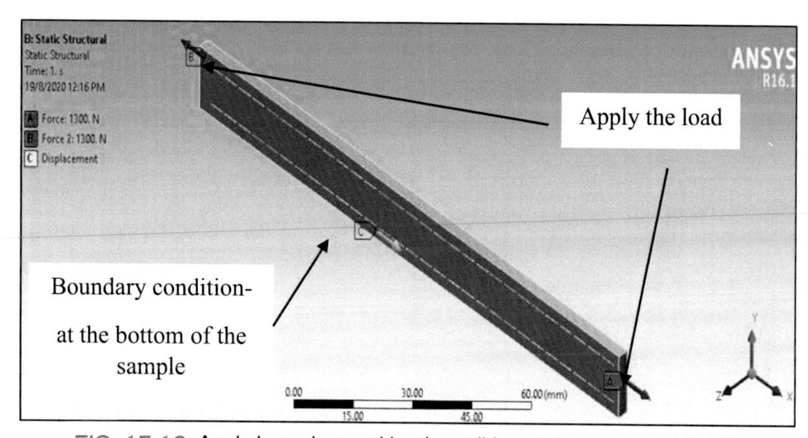

FIG. 15.16 Apply boundary and load condition at the model in ANSYS.

0.05 mm fiber size sample. The maximum deformation for 0.08 and 0.1 mm fiber sizes were 265,820 and 266,080 mm, respectively when subjected to tensile loading as shown in Table 15.2.

Based on the results acquired in Fig. 15.22, the fiber size affected the tensile strength of the samples. For 0.05 mm fiber size, both of loads of 1.3 and 10 kN that were applied to the sample affected the tensile strength of the sample. As observed, as applied load was increased, the displacement gradually increased but suddenly dropped at 1090.4 MPA for 0.10 mm fiber size. From the figure, it can be seen that the increasing size of roselle fiber-reinforced PLA composites applied with both loads showed better tensile strength performance.

The tensile strengths of various roselle fiber sizes does affect the strength of the fiber (Ochi, 2010). The analysis of the tensile testing showed that the composites in their elastic state were brittle but exhibited linear deformation. The 0.08 mm fiber size gave better tensile stress results compared to the other fiber sizes. The results also showed that when the load was increased with the same different size of fiber as above, the tensile strength has the same value as shown in Table 15.3

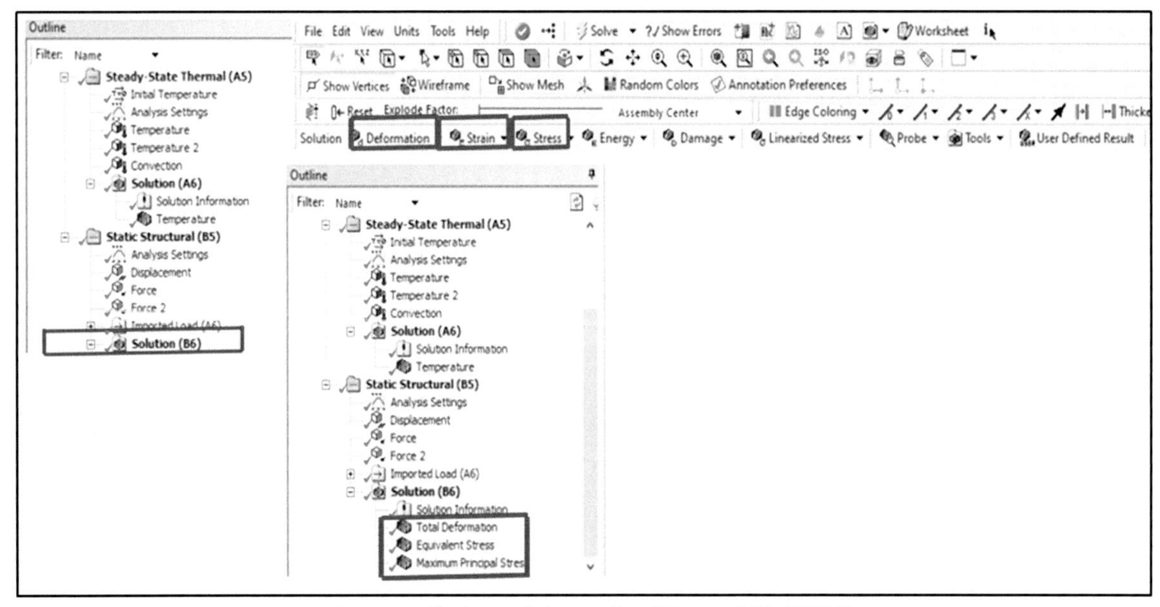

FIG. 15.17 Evaluate all the results of the model in ANSYS.

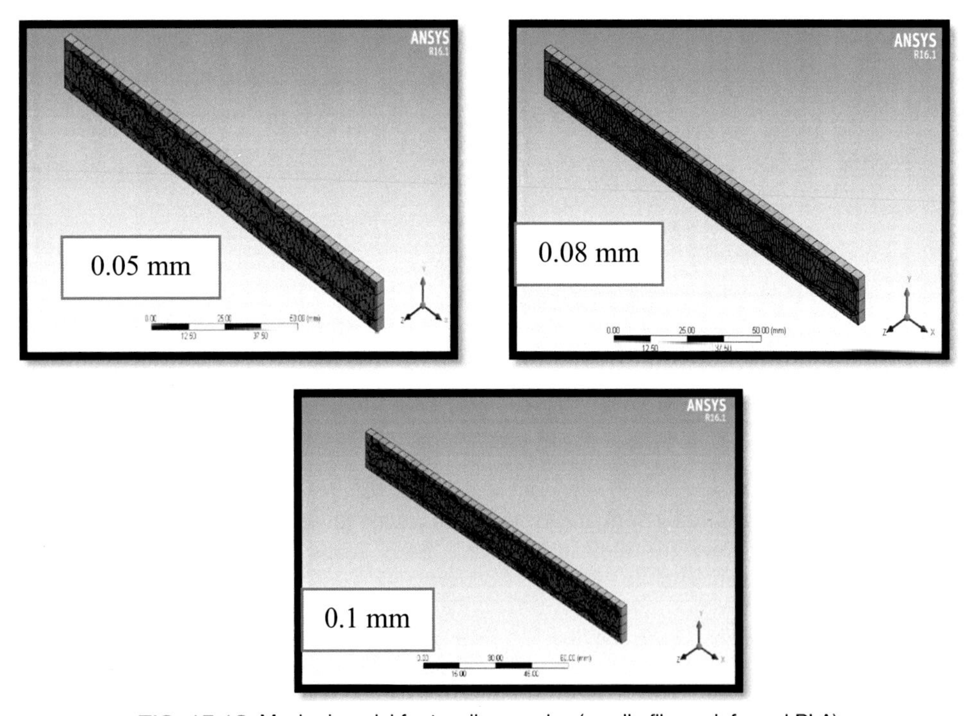

FIG. 15.18 Meshed model for tensile samples (roselle fiber-reinforced PLA).

From the acquired results, it was observed that fiber size of affected shape of the samples when both loads were applied because displacement is proportional to the load exerted within the material's elastic limits. Fig. 15.23 shows the values of total deformation for the three different sizes of roselle fiber. It shows that

0.05mm

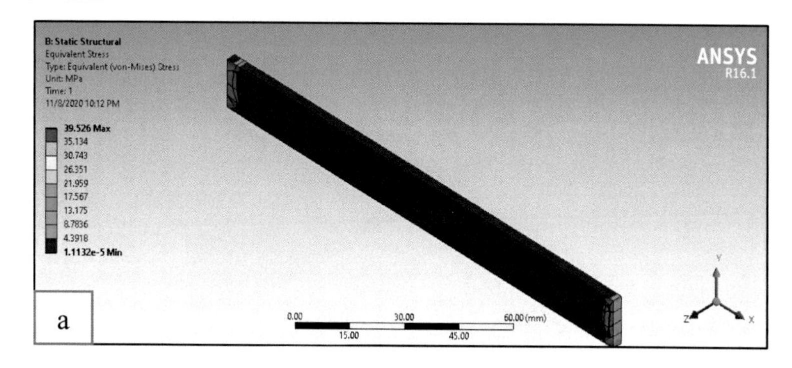

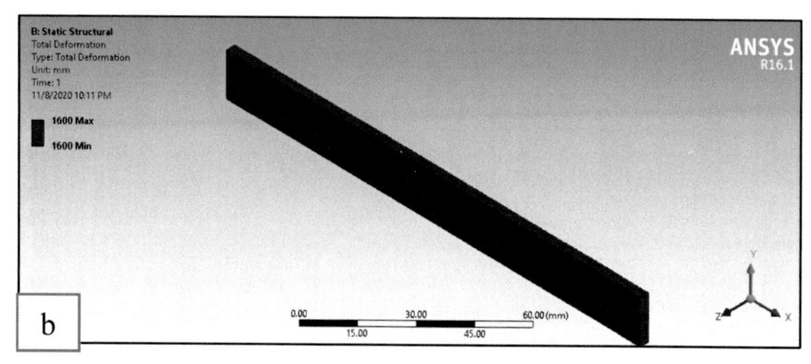

FIG. 15.19 FEA simulation analysis due to both loads. **(A)** stress distribution; **(B)** deformation.

0.08 mm

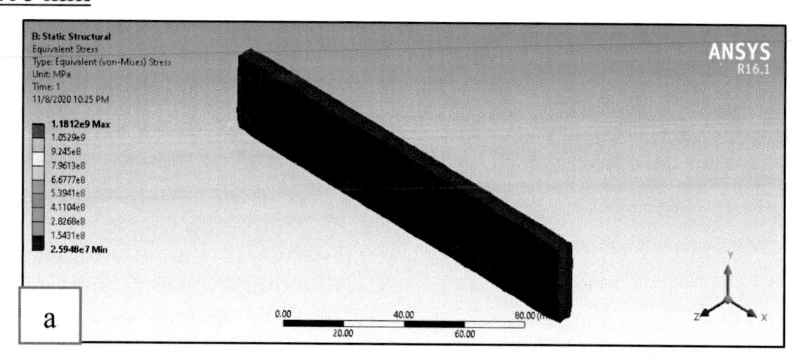

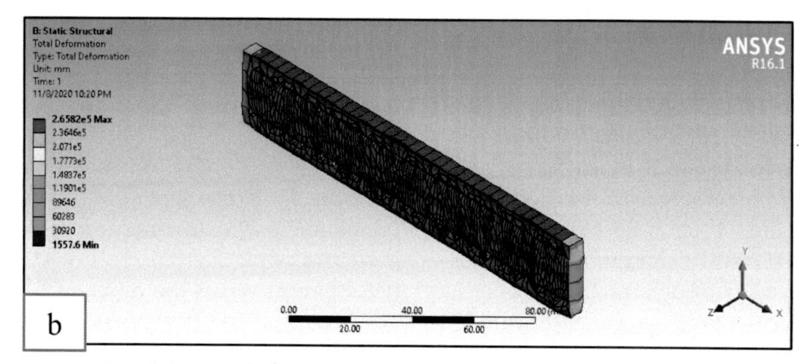

FIG. 15.20 FEA simulation analysis due to both loads. **(A)** stress distribution; **(B)** deformation.

0.10 mm

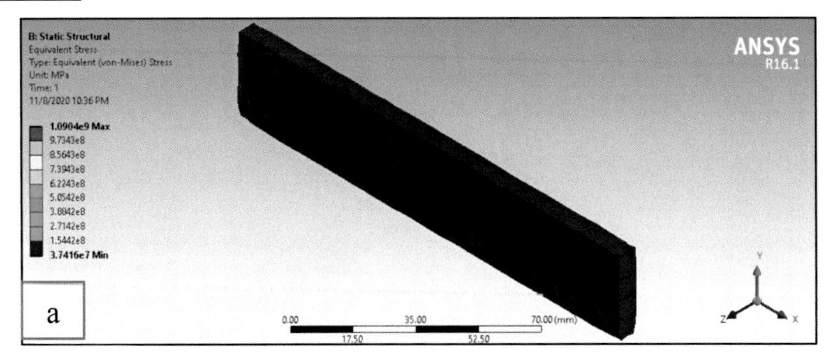

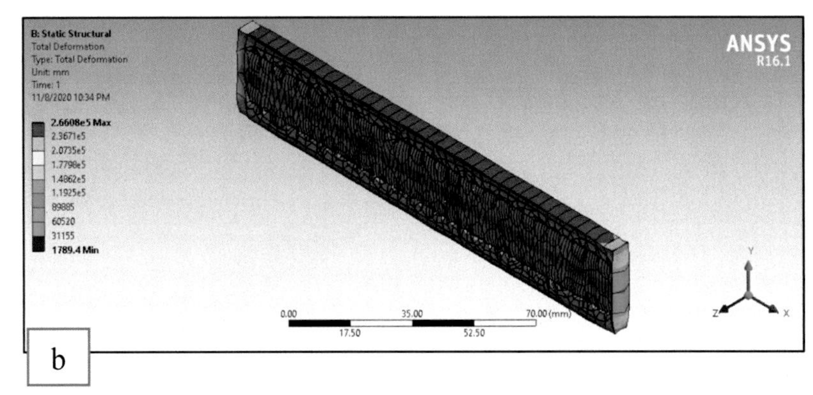

FIG. 15.21 FEA simulation analysis due to both loads. **(A)** stress distribution; **(B)** deformation.

TABLE 15.2
Tensile Strength of the Samples Predicted From ANSYS.

Size of Fiber (mm)	TENSILE STRENGTH (MPA)	
	Force (1.3 kN)	Force (10 kN)
0.05	39.52	304.05
0.08	1181.20	1181.2
0.10	1090.40	1090.4

the sample with 0.05 mm size gave the optimum value for total deformation with 1600 mm, while the 0.10 mm fiber size showed higher total deformation value of 266,080 mm.

Moreover, 0.05 mm fiber size gave better results in load distribution compared to other fiber sizes because the sample only showed a small change in shape. This was because when the value of deformation is bigger, the tendency of the structure to fail is higher. In various applications, the deformation of materials is important for industries. The results also showed that when the load was increased, it will not affect the deformation of the samples with the same different size of fiber.

3.2 Thermal Analysis

3.2.1 FEA result analysis

Fig. 15.24 shows the composites that were obtained under interfacial conditions between the matrix and fiber. Based on the pictorial pictures, it can be seen that the shape of the samples became longer according to changes in temperature. As the temperature was increased, the shape of the sample also changed horizontally. The analysis of the thermal testing indicated that the samples showed color changes with the different fiber sizes. From the figure, it is obviously shown that the bigger fiber size did affect the color change in this simulation analysis.

Table 15.4 represents the thermal deformation obtained for roselle fiber-reinforced PLA with different fiber sizes, while Fig. 15.25 shows a graph for thermal deformation versus the different fiber sizes of fiber for all three studied parameters. When applied with the same temperature of 150°C to all three different fiber

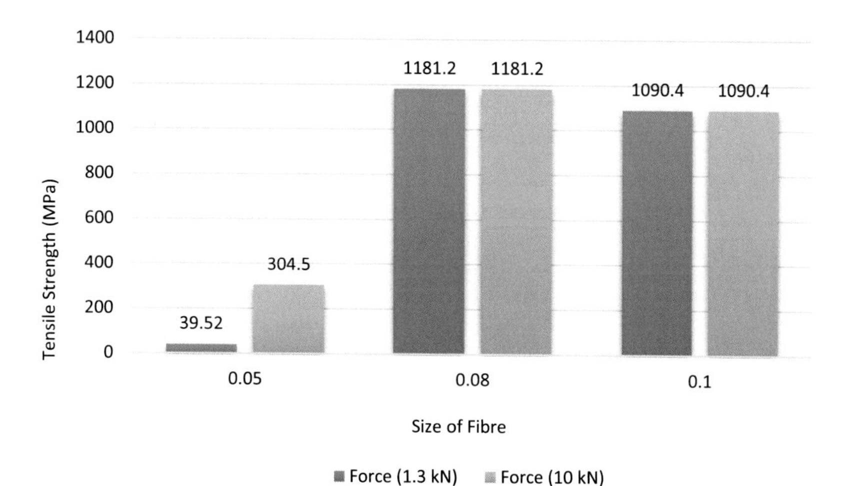

FIG. 15.22 Tensile strength of the samples predicted from ANSYS.

TABLE 15.3
Deformation of the Samples Predicted From ANSYS.

	DEFORMATION (MM)	
Size of Fiber (mm)	**Force (1.3 kN)**	**Force (10 kN)**
0.05	1600	1600
0.08	265,820	265,820
0.10	266,080	266,080

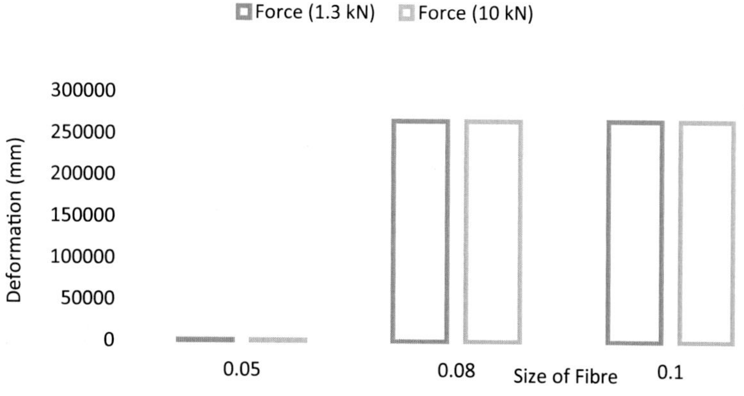

FIG. 15.23 Deformation of the samples predicted from ANSYS.

sizes, 0.10 mm fiber size showed the lowest thermal deformation value of 170.41°C among the different fiber sizes. In this thermal analysis, 0.08 mm fiber size easily degraded and the sample simply burned because it could not withstand high temperatures. Therefore, it can be seen that the fine fiber size of 0.1 mm fiber obtained the best thermal analysis results. The analysis is in good agreement with a previous research that found that the finer the fiber size, the thermal behavior performance significantly improved.

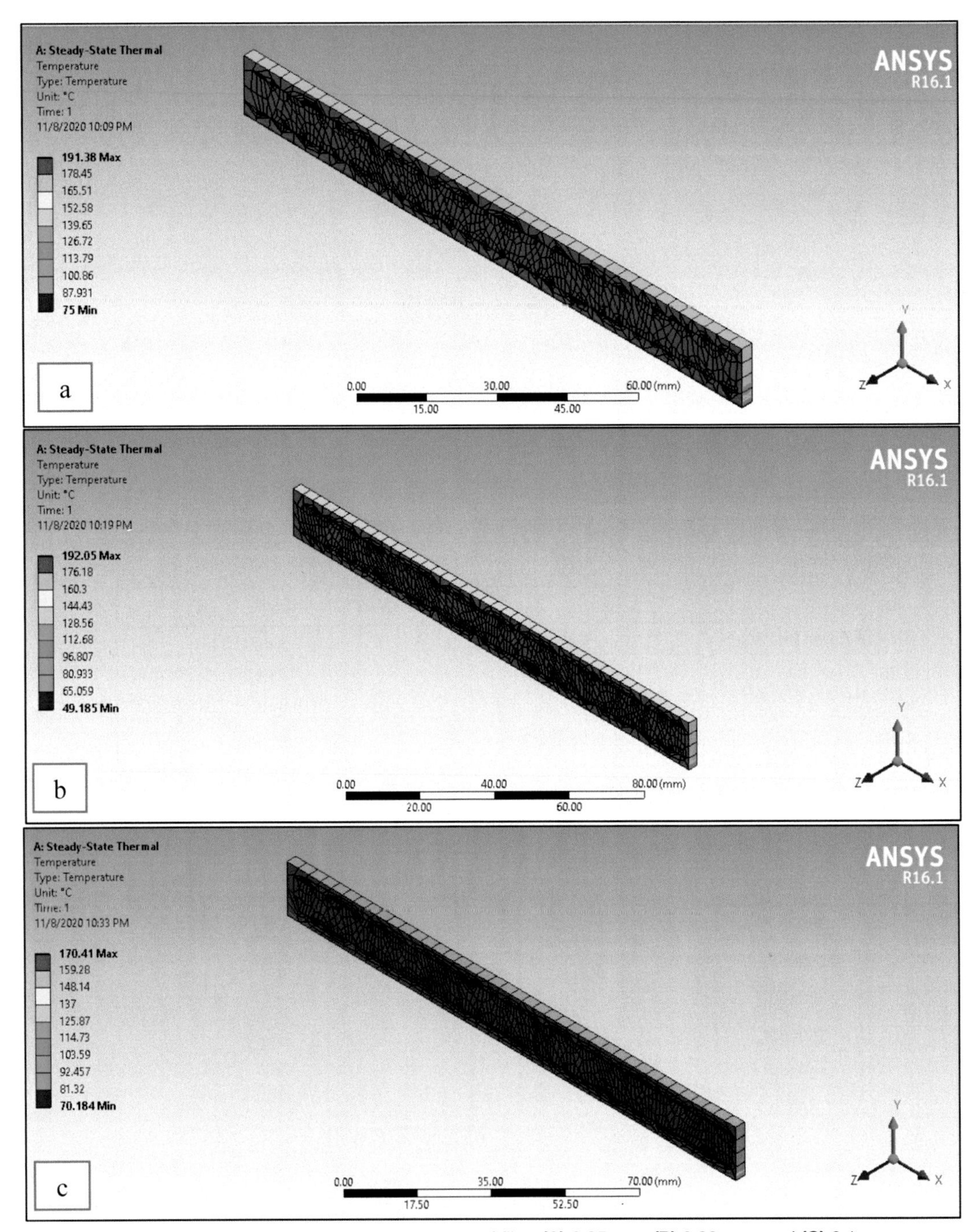

FIG. 15.24 Thermal analysis for different size of fiber **(A)** 0.05 mm; **(B)** 0.08 mm; and **(C)** 0.1 mm.

The optimum fiber size for this analysis was 0.10 mm as it did not easily degrade at high temperature. The deterioration of mechanical properties of the fiber occurred according to the degradation of natural fibers. Thermal degradation is indeed a key factor in the production of NFC, as it has a direct effect on the highest possible temperature to be used for composites manufacturing (Jesuarockiam et al., 2019; Tajvidi &

TABLE 15.4
Thermal Deformation of the Samples Predicted From ANSYS.

Size of Fiber (mm)	Thermal Deformation (°C)
0.05	191.38
0.08	192.05
0.10	170.41

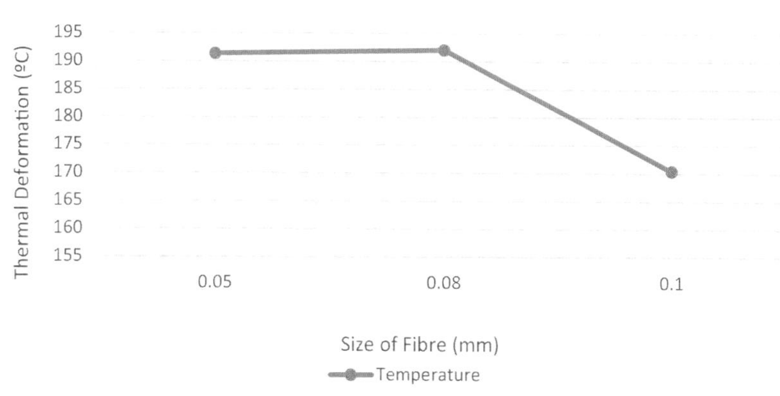

FIG. 15.25 Thermal deformation of the samples predicted from ANSYS.

Takemura, 2010). In composites, the use of natural fibers to PLA reduces the deterioration temperature (Siakeng et al., 2019; Yussuf et al., 2010).

4 CONCLUSION AND RECOMMENDATIONS

The mechanical and thermal properties of roselle fiber-reinforced polylactic acid composites were analyzed using FEA. Natural fiber has an enthusiastic impending to be used as reinforcement in a composite. In this project, roselle fiber was chosen as reinforcement material because of its outstanding properties and environmental considerations acquired from previous research. PLA is one of the eco biodegradable polymers and the characteristic shows it is the top favorable candidate for future improvements in green technology. Based on the tensile strength results, the optimum fiber size was 0.08 mm because it showed the highest tensile strength compared to the other fiber sizes. The analysis of the tensile testing showed that the composites in their elastic state were very brittle but exhibited linear deformation. The sample with 0.05 mm fiber size gave the optimum value for the total deformation with 1600 mm. The tensile strengths of various roselle fiber sizes affected the strength of the fiber. For the thermal analysis, the result showed that the optimum fiber size of fiber was 0.10 mm as it did not easily degrade at high temperatures. The thermal degradation of natural fiber is essential for the application of the composite materials and need to be implemented when considering the use of composite material for the composite industry. As the samples reached the elevated temperature and underwent thermal degradation, the performance of mechanical properties was affected. As seen from the results in tensile strength analysis and thermal analysis, the results were inconsistent in terms of the fiber size because of a few inaccuracies that may be resulted from the meshing or certain conditions when doing analysis simulation in ANSYS software. After analyzing the models or samples of RS/PLA in FEA, there are several improvements for future works that can be recommended. There were a few inaccuracies that may result from the meshing or certain conditions when doing analysis simulation in ANSYS software and need more practice on conducting the software. As for future recommendations, the different fiber content of roselle fiber-reinforced polylactic acid composite should be studied.

REFERENCES

Athijayamani, A., Thiruchitrambalam, M., Winowlin Jappes, J. T., & Alavudeen, A. (2011). Effects of fiber content on the mechanical properties of short roselle/sisal fiber polyester hybrid composite. *International Journal of Computer Aided Engineering and Technology, 3*(5/6), 538–546.

Azammi, A. M. N., Ilyas, R. A., Sapuan, S. M., Ibrahim, R., Atikah, M. S. N., Asrofi, M., & Atiqah, A. (2019). Characterization studies of biopolymeric matrix and cellulose fibres based composites related to functionalized fiber-matrix interface. In *Interfaces in particle and fibre reinforced composites- from macro to nano scales* (1st ed., pp. 1–68). London: Woodhead Publishing. https://doi.org/10.1016/B978-0-08-102665-6

Balla, V. K., Kate, K. H., Satyavolu, J., Singh, P., & Tadimeti, J. G. D. (2019). Additive manufacturing of natural fiber reinforced polymer composites: Processing and prospects. *Composites Part B: Engineering, 174*(October), 106956. https://doi.org/10.1016/j.compositesb.2019.106956

Barbero, E. J., & Shahbazi, M. (2017). Determination of material properties for ANSYS progressive damage analysis of laminated composites. *Composite Structures, 176*, 768–779. https://doi.org/10.1016/j.compstruct.2017.05.074

Castro-Aguirre, E., Auras, R., Selke, S., Rubino, M., & Marsh, T. (2018). Impact of nanoclays on the biodegradation of poly(lactic acid) nanocomposites. *Polymers, 10*(2). https://doi.org/10.3390/polym10020202

Cellulose Fibers: Bio- and Nano-Polymer Composites. (2011). *Cellulose fibers: Bio- and nano-polymer composites*. https://doi.org/10.1007/978-3-642-17370-7

Chandramohan, D., & Marimuthu, K. (2011). A review on natural fibers. *International Journal of Research and Reviews in Applied Sciences, 8*(2), 194–206.

Chauhan, A., & Chaudan, P. (2013). Natural fibers reinforced advanced materials. *Chemical Engineering & Process Technology*, 1–3.

Chauhan, A., & Kaith, B. (2011). Development and evaluation of novel roselle graft copolymer. *Malaysia Polymer Journal, 6*(2), 176–188.

Chauhan, A., & Kaith, B. (2012a). Accreditation of novel roselle grafted fiber reinforced bio-composites. *Journal of Engineered Fibers and Fabrics, 7*(2), 66–75. https://doi.org/10.1177/155892501200700210

Chauhan, A., & Kaith, B. (2012b). Versatile roselle graft-copolymers: XRD studies and their mechanical evaluation after use as reinforcement in composites. *Journal of the Chilean Chemical Society, 57*(3), 1262–1266. https://doi.org/10.4067/s0717-97072012000300014

Dhannush, S., Aushwin, S., Arunagiri, A. R., & Senthamaraikannan, C. (2020). Investigation on mechanical behavior of sisal fiber reinforced polylactic acid and sisal/epoxy composites. *Key Engineering Materials, 841*, 322–326. https://doi.org/10.4028/www.scientific.net/kem.841.322

Drahansky, M., Paridah, M. T., Amin, M., Mohamed, A. Z., Taiwo, O. F. A., Asniza, M., & Shawkataly, H. P. A. K. (2016). We are IntechOpen, the world's leading publisher of open access books built by scientists, for scientists TOP 1%. *Intech i (Tourism)*, 13. https://doi.org/10.5772/57353

El-Shekeil, Y. A., Sapuan, S. M., Abdan, K., & Zainudin, E. S. (2012). Influence of fiber content on the mechanical and thermal properties of kenaf fiber reinforced thermoplastic polyurethane composites. *Materials and Design, 40*, 299–303. https://doi.org/10.1016/j.matdes.2012.04.003

El-Shekeil, Y. A., Sapuan, S. M., Jawaid, M., & Al-Shuja'a, O. M. (2014). Influence of fiber content on mechanical, morphological and thermal properties of kenaf fibers reinforced poly(vinyl chloride)/thermoplastic polyurethane poly-blend composites. *Materials and Design, 58*, 130–135. https://doi.org/10.1016/j.matdes.2014.01.047

Espinach, F. X., Boufi, S., Delgado-Aguilar, M., Julián, F., Mutjé, P., & Méndez, J. A. (2018). Composites from poly(-lactic acid) and bleached chemical fibres: Thermal properties. *Composites Part B: Engineering, 134*, 169–176. https://doi.org/10.1016/j.compositesb.2017.09.055

Frone, A. N., Berlioz, S., Chailan, J. F., & Panaitescu, D. M. (2013). Morphology and thermal properties of PLA-cellulose nanofibers composites. *Carbohydrate Polymers, 91*(1), 377–384. https://doi.org/10.1016/j.carbpol.2012.08.054

Hassan, N. A. A., Chen, R. S., & Ahmad, S. (2019). Expanded biocomposite based on polylactic acid and kenaf fiber. *AIP Conference Proceedings, 2111*(June). https://doi.org/10.1063/1.5111234

Ilyas, R. A., & Sapuan, S. M. (2020a). Biopolymers and biocomposites: Chemistry and technology. *Current Analytical Chemistry, 16*(5), 500–503. https://doi.org/10.2174/157341101605200603095311

Ilyas, R. A., & Sapuan, S. M. (2020b). The preparation methods and processing of natural fiber bio-polymer composites. *Current Organic Synthesis, 16*(8), 1068–1070. https://doi.org/10.2174/157017941608200120105616

Ilyas, R. A., Sapuan, S. M., Ibrahim, R., Abrald, H., Ishak, M. R., Zainudin, E. S., Atikah, M. S. N., Mohd Nurazzia, N., Atiqah, A., Ansari, M. N. M., Syafri, E., Asrofi, M., Sari, N. H., & Jumaidin, R. (2019). Effect of sugar palm nanofibrillated celluloseconcentrations on morphological, mechanical andphysical properties of biodegradable films basedon agro-waste sugar palm (*Arenga pinnata*(*Wurmb.*) *Merr*) starch. *Journal of Materials Research and Technology, 8*(5), 4819–4830. https://doi.org/10.1016/j.jmrt.2019.08.028

Ilyas, R. A., Sapuan, S. M., Ishak, M. R., & Zainudin, E. S. (2017). Effect of delignification on the physical, thermal, chemical, and structural properties of sugar palm fiber. *BioResources, 12*(4), 8734–8754. https://doi.org/10.15376/biores.12.4.8734-8754

Ilyas, R. A., Sapuan, S. M., Ishak, M. R., & Zainudin, E. S. (2018). Development and characterization of sugar palm nanocrystalline cellulose reinforced sugar palm starch bionanocomposites. *Carbohydrate Polymers, 202*, 186–202. https://doi.org/10.1016/j.carbpol.2018.09.002

Jayanth, D., Sathish Kumar, P., Chandra Nayak, G., Kumar, J. S., Pal, S. K., & Rajasekar, R. (2018). A review on biodegradable polymeric materials striving towards the attainment of green environment. *Journal of Polymers and the Environment, 26*(2), 838–865. https://doi.org/10.1007/s10924-017-0985-6

Jesuarockiam, N., Jawaid, M., Zainudin, E. S., Sultan, M. T. H., & Yahaya, R. (2019). Enhanced thermal and dynamic mechanical properties of synthetic/natural hybrid composites with graphene nanoplateletes. *Polymers, 11*(7). https://doi.org/10.3390/polym11071085

Jeyasekaran, A. S., Kumar, K. P., & Rajarajan, S. (2016). Numerical and experimental analysis on tensile properties of banana and glass fibers reinforced epoxy composites. *Sadhana - Academy Proceedings in Engineering Sciences, 41*(11), 1357–1367. https://doi.org/10.1007/s12046-016-0554-z

Kargarzadeh, H., Huang, J., Lin, N., Ahmad, I., Mariano, M., Dufresne, A., Thomas, S., & Gał, A. (2018). Recent developments in nanocellulose-based biodegradable polymers, thermoplastic polymers, and porous nanocomposites. *Progress in Polymer Science, 87*, 197–227. https://doi.org/10.1016/j.progpolymsci.2018.07.008

Murariu, M., Dechief, A. L., Paint, Y., Peeterbroeck, S., Bonnaud, L., & Dubois, P. (2012). Polylactide (PLA)-halloysite nanocomposites: Production, morphology and key-properties. *Journal of Polymers and the Environment, 20*(4), 932–943. https://doi.org/10.1007/s10924-012-0488-4

Nazrin, A., Sapuan, S. M., Zuhri, M. Y. M., Ilyas, R. A., Syafiq, R., & Sherwani, S. F. K. (2020). Nanocellulose reinforced thermoplastic starch (TPS), polylactic acid (PLA), and polybutylene succinate (PBS) for food packaging applications. *Frontiers in Chemistry, 8*(213), 1–12. https://doi.org/10.3389/fchem.2020.00213

Ochi, S. (2010). Tensile properties of kenaf fiber bundle. *SRX Materials Science, 2010*, 1–6. https://doi.org/10.3814/2010/152526

Ochi, S. (2012). Tensile properties of bamboo fiber reinforced biodegradable plastics. *International Journal of Composite Materials, 2*(1), 1–4. https://doi.org/10.5923/j.cmaterials.20120201.01

Pandey, K. (2015). *Natural fiber composites for 3D printing*.

Pickering, K. L., Aruan Efendy, M. G., & Le, T. M. (2016). A review of recent developments in natural fiber composites and their mechanical performance. *Composites Part A: Applied Science and Manufacturing, 83*, 98–112. https://doi.org/10.1016/j.compositesa.2015.08.038

Raman Bharath, V. R., Vijaya Ramnath, B., & Manoharan, N. (2015). Kenaf fiber reinforced composites: A review. *ARPN Journal of Engineering and Applied Sciences, 10*(13), 5483–5485. https://doi.org/10.1016/j.matdes.2011.04.008

Ramli, S. N. R., Fadzullah, S. H. S. M., & Mustafa, Z. (2017). The effect of alkaline treatment and fiber length on pineapple leaf fiber reinforced poly lactic acid biocomposites. *Jurnal Teknologi, 79*(5–2), 111–115. https://doi.org/10.11113/jt.v79.11293

Ramu, P., & Sakthivel, G. V. R. (2013). Preparation and characterization of roselle fiber polymer reinforced composites. *International Science and Research Journals, 1*.

Roslan, M. N., Ismail, A. E., Hashim, M. Y., Zainulabidin, M. H., & Khalid, S. N. A. (2014). Modelling analysis on mechanical damage of kenaf reinforced composite plates under oblique impact loadings. *Applied Mechanics and Materials, 465–466*, 1324–1328. https://doi.org/10.4028/www.scientific.net/AMM.465-466.1324

Rujnić-Sokele, M., & Pilipović, A. (2017). Challenges and opportunities of biodegradable plastics: A mini review. *Waste Management & Research, 35*(2), 132–140. https://doi.org/10.1177/0734242X16683272

Sani, M. S. M., Abdullah, N. A. Z., Zahari, S. N., Siregar, J. P., & Rahman, M. M. (2016). Finite element model updating of natural fiber reinforced composite structure in structural dynamics. *MATEC Web of Conferences, 83*. https://doi.org/10.1051/matecconf/20168303007

Sanjay, M. R., Madhu, P., Jawaid, M., Senthamaraikannan, P., Senthil, S., & Pradeep, S. (2017). Characterization and properties of natural fiber polymer composites: A comprehensive review. *Journal of Cleaner Production, 172*, 566–581. https://doi.org/10.1016/j.jclepro.2017.10.101

Sanjay, M. R., Siengchin, S., Parameswaranpillai, J., Jawaid, M., Pruncu, C. I., & Khan, A. (2019). A comprehensive review of techniques for natural fibers as reinforcement in composites: Preparation, processing and characterization. *Carbohydrate Polymers, 207*(March), 108–121. https://doi.org/10.1016/j.carbpol.2018.11.083

Sari, N. H., Pruncu, C. I., Sapuan, S. M., Ilyas, R. A., Catur, A. D., Suteja, S., Sutaryono, Y. A., & Pullen, G. (2020). The effect of water immersion and fiber content on properties of corn husk fibres reinforced thermoset polyester composite. *Polymer Testing, 91*(November), 106751. https://doi.org/10.1016/j.polymertesting.2020.106751

Siakeng, R., Jawaid, M., Ariffin, H., Sapuan, S. M., Asim, M., & Saba, N. (2019). Natural fiber reinforced polylactic acid composites: A review. *Polymer Composites, 40*(2), 446–463. https://doi.org/10.1002/pc.24747

Tajvidi, M., & Takemura, A. (2010). Thermal degradation of natural fiber-reinforced polypropylene composites. *Journal of Thermoplastic Composite Materials, 23*(3), 281–298. https://doi.org/10.1177/0892705709347063

Thiruchitrambalam, M., Athijayamani, A., & Sathiyamurthy, S. (2010). A review on the natural fiber- reinforced polymer composites for the development of roselle fiber-reinforced polyester composite. *Journal of Natural Fibers, 7*, 307–323.

Tholibon, D., Tharazi, I., Sulong, A. B., Muhamad, N., Ismial, N. F., Radzi, M. K. F. M., Mohd Radzuan, N. A., & Hui, D. (2019). Kenaf fiber composites : A review on synthetic and biodegradable polymer matrix. *Journal of Engineering, 31*(1), 65–76.

Touri, M., Kabirian, F., Saadati, M., Ramakrishna, S., & Mozafari, M. (2019). Additive manufacturing of biomaterials – the evolution of rapid prototyping. *Advanced Engineering Materials, 21*(2). https://doi.org/10.1002/adem.201800511

Venkateshwaran, N., Elaya Perumal, A., & Arunsundaranayagam, D. (2013). Fiber surface treatment and its effect on mechanical and visco-elastic behaviour of banana/epoxy composite. *Materials and Design, 47*, 151–159. https://doi.org/10.1016/j.matdes.2012.12.001

Wester, P. J. (1907). Roselle : Its culture and uses. *U.S. Department of Agriculture, 312*(October), 1–16.

Yussuf, A. A., Massoumi, I., & Hassan, A. (2010). Comparison of polylactic acid/kenaf and polylactic acid/rise husk composites: The influence of the natural fibers on the mechanical, thermal and biodegradability properties. *Journal of Polymers and the Environment, 18*(3), 422–429. https://doi.org/10.1007/s10924-010-0185-0

FURTHER READING

Christiyan, K. G. J., Chandrasekhar, U., & Venkateswarlu, K. (2019). Investigation on the mechanical properties of PLA & its composite fabricated through advanced fusion plastic modelling. *Journal of Mechanical Engineering Research and Developments, 42*(3), 47–54. https://doi.org/10.26480/jmerd.03.2019.47.54

Jawaid, M., & Abdul Khalil, H. P. S. (2011). Cellulosic/synthetic fiber reinforced polymer hybrid composites: A review. *Carbohydrate Polymers, 86*(1), 1–18. https://doi.org/10.1016/j.carbpol.2011.04.043

Managooli, V. A. (2009). *Dyeing mesta* (Hibiscus sabdariffa) *fiber with natural colourant.*

Phattanaphibul, T., Opaprakasit, P., Koomsap, P., & Tangwarodomnukun, V. (2007). Preparing biodegradable PLA for powder-based rapid prototyping. In *Asia pacific industrial engineering & management systems conference, no. 2006.*

CHAPTER 16

A Review of the Mechanical Properties of Roselle Fiber-Reinforced Polymer Hybrid Composites

A.M. RADZI • S.M. SAPUAN • M.R.M. HUZAIFAH • A.M. NOOR AZAMMI • R.A. ILYAS • R. NADLENE

1 INTRODUCTION

Privileged use of natural fiber composites can reduce the cost of materials, characterized by good mechanical, biodegradable and renewable. Besides, the production of products using this natural fiber is almost used in various applications such as furniture, automotive parts, medical, textiles, and more (Sathishkumar et al., 2014).

The uniqueness of this natural fiber is in terms of the low cost of materials, easy to obtain, and very low energy consumption compared to conventional fiber. In addition, roselle fiber is also one of the readily available fibers and also have good properties and is closely related to flax fiber to be used as a reinforcing material in polymer composites.

Roselle or *Hibiscus sabdariffa* L. belongs to the hibiscus group (Venkateshwaran et al., 2011). In addition, this roselle tree will produce fruit and the result is used as a food and medical flavor. In fact, the fiber found in the roselle tree can be used as products such as rope and textiles (Sapuan et al., 2013). Roselle fiber consumption is due to good mechanical properties (tensile and impact) and low water immersion when the fiber is used as reinforced polymer composites (Nadlene et al., 2016a).

1.1 Natural Resources

For more than 3000 years, natural fibers were used as reinforcement and to produce products (Taj et al., 2007). With current technology, many researchers have found something valuable on these fibers by combining natural fibers and polymers (Azwa et al., 2013; Owolabi et al., 2020). Various types of natural fibers have been used to study and produce products such as palm sugar, palm oil, rice husk, sugarcane, ramie, sisal, wood, and animal (Nguong et al., 2013). Natural fibers vary in uniqueness and therefore make it an attraction for many researchers to study in depth the users of this fiber. the use of these fibers can reduce forest ecosystems where tree felling can be reduced and reduce open burning of unused fiber (Ishak et al., 2013; Jawaid & Khalil, 2015; Joshi et al., 2004; Yusriah et al., 2012). The fibers which had not been used and burned by farmers, these fibers can be applied in the form of board composites or various forms of the appropriate product and positive impact on the environment (Kalia & Kaur, 2011; Sathishkumar et al., 2014).

1.2 Natural Fibers

Natural fibers have good mechanical and physical properties and can compete with other conventional fibers (El-Shekeil et al., 2014; Ramesh et al., 2013). This material source is very easily available and can be purchased in bulk, low-cost equipment for handling this fiber and nonabrasiveness (Ku et al., 2011). Natural fiber cultivation distribution is found around southeast Asia(Ishak et al., 2013). Demand for palm oil, kenaf, flax, and banana fibers is increasing in the manufacturing industry. The production of these natural fiber-based products gives high profit to factory operators because the products are more environmentally friendly, easy to recycle, and renewable (Manickam et al., 2015; Nadlene et al., 2016b; Razali et al., 2015).

1.3 Synthetic Fibers

Synthetic fibers or artificial fiber make up approximately half of all fiber utilization, with applications in each field of fiber and material innovation. In spite of the fact that numerous classes of fiber-based on engineered polymers have been assessed as possibly

Roselle. https://doi.org/10.1016/B978-0-323-85213-5.00017-2

important commercial items, four of them—nylon, polyester, acrylic, and polyolefin. These four account for around 98% by volume of manufactured fiber generation, with polyester alone taking up around 60% (McIntyre, 2005).

Among the synthetic fibers such as nylon, polyester, and acrylic, polyolefin gives a brief history of the early assessments that are driven to this circumstance. There are several technique processes that involve synthesizing of chemical intermediates, texturing techniques, polymerization methods, orientation technology, fiber spinning and production of microfibres, and chemical variants, and effect for modified dyeability.

1.4 Polymer Composites

Fabric engineers have utilized characteristic filaments to strengthen polymer composites such as the decrease of timber utilization and debasement of the unused characteristic strands. Other preferences incorporate their moo taken a toll, great mechanical properties, plenteous accessibility, fabric renewability, biodegradability, and rough nature for ease of reusing (Azwa et al., 2013; Joshi et al., 2004; Ticoalu et al., 2010). These unused normal filaments can be handled into composite sheets or other shapes reasonable for different applications whereas protecting the environment. In spite of the fact that common strands have a few positive characteristics, they too have undesirable characteristics such as a tall partiality to water, conflicting temperature, and contradiction with polymer lattices that show hydrophobic traits (Georgopoulos et al., 2005; John & Thomas, 2008; Saheb & Jog, 1999). Chemical treatments have been performed for reasons such as anticipating the partiality of water and decreasing the smoothness of the surface (Edeerozey et al., 2007). Analysts distributed a compressive audit article that proposed a few pretreatment strategies such as mercerization, isocyanate, acrylation, permanganate treatment, acetylation, and a silane coupling specialist to make strides in the compatibility of fibers and the matrix (Kalia et al., 2009).

2 ROSELLE PLANTS

2.1 Roselle Plants and Application

Roselle is additionally known as Sabdariffa, whose title was determined from the Turkish words; it is also known as *Hibiscus sabdariffa* L. This tree is classified as a hibiscus and is easily found in tropical regions (Nadlene et al., 2016a). Roselle plant is not only used for fruit (for nourishment or restorative purposes) but its bast fiber is used to deliver the ropes, jute, and material (Navaneethakrishnan & Athijayamani, 2016; Razali et al., 2015).

Roselle is found inexhaustibly in tropical zones. They are commonly utilized as a mixture and to create bast fiber. They are variously utilized by roselle. The natural product is commonly utilized in a therapeutic application (Hudson, n.d.; Mungole & Chaturvedi, 2011), and within the nourishment industry (Wilson, 2009), whereas the fiber is utilized as material (Kalita et al., 2019).

Customarily, roselle seeds are utilized as feedstocks for chicken. Over the past few decades, roselle strands have been utilized for ropes, heavy-duty cables, and the naval force and vendor marine due to their supportability and use in water. Angling nets made from roselle fiber are too amazingly safe and break as they were beneath extraordinary strain (Konar et al., 2018; Nadlene et al., 2016a). It has been illustrated that roselle is rich in niacin, vitamins, riboflavin, carbohydrates, flavonoids, acids, calcium, minerals, and protein (Ismail et al., 2008). Hence, roselle is exceptionally valuable in therapeutic applications (Hudson, n.d.; Mungole & Chaturvedi, 2011). The plant has been detailed to have antihypertensive, hepato-protective, antihyperlipidemic, anticancer, and antioxidant properties (Khare, 2007). In expansion, roselle calyces have moreover been utilized as common nourishment colorants. New roselle or dry calyces are utilized to create juices (Wilson, 2009; Mohamed et al., 2012). As of late, roselle filaments have been broadly utilized as a fortification fabric for composites and material designing (Muthu & Gardetti, 2020; Mwasiagi et al., 2014).

3 ROSELLE FIBERS

Various types of natural fibers are easily obtained from agricultural or forest areas that have the potential to be fiber such as kenaf, banana, pineapple, roselle, sugar palm, palm oil, and many more. These fibers have their uniqueness where they can be used as reinforcement on polymer materials. One of the fibers or trees that are the focus of researchers is the roselle tree.

This tree is closely related to jute (Croschoruscapsularis L.) and is used as reinforcement on polymer materials. This roselle fiber has good ability in mechanical properties. It also has the potential to produce a variety of products for daily use and even in the manufacturing industry such as in the medical, automotive, furniture, aerospace, and souvenir sectors. This fiber is used because its properties can rival or be comparable to other natural fibers such as kenaf, ramie, and sisal. Table 16.1 shows the mechanical properties of roselle fiber carried out by several researchers.

TABLE 16.1
Mechanical Properties of Roselle Fiber.

Density (g/cm^3)	Tensile (MPa)	Tensile Modulus (GPa)	Elongation (%)	Ref
1.41	130–562	–	–	Nadlene et al. (2015)
1.31	165	–	0.65	Manickam et al. (2015)
1.31	80.193–235.019	7.46–18.82	–	Praveen et al. (2016)

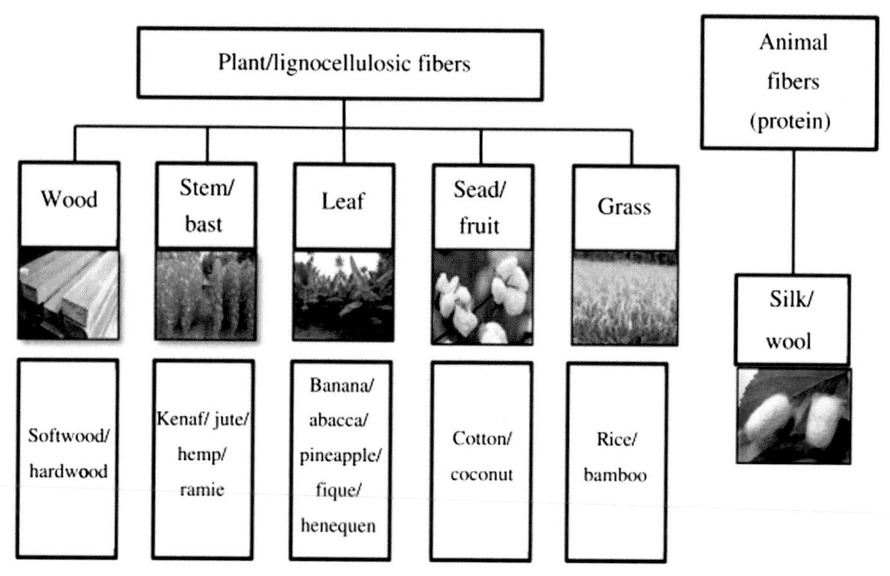

FIG. 16.1 Types of natural fibers (Radzi et al., 2017) (with permission).

To date, even though these roselle fibers have had a positive impact in the manufacturing industry, studies on them are still limited. Therefore, this review was performed to explore in depth the recent breakthroughs on this subject.

4 NATURAL FIBER-REINFORCED POLYMER COMPOSITES

Natural fiber-reinforced polymer-based products have grown since it was introduced into the industry. Natural fibers have two types such as plant and animal fibers. Plant fibers used for various applications are flax, sisal, coconut, kenaf, sugar palm, and abaca. The animal fiber such as silk and wool that is often used since ancient times. Fig. 16.1 shows the types of plant and animal fibers.

The difference between then and now is used through innovation in property modification, design, or manufacturing applications. Natural fiber-reinforced polymer composites are used in a variety of applications such as transporting medical implants, furniture, paper industry, mining, and textiles. Natural fibers also a source of economic income, which depends on the production of the fiber of kenaf, abaca, banana, jute, sisal, and coconut (Kumre et al., 2017). Fig. 16.2 shows type fiber production such as banana, coir (coconut), jute, kenaf, and sisal.

Apart of natural fibers, matrix materials also play an important role in composite systems. Polymers have two different types, namely, thermoset and thermoplastic, which are often used as a matrix for composites. These two types of polymers have their advantages, depending on the desired application. Thermoset

FIG. 16.2 The type fiber production of banana, coconut, jute, kenaf and sisal (Kumre et al., 2017) (With Permission).

TABLE 16.2
Research reported on Natural Fiber composites.

Fiber	Polymer Matrix	Reference
Roselle	TPU Vinyl Easter	Nadlene et al. (2016b), Radzi et al. (2018a)
Sugar palm	Phenolic Vinyl easter TPU	Ammar et al. (2018), Atiqah et al. 2017, Rashid et al. (2016)
Coir	Polyester Polyurethane	Azmi et al. (2012), Prasad et al. (2017)
Sisal	PE Poly (lactic acid)	Orue et al. (2015), Sreekumar et al. (2007)
Pineapple	PP Polycarbonate	Arib et al. (2006), Threepopnatkul et al. (2009)
Kenaf	Epoxy PLA	Ochi (2008), Yousif et al. (2012)
Banana	Polyester Epoxy	Bhoopathi et al. (2014), Idicula et al. (2005)
Silk	Epoxy	Eshkoor et al. (2013)

materials, the preparation of this material is very complicated because it requires several materials such as curing agents, catalysts, and hardeners. This thermoset is chemically cured into a highly cross-linked structure where it can give to a composite that is highly solvent-resistant and creep resistant. The thermoplastic is an easy-to-use material and costs very cheap compared to the thermoset. In addition, thermoplastic properties are easy to design, flexible, durable and have good mechanical properties (Saheb & Jog, 1999). Table 16.2 shows research work on natural fibers composites.

5 NATURAL FIBER HYBRID COMPOSITES

Hybrid composite is a combination of two different matrices or more than one filler material to be used as a reinforcing material (Jawaid & Khalil, 2015). The hybrid composites, containing more than one type of

TABLE 16.3
Research Work on Hybrid Composites.

Natural Fiber	Natural Fiber	Artificial Fiber	Matrix	Reference
Roselle	Sugar palm		TPU	Radzi et al. (2018b)
Wood floor		Glass	PP	Kord and Kiakojouri (2011)
Pineapple	sisal		Polyester	Mishra et al. (2003)
Oil palm		Glass		Mishra et al. (2003)
Sisal		Glass	Epoxy	Rana et al. (2017)
Banana	Sisal		Epoxy	Venkateshwaran et al. (2011)
Jute	Banana		Epoxy	Venkateshwaran and ElayaPerumal (2012)
Jute	Coir		Epoxy	Saw et al. (2012)
Oil palm	Jute		Epoxy	Jawaid et al. (2012)
Hemp		Glass	Pp	Panthapulakkal and Sain (2007)
Kenaf		Kevlar	Epoxy	Yahaya et al. (2014)
Sisal	Oil palm		Natural rubber	Jacob et al. (2004)

fiber, provide advantages or complement with other disadvantages. This hybrid composite consists of two types of fiber, namely, inorganic and organic sources. Organic fiber is a natural fiber that is easily found around and inorganic fiber is like glass, ceramic, and carbon fiber. This combination of fibers aims to increase fatigue life and good mechanical properties compared to single fibers. The strength of this hybrid composite also depends on the properties of the fibers, orientation, length, the interface between fiber and matrix, and failure strain of individual fiber. Various studies have made with a combination of artificial/inorganic fibers showing an increase in mechanical and thermal properties. Therefore, these fibers have a detrimental effect on the environment and humans (Nunna et al., 2012). Due to its adverse effects, many researchers are turning to natural fibers, which are better for humans and the environment, cheaper, and more accessible resources. The use of natural fiber on the composite shows good strength and stiffness value, and it can compete with artificial/inorganic fiber (Goswami et al., 2008; Nunna et al., 2012). Table 16.3 shows research work on hybrid composites.

6 MECHANICAL PROPERTIES OF NATURAL FIBER COMPOSITES

Many researchers reported the potential mechanical properties of composites such as strength, stiffness, rigidity, and specific strength compared to the artificial composites (Pickering et al., 2016). Depending on the specific modulus where the value is comparable or more, it indicates natural fibers have the potential to compete with fiberglass. Therefore, material strength is very important especially from a mechanical point before it becomes the final product. The addition of two types of fiber (natural or synthetic) can enhance the mechanical properties of polymers by enhancing the surface characteristics of natural fibers. However, several factors can affect the mechanical and characterization properties of natural fibers hybrid composites such as fiber loading/volume friction, stacking sequence, fiber treatment, and environmental conditions.

6.1 Fiber Percentage or Weight Fraction of Natural Fiber Composites

Many researchers have studied the mechanical properties of hybrid composites by varying the fiber loading/percentage into the polymer for observation of the positive and negative effects on the material properties. Abdul Khalil et al. (2007) states that the increased fiber percentage between palm oil and glass fiber causes the interaction difficulty of both the fiber and the dispersion matrix is disrupted which can lead to a decrease in tensile and flexural strength. Besides, the fiber percentage also affects the impact test. This effect makes the effectiveness of stress transfer becomes low because the interaction interface between fibers becomes weak. Idicula et al. (2005) reported that hybridization between banana and sisal fiber had a positive effect on tensile, bending and impact tests. Tensile strength and impact at 0.50 friction volume give a high strength value of banana/sisal fiber-reinforced polyester hybrid composite. The increased mechanical strength is due to the high interaction of fiber and matrix dispersion, which the stress transfer through between fiber and matrix can be propagated without failure to the composite.

6.2 Stacking Sequence of Layers

The stacking sequence is the effect of fiber arrangement in a composite hybrid in which the various position of the fiber layer gives an impact on the mechanical properties of the hybrid composites that have been studied by researchers. Naresh et al. (2016) reported the effect of different stacking sequences on tensile strength of glass/hemp/jute reinforced epoxy hybrid composites. Observation shows that the effect of the stacking sequence of fiber has a significant effect on the tensile properties of composites. The fiberglass affects this stacking sequence in enhancing mechanical properties. Fiberglass as extreme plies on both sides showed good improvement in properties. Besides, the continuity of all these fibers indicates a good load-carrying capacity. Jawaid et al. (2011) reported a comparative study of oil palm empty fruit bunches/woven jute reinforced epoxy hybrid composites. The observations have been made on the mechanical properties such as tensile and flexural strength of the hybrid composites. The high fiber content in the composite will affect the mechanical properties. Therefore, in combination with the hybrid layer of jute and the oil palm fiber improved the tensile and flexural strength due to the high content of jute fiber on hybrid composites. Additionally, the strength of jute fiber is higher compared to oil palm fiber.

6.3 Treatment of Fibers

The mechanical strength for hybrid composites depends on the interface bond between the fiber and the matrix. Many studies have been done to increase the strength of the interface by performing fiber treatments as one of the methods to improve and achieve good adhesion fiber surface. This fiber treatment is to further enhance the interaction between the fiber and matrix on increased mechanical properties of the hybrid composite. Atiqah, Jawaid, Ishak, and Sapuan (2017) studied the effect of alkali and silane treatment on mechanical properties of sugar palm and glass fiber-reinforced polyurethane hybrid composites. Their studies show that silane and alkali treatments show optimal tensile strength when compared to untreated. This phenomenon occurs due to the good interface bond between the fiber and the matrix after fiber treated. Fiber surface changes after surface modification of fiber improved the mechanical properties (tensile, flexural, and impact strength) of hybrid composites. Silane and alkaline treatment produce moisture resistance and roughness on the fiber surface, which makes good interaction between the fiber and matrix composites. Singh et al. (2015) investigated the effect of alkaline with maleic anhydride fiber treatment on sisal and hemp-reinforced high-density polyethylene hybrid composites. The fiber treatment of hybrid composites showed good tensile strength compared with single fiber composites. The treatment enhanced the adhesion surface of the fiber and matrix. In this study, they found that the decrease occurred due to maleic anhydride treatment breaking of bundled fibers structure into fine-sized and leads to poor adhesion between the hemp fibers and matrix.

7 MECHANICAL PROPERTIES OF ROSELLE FIBER COMPOSITES

7.1 Tensile Properties

A study on the mechanical properties of hybrid woven jute-/roselle-reinforced polyester composites was done by Hamdan et al. (2019). A few samples were prepared with different layering sequences using jute and roselle fiber. From the result, the addition of jute (JJJJ) and roselle (RRRR) fiber into the composites does not increase the tensile strength. A similar trend was observed for the hybrid composites that show the highest tensile strength was unsaturated polyester (UPE) followed by JJRR, JRJR, JRRJ, and RJJR, respectively. However, the single jute composite shows the highest tensile strength compared to the hybrid jute-roselle composite event do they shared the same volume weight fraction. This result showed that layering selection affects the strength of the composites.

Similar studies also have been conducted by Radzi et al. (2018b) to determine the mechanical performance of roselle/sugar palm fiber hybrid reinforced polyurethane composites. In the study, the ratio of the fiber and matrix that were used is 40:60 weight percentage, which is the percentage the matrix was maintained. Five samples were prepared and labeled as roselle fiber-reinforced polyurethane (RF-T), sugar palm fiber-reinforced polyurethane (SPF-T), and roselle/sugar palm fiber-reinforced polyurethane with a different combination, as shown in Table 16.4.

TABLE 16.4
Composition of Roselle/Sugar Palm Fiber-Reinforced Polyurethane (Radzi et al., 2018b).

RF (%)	SPF (%)	Designation
75	25	RS-1
50	50	RS-2
25	75	RS-3

From the study, RF-T composites have the highest tensile strength and tensile modulus. In contrast, the lowest tensile strength and tensile modulus were SPF-T composites. From Fig. 16.2, hybrid composites (RS-1, RS-2, RS-3) shows the decrease in their tensile strength and modulus after the addition of SPF into the RF composites. The decreased in tensile strength may be due to poor adhesion, high void, fiber pull-outs, incompatibility, or agglomeration, which may have led to low load transfer between the fiber and the matrix.

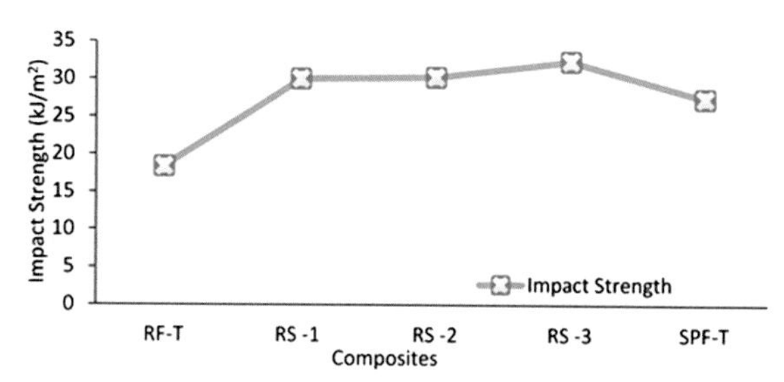

FIG. 16.3 Impact strength of RF/SPF/TPU hybrid composites (Radzi et al., 2018b). (With Permission).

7.2 Flexural Properties

Flexural properties are one of the essential measurements used in a structural application. According to Huzaifah et al. (2019b), the flexural strength of composites material is the capability of composites to endure the stress applied perpendicular to its longitudinal axis. Another study done by Mohamad Hamdan et al. (2019) shows that all composites samples demonstrated superior flexural properties compared to the pure UPE polymer. It indicates that the addition of fiber into UPE matrix increased the flexural strength of the composites. Also, it had better potential to be used as a reinforcing agent in UPE matrix composites for manufacturing products that required more flexural strength.

7.3 Impact Properties

A similar trend with flexural strength was shown by impact strength by Mohamad Hamdan et al. (2019). The addition of fiber into UPE matrix increased the impact strength of all composite samples. From the result also, hybrid composite has better impact strength compared to single fiber composites. According to Huzaifah et al. (2019a), the impact test is to determine the ability of a material to absorb energy under certain pressures. The relationship of interface adhesion, fiber distribution to the composite, the type of fiber used, and the geometry of the fiber are factors that can influence the impact test whether it has a positive or negative effect on the composite material. The failure or low value of the impact test on the material tested is caused by fiber breakage, percentage of fiber in the composite, fiber fracture, and matrix debonding.

A study done by Radzi et al. shows a similar trend which is hybrid composites (RS/SPF/TPU) have a better impact strength compared to composites sample consist only roselle fiber. From Fig. 16.3, RF-T had the lowest impact strength compared to SPF-T. The hybrid sample which are RS-1, RS-2 and RS-3 shows that the combination of SPF and RS fiber has enhanced the impact strength of hybrid composite compared to RF-T composites (Radzi et al., 2018b).

8 CONCLUSIONS

Roselle-reinforced hybrid composites have increased in use lately because of their good properties and safety. A brief discussion on the mechanical properties of roselle fiber hybrid composites was given, along with a review of the previous studies. The mechanical performance of roselle hybrid composites are presented for researchers to see the potential of roselle and other fibers (natural source or conventional fiber) combined to further enhance the properties of polymer composites for use in various product applications. Indirectly, this review can provide various information especially in terms of mechanical properties to study roselle hybrid composites in depth, and some studies have used various natural sources combined with several other types of fiber such as flax, kenaf, banana fiber, conventional fiber, and others. Future research is to produce roselle hybrid composites by combining various fibers, various types of fiber treatment, and performing various mechanical tests (industry scale) to improve the mechanical properties of composite hybrids to produce real products.

REFERENCES

Abdul Khalil, H. P. S., Hanida, S., Kang, C. W., & Nik Fuaad, N. A. (2007). Agro-hybrid composite: The effects on mechanical and physical properties of oil palm fiber (EFB)/glass hybrid reinforced polyester composites. *Journal of Reinforced Plastics and Composites, 26*(2), 203–218. https://doi.org/10.1177/0731684407070027

Ammar, I. M., Huzaifah, M. R. M., Sapuan, S. M., Ishak, M. R., & Leman, Z. B. (2018). Development of sugar palm fiber reinforced vinyl ester composites. In *Natural fibre reinforced vinyl ester and vinyl polymer composites* (pp. 211–224). Elsevier. https://doi.org/10.1016/b978-0-08-102160-6.00011-1

Arib, R. M. N., Sapuan, S. M., Ahmad, M. M. H. M., Paridah, M. T., & Khairul Zaman, H. M. D. (2006). Mechanical properties of pineapple leaf fibre reinforced polypropylene composites. *Materials and Design, 27*(5), 391–396. https://doi.org/10.1016/j.matdes.2004.11.009

Atiqah, A., Jawaid, M., Ishak, M. R., & Sapuan, S. M. (2017). Moisture absorption and thickness swelling behaviour of sugar palm fibre reinforced thermoplastic polyurethane. *Procedia Engineering, 184*, 581–586. https://doi.org/10.1016/j.proeng.2017.04.142

Atiqah, A., Jawaid, M., Sapuan, S. M., & Ishak, M. R. (2017). Effect of surface treatment on the mechanical properties of sugar palm/glass fiber-reinforced thermoplastic polyurethane hybrid composites. *BioResources, 13*(1), 1174–1188. https://doi.org/10.15376/biores.13.1.1174-1188

Azmi, M. A., Yusoff, M. F. C., Abdullah, H. Z., & Idris, M. I. (2012). Rigid polyurethane foam reinforced coconut coir fiber properties. *International Journal of Integrated Engineering, 4*(1), 11–15.

Azwa, Z. N., Yousif, B. F., Manalo, A. C., & Karunasena, W. (2013). A review on the degradability of polymeric composites based on natural fibres. *Materials and Design, 47*, 424–442. https://doi.org/10.1016/j.matdes.2012.11.025

Bhoopathi, R., Ramesh, M., & Deepa, C. (2014). Fabrication and property evaluation of banana-hemp-glass fiber reinforced composites. *Procedia Engineering, 97*(December), 2032–2041. https://doi.org/10.1016/j.proeng.2014.12.446

Edeerozey, A. M. M., Akil, H. M., Azhar, A. B., & Ariffin, M. I. Z. (2007). Chemical modification of kenaf fibers. *Materials Letters, 61*(10), 2023–2025. https://doi.org/10.1016/j.matlet.2006.08.006

El-Shekeil, Y. A., Sapuan, S. M., Jawaid, M., & Al-Shuja'a, O. M. (2014). Influence of fiber content on mechanical, morphological and thermal properties of kenaf fibers reinforced poly(vinyl chloride)/thermoplastic polyurethane polyblend composites. *Materials & Design, 58*, 130–135. https://doi.org/10.1016/j.matdes.2014.01.047

Eshkoor, R. A., Oshkovr, S. A., Sulong, A. B., Zulkifli, R., Ariffin, A. K., & Azhari, C. H. (2013). Effect of trigger configuration on the crashworthiness characteristics of natural silk epoxy composite tubes. *Composites Part B: Engineering, 55*(1), 5–10. https://doi.org/10.1016/j.compositesb.2013.05.022

Georgopoulos, S. T., Tarantili, P. A., Avgerinos, E., Andreopoulos, A. G., & Koukios, E. G. (2005). Thermoplastic polymers reinforced with fibrous agricultural residues. *Polymer Degradation and Stability, 90*(2 SPEC. ISS.), 303–312. https://doi.org/10.1016/j.polymdegradstab.2005.02.020

Goswami, D. N., Ansari, M. F., Day, A., Prasad, N., & Baboo, B. (2008). Jute-fibre glass-plywood/particle board composite. *Indian Journal of Chemical Technology, 15*(4), 325–331.

Hudson, T. (n.d.). A research review ON the use OF HIBISCUS sabdariffa A research review on the use of Hibiscus sabdariffa 2 background and uses. Gaia Herbs. Professional Solution.

Huzaifah, M. R. M., Sapuan, S. M., Leman, Z., & Ishak, M. R (2019a). Effect of fibre loading on the physical, mechanical and thermal properties of sugar palm fibre reinforced vinyl ester composites. *Fibers and Polymers*. https://doi.org/10.1007/s12221-019-1040-0

Huzaifah, M. R. M., Sapuan, S. M., Leman, Z., & Ishak, M. R (2019b). Effect of soil burial on physical, mechanical and thermal properties of sugar palm fibre reinforced vinyl ester composites. *Fibers and Polymers*. https://doi.org/10.1007/s12221-019-9159-6

Idicula, M., Neelakantan, N. R., Oommen, Z., Joseph, K., & Thomas, S. (2005). A study of the mechanical properties of randomly oriented short banana and sisal hybrid fiber reinforced polyester composites. *Journal of Applied Polymer Science, 96*(5), 1699–1709. https://doi.org/10.1002/app.21636

Ishak, M. R., Sapuan, S. M., Leman, Z., Rahman, M. Z. A., Anwar, U. M. K., & Siregar, J. P. (2013). Sugar palm (*Arenga pinnata*): Its fibres, polymers and composites. *Carbohydrate Polymers, 91*(2), 699–710. https://doi.org/10.1016/j.carbpol.2012.07.073

Ismail, A., Hainida, E., Ikram, K., Saadiah, H., & Nazri, M. (2008). Roselle (*Hibiscus sabdariffa* L.) seeds – nutritional composition, protein quality and health benefits. *Food, 2*(1), 1–16.

Jacob, M., Thomas, S., & Varughese, K. T. (2004). Mechanical properties of sisal/oil palm hybrid fiber reinforced natural rubber composites. *Composites Science and Technology, 64*(7–8), 955–965. https://doi.org/10.1016/S0266-3538(03)00261-6

Jawaid, M., Abdul Khalil, H. P. S., & Abu Bakar, A. (2011). Woven hybrid composites: Tensile and flexural properties of oil palm-woven jute fibres based epoxy composites. *Materials Science and Engineering A, 528*(15), 5190–5195. https://doi.org/10.1016/j.msea.2011.03.047

Jawaid, M., Abdul Khalil, H. P. S., & Alattas, O. S. (2012). Woven hybrid biocomposites: Dynamic mechanical and thermal properties. *Composites Part A: Applied Science and Manufacturing, 43*(2), 288–293. https://doi.org/10.1016/j.compositesa.2011.11.001

Jawaid, M., & Khalil, H. P. S. A. (2015). Cellulosic/synthetic fibre reinforced polymer hybrid composites: A review. *Carbohydrate Polymers, 86*(1), 1–18. https://doi.org/10.1016/j.carbpol.2011.04.043

John, M., & Thomas, S. (2008). Biofibres and biocomposites. *Carbohydrate Polymers, 71*(3), 343–364. https://doi.org/10.1016/j.carbpol.2007.05.040

Joshi, S., Drzal, L., Mohanty, A., & Arora, S. (2004). Are natural fiber composites environmentally superior to glass fiber reinforced composites? *Composites Part A: Applied Science and Manufacturing, 35*(3), 371–376. https://doi.org/10.1016/j.compositesa.2003.09.016

Kalia, S., Kaith, B. S., & Kaur, I. (2009). Pretreatments of natural fibers and their application as reinforcing material in polymer composites-a review. *Polymer Engineering & Science, 49*(7), 1253–1272. https://doi.org/10.1002/pen.21328

Kalia, S., & Kaur, B. S. K. (2011). In S. Kalia, B. S. Kaith, & I. Kaur (Eds.), *Cellulose fibers: Bio- and nano-polymer composites*. Berlin, Heidelberg: Springer Berlin Heidelberg. https://doi.org/10.1007/978-3-642-17370-7.

Kalita, B. B., Jose, S., Baruah, S., Kalita, S., & Saikia, S. R. (2019). *Hibiscus sabdariffa* (roselle): A potential source of bast fiber. *Journal of Natural Fibers, 16*(1), 49–57. https://doi.org/10.1080/15440478.2017.1401504

Khare, C. P. (2007). *Hibiscus sabdariffa* Linn. *Indian Medicinal Plants, 8*(1), 1. https://doi.org/10.1007/978-0-387-70638-2_749

Konar, S., Karmakar, J., Ray, A., Adhikari, S., & Bandyopadhyay, T. K. (2018). Regeneration of plantlets through somatic embryogenesis from root derived calli of *Hibiscus sabdariffa* L. (Roselle) and assessment of genetic stability by flow cytometry and ISSR analysis. *PloS One, 13*(8), 1–17. https://doi.org/10.1371/journal.pone.0202324

Kord, B., & Kiakojouri, S. (2011). Effect of nanoclay dispersion on physical and mechanical properties of wood flour/polypropylene/glass fibre hybrid composites. *BioResources, 6*(2), 1741–1751.

Kumre, A., Rana, R. S., & Purohit, R. (2017). A review on mechanical property of sisal glass fiber reinforced polymer composites. *Materials Today: Proceedings, 4*(2), 3466–3476. https://doi.org/10.1016/j.matpr.2017.02.236

Ku, H., Wang, H., Pattarachaiyakoop, N., & Trada, M. (2011). A review on the tensile properties of natural fiber reinforced polymer composites. *Composites Part B: Engineering, 42*(4), 856–873. https://doi.org/10.1016/j.compositesb.2011.01.010

Manickam, C., Kumar, J., Athijayamani, A., & Easter Samuel, J. (2015). Effect of various water immersions on mechanical properties of roselle fiber-vinyl ester composites. *Polymer Composites, 36*(9), 1638–1646. https://doi.org/10.1002/pc.23072

McIntyre, J. E. (Ed.). (2005). *Synthetic fibres: nylon, polyester, acrylic, polyolefin*. Taylor & Francis US.

Mishra, S., Mohanty, A. K., Drzal, L. T., Misra, M., Parija, S., Nayak, S. K., & Tripathy, S. S. (2003). Studies on mechanical performance of biofibre/glass reinforced polyester hybrid composites. *Composites Science and Technology, 63*(10), 1377–1385. https://doi.org/10.1016/S0266-3538(03)00084-8

Mohamad Hamdan, M. H., Siregar, J. P., Thomas, S., Jacob, M. J., Jaafar, J., & Tezara, C. (2019). Mechanical performance of hybrid woven jute–roselle-reinforced polyester composites. *Polymers and Polymer Composites, 27*(7), 407–418. https://doi.org/10.1177/0967391119847552

Mohamed, B. B., Sulaiman, A. A., & Dahab, A. A. (2012). Roselle (*Hibiscus sabdariffa* L.) in Sudan, cultivation and their uses. *Bulletin of Environment, Pharmacology and Life Sciences, 1*(6), 48–54.

Mungole, A., & Chaturvedi, A. (2011). *Hibiscus sabdariffa* L a rich source of secondary metabolites. *International Journal of Pharmaceutical Sciences Review and Research, 6*(1), 83–87.

Muthu, S. S., & Gardetti, M. A. (2020). *Sustainability in the textile and apparel industries: Sourcing synthetic and novel alternative raw materials*. Retrieved from http://www.springer.com/series/16490.

Mwasiagi, J. I., Yu, C. W., Phologolo, T., Waithaka, A., Kamalha, E., & Ochola, J. R. (2014). Characterization of the Kenyan Hibiscus sabdariffa L.(Roselle) bast fibre. *Fibres & Textiles in Eastern Europe.*

Nadlene, R., Sapuan, S. M., Jawaid, M., & Ishak, M. R. (2015). Mercerization effect on morphology and tensile properties of roselle fibre. *Applied Mechanics and Materials, 754–755*, 955–959. https://doi.org/10.4028/www.scientific.net/AMM.754-755.955

Nadlene, R., Sapuan, S. M., Jawaid, M., Ishak, M. R., & Yusriah, L. (2015). Material characterization of roselle fibre (*Hibiscus sabdariffa*L.) as potential reinforcement material for polymer composites. *Fibres and Textiles in Eastern Europe, 23*(6(114)), 23–30. https://doi.org/10.5604/12303666.1167413

Nadlene, R., Sapuan, S. M., Jawaid, M., Ishak, M. R., & Yusriah, L. (2016a). A review on roselle fiber and its composites. *Journal of Natural Fibers, 13*(1), 10–41. https://doi.org/10.1080/15440478.2014.984052

Nadlene, R., Sapuan, S. M., Jawaid, M., Ishak, M. R., & Yusriah, L. (2016b). The effects of chemical treatment on the structural and thermal, physical, and mechanical and morphological properties of roselle fiber-reinforced vinyl ester composites. *Polymer Composites, 39*(1), 274–287. https://doi.org/10.1002/pc.23927

Naresh, C., Kumar, Y. R., & Manikantesh, K. (2016). Effect of stacking sequence and orientation on tensile response of natural fiber reinforced hybrid composites: Fibrous - glass/hemp / jute / epoxy composite plates. *International Journal of Engineering Research & Technology, 5*(04), 161–167.

Navaneethakrishnan, S., & Athijayamani, A. (2016). Mechanical properties and absorption behavior of CSP filled roselle fiber reinforced hybrid composites. *Materials and Environment Science, 7*(5), 1674–1680.

Nguong, C. W., Lee, S. N. B., & Sujan, D. (2013). A review on natural fibre polymer composites. *International Journal of Materials and Metallurgical Engineering, 1*, 52–59. https://doi.org/10.5281/zenodo.1332600

Nunna, S., Chandra, P. R., Shrivastava, S., & Jalan, a. (2012). A review on mechanical behavior of natural fiber based hybrid composites. *Journal of Reinforced Plastics and Composites, 31*(11), 759–769. https://doi.org/10.1177/0731684412444325

Ochi, S. (2008). Mechanical properties of kenaf fibers and kenaf/PLA composites. *Mechanics of Materials, 40*(4–5), 446–452. https://doi.org/10.1016/j.mechmat.2007.10.006

Orue, A., Jauregi, A., Peña-Rodriguez, C., Labidi, J., Eceiza, A., & Arbelaiz, A. (2015). The effect of surface modifications on sisal fiber properties and sisal/poly (lactic acid) interface adhesion. *Composites Part B: Engineering, 73*, 132–138. https://doi.org/10.1016/j.compositesb.2014.12.022

Owolabi, F. A. T., Deepu, A. G., Thomas, S., Shima, J., Rizal, S., Sri Aprilia, N. A., & Abdul Khalil, H. P. S. (2020). Green composites from sustainable cellulose nanofibrils. *Encyclopedia of Renewable and Sustainable Materials*, 81–94. https://doi.org/10.1016/b978-0-12-803581-8.11422-5

Panthapulakkal, S., & Sain, M. (2007). Studies on the water absorption properties of short hemp—glass fiber hybrid polypropylene composites. *Journal of Composite Materials, 41*(15), 1871–1883. https://doi.org/10.1177/0021998307069900

Pickering, K. L., Efendy, M. G. A., & Le, T. M. (2016). A review of recent developments in natural fibre composites and their mechanical performance. *Composites Part A, 83*, 98–112. https://doi.org/10.1016/j.compositesa.2015.08.038

Prasad, G. L. E., Gowda, B. S. K., & Velmurugan, R. (2017). A study on impact strength characteristics of coir polyester composites. *Procedia Engineering, 173*, 771–777. https://doi.org/10.1016/j.proeng.2016.12.091

Praveen, V., Saran, C., Sasikumar, M., Praveen, K., Manickam, P. C., & Kumarasamy, M. (2016). Study of mechanical behaviour of natural hybrid fibre reinforced polymer matrix composite by using Roselle and Luffa fibre abstract. *International Journal for Innovative Research in Science & Technology, IV*, 99–105.

Radzi, A. M., Sapuan, S. M., Jawaid, M., & Mansor, M. R. (2017). Influence of fibre contents on mechanical and thermal properties of roselle fibre reinforced polyurethane composites. *Fibers and Polymers, 18*(7), 1353–1358.

Radzi, A. M., Sapuan, S. M., Jawaid, M., & Mansor, M. R. (2018a). Mechanical and thermal performances of roselle fiber-reinforced thermoplastic polyurethane composites. *Polymer-Plastics Technology and Engineering, 57*(7), 601–608. https://doi.org/10.1080/03602559.2017.1332206

Radzi, A. M., Sapuan, S. M., Jawaid, M., & Mansor, M. R. (2018b). Mechanical performance of roselle/sugar palm fiber hybrid reinforced polyurethane composites. *BioResources, 13*(3), 6238–6249. https://doi.org/10.15376/biores13.3.6238-6249

Ramesh, M., Palanikumar, K., & Reddy, K. H. (2013). Comparative evaluation on properties of hybrid glass fiber-sisal/jute reinforced epoxy composites. *Procedia Engineering, 51*(NUiCONE 2012), 745–750. https://doi.org/10.1016/j.proeng.2013.01.106

Rana, R. S., Kumre, A., Rana, S., & Purohit, R. (2017). Characterization of properties of epoxy sisal/glass fiber reinforced hybrid composite. *Materials Today: Proceedings, 4*(4), 5445–5451. https://doi.org/10.1016/j.matpr.2017.05.056

Rashid, B., Leman, Z., Jawaid, M., Ghazali, M. J., & Ishak, M. R. (2016). The mechanical performance of sugar palm fibres (ijuk) reinforced phenolic composites. *International Journal of Precision Engineering and Manufacturing, 17*(8), 1001–1008. https://doi.org/10.1007/s12541-016-0122-9

Razali, N., Salit, M. S., Jawaid, M., Ishak, M. R., & Lazim, Y. (2015). A study on chemical composition, physical, tensile, morphological, and thermal properties of roselle fibre: Effect of fibre maturity. *BioResources, 10*(1), 1803–1823.

Saheb, D. N., & Jog, J. P. (1999). Natural fiber polymer composites: A review. *Advances in Polymer Technology, 18*(4), 351–363. https://doi.org/10.1002/(SICI)1098-2329(199924)18

Sapuan, S. M., Pua, F., El-Shekeil, Y. A., & AL-Oqla, F. M. (2013). Mechanical properties of soil buried kenaf fibre reinforced thermoplastic polyurethane composites. *Materials & Design, 50*, 467–470. https://doi.org/10.1016/j.matdes.2013.03.013

Sathishkumar, T. P., Navaneethakrishnan, P., Shankar, S., & Rajasekar, R. (2014). Mechanical properties and water absorption of short snake grass fiber reinforced isophthallic polyester composites. *Fibers and Polymers, 15*(9), 1927–1934. https://doi.org/10.1007/s12221-014-1927-8

Saw, S. K., Sarkhel, G., & Choudhury, A. (2012). Preparation and characterization of chemically modified Jute-Coir hybrid fiber reinforced epoxy novolac composites. *Journal of Applied Polymer Science, 125*(4), 3038–3049. https://doi.org/10.1002/app.36610

Singh, N. P., Aggarwal, L., & Gupta, V. K. (2015). Tensile behavior of sisal/hemp reinforced high density polyethylene hybrid composite. *Materials Today: Proceedings, 2*(4–5), 3140–3148. https://doi.org/10.1016/j.matpr.2015.07.102

Sreekumar, P. A., Joseph, K., Unnikrishnan, G., & Thomas, S. (2007). A comparative study on mechanical properties of sisal-leaf fibre-reinforced polyester composites prepared by resin transfer and compression moulding techniques. *Composites Science and Technology, 67*(3–4), 453–461. https://doi.org/10.1016/j.compscitech.2006.08.025

Taj, S., Munawar, M. A., & Khan, S. (2007). Natural fiber-reinforced polymer composites. *Carbon, 44*(2), 129–144. Retrieved from http://apps.isiknowledge.com/InboundService.do?product=WOS&action=retrieve&SrcApp=Papers&UT=000258388200018&SID=2E7cMAE37HiPjHo@oAg&SrcAuth=mekentosj&mode=FullRecord&customersID=mekentosj&DestFail=http://access.isiproducts.com/custom_images/wok_failed_aut.

Threepopnatkul, P., Kaerkitcha, N., & Athipongarporn, N. (2009). Effect of surface treatment on performance of pineapple leaf fiber-polycarbonate composites. *Composites Part B: Engineering, 40*(7), 628–632. https://doi.org/10.1016/j.compositesb.2009.04.008

Ticoalu, A., Aravinthan, T., & Cardona, F. (2010). A review of current development in natural fiber composites for structural and infrastructure applications. In *Southern region engineering conference 2010, SREC 2010 - incorporating the 17th annual international conference on mechatronics and machine vision in practice, M2VIP 2010, (November)* (pp. 113–117).

Venkateshwaran, N., & ElayaPerumal, A. (2012). Mechanical and water absorption properties of woven jute/banana hybrid composites. *Fibers and Polymers, 13*(7), 907–914. https://doi.org/10.1007/s12221-012-0907-0

Venkateshwaran, N., ElayaPerumal, A., Alavudeen, A., & Thiruchitrambalam, M. (2011). Mechanical and water absorption behaviour of banana/sisal reinforced hybrid composites. *Materials & Design, 32*(7), 4017–4021. https://doi.org/10.1016/j.matdes.2011.03.002

Wilson, W. (2009). *Discover the Many Uses of the Roselle Plant!*. Singapore: National Parks (NParks).

Yahaya, R., Sapuan, S. M., Jawaid, M., Leman, Z., & Zainudin, E. S. (2014). Mechanical performance of woven kenaf-Kevlar hybrid composites. *Journal of Reinforced Plastics and Composites, 33*(24), 2242–2254. https://doi.org/10.1177/0731684414559864

Yousif, B. F., Shalwan, A., Chin, C. W., & Ming, K. C. (2012). Flexural properties of treated and untreated kenaf/epoxy composites. *Materials and Design, 40*, 378–385. https://doi.org/10.1016/j.matdes.2012.04.017

Yusriah, L., Sapuan, S. M., Zainudin, E. S., & Mariatti, M. (2012). Exploring the potential of betel nut husk fiber as reinforcement in polymer composites: Effect of fiber maturity. *Procedia Chemistry, 4*, 87–94. https://doi.org/10.1016/j.proche.2012.06.013

CHAPTER 17

A Review of the Physical and Thermal Properties of Roselle Fiber-Reinforced Polymer Hybrid Composites

A.M. RADZI • S.M. SAPUAN • M.R.M. HUZAIFAH • A.M. NOOR AZAMMI • R.A. ILYAS • R. NADLENE • J.M. KHIR

1 INTRODUCTION

Recently, many researchers have shifted to the consumption of natural fibers in their research, and it was ideal economically to produce good quality fiber-reinforced polymer composites for furniture, building, structural, and other needs (Huzaifah et al., 2017; Sahari et al., 2014; Sastra et al., 2006). Roselle is one of the fibers derived from natural sources, and it is an alternative source for the researcher to use as reinforced polymer composites. In addition, it is also one of the fibers that have environmentally friendly characteristics, biodegradability, and low material cost (Harish et al., 2015; Junkasem et al., 2006; Radzi et al., 2019a). Within the assortment of common strands, roselle fiber has been utilized customarily within the generation of high-strength material, ropes, and floor mats (Chauhan & Kaith, 2012; Nadlene et al., 2016; Navaneethakrishnan & Athijayamani, 2016; Soundhar et al., 2019). It includes an interesting intrigued in composite frameworks since it has solid ductile properties such as kenaf, jute, and moderates to affect quality compared with other normal filaments. Therefore, a review paper was concentrated on the mechanical properties of roselle fiber-reinforced polymer hybrid composites to provide a perfect source of literature for doing further research.

1.1 Natural Resources and Fibers

The natural fiber is chosen to be commercialized in the industry because of its density compared with manufactured fiber, which leads to the making of a lightweight composite (Thyavihalli Girijappa et al., 2019). As a result, there's an increment within the request for the commercial use of characteristic fiber-based composites in different mechanical segments. Subsequently, common filaments such as hemp, jute, sisal, banana, coir, and kenaf are broadly utilized within the generation of the lightweight composites (Sreekumar et al., 2009; Thakur et al., 2014).

The characteristic fiber-based composites have been utilized in car interior linings (roof, raise divider, side panel lining), furniture, development, bundling, shipping pallets, and so forth (Lau et al., 2018; Oksman, 2001). Common filaments are extracted from diverse plants and animals (chicken feathers, hair, etc.; Aziz & Ansell, 2004; Huda et al., 2006).

The plant filaments are made up of constituents like cellulose, lignin, hemicellulose, pectin, waxes, and water-soluble substances. The nearness of cellulose, which is hydrophilic in nature, influences the interfacial holding between the polymer lattice and the strands since the framework is hydrophobic. Chemical treatment of the common filaments is one of the ways to optimize the interaction between the strands and polymer framework. Because it decreases the OH useful bunches show on the fiber surface additionally it increments the surface harshness and thus upgrades the interfacial interaction between the network and the strands (Liu et al., 2005; Mahjoub et al., 2014; Noor Azammi et al., 2018b), the study of natural fibers as primary research is very essential to create ecofriendly composites.

1.2 Roselle Plant and Application

Roselle is additionally known as Sabdariffa, whose title was inferred from Turkish. Roselle is formally known as *Hibiscus sabdariffa* L. This tree is classified with several hibiscus trees and is easily found in tropical ranges (Nadlene et al., 2016). The roselle plant is not used for its fruit (for nourishment or therapeutic purposes) but for its bast fiber, and it is additionally utilized for

Roselle. https://doi.org/10.1016/B978-0-323-85213-5.00004-4

rope, jute, and other material (Nadlene et al., 2016; Navaneethakrishnan & Athijayamani, 2016; Razali et al., 2015).

Roselle is found abundantly in tropical zones. They are commonly utilized as implantation and to form bast fiber. The characteristic item is commonly utilized in restorative applications and inside the food industry, while the fiber is utilized as the fabric (Tori Hudson, 2011; Mungole & Chaturvedi, 2011).

Generally, roselle seeds are utilized as feedstock for chickens. Over the past few decades, roselle fibers have been utilized for ropes, heavy-duty cables, and maritime drive and merchant marine purposes due to their practicality and usefulness in water. Using nets made from roselle fiber is secure and is less likely to break under strain (Crane, 1949). The roselle nutrient contained are protein, vitamin, high calcium, carbohydrate, mineral, riboflavin, niacin, and flavonoid (Mohamad et al., 2002). Subsequently, roselle is especially profitable in restorative applications (Mungole & Chaturvedi, 2011). Fig. 17.1 shows the roselle plant and its fruit.

In addition to the benefits of the fruit, this roselle can protect humans who eat roselle fruit due to its antioxidant, cancer, hypertensivity, and hepatic properties (Mahadevan & Kamboj, 2007). In development, roselle calyces have been utilized as common food colorants. Unused roselle or dry calyces are utilized to form juices (Abdul Khalil et al., 2012). As of late, roselle strands have been broadly utilized as a bolster fabric for composites and material plans (Wilson, 2009; Das Gupta, 1959).

FIG. 17.1 Roselle plant.

1.3 Roselle Fiber

There's an assortment of common (Nadlene et al., 2015) filaments that can be obtained from the environment that has been explored to fill within the polymer system as reinforces. One of these strands may be a sort of roselle fiber (*Hibiscus sabdariffa* L) and was closely related to jute fiber to fortify fabric. The word *Sabdariffa* is inferred from Turkish, which means the root of the tree. In fact, it is also characterized as a hibiscus group and easy to grow in Southeast Asia (Nadlene et al., 2015). Fig. 17.1 shows the roselle plant.

Roselle plant is utilized as a natural product for nourishment or therapeutics, but it too utilized the bast fiber to create rope, jute, and other material (Junkasem et al., 2006; Navaneethakrishnan & Athijayamani, 2016). The roselle has the potential to apply within the car portion, biomedical, and aviation industry since roselle fiber is comparable to other characteristic fiber such as jute, ramie, kenaf, and sisal fiber. Fig. 17.2 shows roselle fiber after the water-retting process.

1.4 Natural Resources and Synthetic Fibers Composites

The thermal durability characterization on the filaments become priorities in the handling of composite materials to produce products of fiber-reinforced very limited especially on polypropylene and polyethylene,

FIG. 17.2 Roselle fiber.

it is caused by the high thermal for designing plastics such as nylon and polyethylene which will cause deterioration of heat naturally (Tajvidi et al., 2009). Be that as it may, certain under-the-hood applications within the vehicle industry are unforgiving for commodity plastics to resist (Johnson et al., 2004). Plastic materials such as nylon play an important role in the automotive industry, especially in the lid part because it has a high resistance to oils, high temperatures, and chemicals that can interfere with quality and modulus. Moreover, it is easily shaped according to the desired specifications and easily combined with other parts (Graff, 2005).

The preferences of normal strands over traditional reinforcing materials such as glass filaments, powder, and mica are satisfactory particular quality properties, low cost, low thickness, sturdiness, great warm properties, diminished instrument wear, decreased dermal, and respiratory disturbance. ease of division, upgraded vitality recuperation, and biodegradability. It has been illustrated that wood-fiber strengthened PP composites have properties comparable to conventional glass fiber fortified PP composites (Krishnan & Narayan, 1992). Lignocellulosic-plastic composites have been checked on by Kowell, Youngquist, and Narayan (Kennedy, 1993). The poor combination of filaments with other materials, moisture absorption, and dimensional changes are things that need full attention when using these fibers. A survey by Maldas and Kokta (1993) covers the complexities included with the compatibilization of these materials and the distinctive methods utilized to understand the interfacial interaction. The effectiveness of a fiber-reinforced composite depends on the fiber-matrix interface and the capacity to exchange stretch from the framework to the fiber.

1.5 Physical Properties

Inquiries into the age of roselle fiber has been conducted by Razali et al. (2015). The study investigated three diverse fiber ages: three, six, and nine months. Measurements showed the three-month-old fiber displayed the littlest run of fiber breadth in the range of 40–100 μm. The results of the study found that the best natural fiber used to produce various products for three-month-old fiber. In any case, the cross-section of roselle fiber shifts the bundle of single strands. It is troublesome to initiate a single fiber estimation with uncovered eyes. From the observation, there was a refinement within the cross-section between the three unmistakable age types of fiber, and the separate over of roselle fiber extended from three to nine months. From the study accomplished, in conclusion, fiber separation increases with age since the cell divider gets thicker since it creates. The components impacting the particular physical properties of roselle fiber are the cell divider thickness, breadth, and length of the tracheid inside the fibers. The comes about that appeared were taken from 15 tests (Rowell, Han, & Rowell, 2000). Roselle fiber may be a fine bast fiber. The physical properties of characteristic filaments depend on several components. Actually, it is difficult to urge the consistency properties of common fiber (Chandramohan & Marimuthu, 2011), and moreover, the nature of fiber depends on the cultivation, crop development, condition plant, extraction, and resources obtained (Reddy & Yang, 2005).

The density of the Roselle for three, six, and nine months were found to be 1.332, 1.419, and 1.421 g/cm^3, respectively. It is obvious that the thickness grows as the plant develops. The thickness of roselle fiber is generally low. This is contributed by the presence of a lumen (Vilay et al., 2008). This characteristic contributes to fiber delicacy, one of the foremost alluring variables of normal fiber as a support fabric for composite items.

Without a doubt, although characteristic strands have different centres of interested relative to them utilize as a back surface such as being regularly accepted and promising, for the foremost portion, comparable properties as made fiber, there are still defect in standard fiber. The hydrophilic behavior present in natural fibers usually has a combined weakness between the fiber and matrix. This contributes to high water absorption and weakens the composite properties in various applications (Nguong, Lee & Sujan, 2013). In any case, surface treatment of fiber is one method of repairing existing weaknesses (Xie et al., 2010). From the same soil that collected, rates of water maintenance of roselle fiber three, six, and nine months were averagely tall. The lumen structure joins astounding preference toward the water. The more lumen exists, the more water is absorbed by roselle strands. These consider other than exist due to the cellulose substance in characteristic strands in common and roselle fiber. Another rate of cellulose substance increases free hydroxyl (Yusriah et al., 2014). In this examination, it was found that a cellulose substance was the foremost growth for a three-month-old plant. This appears that the water uptake comes around were in unimaginable understanding with other scattered composing. The water maintenance of common fiber must be lessened to produce a tall quality composite. In advancement, fiber and framework association can be improvement made strides by fortifying the composite.

1.6 Thermal Properties

An examination was conducted on the characteristic information of the warm behavior of roselle fiber at tall temperatures (Razali et al., 2015). Around 5 mg of roselle fiber was utilized to assess the thermal conduct. For the most part, there are four stages within the warm degradation of normal fiber (Sathishkumar et al., 2013; Shahzad, 2013).

The primary debasement is dampness dissipation, taken after by the decay of hemicelluloses, cellulose, and lignin, clearing out fiery debris as the ultimate buildup (Ishak et al., 2012). The primary debasement of normal strands happens between 30°C and 110°C (De Rosa et al., 2010). Ordinarily due to the vanishing of moistness substance inside the fiber. Inside the case of roselle fiber, dissemination of dampness amplified from 30°C to 110°C. As the temperature of fiber increases, while it is warmed, the fibers ought to be lighter due to the dissemination of bound water and unsteady extractives. Indeed, in spite of the fact that less unsteady extractives have appeared, it will move toward the outer surface of the fiber stem surface. This improvement of unsteady extractives happens due to the water advancement from the internal to the outside parcel of the fiber stem surface that has water on the outside parcel disseminates. The unsteady extractives coalesce and move to the fiber surface (the outside parcel of the fiber stem). It can be seen that nine-month-old roselle fiber shows the slightest rate of mass misfortune, which 4.1%, whereas the three-month and six-month cases showed up 10.28% and 8.25%, exclusively. This mass misfortune reflects the clammy substances of the roselle fiber.

The lignocellulose component breaks down inside the run of 200°C–520°C. The minute arrangement and warm debasement of roselle fiber is due to the thermochemical changes of hemicellulose substances inside the fiber, caused by high temperatures that damage the cells. For three-month-old roselle fiber, hemicelluloses will break down at 220°C–350°C. The six- and nine-month-old roselle fiber degrades at 200°C–315°C and 210°C–320 °C, independently. Hemicelluloses degenerate earlier compared with the lignocellulosic, cellulose, and lignin components. Cellulose is more heat resistant and unfaltering compared to hemicelluloses. Ordinarily, due to the reality that the hemicellulose component contains heterogeneous polysaccharides such as glucose, mannose, galactose, and xylose, such polysaccharides are especially indistinct, which licenses them to viably migrate from the largest stem. Unavoidably, the hemicellulosic saccharides finished up volatiles at lower temperatures (Yang et al., 2007).

The minute organization includes the cellulose structure. The debasement of cellulose will start after hemicellulose rot is added up. The foremost reason behind the higher substance of crystalline chain compared is unclear. In this stage, cellulose is more thermally steady (Ishak et al., 2012). For the most part, for all unmistakable ages of roselle fiber, cellulose breaks down at a temperature range of 315°C and 400°C. Cellulose degenerates at a temperature range of 315°C (Yang et al., 2007). Once the desired temperature is fulfilled, the rot starts and the mass misfortune rate is fast. The rate of weight hardship for three, six, and nine-month-old roselle fiber is 76.36%, 62.27%, and 63.69%, separately. The three-month-old cases appeared up the preeminent basic weight occurrence in this temperature amplify since of the first basic cellulose substance. No change in all three ages of roselle fiber. Overall, the decrease occurs at the same temperature of hemicellulose and cellulose components. The probability occurs due to insignificant changes in chemical substances between the ages of the fiber. The chemical substance of roselle fiber is related to the warm behavior of the fiber.

2 FACTORS AFFECTING PHYSICAL AND THERMAL BEHAVIOR OF HYBRID COMPOSITES

This chapter will explore research activities on the effect of hybrid composites: either a single type of fiber with combined composite or the other way around toward the thermal and physical effect.

2.1 Fiber Percentage or Weight Fraction

Studies on hybrid natural fiber with various composite has been explored by many researchers, for example, kenaf fiber with combined composites of natural rubber and thermoplastic (Noor Azammi et al., 2018a). The varies of fiber and composites composition has resulted in different physical and thermal results.

The thermal effect of the 1KF4NR3TPU frameworks has the least char buildup of 4.9% compared to other frameworks. The kenaf cellulose began to break down at 300°C–350°C and after 435°C the polymer was damaged in the shape of char buildup. The most noteworthy substance of elastic within the composite allows the highest warm resistance within the polymer composite (Azwa & Yousif, 2013; Sarifuddin & Ismail, 2013; Srinivasan et al., 2015). The same substance sum of elastic moreover appears the least buildup or char buildup.

The physical effect of water absorption, the water assimilation esteem was found to extend with the increase of the TPU within the polymer composites. In the interim, the polymer composites with the next sum of common elastic T143 appeared lower esteem of water retention. The T116 composition of treated kenaf fiber-filled tended to retain dampness more than other tests (Azammi et al., 2020). This clearly shown the increase of the dampness retention was due to destitute holding or grip between TPU and treated kenaf fiber. On the other hand, NR had superior holding and attachment (Ahad et al., 2014; Anuar et al., 2010) for treated kenaf fiber, which had less dampness to assimilate for the T143 test.

2.2 Treatment of Fibers

A study on alkaline treated on kenaf fiber has been conducted by using NaOH as a treatment on kenaf fiber utilized in TPUR/NR composites on their mechanical properties was explored. A critical change within the impact characteristics (increments that affect quality by up to 127%) was watched for the composites with alkali-treated kenaf strands in comparison with those containing untreated strands (Noor Azammi et al., 2018b). This appears that the NaOH treatment of the kenaf filaments includes a positive impact on the holding between the fiber and the polymer framework. Concurring to other ponders (Ismail et al., 2010) antacid treatment cleans kenaf fiber surface and may make strides in the fiber-matrix attachment.

2.3 Effect of Environmental Conditions

The dampness aspect of normal fiber is a vital criterion that must be considered in choosing natural fiber as support fabric. Dampness impacts dimensional consistent quality, electrical resistivity, moldable quality, porosity, and swelling behavior of common fiber in a composite texture. From other dispersed composing, it was found that the moo moistness substance of the typical fiber is the first appealing criteria for polymer composites in organizing to overcome the issues (Jawaid & Khalil, 2015). Composites combined with moistness substance fiber are less likely to spoil in separation to composites combined with tall clamminess substances. As a rule, it may be due to the ability of the fiber to hold water in the composite, which can cause composite damage (Rowell et al., 2000). The result of clamminess substances of roselle fiber illustrated that plant age of nine months antiquated roselle strands was slightest. In this way, nine-month-old roselle fiber is most fitting in making composites to have tall dimensional soundness and good quality of the thing. By and expansive, fiber interior these three assorted plant ages is palatable to be utilized as a common fiber in composites product in terms of moistness substance and water maintenance since the comes approximately are comparable to other built up characteristic fiber such as kenaf and jute, where their amplify is 3%–5% and ~200%, separately (Saheb & Run, 1999).

3 PHYSICAL PROPERTIES OF ROSELLE FIBER AND HYBRID COMPOSITES

3.1 Water Absorption Properties

The drawbacks of utilizing characteristic strands in polymer networks included the tall dampness retention of the filaments and composites. The dampness assimilation propensity in polymer matrix composites has the most impact on their mechanical properties. According to Wang et al. (2010), composite materials that contained natural fibers have several adverse effects on their properties and thus affect their long-term performance due to high moisture absorption. Modibbo et al. (2007) have explored the dampness assimilation characteristics of four cellulosic bast filaments: roselle, okra, baobab, and kenaf with and without treatment. The dampness assimilation propensities for the treated materials were: roselle–okra–baobab–kenaf. Subsequently, these filaments were appropriate to be utilized as an elective to inorganic/mineral-based fortifying strands.

Generally, moisture absorption increases with increasing fiber loading (Huzaifah et al., 2019). The dampness retention of pineapple-leaf fiber-reinforced low-density polyethylene composites was explored. It was found that the dampness retention expanded nearly straightly with the fiber stacking (Huzaifah et al., 2019). All polymers and their composites assimilate dampness in a muggy environment and when submerged in water. Common strands assimilate more dampness compared with manufactured strands. The impact of this ingested dampness was to corrupt the properties such as ductile quality.

Nadlene et al., (2016) have done investigate the impacts of chemical treatment on the auxiliary and heat, physical, and mechanical and morphological properties of roselle fiber-fortified vinyl ester composites. Four diverse medications for roselle fiber were submerged into three concentrations of NaOH (3%, 6%, 9%) at room temperature. In the silane coupling treatment, the 6% NaOH-treated filaments were encourage submerged in silane arrangements for 24 h. To arrange the composites test, a damp hand lay-up preparation was utilized by blending 5 wt% of roselle fiber and 92.5 wt % of vinyl ester lattice. All tests show that the water assimilation happens in increments, most likely due to the hydrophilic of the plant-based fiber. The water retention rate expanded with inundation time, coming to certain esteem at an immersion level, past which no more water was ingested, and the water substance ingested by the composite examples remained steady.

Navaneethakrishnan and Athijayamani (2016) have done a study on mechanical properties and absorption behavior of coconut shell particulates (CSP) filled roselle fiber-reinforced hybrid composites. Fiber-reinforced vinyl ester composites were prepared by incorporating coconut shell at four different filler contents viz. 0%, 5%, 10%, and 15%. From the results, the water absorption goes on increasing with the increase of time duration and CSP filler content. The maximum water absorption is obtained from the composite prepared with 15% CSP filler.

3.2 Thickness Swelling Properties

The thickness swelling was calculated on the thickness examples sometime recently and after water submersion for 24 h. This test was conducted to analyze the changes within the dimensional solidness of the composites (Radzi et al., 2019b). Regularly, thickness swelling comes about to have comparable slant with water retention properties. Radzi et al. (2019b) explored water assimilation, thickness swelling, and warm properties of roselle/sugar palm fiber strengthened thermoplastic polyurethane cross-breed composites. The hybridized forms of roselle and sugar palm fiber were arranged at distinctive weight apportions through softening blending and hot compression strategy. The thickness of swelling comes about uncovered that an increment in roselle fiber substance driven to an increment in water take-up and thickness of swelling of the roselle fiber (RF)/sugar palm fiber (SPF) cross breed composites.

4 THERMAL PROPERTIES OF ROSELLE FIBER AND HYBRID COMPOSITES

The execution of roselle fiber-fortified polymer composites are more often than not measured by their physical and warm properties, such as water assimilation, pliability, thermalgravimetric examination (TGA), differential filtering calorimetry (DSC), and energetic mechanical investigation (DMA). The thermal properties of roselle composites, as it has been proven from previous studies, have a significant influence on the behavior between fibers and composites. Singhaa and Thakura (Singhaa & Thakura, 2008), investigated the effect of thermal properties of single RF, phenol-formaldehyde (PF), and RF/PF composites with 10% fiber loading. TGA thermal studies were performed on PF, RF, and roselle/PF supported by differential thermal analysis (DTA) and thermogravimetric differential (DTG). The result show thermal changes when the PF matrix is reinforced with RF. Overall, the author concludes that the PF matrix was slightly thermally stable compared to the polymer composites, possibly caused by the presence of cellulose of RF on the polymer matrix disrupting the original crystal lattice structure of PF matrix (Sgriccia & Hawley, 2007; Thakur & Singha, 2008).

Radzi et al. (2018) studied the effect of different sizes on the mechanical and thermal properties of RF/thermoplastic polyurethane (TPU) composites. The RF/TPU composites size used in this study is 125 μm and lower; 123–300 μm and 300–425 μm were prepared using the hot press machine. Figs. 17.3 and 17.4 shows TGA and DTG for Neat TPU, RF/TPU composites with different sizes (<125, 125–300 and 300–425 μm). In these thermal properties, there are three main stages involved, namely, thermal degradation related to the

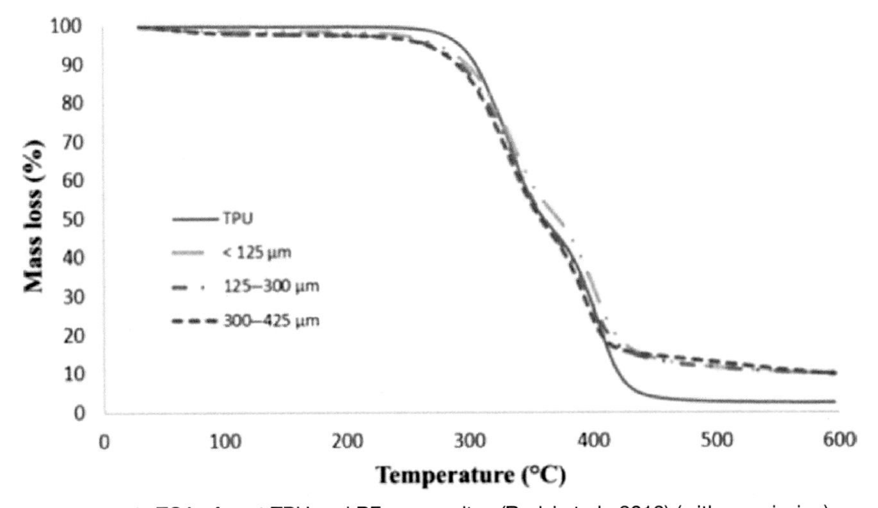

FIG. 17.3 TGA of neat TPU and RF composites (Radzi et al., 2018) (with permission).

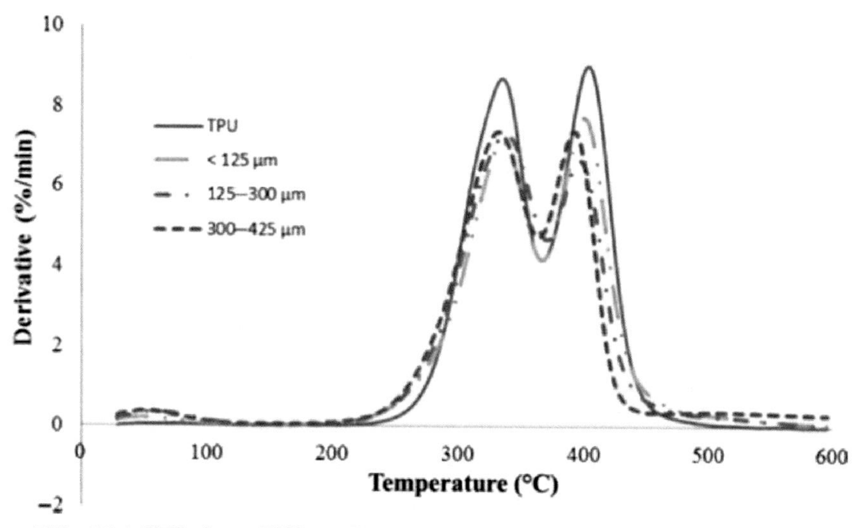

FIG. 17.4 DTG of neat TPU and RF composites (Radzi et al., 2018) (with permission).

moisture content of RF composites between temperature of 30°C–100°C. The second phase is the change in hemicellulose within fiber in the temperature range of 330°C–338°C. The third phase involves the decomposition of lignin followed by cellulose and ash in a temperature range of 392°C–403°C. From the observation, they believe that the size factor does not show a significant effect on the thermal properties of the three size parameters. The results of this test can be concluded that these RF composites have better thermal stability compared to TPU.

Radzi et al. (2017) focused on different fiber content on the mechanical and thermal properties of RF-reinforced TPU composites. The different RF contents prepared were 10, 20, 30, 40, and 50wt% using the hot compression method. TGA properties were carried out using Q series thermal analysis machine model TA instrument (TGA Q500). The author argues that the 40wt% RF content indicates high thermal stability compared to others and it also confirms that 40% of roselle fiber weight is the most suitable fiber content to be used in TPU composites. Research by Razali et al. (2016), investigated the mechanical and thermal stability properties of different fiber loading of RF reinforced vinyl ester composites (RFVE). Figs. 17.5 and 17.6 show TGA and DTG curves for Neat VE, RF/VE composites with different fiber loadings. RF composites were prepared by four parameters; 10, 20, 30, and 40 wt %, and neat polymer using the hand layout method. They investigated thermal stability using TGA according to ASTM D3850, which was used in this analysis.

From the results, TGA and DTG showed three-phase degradation of the RFVE, which were loss of moisture content, degradation of hemicellulose, degradation of cellulose/lignin and residual ash. From observation, they state that the increasing of RF into the VE composite will contribute to the increase in thermal stability compared to the neat VE. Besides, the increase of RF in VE will also increase the residual ash. Weight loss on neat VE and RF will affect the formation of residual char in the VE composite during the thermal testing.

The research studied on thermal properties of roselle/SPF hybrid composites, they observed the effect of these two types of fiber with different fiber loading (Radzi et al., 2019b). Figs. 17.7 and 17.8 shows the TGA and DTG curves of RF composites (RFT), RF/SPF hybrid with different fiber contents (RST 1–75:25, RST 2–50:50, and RST 3–25:75) and SPF composites (SPFT) with varying temperature TGA. From observation, the curve trends show a similar degradation in each composition. All the composition of these composites undergoes four core phases, namely, decomposition of lignocellulosic and cellulosic components, lignin, and ash. In the TGA curves it was shown that RFT3 had higher thermal stability. The addition of RF in pure SPF hybrid composites increased thermal stability compared with other compositions.

Radzi et al. (Radzi et al., 2019a) investigated the thermal properties of alkaline treatment of RF/SPF reinforced thermoplastic polyurethane hybrid composites. Figs. 17.9 and 17.10 shows the TGA and DTG curves of RF/SPF hybrid composites with untreated and treated hybrid fibers using sodium hydroxide (NaOH) with different concentration (3%, 6%, and 9%).

From observation, both figures show no significant difference between RF/SPF hybrid composites untreated

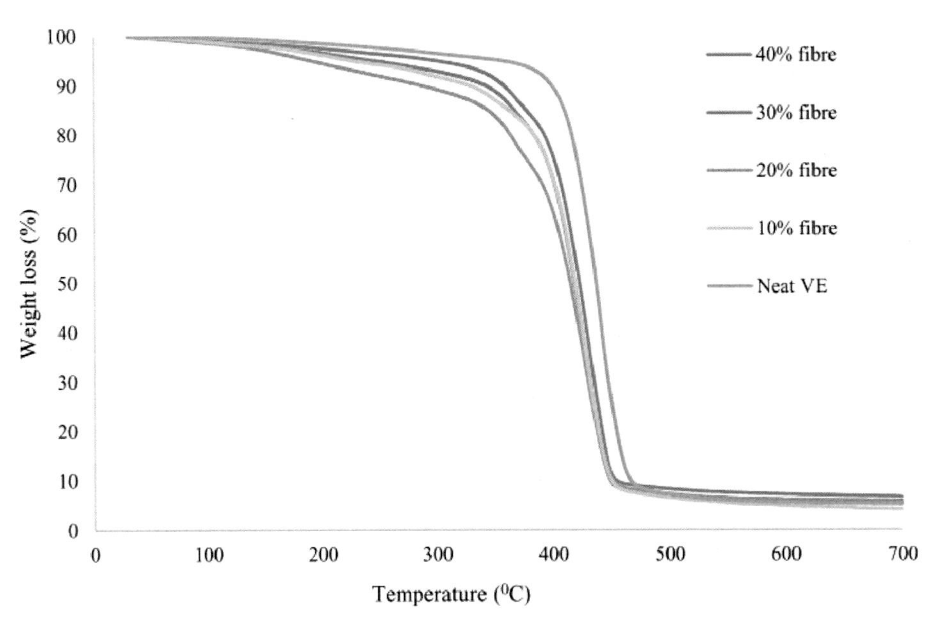

FIG. 17.5 TGA of neat VE and RF composites (Razali et al., 2016).

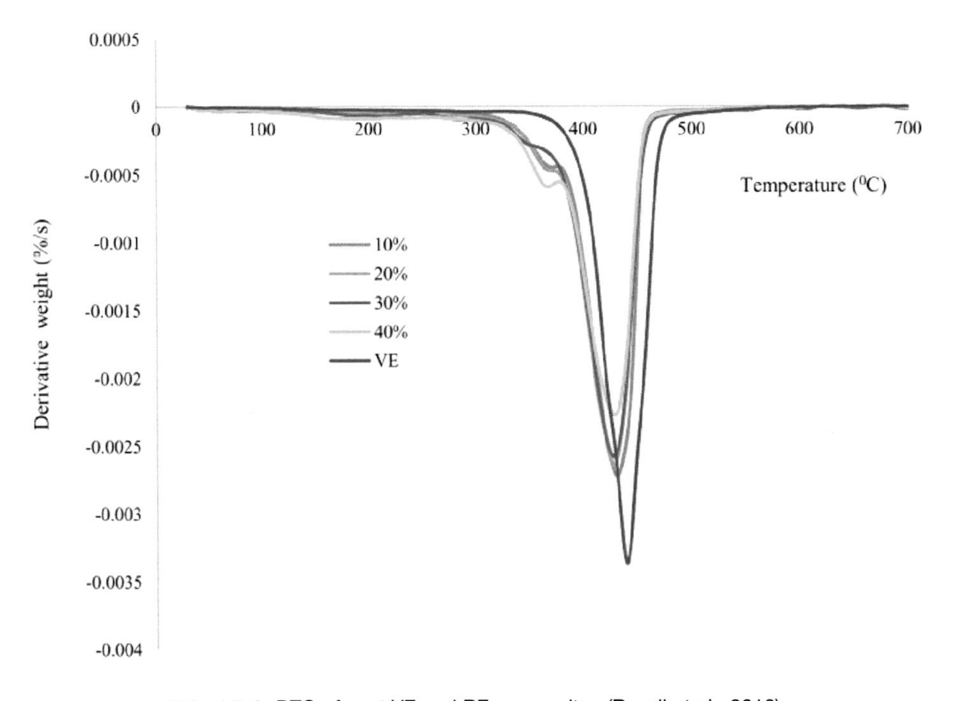

FIG. 17.6 DTG of neat VE and RF composites (Razali et al., 2016).

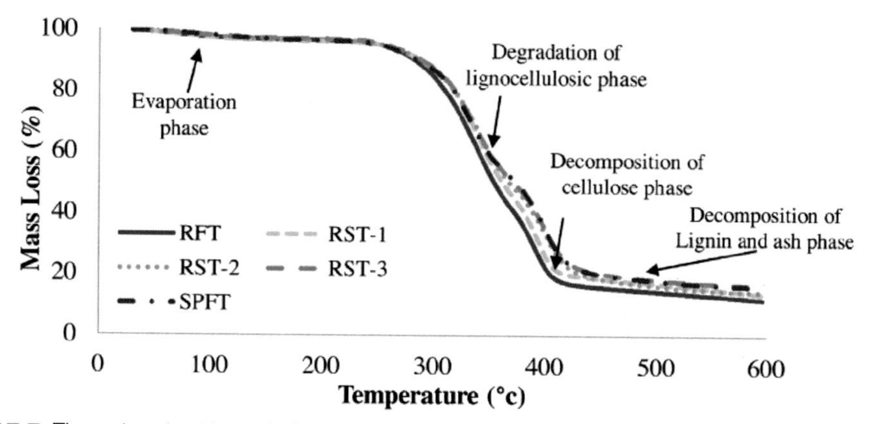

FIG. 17.7 Thermalgravimetric analysis of RF/SPF hybrid composites (Radzi et al., 2019b) (with Permission).

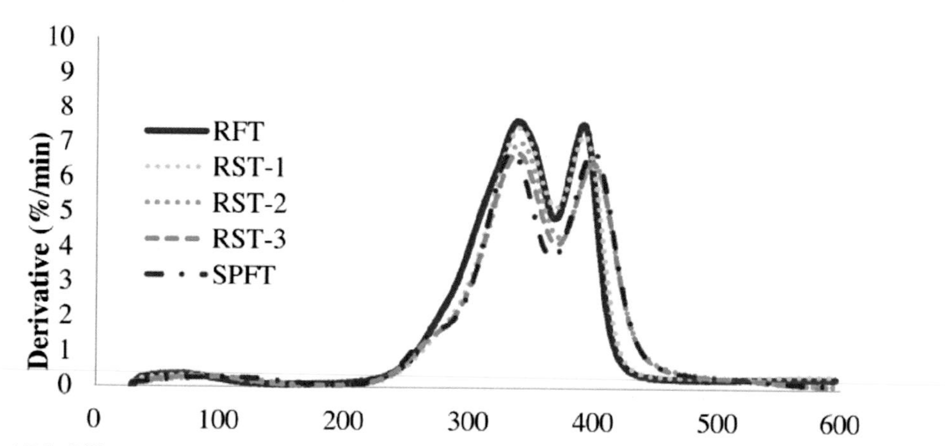

FIG. 17.8 Differential Thermogravimetric of RF/SPF hybrid composites (Radzi et al., 2019b) (with Permission).

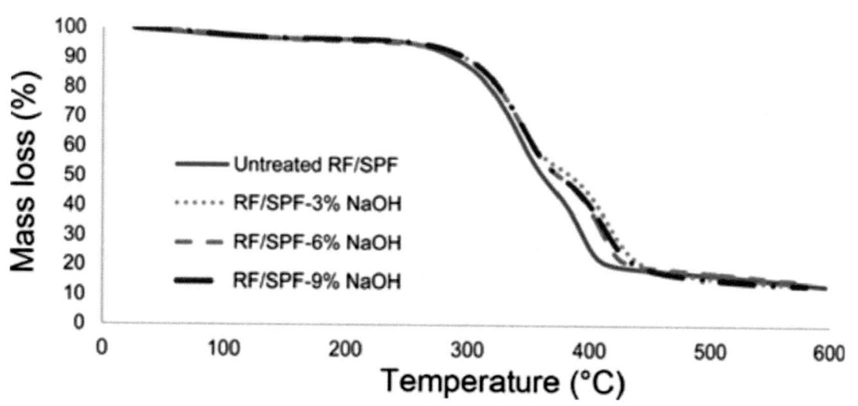

FIG. 17.9 Thermogravimetric analysis of RF/SPF reinforced thermoplastic polyurethane hybrid composites with different NaOH concentrations (Radzi et al., 2019a) (with permission).

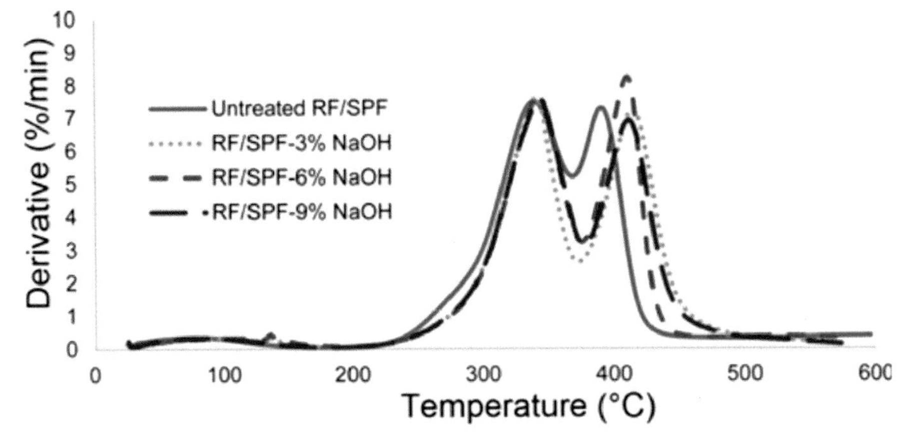

FIG. 17.10 Differential Thermogravimetric of RF/SPF reinforced thermoplastic polyurethane hybrid composites with different NaOH concentrations (Radzi et al., 2019a) (with permission).

and alkaline treated. All composites undergo four core phases, namely, decomposition of lignocellulosic, cellulose components, lignin, and ash. The TGA and DTG curve showed that the NaOH-treated hybrid composites had better thermal stability compared with untreated fiber hybrid composites. This happens because alkaline treatment removes all impurities on the surface of fiber, wax, and hemicellulose (Kaewkuk et al., 2013; Joseph et al., 2003). Comparison of three differences in NaOH concentration, NaOH-6% showed excellent thermal stability compared to the other's concentration.

5 CONCLUSIONS

Studies on natural hybrid composites have increased due to being environmentally friendly and with low-cost materials for the various product applications. A brief discussion of the thermal and physical properties of roselle fiber hybrid composites from a previous study were given. Roselle hybrid composites are presented for researchers to see the potential of roselle combined with other fibers to improve the performance of composites. This review paper can provide information for future studies on the thermal and physical properties of roselle hybrid composites. Future research is to produce roselle hybrid composites with combinations of various conventional fibers and types of fiber treatment to improve the thermal and physical properties of the hybrid composites.

REFERENCES

Abdul Khalil, H. P. S., Bhat, H., & Ireana Yusra, F. (2012). Green composites from sustainable cellulose nanofibrils: A review. *Carbohydrate Polymers, 87*(2), 963–979.

Ahad, N. A., Ahmad, S. H., & Zain, N. M. (2014). The compatibilty of TPU and NR blends. *Advanced Materials Research, 879*, 107–111. https://doi.org/10.4028/www.scientific.-net/AMR.879.107

Anuar, H., Surip, S. N., & Adilah, A. (June 2010). Development of reinforced thermoplastic elastomer with kenaf bast fibre for automotive component. In *14th European conference on composite materials* (pp. 1–5).

Azammi, A. M. N., Sapuan, S. M., Ishak, M. R., & Sultan, M. T. H. (2020). Physical and damping properties of kenaf fibre filled natural rubber/thermoplastic polyurethane composites. *Defence Technology, 16*, 29–34. https://doi.org/10.1016/j.dt.2019.06.004

Aziz, S. H., & Ansell, M. P. (2004). The effect of alkalization and fibre alignment on the mechanical and thermal properties of kenaf and hemp bast fibre composites: Part 1 - polyester resin matrix. *Composites Science and Technology, 64*(9), 1219–1230. https://doi.org/10.1016/j.compscitech.2003.10.001

Azwa, Z. N., & Yousif, B. F. (2013). Characteristics of kenaf fibre/epoxy composites subjected to thermal degradation. *Polymer Degradation and Stability, 98*(12). https://doi.org/10.1016/j.polymdegradstab.2013.10.008

Chandramohan, D., & Marimuthu, K. (2011). A review on natural fibers. *International Journal of Research and Reviews in Applied Sciences, 8*(2), 194–206. Retrieved from http://www.arpapress.com/Volumes/Vol8Issue2/IJRRAS_8_2_09.pdf.

Chauhan, A., & Kaith, B. (2012). Accreditation of novel Roselle grafted fiber reinforced bio-composites. *Journal of Engineered Fibers and Fabrics, 7*(2), 66–75. https://doi.org/10.1177/155892501200700210

Crane, J. C. (1949). Roselle—a potentially important plant fiber. *Economic Botany, 3*(1), 89–103. https://doi.org/10.1007/BF02859509

Das Gupta, P. C. (1959). The hemicelluloses of roselle fiber (*Hibiscus sabdariffa*). *Textile Research Journal, 30*(3), 237.

Graff, G. (2005). *Under-hood Applications of Nylon Accelerate.* retrieved from Omnexus by SpecialChem: http://www.omnexus.com/resources/articles/article.aspx?id¼9660.

Harish, R., Muppal, D., & Kumar, M. (2015). Investigations on mechanical properties in FRP composites using woven roselle and sisal fibers. *International Journal of Computer Science Information and Engineering Technologies ISSN, 2*, 1–8.

Huda, M. S., Drzal, L. T., Mohanty, A. K., & Misra, M. (2006). Chopped glass and recycled newspaper as reinforcement fibers in injection molded poly(lactic acid) (PLA) composites: A comparative study. *Composites Science and Technology, 66*(11–12), 1813–1824. https://doi.org/10.1016/j.compscitech.2005.10.015

Huzaifah, M. R. M., Sapuan, S. M., Leman, Z., & Ishak, M. R. (2017). Comparative study on chemical composition, physical, tensile, and thermal properties of sugar palm fiber (*Arenga pinnata*) obtained from different geographical locations. *BioResources, 12*, 9366–9382. https://doi.org/10.15376/biores.12.4.9366-9382

Huzaifah, M. R. M., Sapuan, S. M., Leman, Z., & Ishak, M. R. (2019). Effect of soil burial on physical, mechanical and thermal properties of sugar palm fibre reinforced vinyl ester composites. *Fibers and Polymers*. https://doi.org/10.1007/s12221-019-9159-6

Ishak, M. R., Sapuan, S. M., Leman, Z., Rahman, M. Z. A., & Anwar, U. M. K. (2012). Characterization of sugar palm (*Arenga pinnata*) fibres tensile and thermal properties. *Thermal Analysis and Calorimetry, 109*(2), 981–989. https://doi.org/10.1007/s10973-011-1785-1

Ismail, H., Norjulia, A. M., & Ahmad, Z. (2010). The effects of untreated and treated kenaf loading on the properties of kenaf fibre-filled natural rubber compounds. *Polymer-Plastics Technology and Engineering, 49*(5), 519–524. https://doi.org/10.1080/03602550903283117

Jawaid, M., & Khalil, H. P. S. A. (2015). Cellulosic / synthetic fibre reinforced polymer hybrid composites : A review. *Carbohydrate Polymers, 86*(1), 1–18. https://doi.org/10.1016/j.carbpol.2011.04.043

Johnson, R. W., Evans, J. L., Jacobsen, P., Thompson, J. R., & Christopher, M. (2004). The changing automotive environment: High-temperature electronics. *IEEE Transactions on Electronics Packaging Manufacturing, 27*(3), 164–176. https://doi.org/10.1109/TEPM.2004.843109

Joseph, P. V., Joseph, K., Thomas, S., Pillai, C. K. S., Prasad, V. S., Groeninckx, G., & Sarkissova, M. (2003). The thermal and crystallisation studies of short sisal fibre reinforced polypropylene composites. *Composites Part A: Applied Science and Manufacturing, 34*(3), 253–266. https://doi.org/10.1016/S1359-835X(02)00185-9

Junkasem, J., Menges, J., & Supaphol, P. (2006). Mechanical properties of injection-molded isotactic polypropylene/roselle fiber composites. *Journal of Applied Polymer Science, 101*(5), 3291–3300. https://doi.org/10.1002/app.23829

Kaewkuk, S., Sutapun, W., & Jarukumjorn, K. (2013). Effects of interfacial modification and fiber content on physical properties of sisal fiber/polypropylene composites. *Composites Part B: Engineering, 45*(1), 544–549. https://doi.org/10.1016/j.compositesb.2012.07.036

Kennedy, J. F. (1993). Emerging technologies for materials and chemicals from biomass. *Polymer International, 30*(4), 1993.

Krishnan, M., & Narayan, R. (1992). Compatibilization of biomass fibers with hydrophobic materials. *MRS Proceedings, 266*(ii), 93–104. https://doi.org/10.1557/proc-266-93

Lau, K. tak, Hung, P. yan, Zhu, M. H., & Hui, D. (2018). Properties of natural fibre composites for structural engineering applications. *Composites Part B: Engineering, 136*(November 2017), 222–233. https://doi.org/10.1016/j.compositesb.2017.10.038

Liu, W., Mohanty, A. K., Drzal, L. T., & Misra, M. (2005). Novel biocomposites from native grass and soy based bioplastic: Processing and properties evaluation. *Industrial and Engineering Chemistry Research, 44*(18), 7105–7112. https://doi.org/10.1021/ie050257b

Mahadevan, N., S., & Kamboj, P. (2007). Hibiscus sabdariffa Linn. *Indian Medicinal Plants, 8*(1), 1–1. https://doi.org/10.1007/978-0-387-70638-2_749

Mahjoub, R., Yatim, J. M., Mohd Sam, A. R., & Hashemi, S. H. (2014). Tensile properties of kenaf fiber due to various conditions of chemical fiber surface modifications. *Construction and Building Materials, 55*, 103–113. https://doi.org/10.1016/j.conbuildmat.2014.01.036

Maldas, D., & Kokta, V. (1993). Interfacial adhesion of lignocellulosic materials in polymer composites: An overview. *Composite Interfaces, 1*(1), 87–108. https://doi.org/10.1163/156855493X00338

Modibbo, U. U., Aliyu, B. A., Nkafamiya, I. I., & Manji, A. J. (2007). The effect of moisture imbibition on cellulosic bast fibers as industrial raw materials. *International Journal of Physical Sciences, 2*(7), 163–168.

Mohamad, O., Nazir, B. M., Rahman, M. A., & Herman, S. (2002). Roselle: A new crop in Malaysia. In *Buletin Persatuan Genetik Malaysia* (pp. 12–13).

Mungole, A., & Chaturvedi, A. (2011). Hibiscus sabdariffa L. - A rich source of secondary metabolism. *International Journal of Pharmaceutical Sciences Review and Research, 6*(1), 83–87.

Nadlene, R., Sapuan, S. M., Jawaid, M., Ishak, M. R., & Yusriah, L. (2015). Material characterization of roselle fibre (*Hibiscus sabdariffa* L.) as potential reinforcement material for polymer composites. *Fibres and Textiles in Eastern Europe*, 23(6(114)), 23–30.

Nadlene, R., Sapuan, S. M., Jawaid, M., Ishak, M. R., & Yusriah, L. (2016). A review on roselle fiber and its composites. *Journal of Natural Fibers, 13*(1), 10–41. https://doi.org/10.1080/15440478.2014.984052

Navaneethakrishnan, S., & Athijayamani, A. (2016). Mechanical properties and absorption behavior of CSP filled roselle fiber reinforced hybrid composites. *Materials and Environment Science, 7*(5), 1674–1680.

Nguong, C. W., Lee, S. N. B., & Sujan, D. (2013). A review on natural fibre reinforced polymer composites. In *World Academy of Science, Engineering and Technology* (pp. 1123–1130).

Noor Azammi, A. M., Sapuan, S. M., Ishak, M. R., & Sultan, M. T. H. (2018a). Mechanical and thermal properties of kenaf reinforced thermoplastic polyurethane (TPU)-Natural rubber (NR) composites. *Fibers and Polymers, 19*(2), 446–451. https://doi.org/10.1007/s12221-018-7737-7

Noor Azammi, A. M., Sapuan, S. M., Ishak, M. R., & Sultan, M. T. H. (2018b). Mechanical properties of kenaf fiber thermoplastic polyurethane-natural rubber composites. *Polimery, 63*(7/8), 524–530. https://doi.org/10.14314/polimery.2018.7.6

Oksman, K. (2001). High quality flax fibre composites manufactured by the resin transfer moulding process. *Journal of Reinforced Plastics and Composites, 20*(7), 621–627. https://doi.org/10.1177/073168401772678634

Radzi, A. M., Sapuan, S. M., Jawaid, M., & Mansor, M. R. (2017). Influence of fibre contents on mechanical and thermal properties of roselle fibre reinforced polyurethane composites. *Fibers and Polymers, 18*(7), 1353–1358. https://doi.org/10.1007/s12221-017-7311-8

Radzi, A. M., Sapuan, S. M., Jawaid, M., & Mansor, M. R. (2018). Mechanical and thermal performances of roselle fiber-reinforced thermoplastic polyurethane composites. *Polymer-Plastics Technology and Engineering, 57*(7), 601–608. https://doi.org/10.1080/03602559.2017.1332206

Radzi, A. M., Sapuan, S. M., Jawaid, M., & Mansor, M. R. (2019a). Effect of alkaline treatment on mechanical, physical and thermal properties of roselle/sugar palm fiber reinforced thermoplastic polyurethane hybrid composites. *Fibers and Polymers, 20*(4), 847–855. https://doi.org/10.1007/s12221-019-1061-8

Radzi, A. M., Sapuan, S. M., Jawaid, M., & Mansor, M. R. (2019b). Water absorption, thickness swelling and thermal properties of roselle/sugar palm fibre reinforced thermoplastic polyurethane hybrid composites. *Journal of Materials Research and Technology, 8*(5), 3988–3994. https://doi.org/10.1016/j.jmrt.2019.07.007

Razali, N., Salit, M. S., Jawaid, M., Ishak, M. R., & Lazim, Y. (2015). A study on chemical composition, physical, tensile, morphological, and thermal properties of roselle fibre: Effect of fibre maturity. *BioResources, 10*(1), 1803–1823.

Razali, N., Sapuan, S. M., Jawaid, M., Ishak, M. R., & Lazim, Y. (2016). Mechanical and thermal properties of roselle fibre reinforced vinyl ester composites. *BioResources, 11*(4), 9325–9339. https://doi.org/10.15376/biores.11.4.9325-9339

Reddy, N., & Yang, Y. (2005). Biofibers from agricultural byproducts for industrial applications. *Trends in Biotechnology, 23*(1), 22–27. https://doi.org/10.1016/j.tibtech.2004.11.002

Rosa, I. M. D., Kenny, J. M., Puglia, D., Santulli, D., & Sarasini, F. (2010). Morphological, thermal and mechanical characterization of okra (Abelmoschus esculentus) fibres as potential reinforcement in polymer composites. *Composites Science and Technology, 70*(1), 116–122.

Rowell, R. M., Han, J. S., & Rowell, J. S. (2000). Characterization and factors effecting fiber properties. *Natural Polymers an Agrofibers Composites*, (January), 115–134.

Sahari, J., Sapuan, S. M., Zainudin, E. S., & Maleque, M. A. (2014). Physico-chemical and thermal properties of starch derived from sugar palm tree (*Arenga pinnata*). *Asian Journal of Chemistry, 26*(4), 955–959. https://doi.org/10.14233/ajchem.2014.15652

Saheb, D. N., & Jog, J. P. (1999). Natural fiber polymer composites: a review. *Advances in Polymer Technology,* 18(4), 351–363. https://doi.org/10.1002/(SICI)1098-2329(199924)18

Sarifuddin, N., & Ismail, H. (2013). Comparative study on the effect of bentonite or feldspar filled low-density polyethylene/thermoplastic sago starch/kenaf core fiber composites. *Journal of Physical Science, 24*(2), 97–115.

Sastra, H. Y., Siregar, J. P., Sapuan, S. M., & Hamdan, M. M. (2006). Tensile properties of *Arenga pinnata* fiber-reinforced epoxy composites. *Polymer-Plastics Technology and Engineering, 45*(1), 149–155. https://doi.org/10.1080/03602550500374038

Sathishkumar, T. P., Navaneethakrishnan, P., Shankar, S., Rajasekar, R., & Rajini, N. (2013). Characterization of natural fiber and composites – a review. *Journal of Reinforced Plastics and Composites, 32*(19), 1457–1476. https://doi.org/10.1177/0731684413495322

Sgriccia, N., & Hawley, M. C. (2007). Thermal, morphological, and electrical characterization of microwave processed natural fiber composites. *Composites Science and Technology, 67*(9), 1986–1991. https://doi.org/10.1016/j.compscitech.2006.07.031

Shahzad, A. (2013). A study in physical and mechanical properties of hemp fibres. *Advances in Materials Science and Engineering, 2013*, 1–9. https://doi.org/10.1155/2013/325085

Singhaa, A. S., & Thakura, V. K. (2008). Fabrication and study of lignocellulosic *Hibiscus sabdariffa* fiber reinforced polymer composites. *BioResources, 3*(4), 1173–1186. https://doi.org/10.15376/biores.3.4.1173-1186

Soundhar, A., Rajesh, M., Jayakrishna, K., Sultan, M. T. H., & Shah, A. U. M. (2019). Investigation on mechanical properties of polyurethane hybrid nanocomposite foams reinforced with roselle fibers and silica nanoparticles. *Nanocomposites, 5*(1), 1–12. https://doi.org/10.1080/20550324.2018.1562614

Sreekumar, P. A., Thomas, S. P., Saiter, J marc, Joseph, K., Unnikrishnan, G., & Thomas, S. (2009). Effect of fiber surface modification on the mechanical and water absorption characteristics of sisal/polyester composites fabricated by resin transfer molding. *Composites Part A: Applied Science and Manufacturing, 40*(11), 1777–1784. https://doi.org/10.1016/j.compositesa.2009.08.013

Srinivasan, V. S., Boopathy, S. R., & Ramnath, B. V. (2015). Thermal behaviour of flax kenaf hybrid natural fiber composite. *International Journal of Advanced Research in Science, Engineering and Technology, 2*, 883–888.

Tajvidi, M., Feizmand, M., Falk, R. H., & Felton, C. (2009). Effect of cellulose fiber reinforcement on the temperature dependent mechanical performance of nylon 6. *Journal of Reinforced Plastics and Composites, 28*(22), 2781–2790. https://doi.org/10.1177/0731684408093875

Thakur, V. K., & Singha, A. (2008). Fabrication of *Hibiscus sabdariffa* fibre reinforced polymer composites. *Computer, 17*(7), 541–553.

Thakur, V. K., Thakur, M. K., & Gupta, R. K. (2014). Review: Raw natural fiber-based polymer composites. *International Journal of Polymer Analysis and Characterization, 19*(3), 256–271. https://doi.org/10.1080/1023666X.2014.880016

Thyavihalli Girijappa, Y. G., Mavinkere Rangappa, S., Parameswaranpillai, J., & Siengchin, S. (2019). Natural fibers as sustainable and renewable resource for development of eco-friendly composites: A comprehensive review. *Frontiers in Materials, 6*(September), 1–14. https://doi.org/10.3389/fmats.2019.00226

Tori Hudson, N. D. (2011). A research review on the use of Hibiscus sabdariffa. In *Better Medicine - National Network of Holistic Practitioner Communities*. Available at www.todaysdietitian.com/ whitepapers/Hibiscus_Sabdariffa.pdf.

Vilay, V., Mariatti, M., Mat Taib, R., & Todo, M. (2008). Effect of fiber surface treatment and fiber loading on the properties of bagasse fiber-reinforced unsaturated polyester composites. *Composites Science and Technology, 68*, 631–638. https://doi.org/10.1016/j.compscitech.2007.10.005

Wang, W. M., Cai, Z. S., & Yu, J. Y. (2010). Study on the chemical modification process of jute fiber. *Tappi Journal, 9*(2), 23–29. https://doi.org/10.32964/tj9.2.23

Wilson, W. (2009). *Discover the Many uses of the Roselle Plant*. NParks. Available at: http://mygreenspace.nparks.gov.sg/discover-the-many-uses-of-the-roselle-plant/.

Xie, Y., Hill, C. A. S., Xiao, Z., Militz, H., & Mai, C. (2010). Composites: Part A. Silane coupling agents used for natural fiber/polymer composites: A review. *Composites Part A, 41*(7), 806–819. https://doi.org/10.1016/j.compositesa.2010.03.005

Yang, H., Yan, R., Chen, H., Lee, D. H., & Zheng, C. (2007). Characteristics of hemicellulose, cellulose and lignin pyrolysis. *Fuel, 86*(12–13), 1781–1788. https://doi.org/10.1016/j.fuel.2006.12.013

Yusriah, L., Sapuan, S. M., Zainudin, E. S., & Mariatti, M. (2014). Characterization of physical, mechanical, thermal and morphological properties of agro-waste betel nut (Areca catechu) husk fibre. *Journal of Cleaner Production, 72*, 174–180. https://doi.org/10.1016/j.jclepro.2014.02.025

CHAPTER 18

Development and Characterization of Roselle Nanocellulose and Its Potential in Reinforced Nanocomposites

R.A. ILYAS • S.M. SAPUAN • M.M. HARUSSANI • M.S.N. ATIKAH • R. IBRAHIM • M.R.M. ASYRAF • A.M. RADZI • R. NADLENE • LAU KIA KIAN • SUZANA MALI • MOCHAMAD ASROFI • SANJAY MAVIKERE RANGAPPA • SUCHART SIENGCHIN

1 INTRODUCTION

According to a reference module in the 2020 edition of *Natural Fiber Composites* written by Mohamed Zakriya and Ramakrishnan (2020), natural fibers from the plant-based kingdom are renewable resources with several advantages. They provide the composite with high specific stiffness and strength and lightweight feature, are available at low cost, have a desirable fiber aspect ratio, possess excellent mechanical properties, are biodegradable, and are abundantly and readily available from natural sources. However, artificial or synthetic fibers are petroleum-based products. As one of the fossil fuels, petroleum is nonrenewable and exploration and drilling of petroleum have many adverse effects on the environment. Thus these advantages of natural plant fibers have allowed scientists and engineers to use them to reinforce with polymer composites that help reduce the forest source utilization as well as minimize the surplus of natural fibers (Abral, Atmajaya, et al., 2020; Abral, Basri, et al., 2019; Halimatul et al., 2019a; Nadlene et al., 2016). Plant natural fibers can be used as reinforcements in polymer composite materials, where the orientation of fibers would improve the mechanical properties of the composite. Currently, natural fibers, such as cassava (Ibrahim, Edhirej, et al., 2020), cogon (Jumaidin et al., 2020; Jumaidin, Saidi, et al., 2019), corn (Ibrahim, Sapuan, et al., 2020; Sari et al., 2020), flax (Akonda et al., 2020), ginger (Abral, Ariksa, et al., 2019; 2020), hemp (Arulmurugan et al., 2019), jute (Islam et al., 2019), kenaf (Aisyah et al., 2019; Mazani et al., 2019), oil palm (Ayu et al., 2020), ramie (Syafri, Kasim, et al., 2019), sisal (Yorseng et al., 2020), sugarcane (Asrofi, Sapuan, et al., 2020; Asrofi, Sujito, et al., 2020; Jumaidin, Ilyas, et al., 2019), sugar palm (Atiqah et al., 2019; Halimatul et al., 2019b; Hazrol et al., 2020; Maisara et al., 2019; Norizan et al., 2020; Nurazzi, Khalina, Sapuan, & Ilyas, 2019, Nurazzi, Khalina, Sapuan, Ilyas, Rafiqah, & Hanafee, 2019; Nurazzi et al., 2020; Sapuan et al., 2020), and water hyacinth (Syafri, Sudirman, et al., 2019), had been used as reinforcing agents in polymer composites. This is owing to their benefits, such as excellent mechanical properties, ease of processing, and low-cost manufacturing, compared to synthetic fibers (Ilyas & Sapuan, 2020a, 2020b; Nadlene et al., 2018; Nazrin et al., 2020). Plant natural fibers are mainly composed of cellulose, hemicellulose, lignin, pectin, and other waxy substances (Ilyas et al., 2017; Ilyas, Sapuan, Atikah, Asyraf, et al., 2020; Ilyas, Sapuan, Atiqah, Ibrahim, et al., 2020; Ilyas, Sapuan, et al., 2020; Ilyas, Sapuan, Ishak, Zainudin, et al., 2018; Ilyas, Sapuan, & Ishak, 2018; Ilyas, Sapuan, Ibrahim, et al., 2019; Ilyas, Sapuan, Ishak, et al., 2019). Reinforcement of natural fibers and polymers that have emerged in the field of polymer science has drawn attention for use in a variety of advanced applications, ranging from plastic packaging to the automotive industry (Atikah et al., 2019; Ilyas, Sapuan, Atiqah, et al., 2020; Ilyas, Sapuan, Ibrahim, et al., 2020, 2019; Ilyas, Sapuan, Sanyang, et al., 2018; Ilyas, Sapuan, Ishak, & Zainudin, 2018a, 2018b, 2018c; Sanyang et al., 2018).

The usage of natural fibers as reinforcement in polymer has garnered attention during the past few decades owing to their low cost, their density and weight, and the less pollution during production resulting in minimal health hazards, as well as their eco-friendly nature (Aisyah et al., 2019; Asyraf et al., 2020; Ayu et al., 2020; Nazrin et al., 2020; Nurazzi, Khalina, Sapuan,

[illegible]

Ilyas, et al., 2019; Sari et al., 2020). Natural fibers from plant consist of three main components: cellulose, hemicellulose, and lignin. Cellulose is one of the most abundant biopolymers in the Earth (Ilyas & Sapuan, 2020a, 2020b; Jumaidin et al., 2020; Jumaidin, Ilyas, et al., 2019; Jumaidin, Saidi, et al., 2019). Cellulose is a structural component in lignocellulosic materials, natural fibers, and wood (Ilyas et al., 2017). It accounts for 40%–60% of the mass of the wood and can be extracted in the form of 20- to 40-μm-thick fibers upon the pulping process. The term "nanocellulose" has been used extensively for describing a range of quite different cellulose-based nanomaterials that have a single dimension in the range of nanometers, such as nanocrystalline cellulose (NCC), whiskers, nanofibrillated cellulose (NFC), nanofibrils, cellulose microfibrils, and bacterial nanocellulose. Nanocellulose can be classified into three types of materials (Abitbol et al., 2016; Ilyas, Sapuan, Sanyang, et al., 2018): (1) NCC, often named as crystallites, whiskers, nanowhiskers, rodlike cellulose microcrystals, cellulose nanocrystal, and cellulose nanowhisker (CNW); (2) NFC, often named as cellulose nanofibril (CNF), cellulose microfibrils, nanofiber, and nanofibrils; and (3) bacterial nanocellulose, also referred to as bacterial cellulose, biocellulose, and microbial cellulose. Extraction and isolation processes, characterization, and potential applications of novel forms of nanocellulose, such as NFC and NCC, have created substantial interest in the scientific and industrial fields. NCC (Fig. 18.1A) and NFC (Fig. 18.2B) are renowned not only for their biodegradation ability, superb properties, unique structures, low density, excellent mechanical performance, high surface area, high aspect ratio, biocompatibility, and natural abundance but also for their abundant hydroxyl groups that make the modification of their surface possible, which consequently leads to enhancement in their nanoreinforcement compatibility with other polymers. Because of these excellent properties, NCC and NFC have demonstrated numerous potential advanced applications, including automotive, optically transparent materials, drug supply, coating films, tissue technology, biomimetic materials, aerogels, sensor, 3D printing, rheology modifiers, energy harvesters, filtration, textiles, printed and flexible electronics, composites, paper and board, packaging, oil and gas, and medical and healthcare (Abral, Basri, et al., 2019; Abral, Atmajaya, et al., 2020; Criado et al., 2018; Ferreira et al., 2019; Picheth et al., 2017; Rajinipriya et al., 2018). In consequence, the number of patents and publications on nanocellulose has significantly increased from 300 in 2000 to 2500 in 2020 (Fig. 18.1C).

Natural fibers such as roselle fibers (*Hibiscus sabdariffa*) are found in abundance in nature and cultivated in Malaysia, India, Panama, Indonesia, Jamaica, Mexico, Guatemala, Australia, Philippines, Kenya, Madagascar, Mozambique, Malawi, Uganda, Somalia, Tanzania, Djibouti, Cambodia, Vietnam, Namibia, Gabon, Congo, Burundi, Rwanda, DR Congo, Myanmar, Thailand, Belize, China, Sudan, South Sudan, Egypt, Gambia, Senegal, Saudi Arabia, Bangladesh, Laos, Sri Lanka, Ghana, Nigeria, Brazil, and Cuba. In Malaysia, after a year, the roselle plant is cut and disposed as agro-waste, because of the deterioration of the fruit quality beyond this period. For efficient use of this plant, the fiber can be utilized as reinforcing material in polymer composites owing to the high cellulose content in its nanocellulose (Table 18.1). This chapter focuses on the top-down techniques, such as acid hydrolysis, used to prepare roselle nanocellulose from agro-waste roselle fiber. This chapter also elaborates the development and characterization of roselle nanocellulose and its potential reinforcement in nanocomposites.

2 EXTRACTION PROCESS OF NANOCRYSTALLINE CELLULOSE AND NANOFIBRILLATED CELLULOSE

NCC and NFC can be extracted from various sources of (1) **hardwood and softwood**, such as roselle, bleached softwood kraft pulp, eucalyptus, *Acacia mangium*, wood sawdust, mengkuang, and wood pulp, and (2) **plant fiber**, such as sugar palm fiber, sugar palm frond, dunchi fiber stalks, broomcorn stalks, cotton linters, ginger fiber, kenaf, sugarcane bagasse, rice straw, kraft pulp, water hyacinth, blue agave leaf, grass fibers, oil palm empty fruit bunch, oil palm trunk, ramie, sago seed shell, banana pseudostem, cornhusk, groundnut shell, *Humulus japonicus* stem, wheat straw, arecanut husk fiber, corncob, ushar seed fiber, *H. sabdariffa* fibers, kenaf core wood, *Phormium tenax* fiber, flax fiber, soy hulls, bamboo, potato peel waste, sesame husk, coconut husk, colored cotton, curaua fiber, cassava bagasse, grass fibers, mulberry, ramie, and sisal fiber, as shown in Tables 18.2 and 18.3. Among all sources, agro-industrial wastes are seen as the outstanding ones to be extensively utilized for producing NCC and NFC. This might be due to the higher NCC and NFC productivities and large-scale accessibility, as well as the abundance of agro-industrial waste resources. Both NCC and NFC are cellulose-based nanomaterials that can be extracted and isolated from plant sources via top-down methods that include chemical, mechanical, chemomechanical, and enzymatic treatments for their isolation from lignocellulosic residues.

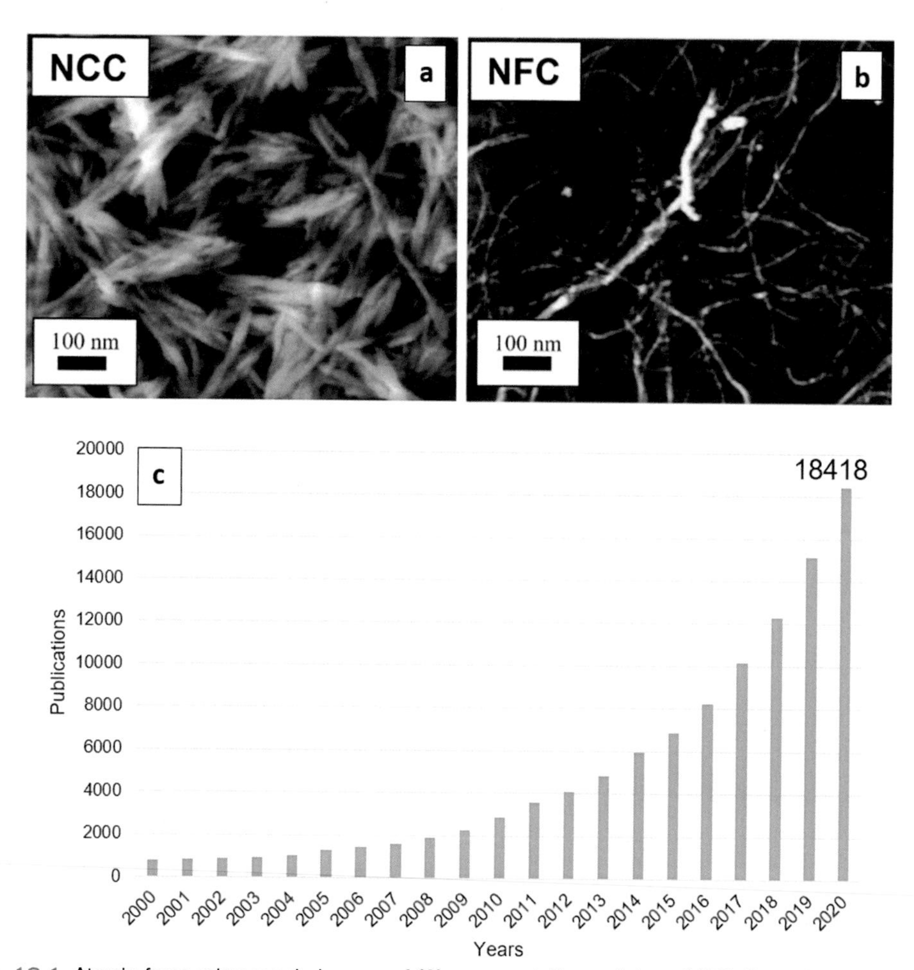

FIG. 18.1 Atomic force microscopic images of **(A)** nanocrystalline cellulose (NCC) (Ilyas, Sapuan, Atikah, et al., 2020) and **(B)** nanofibrillated cellulose (NFC) (Ilyas, Sapuan, Ishak, et al., 2019). **(C)** The number of publications on nanocellulose in the past decade indicating increasing interest in nanocellulosic materials (Scopus, **December 2020**, Search: "nanocellulose" OR "nanofibril" OR "nanocrystalline cellulose" OR "nanofibrillated cellulose" OR "microcrystalline cellulose" OR "microfibrillated cellulose" OR "cellulose nanocrystals" OR "cellulose whiskers" OR "cellulose fibers" OR "cellulose fibrils"). (**(A)** Reproduced with copyright permission from Ilyas, R. A., Sapuan, S. M., Ibrahim, R., Abral, H., Ishak, M. R., Zainudin, E. S., Atiqah, A., Atikah, M. S. N., Syafri, E., Asrofi, M., & Jumaidin, R. (2020). Thermal, biodegradability and water barrier properties of bio-nanocomposites based on plasticised sugar palm starch and nanofibrillated celluloses from sugar palm fibres. *Journal of Biobased Materials and Bioenergy, 14*(2), 234–248. https://doi.org/10.1166/jbmb.2020.1951, SAGE Publications. **(B)** Reproduced with copyright permission from Ilyas, R. A., Sapuan, S. M., Ibrahim, R., Abral, H., Ishak, M. R., Zainudin, E. S., Atikah, M. S. N., Mohd Nurazzi, N., Atiqah, A., Ansari, M. N. M., Syafri, E., Asrofi, M., Sari, N. H., & Jumaidin, R. (2019). Effect of sugar palm nanofibrillated cellulose concentrations on morphological, mechanical and physical properties of biodegradable films based on agro-waste sugar palm (*Arenga pinnata* (Wurmb.) Merr) starch. *Journal of Materials Research and Technology, 8*(5), 4819–4830. https://doi.org/10.1016/j.jmrt.2019.08.028, Elsevier.)

NCC are crystalline ricelike nanoparticles that are 5–80 nm in diameter and 100–250 nm in length, as shown in Fig. 18.1A (Ilyas, Sapuan, Atikah, et al., 2020). NCC can be extracted and isolated with different reagents, chemicals, and processes directly from lignocellulose and wood. From Table 18.2, it can be seen that the preferred methods used by researchers were H_2SO_4 hydrolysis treatment, followed by HCl

FIG. 18.2 Extraction of roselle fiber: **(A)** roselle plant, **(B)** stalks in bundle form, **(C)** water retting process, **(D)** removing the fibers from the stalks, and **(E)** final form of roselle fibers (Athijayamani et al., 2009; Nadlene et al., 2016).

hydrolysis, TEMPO oxidation, sonochemical acid hydrolysis, enzymatic hydrolysis, KOH hydrolysis, formic acid hydrolysis, irradiation oxidation, and organosolv solubilization treatments.

On the other hand, NFC are semicrystalline thread-like nanoparticles that are 5–70 nm in diameter and a few micrometers in length, as shown in Fig. 18.1B (Ilyas, Sapuan, Ishak, et al., 2019). A wide variety of

TABLE 18.1
Chemical Composition of Agro-Waste Fibers and Forest By-Products From Different Plants and Their Parts.

Fibers	HOLOCELLULOSE (WT%)		Lignin (wt %)	Ash (wt%)	Extractives (wt%)	Crystallinity (%)	Ref.
	Cellulose (wt%)	Hemicellulose (wt%)					
Roselle fiber	58–64	16–20	6–10	1–2	–	–	Razali et al. (2015)
Roselle fiber	60	10	15	–	–	–	Thiruchitrambalam et al. (2010)
Roselle skin	40.10		23.82	5.30	–	–	Ghalehno and Nazerian (2013)
Roselle pith	44.73		22.67	1.92	–	–	Ghalehno and Nazerian (2013)
Sugar palm fiber	43.88	7.24	33.24	1.01	2.73	55.8	Ilyas et al. (2018)
Sugar palm frond	66.49	14.73	18.89	3.05	2.46	–	Sanyang et al. (2016)
Sugar palm bunch	61.76	10.02	23.48	3.38	2.24	–	Sanyang et al. (2016)
Sugar palm trunk	40.56	21.46	46.44	2.38	6.30	–	Sanyang et al. (2016)
Wheat straw fiber	43.2 ± 0.15	34.1 ± 1.2	22.0 ± 3.1	–	–	57.5	Alemdar and Sain (2008c)
Soy hull fiber	56.4 ± 0.92	12.5 ± 0.72	18.0 ± 2.5	–	–	59.8	Alemdar and Sain (2008c)
Arecanut husk fiber	34.18	20.83	31.60	2.34	–	37	Julie Chandra et al. (2016)
Helicteres isora plant	71 ± 2.6	3.1 ± 0.5	21 ± 0.9	–	–	38	Chirayil et al. (2014)
Pineapple leaf fiber	81.27 ± 2.45	12.31 ± 1.35	3.46 ± 0.58	–	–	35.97	Cherian et al. (2010)
Ramie fiber	69.83	9.63	3.98	–	–	55.48	Syafri et al. (2018)
Oil palm mesocarp fiber	28.2 ± 0.8	32.7 ± 4.8	32.4 ± 4.0	–	6.5 ± 0.1	34.3	Megashah et al. (2018)
OPEFB	37.1 ± 4.4	39.9 ± 0.75	18.6 ± 1.3	–	3.1 ± 3.4	45.0	Megashah et al. (2018)
Oil palm frond	45.0 ± 0.6	32.0 ± 1.4	16.9 ± 0.4	–	2.3 ± 1.0	54.5	Megashah et al. (2018)
OPEFB fiber	40 ± 2	23 ± 2	21 ± 1	–	2.0 ± 0.2	40	Jonoobi et al. (2011)
Rubber wood	45 ± 3	20 ± 2	29 ± 2	–	2.5 ± 0.5	46	Jonoobi et al. (2011)
Curaua fiber	70.2 ± 0.7	18.3 ± 0.8	9.3 ± 0.9	–	–	64	Corrêa et al. (2010)
Banana fiber	7.5	74.9	7.9	0.01	9.6	15.0	Tibolla et al. (2014)
Sugarcane bagasse	43.6	27.7	27.7	–	–	76	Teixeira, Bondancia et al. (2011), Teixeira, Lotti, et al. (2011)
Kenaf bast	63.5 ± 0.5	17.6 ± 1.4	12.7 ± 1.5	2.2 ± 0.8	4.0 ± 1.0	48.2	Jonoobi et al. (2009)

Continued

TABLE 18.1
Chemical Composition of Agro-Waste Fibers and Forest By-Products From Different Plants and Their Parts.—cont'd

Fibers	HOLOCELLULOSE (WT%)		Lignin (wt %)	Ash (wt%)	Extractives (wt%)	Crystallinity (%)	Ref.
	Cellulose (wt%)	Hemicellulose (wt%)					
Phoenix dactylifera palm leaflet	33.5	26.0	27.0	6.5	–	50	Bendahou et al. (2009)
P. dactylifera palm rachis	44.0	28.0	14.0	2.5	–	55	Bendahou et al. (2009)
Kenaf core powder	80.26	23.58	–	–	–	48.1	Chan et al. (2013)
Water hyacinth fiber	42.8	20.6	4.1	–	–	59.56	Abral et al. (2018)
Wheat straw	43.2 ± 0.15	34.1 ± 1.2	22.0 ± 3.1	–	–	57.5	Alemdar and Sain (2008a)
Sugar beet fiber	44.95 ± 0.09	25.40 ± 2.06	11.23 ± 1.66	17.67 ± 1.54	–	35.67	Li et al. (2014)
Mengkuang leaf	37.3 ± 0.6	34.4 ± 0.2	24 ± 0.8		2.5 ± 0.02	55.1	Sheltami et al. (2012)

OPEFB, oil palm empty fruit bunch.

TABLE 18.2
Available Processes of Extraction Approaches From Different Sources for Isolation of Nanocrystalline Cellulose.

Source	Process	References
Roselle	H_2SO_4 hydrolysis and ultrasonic treatment	Kian et al. (2017, 2018)
Bleached softwood kraft pulp	H_2SO_4 hydrolysis and sonification	Wang et al. (2021)
Sugar palm fiber	H_2SO_4 hydrolysis	Hazrol et al. (2020), Ilyas, Sapuan and Ishak (2018), Ilyas, Sapuan, Atikah, et al. (2020), Ilyas et al. (2018b, 2018c)
Dunchi fiber and *Sesbania bispinosa* stalks	H_2SO_4 hydrolysis	Khan et al. (2020)
Broomcorn stalks	Hydrolysis and sonication	Langari et al. (2019)
Cotton linters	Single-step ammonium persulfate-assisted swelling, followed by oxidation	Wang et al. (2020)
Ginger fiber	HCl hydrolysis and ultrasonication	Abral, Ariksa, et al. (2019, 2020)
Kenaf	H_2SO_4 hydrolysis and ultrasonic treatment	Barbash and Yashchenko (2020)
Sugarcane bagasse	Enzymatic hydrolysis	De Aguiar et al. (2020)
Sugarcane straw	Enzymatic hydrolysis	De Aguiar et al. (2020)
Sugarcane bagasse	Acid hydrolysis	Plengnok and Jarukumjorn (2020)
Sugarcane bagasse	H_2SO_4 hydrolysis	Asrofi, Sapuan, et al. (2020), Asrofi, Sujito, et al. (2020)
Cotton cellulose powder	High-pressure homogenization	Park et al. (2019)
Kraft pulp	H_2SO_4 hydrolysis	He et al. (2019)
Water hyacinth (*Eichhornia crassipes*)	HCl hydrolysis and ultrasonication	Syafri, Sudirman, et al. (2019)
Bagasse fibers	Sonochemical acid hydrolysis and sonication	Robles et al. (2018)
Blue agave leaves	Sonochemical acid hydrolysis and sonication	Robles et al. (2018)
Eucalyptus pulp	Periodate oxidation route followed by reductive treatment with $NaBH_4$	Errokh et al. (2018)
Eucalyptus hardwood	Irradiation oxidation and organosolv solubilization	Zhang and Liu (2018)
Grass fibers (*Imperata brasiliensis*)	H_2SO_4 hydrolysis	Benini et al. (2018)
Oil palm	Sono-assisted TEMPO oxidation	Wardhono et al. (2018)
Ramie	KOH hydrolysis	Wahono et al. (2018)
Sago seed shells	H_2SO_4 hydrolysis	Naduparambath et al. (2018)
Water hyacinth fiber	HCl hydrolysis	Asrofi et al. (2018)
Wood sawdust	Sonochemical synthesis using acid hydrolysis	Shaheen and Emam (2018)
Acacia mangium	H_2SO_4 hydrolysis	Jasmani and Adnan (2017)
Banana pseudostem	TEMPO-mediated oxidation, formic acid hydrolysis	Faradilla et al. (2017)

Continued

TABLE 18.2
Available Processes of Extraction Approaches From Different Sources for Isolation of Nanocrystalline Cellulose.—cont'd

Source	Process	References
Cornhusk	H_2SO_4 hydrolysis	Yang et al. (2017)
Groundnut shells	H_2SO_4 hydrolysis	Bano and Negi (2017)
Humulus japonicus stem	H_2SO_4 hydrolysis with high-temperature pretreatment	Jiang et al. (2017)
Wheat Straw	H_2SO_4 hydrolysis	Pereira et al. (2017)
Arecanut husk fiber	HCl hydrolysis	Julie Chandra et al. (2016)
Corncob	H_2SO_4 hydrolysis	Liu et al. (2016)
Corncob	Formic acid hydrolysis	Liu et al. (2016)
Ushar (*Calotropis procera*) *seed fiber*	H_2SO_4 hydrolysis	Oun and Rhim (2016)
Cotton stalk	TEMPO-mediated oxidation and H_2SO_4 hydrolysis	Soni et al. (2015)
Oil palm trunk	H_2SO_4 hydrolysis	Lamaming et al. (2015)
Cotton fiber	H_2SO_4 hydrolysis	Pereda et al. (2014)
Sugar palm frond	H_2SO_4 hydrolysis	Sumaiyah et al. (2014)
Cotton	H_3PO_4 hydrolysis	Camarero Espinosa et al. (2013)
Cotton (*Gossypium hirsutum*) linters	H_2SO_4 hydrolysis	Morais et al. (2013)
Hibiscus sabdariffa fibers	Steam explosion H_2SO_4 hydrolysis	Sonia and Priya Dasan (2013)
Kenaf core wood	H_2SO_4 hydrolysis	Chan et al. (2013)
Oil palm empty fruit bunch	H_2SO_4 hydrolysis	Haafiz et al. (2014)
Phormium tenax (harakeke) fiber	H_2SO_4 hydrolysis	Fortunati et al. (2013)
Flax fiber	H_2SO_4 hydrolysis	Fortunati et al. (2013)
Soy hulls	H_2SO_4 hydrolysis	Flauzino Neto et al. (2013)
Bamboo	H_2SO_4 hydrolysis	Brito et al. (2012)
Eucalyptus kraft pulp	H_2SO_4 hydrolysis	Tonoli et al. (2012)
Mengkuang leaves	H_2SO_4 hydrolysis	Sheltami et al. (2012)
Potato peel waste	H_2SO_4 hydrolysis	Chen et al. (2012)
Rice straw	H_2SO_4 hydrolysis	Lu and Hsieh (2012)
Wood pulp	TEMPO oxidation followed by HCl hydrolysis	Salajková et al. (2012)
Industrial bioresidue (sludge)	H_2SO_4 hydrolysis	Herrera et al. (2012)
Sesame husk	H_2SO_4 hydrolysis	Purkait et al. (2011)
Bamboo (*Pseudosasa amabilis*)	H_2SO_4 hydrolysis	Liu et al. (2010)
Banana fiber	$H_2C_2O_4$ hydrolysis	Cherian et al. (2010)
Coconut husk	H_2SO_4 hydrolysis	Rosa et al. (2010)
Colored cotton	H_2SO_4 hydrolysis	de Morais Teixeira et al. (2010)
Curaua fiber	H_2SO_4, H_2SO_4/HCl, HCl hydrolysis	Corrêa et al. (2010)
Industrial bioresidue	H_2SO_4 hydrolysis	Oksman et al. (2010)
Rice husk	H_2SO_4 hydrolysis	Rosa et al. (2010)

TABLE 18.2
Available Processes of Extraction Approaches From Different Sources for Isolation of Nanocrystalline Cellulose.—cont'd

Source	Process	References
Sugarcane bagasse	H_2SO_4 hydrolysis	Teixeira, Bondancia et al. (2011), Teixeira, Lotti, et al. (2011)
Cassava bagasse	H_2SO_4 hydrolysis	Teixeira et al. (2009)
Cotton (cotton wool)	H_2SO_4 hydrolysis	Morandi et al. (2009)
Grass fibers	H_2SO_4 hydrolysis	Pandey et al. (2009)
Mulberry	H_2SO_4 hydrolysis	Li et al. (2009)
Cotton linters	HCl hydrolysis	Braun et al. (2008)
Cotton Whatman filter paper	H_2SO_4 hydrolysis	Paralikar et al. (2008)
Ramie	H_2SO_4 hydrolysis	Habibi and Vignon (2008)
Sisal fiber	H_2SO_4 hydrolysis	Morán et al. (2008)
Microcrystalline cellulose	H_2SO_4 hydrolysis	Bondeson et al. (2006)
Ramie	H_2SO_4 hydrolysis	Lu et al. (2006)
Algae	H_2SO_4 hydrolysis	Imai et al. (2003)
Bacterial cellulose	H_2SO_4 hydrolysis	Grunert and Winter (2002)
Tunicate	H_2SO_4 hydrolysis	Favier et al. (1995)
Valonia ventricosa	HCl hydrolysis	Revol (1982)

techniques, such as microfluidation and high-pressure homogenization (HPH), as well as a combination of chemical and mechanical techniques can yield NFC through mechanical fibrillation of pulp fibers. According to Table 18.3, the most favored process used by researchers is supermasscolloider, followed by ultrasonic treatment, HPH, TEMPO-mediated oxidation, PFI mill, grinding, and a combination of chemomechanical treatments, such as acid hydrolysis and ultrasound/HPH/refining treatments.

3 ROSELLE

Roselle has attained colossal attention as a jute substitute, and attempts are being made to extend its cultivation in areas that are not favorable for jute cultivation. Roselle plant is classified as a rapid-growing plant and renewable natural bioresource.

3.1 Roselle Fiber Production

Several researchers explored roselle bast fiber's potential as reinforcing materials because of its similar characteristics to other known natural fibers, such as jute (Nadlene et al., 2016). The fiber is located under roselle stalk's bark and was extracted via a series of water retting steps, as shown in Fig. 18.2. High-quality fiber is obtainable by harvesting during the bud stage. The stalks were packed and water-retted for 3–4 days. The retted stalks were then washed under flowing water and separated from the stalks, washed, and sundried (Nadlene et al., 2016; Radzi et al., 2018a, 2018b). The final product is illustrated in Fig. 18.2E.

3.2 Chemical Composition and Anatomy of Roselle Fiber

Roselle fiber is one of the natural fibers that draws scientists and engineers to explore its potential as a filler in the composites. Some of the research and review papers had discussed the physical, morphologic, mechanical, and chemical properties of roselle fiber (Chauhan & Kaith, 2012; Kian et al, 2018, 2019; Nadlene et al., 2016; Radzi et al., 2019; Razali et al., 2015). According to Nadlene et al. (2016), Young's modulus and tensile strength of the roselle fiber were increased with cellulose content. Before decisions can be made for their potential applications, it is crucial to know the properties of roselle to overcome the main problems of using natural fiber instead of wood fiber. Studies on the chemical and anatomic properties of roselle were carried out by Nadlene et al. (2016), Ghalehno and Nazerian

TABLE 18.3
Available Processes of Extraction Approaches From Different Sources for NFC Isolation.

NFC Sources	NFC Preparation	Ref.
Bleached softwood kraft pulp	H_2SO_4 hydrolysis, sonification, and homogenization	Wang et al. (2021)
Oil palm fibers	NaOH and H_2O_2	Fahma et al. (2020)
Ginkgo seed shells	Acid hydrolysis and high-pressure homogenization	Ni et al. (2020)
Ramie fibers	Super masscolloider	Marinho et al. (2020)
Cornstalk	Microfluidizer	Suopajärvi et al. (2020)
Wheat straw	Microfluidizer	Suopajärvi et al. (2020)
Rapeseed stem residues	Microfluidizer	Suopajärvi et al. (2020)
Borassus flabellifer leaf stalk	NaClO hydrolysis	Athinarayanan et al. (2020)
Raffia fiber (*Raphia vinifera*)	Supermasscolloider	Stanislas et al. (2020)
Ambarella (*Spondias dulcis*)	Supermasscolloider	Stanislas et al. (2020)
Cassava bagasse (*Manihot esculenta*)	Supermasscolloider	Stanislas et al. (2020)
Banana pseudostem (*Musa* spp.)	Acid hydrolysis	Poiini et al. (2020)
Sugarcane bagasse	Grinding, high-pressure homogenization, and ultrasonic treatment	Zhang et al. (2020)
Pineapple stems	Acid hydrolysis	Rigg-Aguilar et al. (2020)
Oat hull	H_2SO_4 hydrolysis and ultrasonication	Debiagi et al. (2020)
Kelempayan (*Neolamarckia cadamba*)	TEMPO oxidation method, PFI mill, and homogenization	Latifah et al. (2020)
Mixed hardwood pulp	$NaIO_4$ and $NaClO_2$ and refining	Kumar et al. (2020)
Sugar palm	High-pressure homogenization	Atikah et al. (2019), Ilyas, Sapuan, Atiqah, et al. (2020), Ilyas, Sapuan, Ibrahim et al. (2019, 2020); Ilyas et al. (2018a, 2018b)
Stalks of wheat straw (*Triticum paleas*)	H_2SO_4 hydrolysis and ultrasound treatment	Barbash et al. (2017)
Bamboo fiber	–	Llanos and Tadini (2017)
Cornhusk	TEMPO-mediated oxidation	Yang et al. (2017)
Cornhusk	High-intensity ultrasonication	Yang et al. (2017)
Ushar (*Calotropis procera*) seed fiber	TEMPO oxidation	Oun and Rhim (2016)
Banana pseudostem	High-pressure homogenization	Velásquez-Cock et al. (2016)
Corncob	TEMPO-mediated oxidation	Liu et al. (2016)
Corncob	PFI refining	Liu et al. (2016)
Kenaf	Supermasscolloider	Babaee et al. (2015)
Rice straw	Ultrasonication	Nasri-Nasrabadi et al. (2014)
Kenaf	Supermasscolloider	Karimi et al. (2014)
Softwood wood flour	Supermasscolloider	Hietala et al. (2013)
Cotton cellulose	Hydrolyzed in 6.5 M sulfuric acid/ 75 min	Teixeira, Bondancia, et al. (2011), Teixeira, Lotti, et al. (2011)
Wheat straw	High-pressure homogenizer/15 min	Kaushik et al. (2010)
Wheat	–	Azeredo et al. (2009)

NFC, nanofibrillated cellulose

TABLE 18.4
Chemical Properties of Roselle Fiber.

CHEMICAL PROPERTIES OF ROSELLE FIBER FROM SKIN AND PITH (GHALEHNO & NAZERIAN, 2013)			
Property Fiber	Skin	Pith	Fiber
Cellulose	40.1	44.73	–
Lignin (%)	23.82	22.67	–
Alcohol-acetone solubility (%)	8.96	6.87	–
Ash (%)	5.3	1.92	–
CHEMICAL PROPERTIES OF ROSELLE FIBER (THIRUCHITRAMBALAM ET AL., 2010)			
Cellulose (%)	–	–	60
Lignin (%)	–	–	10
Hemicellulose (%)	–	–	15
CHEMICAL PROPERTIES OF ROSELLE FIBER (RAZALI ET AL., 2015)			
Cellulose (%)	–	–	58–64
Hemicellulose (%)	–	–	16–20
Lignin (%)	–	–	6–10
Ash (%)	–	–	1–2

(2013), and Thiruchitrambalam et al. (2010) as a substitute material of solid forest products, because of the decreasing forest wood resources. To determine the chemical composition of roselle fiber, Ghalehno and Nazerian (2013) and Nadlene et al. (2016) used the standard methods of TAPPI (Tappi Test Methods, 1992–93) and neutral detergent fiber, respectively. The result and analysis of the chemical and anatomic properties of the roselle fiber are listed in Tables 18.4 and Table 18.5, respectively. The cellulose chemical compositions analyzed by Nadlene et al. (2015a) and Thiruchitrambalam et al. (2010) were slightly different than those analyzed by Ghalehno and Nazerian (2013), with values of 58%–64%, 60%, and 40%, respectively. The differences in the chemical compounds of roselle is due to the origin, type, parts, and difference in maturity of roselle fiber, as well as the methods to determine the chemical composition of fiber.

3.3 Mechanical Properties of Roselle Fiber

Roselle's chemical properties are similar to those of jute fibers, indicating its ability as a good reinforcing material in polymer materials. Table 18.6 shows the mechanical properties of roselle fiber and other comparative bast natural fibers. It can be summarized that roselle fiber has comparable tensile properties to the group of bast fibers. Thus mechanical properties of roselle fiber proved its potential as a reinforcing material for advanced green engineering composite products. Besides, the application of this composite product can help improve the disposal method for the presently generated roselle agro-waste. Currently, in Malaysia,

TABLE 18.5
Anatomic Properties of Roselle Stem (Ghalehno & Nazerian, 2013).

Property Fiber	Skin	Pith
Fiber length (μm)	3030	1279
Fiber diameter (μm), lumen	17.54	26.7
Lumen (μm)	6.88	10.66
Cell wall thickness (μm)	5.33	8.21

TABLE 18.6
Tensile Strength of Roselle and Bast Fibers.

Fiber	Tensile Strength (MPa)	Reference
Roselle 3 months	414.72	Razali et al. (2015)
Roselle 6 months	252.64	Razali et al. (2015)
Roselle 9 months	228.57	Razali et al. (2015)
Kenaf	18–180	Akil et al. (2011)
Hemp	300–800	Clemons (2010)
Jute	340–384	Xia et al. (2009)
Flax	500–900	Clemons (2010)
Ramie	400–938	Ku et al. (2011)

roselle is harvested for its calyx as the main product, and its stems are usually disposed after a year. Therefore roselle fiber is seen as a potential candidate for composite products.

3.4 Morphologic Properties of Roselle Fiber

One of the factors that influence the interfacial bonding and mechanical properties of natural fiber-reinforced polymer composites is the morphologic structure of the natural fiber. The cross-sectional and surface morphology of roselle fiber is shown in Fig. 18.3A and B, respectively. Bast fiber is naturally composed of a combination of elementary fibers, in which these fibers are firmly bonded together by pectin and other noncellulosic components along the length of the fibers. These compounds give strength to the bast fiber. From Fig. 18.3A, it can be observed that the cross section of roselle fiber has a clear lumen structure in the center. The lumen appeared to be smaller with the increase in plant age, which might be due to the maturity of the fiber (Yusriah et al., 2014). The size of the lumen is correlated with roselle water uptake (Nadlene et al., 2016). Thus the bigger the lumen size, the higher the water uptake of the fiber. According to Nadlene et al. (2015a), the size of the roselle lumen shrinks because of the presence of the thicker cell wall with increasing plant age. Fig. 18.3B displays the surface morphology of roselle fiber.

According to Nadlene et al. (2016), the surface structure of a roselle plant aged 9 months is more desirable to be reinforced with polymer composites, compared with roselle plants at the age of 3 and 6 months. This

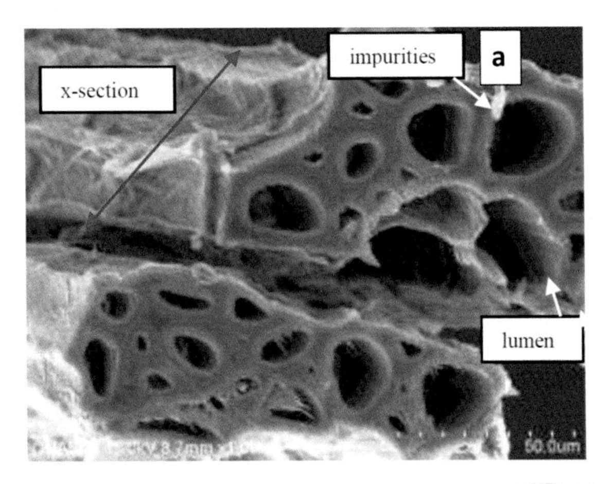

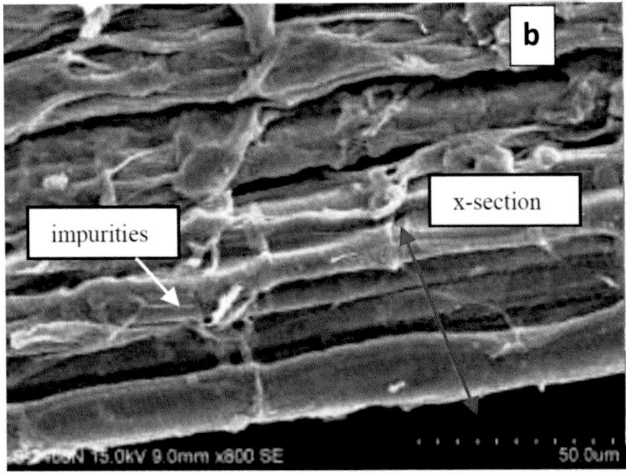

FIG. 18.3 **(A)** Cross section and **(B)** surface structure of roselle fiber (Nadlene et al., 2016).

phenomenon might be due to the condition of the surface, which is rougher and cleaner in a 9-month-old plant. Coarse surface fiber is required for better mechanical interlocking between fiber and polymer matrix, which would result in high mechanical properties of polymer composites (Azammi et al., 2020).

4 ROSELLE NANOCELLULOSE

4.1 Morphology

Roselle-derived fibers mainly contain 59%–65% of cellulose, 16%–20% hemicellulose, and 6%–10% of lignin (Razali et al., 2015). Studies on the advancement of nanoparticles from renewable sources as reinforcing materials are vastly ongoing owing to their excellent thermal stability and tensile strength, biocompatibility, and environment-friendly biodegradability (Capron et al., 2017). Microcrystalline cellulose (MCC) is a possible material produced from natural cellulose. Nowadays, MCC is recognized as an alternative source for the production of nanocellulose nanocomposites or polymer composites (Bandera et al., 2014; Mohamad Haafiz et al., 2013).

According to Kian et al. (2017), the morphology of the MCC derived from roselle fibers had been analyzed, displayed by the scanning electron microscopic (SEM) images in Fig. 18.4. Here, the alkali-treated pulp (R-pulp) exhibited regular-shaped, individualized fibers

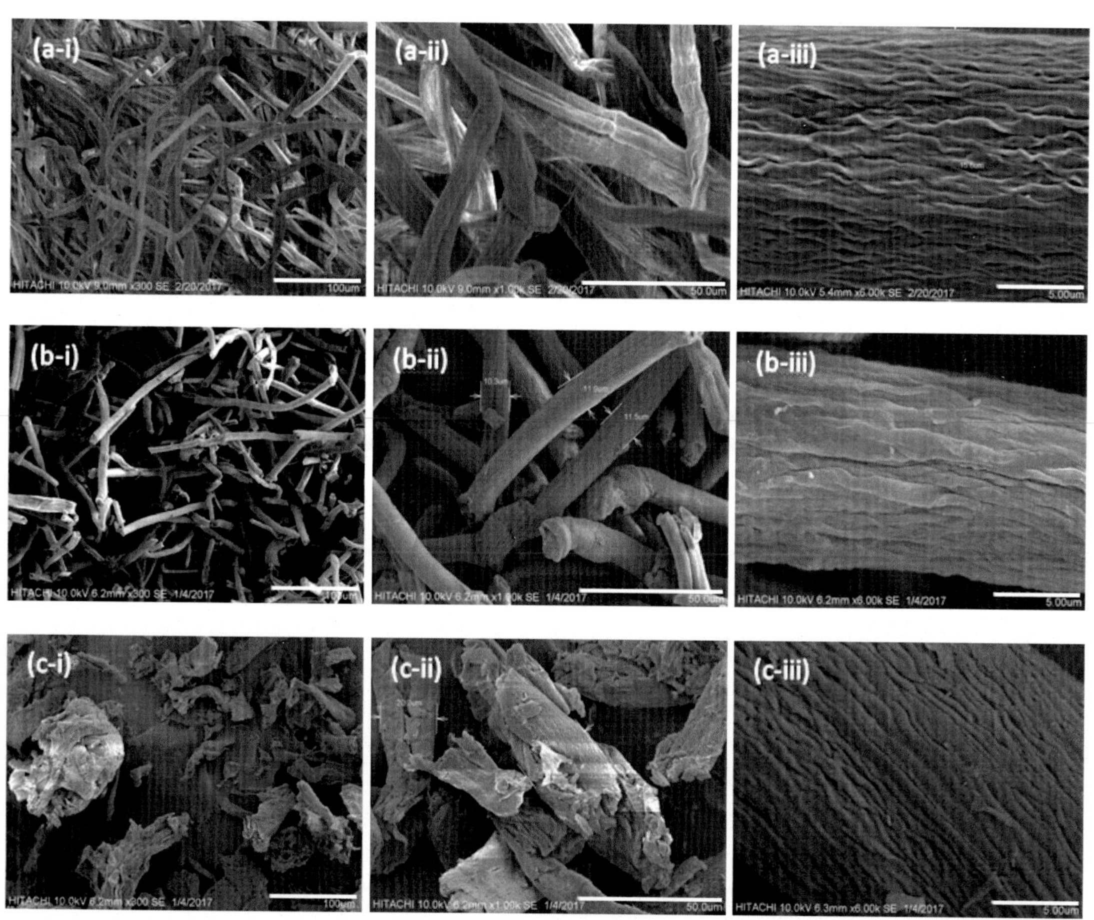

FIG. 18.4 Scanning electron microscopic images of **(A)** alkali-treated pulp (R-pulp), **(B)** roselle microcrystal line cellulose, and **(C)** commercial microcrystalline cellulose, with several magnifications (i) 100×, (ii) 1000×, and (iii) 6000× (Reproduced with copyright permission from Kian, L. K., Jawaid, M., Ariffin, H., & Alothman, O. Y. (2017). Isolation and characterization of microcrystalline cellulose from roselle fibers. *International Journal of Biological Macromolecules*, 103, 931–940. https://doi.org/10.1016/j.ijbiomac.2017.05.135, Elsevier.)

of the normal form (Alemdar & Sain, 2008b; Xiang et al., 2016) because the breakdown of plant components in fiber bundles occurred during bleaching and alkaline treatment. Hence, it contributed to the separation of fiber bundles into separate fibrous strands. In contrast, roselle MCC (R-MCC) showed the modified and uneven shape of microsized fibrils with an irregular base (Kian et al., 2017). HCl exposure toward the MCC led to the fragmentation of the structure of fibrous strands into smaller microcrystallites, owing to acid hydrolysis. Generally, the acid treatment hydrolyzed the glycosidic bonds of the cellulose (Owolabi et al., 2017). From the findings, R-MCC also displayed lower aggregation and thinner microstructure than the commercialized one. The longer microfibrillar configuration of the R-MCC with a recorded high aspect ratio proved its suitability in manufacturing high-tensile-strength biocomposite materials (Adel et al., 2010).

Based on the study by Kian et al. (2018), various sizes of NCC derived from R-MCC samples were observed via field emission scanning electron microscopy (FESEM) and transmission electron microscopy (TEM) for accurate morphologic study. Through further TEM analysis for NCC-I (Fig. 18.5A), the observed NCC-I nanostructures were well-coordinated in bulky bundles. In addition, this situation was presumably due to the short hydrolysis reaction time, thus the MCC microfibrils were partly depolymerized (Elanthikkal et al., 2010). Fig. 18.5B and C show the needle-shaped nanostructures of NCC-II and NCC-III, which operated with higher hydrolysis reaction of time. The visible clusters of nanoparticles are well-align in parallel, for NCC-II and NCC-III, because of the strong lateral bonding of hydrogen between the nanoparticles—a self-assembly mechanism of nanosized particles contributed by the hydrophilic copper grid substrate (Adel et al., 2010).

Besides, from the work of Kian et al. (2019), ultrasonication process was performed in a study on CNWs formed by isolating roselle fibers. CNW-I was applied with 20% amplitude during ultrasonication treatment; thus an entwined whisker-shaped cellulose bundle was clearly observed, as shown in Fig. 18.6A. This represented the occurrence of MCC degradation at low ultrasonic amplitude (Cui et al., 2016). The extended rod-shaped nanoparticles for CNW-II and CNW-III were observed and treated with increased ultrasonication amplitude, as presented in Fig. 18.6B and C,

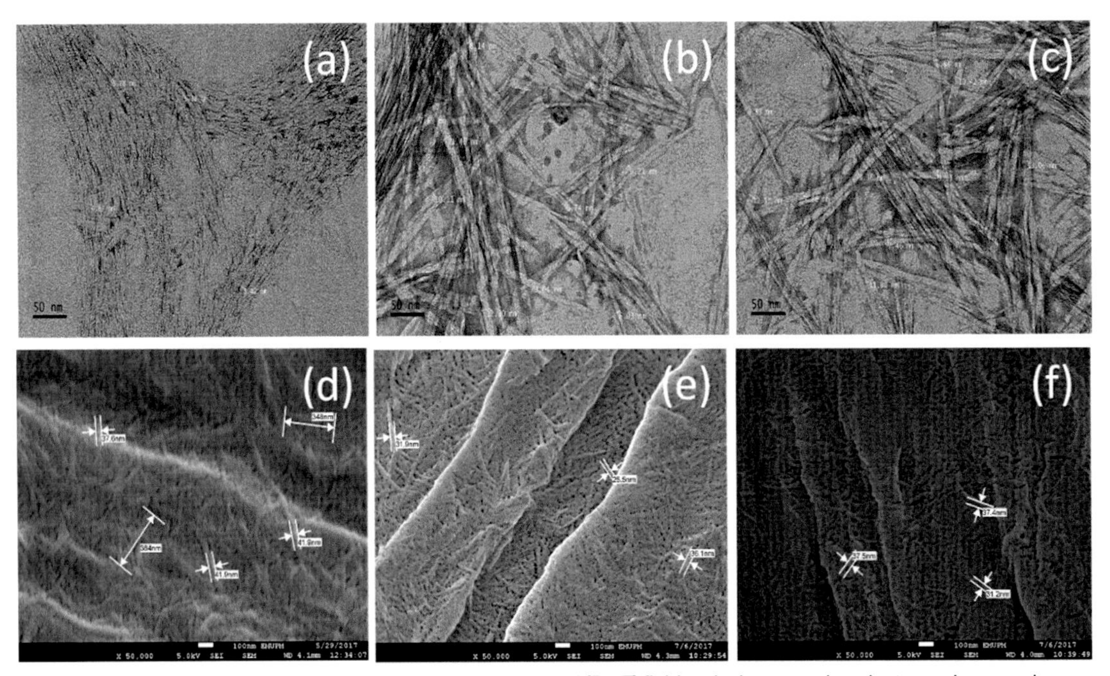

FIG. 18.5 **(A–C)** Transmission electron microscopic and **(D–F)** field emission scanning electron microscopic images of nanocrystalline cellulose (NCC)-I, NCC-II, and NCC-III, respectively. (Reproduced with copyright permission from Kian, L. K., Jawaid, M., Ariffin, H., & Karim, Z. (2018). Isolation and characterization of nanocrystalline cellulose from roselle-derived microcrystalline cellulose. *International Journal of Biological Macromolecules*, 114, 54–63. https://doi.org/10.1016/j.ijbiomac.2018.03.065, Elsevier.)

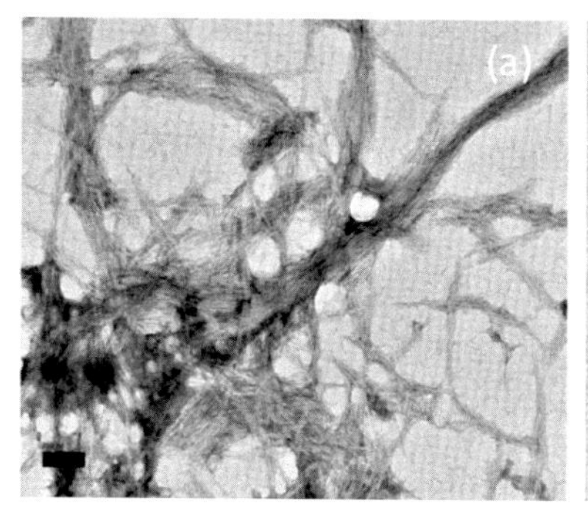

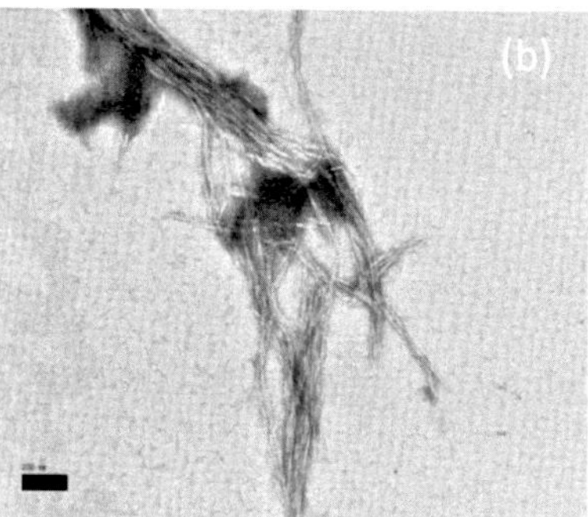

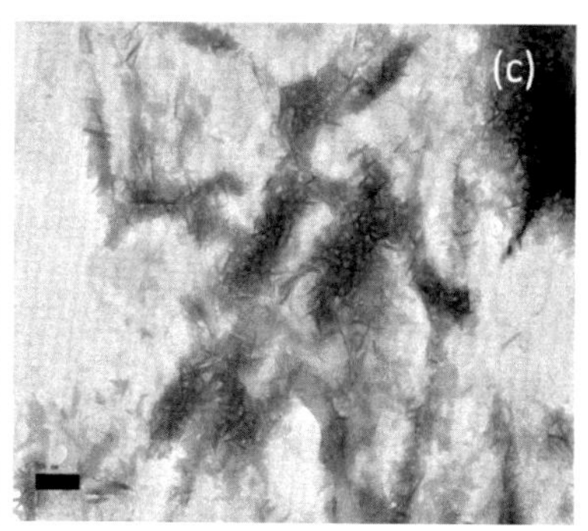

FIG. 18.6 **(A–C)** Transmission electron microscopic images of cellulose nanowhiskers (CNWs), with 20%, 30%, and 40% low amplitude of ultrasonication for CNW-I, CNW-II, and CNW-III, respectively. (Reproduced with copyright permission from Kian, L. K., Jawaid, M. L. C. pd, Ariffin, H., Karim, Z., & Sultan, M. T. H. (2019). Morphological, physico-chemical, and thermal properties of cellulose nanowhiskers from roselle fibers. *Cellulose*, *26*(11), 6599–6613, Springer Nature.)

respectively. Thus the segregation of cellulose bundles into nanostructure of CNWs was aided and facilitated by the higher amplitudes of ultrasonication at 30% and 40% amplitude. Theoretically, the ultrasonication provided efficient shear forces toward the bundles to eradicate the strong cohesion force between the cellulose, encouraged by the freeze-drying treatments prior to FESEM imaging (Azrina et al., 2017). Hence, the nanoaggregation, which took place during CNW-II film formation, affected the entanglement of nanowhiskers (Dongyan Liu et al., 2017).

4.2 Particle Size

According to Kian et al. (2017), from the morphologic analysis, the distribution of the sample particle size had been measured, as represented in Fig. 18.7. The volume-weighted average diameters of sample R-pulp, R-MCC, and commercial MCC (C-MCC) were 231, 44, and 277 μm, respectively. The decrease in the particle size of R-pulp that was caused by acid disintegrations could be observed in Table 18.7. The acid treatment of the pulp hydrolyzed the amorphous cellulose fibrils into shorter R-MCC particles (Wulandari et al., 2016). Moreover, unsymmetric large-scale distribution was detected from Fig. 18.7, which was affected by the chemical treatment of the roselle pulp that strongly led to the reduction in particle size (Salminen et al., 2017).

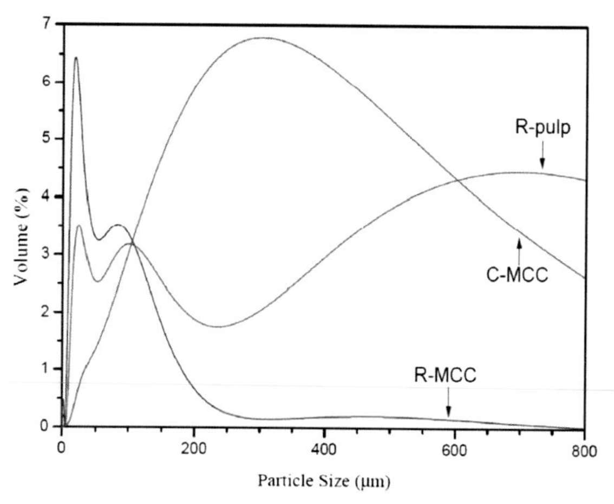

FIG. 18.7 Distribution of alkali-treated pulp (R-pulp), roselle microcrystalline cellulose (R-MCC), and commercial microcrystalline cellulose (C-MCC) sample particle size. (Reproduced with copyright permission from Kian, L. K., Jawaid, M., Ariffin, H., & Alothman, O. Y. (2017). Isolation and characterization of microcrystalline cellulose from roselle fibers. *International Journal of Biological Macromolecules*, 103, 931–940. https://doi.org/10.1016/j.ijbiomac.2017.05.135, Elsevier.)

TEM offered the most precise insight into the morphology of the resulting NCC with a width of 4.67–13.6 nm, as reported by Kian et al. (2018). Atomic force microscopic observations specified the uniformly distributed thickness values of NCCs, as displayed in Fig. 18.8A. Moreover, the average thickness of the samples, NCC-I, NCC-II, and NCC-III, with the hydrolysis reaction time of 30, 45, and 60 min, were about 48, 41, and 34 nm, respectively. However, as shown in Fig. 18.8B, the DLS analysis indicated inconsistent NCC distance distribution, which primarily occurred because of the light scattering deflection toward the NCC particle (El Achaby et al., 2017; Prathapan et al., 2016). NCC-I, NCC-II, and NCC-III had mean lengths of 552.6, 272.6, and 280.9 nm, respectively. Thus the longer the reaction time, the higher the MCC size reduction into nanosized NCCs, both for thickness and length measurements, which had been further verified.

TABLE 18.7
Particle Size Analysis Data of R-Pulp, R-MCC, and C-MCC.

Sample	Surface-Weighted Average Diameter (μm)	Volume-Weighted Average Diameter (μm)
R-pulp	28.41	231.25
R-MCC	11.53	44.28
C-MCC (commercialized)	119.13	277.27

C-MCC, commercial microcrystalline cellulose; *R-MCC*, roselle microcrystalline cellulose; *R-pulp*, alkali-treated pulp.

As recorded by Kian et al. (2019), the mean width of the samples CNW-I, CNW-II, and CNW-III were 14.7, 10.2, and 9.6 nm, respectively. Table 18.8 displays the decreasing CNW height that was affected by the reduction of the CNW length. The minimum aspect ratio (L/W) of CNW recorded from this work was approximately 37 nm, which was greater than the previous NCC particle, 21.5 nm. Thus it was verified that the increasing ultrasonic amplitudes affected the disintegration of MCC into smaller sized CNWs.

As mentioned earlier, different treatments of roselle fiber-derived MCC and process parameters, such as hydrolysis reaction time and ultrasonication amplitude, in deriving the roselle fiber into nanocellulose and nanowhiskers play an important role in the morphologic analysis of the products. The nanoparticle with the highest crystallinity was the remote NCC-III, i.e.,79.5% of crystallinity, with 1 h of hydrolysis time. From the observations, NCC III had shorter mean lengths, which is useful for the mechanical enhancement of the NCC. A higher amplitude of ultrasonication developed excellent colloidal stability in aqueous solution, hence providing better dispersibility of CNW nanoparticles in polymer matrices.

4.3 Thermal Stability

According to Kian et al. (2017), thermogravimetric analysis (TGA) and derived thermogravimetric (DTG; differential scanning calorimetry [DSC]) outcomes specified that the MCC from roselle had better thermal stability than the roselle pulp. TGA was performed for approximately 6 mg of sample that was heated at a

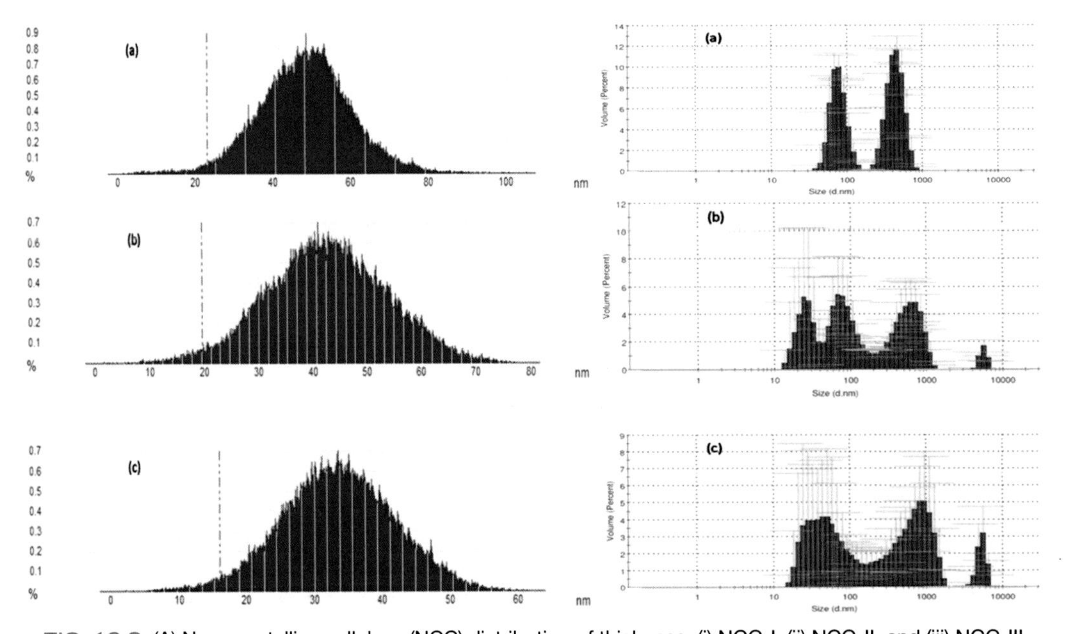

FIG. 18.8 (A) Nanocrystalline cellulose (NCC) distribution of thickness: (i) NCC-I, (ii) NCC-II, and (iii) NCC-III. (B) Distribution of sample length: (i) NCC-I, (ii) NCC-II, and (iii) NCC-III. (Reproduced with copyright permission from Kian, L. K., Jawaid, M., Ariffin, H., & Karim, Z. (2018). Isolation and characterization of nanocrystalline cellulose from roselle-derived microcrystalline cellulose. *International Journal of Biological Macromolecules, 114*, 54–63. https://doi.org/10.1016/j.ijbiomac.2018.03.065, Elsevier.)

TABLE 18.8
Width Average, Thickness Average, and Length Average of CNW-I, CNW-II, and CNW-III.

Samples	Width Size Average, W (nm)	Thickness Size Average, T (nm)	Length Size Average, L (nm)
CNW-I	14.7	101.6	502.1
CNW-II	10.2	33.6	381.4
CNW-III	9.6	29.4	372.5

CNW, cellulose nanowhisker.

temperature range of 30–900°C, at the heating rate of 20°C min^{-1} in a nitrogen environment. In contrast, 10 mg of dry sample was heated at 10°C/min in nitrogen flux with a temperature range from ambient temperature to 350°C for the DSC thermograms. Fig. 18.9A and B represent the TGA and DTG curves for R-pulp, R-MCC, and C-MCC.

Instantly, for R-pulp, its cellulose decomposition was started at 298.14°C, and for R-MCC, it was activated at 315.43°C. Furthermore, the R-MCC displayed higher degradation temperature at 10% weight loss, presumably due to the high cellulose molecular ordering, where more heat energy was essential for the thermal degradation process. The degradation peak temperature of R-MCC and the R-pulp sample was 340.12°C and 335.15°C, respectively. The degradation of cellulosic components occurred as the decarboxylation, depolymerization, and decomposition of the hemicellulose and cellulose fragments took place. Besides, over the temperature of 380°C, the biomass faced aromatization, combustion, disintegration, lignin pyrolysis, and solid char formation. Meanwhile, the weight of the char residue yielded from R-pulp was more than that of the sample R-MCC; hence efficient char formation was contributed by the aid of flame-retardant compounds.

According to Kian et al. (2018), outcomes from TGA studies proved that sulfuric acid hydrolysis treatment could segregate NCC from MCC-derived roselle, suitable to be the pioneer cellulose material for massive NCC production. The shorter hydrolysis time will lead to the production of NCC with high thermal stability, discovered by thermal analysis based on the TGA and DSC observations. Fig. 18.10 shows the TGA curves of NCC samples. The NCC samples were reported to lose weights at 60–150°C temperatures, accredited to the portion of evaporated residual water in the cellulose. While the NCCs were primarily associated with the degradation of the cellulosic chain at the temperature of 200–380°C. Due to the thermal instability of the sulfate group surface area, a defective TGA curve could be explicitly observed for NCC-III samples, which are susceptible to heat degradation.

DTG analysis (Fig. 18.11) showed the degradation of cellulose chain indicated by the distinct peaks from the curve of NCC-III, specifying the relationship between

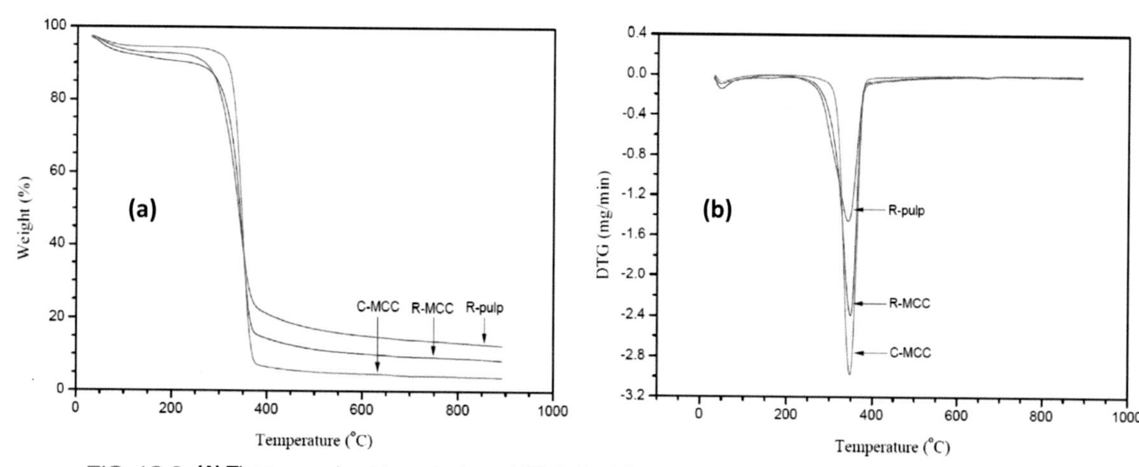

FIG. 18.9 **(A)** Thermogravimetric analysis and **(B)** derived thermogravimetric curves of alkali-treated pulp (R-pulp), roselle microcrystalline cellulose (R-MCC), and commercial microcrystalline cellulose (C-MCC) samples. (Reproduced with copyright permission from Kian, L. K., Jawaid, M., Ariffin, H., & Alothman, O. Y. (2017). Isolation and characterization of microcrystalline cellulose from roselle fibers. *International Journal of Biological Macromolecules*, 103, 931–940. https://doi.org/10.1016/j.ijbiomac.2017.05.135. Elsevier.)

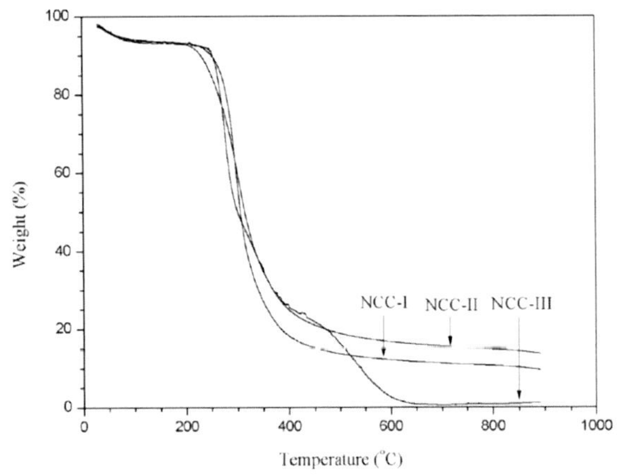

FIG. 18.10 Thermogravimetric analysis curves of nanocrystalline cellulose (NCC)-I, NCC-II, and NCC-III. (Reproduced with copyright permission from Kian, L. K., Jawaid, M., Ariffin, H., & Karim, Z. (2018). Isolation and characterization of nanocrystalline cellulose from roselle-derived microcrystalline cellulose. *International Journal of Biological Macromolecules, 114*, 54–63. https://doi.org/10.1016/j.ijbiomac.2018.03.065, Elsevier.)

the temperature applied and the rate of changes in material weight as heated. The degradation peak at about 277°C denoted the decomposition of cellulose. At 346.21°C, the second degradation peak represented the decomposition of nonchemically modified cellulose. In addition, the degradation peak at 535°C contributed to the sulfur component degradation with high thermal resistant capability. Thus it inhibited the decomposition of carbonaceous residue for materials with lower molecular weight. Conclusively, the NCC-III sample had the highest weight loss plus lowest solid residues, affected by the synergistic effect of high sulfur content, along with large crystalline fraction. The larger the weight loss, the lower the residue formation for the NCCs.

Next, according to Kian et al. (2019), CNWs from roselle fibers possessed high heat resistance when exposed to higher temperatures. Initially, CNWs showed mass loss along the 60–120°C temperature profile associated with the evaporation of water. The CNW-II sample had a lower initial decomposition temperature than CNW-I. It affected the cellulose decomposition at a lower temperature because of the acoustic disruption of CNW-II internal crystal structure. Fig. 18.12 shows the DTG analysis; CNW-III underwent two-stage thermal decomposition at a temperature of 266.16°C, due to the unstable crystalline fraction when excessive heat is applied, and at 357.21°C, which was contributed by a higher thermally stable crystalline fraction. Thus both peaks inferred the thermal instability properties of the nanowhisker samples. Moreover, both CNW-II and CNW-III samples lost around 84% of the weight and were left with about 15% residue weight. This occurred due to the same morphologic feature, as

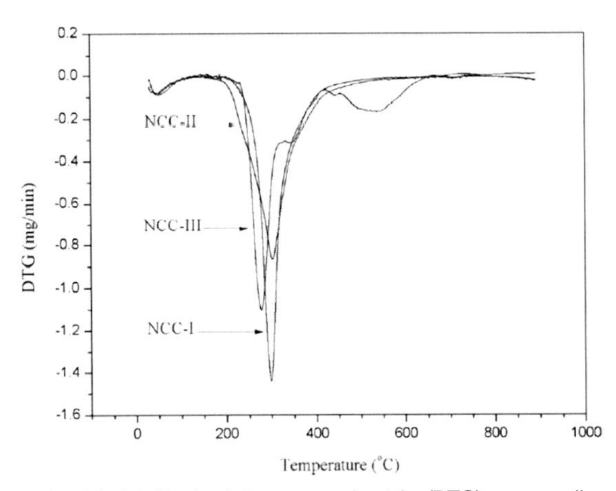

FIG. 18.11 Derived thermogravimetric (DTG) curves displayed from samples of nanocrystalline cellulose (NCC)-I, NCC-II, and NCC-III. (Reproduced with copyright permission from Kian, L. K., Jawaid, M., Ariffin, H., & Karim, Z. (2018). Isolation and characterization of nanocrystalline cellulose from roselle-derived microcrystalline cellulose. *International Journal of Biological Macromolecules, 114*, 54–63. https://doi.org/10.1016/j.ijbiomac.2018.03.065, Elsevier.)

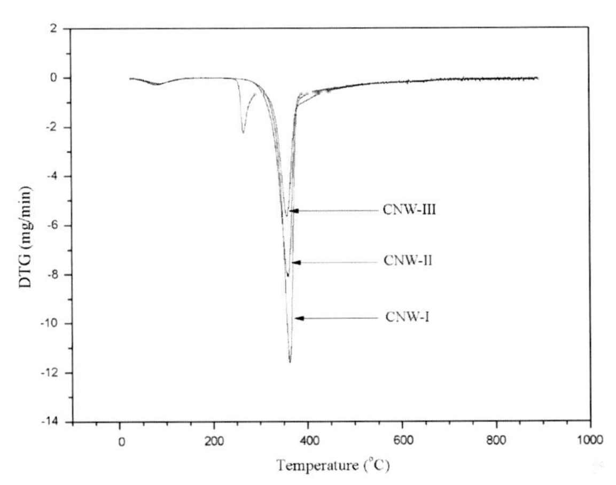

FIG. 18.12 Derived thermogravimetric (DTG) curves of cellulose nanowhisker (CNW-I, CNW-II, and CNW-III samples. (Reproduced with copyright permission from Kian, L. K., Jawaid, M., Ariffin, H., Karim, Z., & Sultan, M. T. H. (2019). Morphological, physico-chemical, and thermal properties of cellulose nanowhiskers from roselle fibers. *Cellulose, 26*(11), 6599–6613, Springer Nature.)

shown by SEM analysis, hence contributing to the comparable mass transfer activity. As summarized in Table 18.9, the thermal resistant ability of CNWs was improved and higher than that of the NCC samples in the previous study.

4.4 Fourier Transform Infrared Spectroscopy

Fig. 18.13 displays the Fourier transform infrared (FTIR) spectra of R-pulp, R-MCC, and C-MCC. All samples had similar spectrum, which showed that no change in the chemical composition of the samples was achieved by the process of extraction. The spectrum of all samples contained two prominent absorbance regions within the wavelength range of 500–1700 cm^{-1} and between 2800 and 3500 cm^{-1}. Table 18.10 summarized the FTIR bands observed for the R-pulp, R-MCC, and C-MCC samples, and these results were supported by works done by Ilyas et al. (2017), Ilyas et al. (2018), Ilyas, Sapuan, Atikah, Asyraf, et al. (2020), Ilyas, Sapuan, Atiqah, Ibrahim, et al. (2020), Ilyas, Sapuan, et al. (2020). From the result obtained by Kian et al. (2017), the reduction in the absorbance region broadening ranging from 3300 to 3500 cm^{-1} (-OH groups) of R-pulp to that of R-MCC might be due to the cellulose chain separation during acid hydrolysis treatment. Besides that, the increase in the intensity of absorbance bands located at 1739 and 1740 cm^{-1} (uronic ester or acetyl groups in hemicellulose) from the spectrum of R-pulp to that of R-MCC might be attributed to the hemicellulose remaining after the process of hydrolysis using hydrochloric acid. This phenomenon also occurred at the region from 1509 to 1609 cm^{-1} (C=C stretching of aromatic skeletal vibrations), where this

TABLE 18.9
Summary of TGA Performances of Different Nano-Roselle Compositions.

Samples	TGA			
	T_{onset} (°C)	T_{peak} (°C)	W_{loss} (%)	$W_{residue}$ (%)
R-MCC	315.43	340.12	81.23	8.48
NCC-III	257.77	277.48	92.23	1.13
CNW-II	318.33	360.26	84.75	15.14

CNW, cellulose nanowhisker; *NCC*, nanocrystalline cellulose; *R-MCC*, roselle microcrystalline cellulose; *TGA*, thermogravimetric analysis.

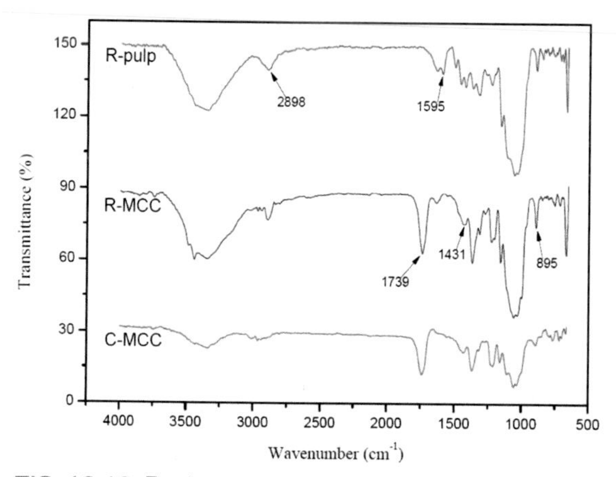

FIG. 18.13 Fourier transform infrared spectra of alkali-treated pulp (R-pulp), roselle microcrystalline cellulose (R-MCC), and commercial microcrystalline cellulose (C-MCC) samples. (Reproduced with copyright permission from Kian, L. K., Jawaid, M., Ariffin, H., & Alothman, O. Y. (2017). Isolation and characterization of microcrystalline cellulose from roselle fibers. *International Journal of Biological Macromolecules, 103*, 931–940. https://doi.org/10.1016/j.ijbiomac.2017.05.135, Elsevier).

TABLE 18.10
Summary of the FTIR Bands Observed for the R-Pulp, R-MCC, C-MCC, NCC-I, NCC-II, and NCC-III Samples.

Peak Assigned	Wavelength (cm^{-1})	Structural Polymer
–	3300–3500	-OH groups
–	2800–2900	C-H stretching
Cellulose	2903	(C\\H symmetric stretching
–	1632–1640	-OH groups
Hemicellulose	1739–1740	Acetyl or uronic ester groups
–	1430–1431	CH_2 bending vibration
Lignin	1509–1609	C=C stretching of aromatic skeletal vibrations
Cellulose	1368–1369	C-H asymmetric deformations
Cellulose	1058, 1157, 1163, and 1159	C-O-C pyranose ring skeletal vibration
Cellulose	895	C–H rocking vibrations

region represented lignin properties. The decrease in the intensity of the band from R-pulp to that of R-MCC was due to the elimination of lignin by acid hydrolysis treatment (Kian et al., 2017).

For the FTIR of roselle NCC in Fig. 18.14, the intensity of the broad band at wavelength 3347 cm^{-1}, which was assigned for O-H bond stretching, was observed to be more intense with the increase in hydrolysis time from 30 min (NCC-I) to 60 min (NCC-III). Besides, the FTIR band located at 1058 cm^{-1} (C-O-C pyranose ring vibration) became narrower and sharper because of the high cellulose content. The properties of cellulose at the wavelengths of 2903, 1163, 1110, and 898 cm^{-1} were more visible after the hydrolysis reaction time was increased. Besides, the FTIR band associated with water absorption at the wavelength of 1637 cm^{-1} (H-O-H stretching vibration) was observed to rise from NCC-I to NCC-III.

4.5 X-ray Powder Diffraction

The nanocellulose crystallinity is linked with its structural rigidity that not only contributes to the structural integrity of nanocellulose but also plays an essential role for the final composite products in terms of mechanical properties. The higher the structural rigidity of nanocellulose, the higher the mechanical properties of the composite. The X-ray diffraction (XRD) patterns of R-pulp, R-MCC, and C-MCC samples are shown in Fig. 18.15, and the XRD patterns of NCC-I, NCC-II, and NCC-III samples are shown in Fig. 18.16. Their crystallinity degrees are tabulated in Table 18.11. All XRD patterns display three main reflections at

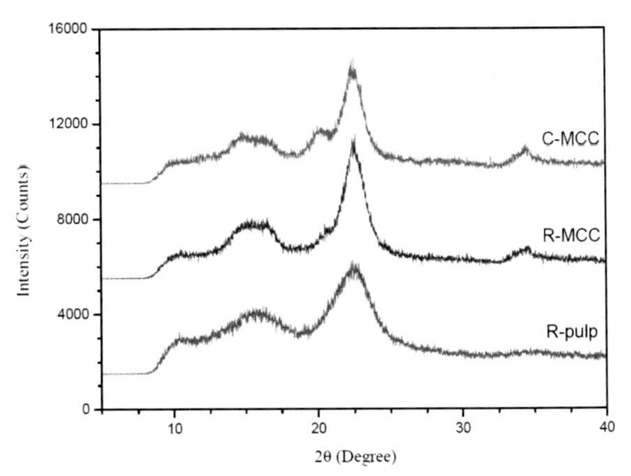

FIG. 18.15 X-ray diffractograms of alkali-treated pulp (R-pulp), roselle microcrystalline cellulose (R-MCC), and commercial microcrystalline cellulose (C-MCC) samples. (Reproduced with copyright permission from Kian, L. K., Jawaid, M., Ariffin, H., & Alothman, O. Y. (2017). Isolation and characterization of microcrystalline cellulose from roselle fibers. *International Journal of Biological Macromolecules, 103*, 931–940. https://doi.org/10.1016/j.ijbiomac.2017.05.135, Elsevier.)

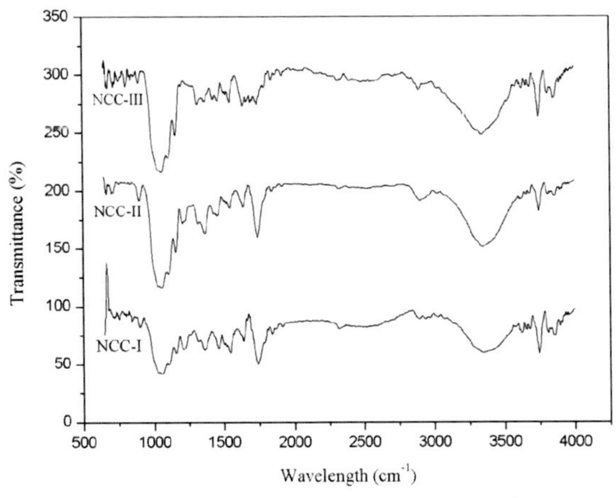

FIG. 18.14 Fourier transform infrared spectra of nanocrystalline cellulose (NCC)-I, NCC-II, and NCC-III samples. (Reproduced with copyright permission from Kian, L. K., Jawaid, M., Ariffin, H., & Karim, Z. (2018). Isolation and characterization of nanocrystalline cellulose from roselle-derived microcrystalline cellulose. *International Journal of Biological Macromolecules, 114*, 54–63. https://doi.org/10.1016/j.ijbiomac.2018.03.065, Elsevier.)

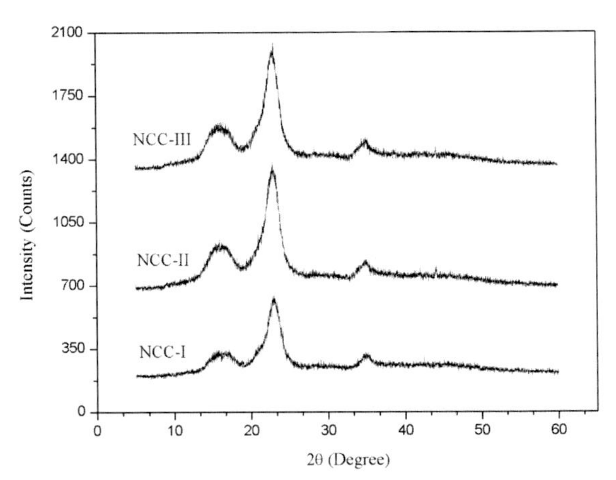

FIG. 18.16 X-ray diffractograms of nanocrystalline cellulose (NCC)-I, NCC-II, and NCC-III. (Reproduced with copy right permission from Kian, L. K., Jawaid, M., Ariffin, H., & Karim, Z. (2018). Isolation and characterization of nanocrystalline cellulose from roselle-derived microcrystalline cellulose. *International Journal of Biological Macromolecules, 114*, 54–63. https://doi.org/10.1016/j.ijbiomac.2018.03.065, Elsevier.)

TABLE 18.11
XRD Analysis Data of R-Pulp, R-MCC, and C-MCC.

Samples	Crystallinity Index (%)	References
R-pulp	63	Kian et al. (2017)
R-MCC	78	Kian et al. (2017)
C-MCC	74	Kian et al. (2017)
NCC-I	76.2	Kian et al. (2018)
NCC-II	77.5	Kian et al. (2018)
NCC-III	79.5	Kian et al. (2018)
CNW-I	78.2	Kian et al. (2019)
CNW-II	79.9	Kian et al. (2019)
CNW-III	79.5	Kian et al. (2019)

C-MCC, commercial microcrystalline cellulose; *CNW*, cellulose nanowhisker; *NCC*, nanocrystalline cellulose; *R-MCC*, roselle microcrystalline cellulose; *R-pulp*, alkali-treated pulp; *XRD*, X-ray diffraction.

$2\theta = 15.8-16.1$ degrees, $22.6-22.8$ degrees, and $34.5-34.9$ degrees, which are assigned to the crystallographic planes (110), (200), and (004), respectively. According to Kian et al. (2017), these type of crystallographic patterns represent the cellulose Iβ type. The crystallinity index value of R-MCC was observed higher with a value of 78% compared to R-pulp, which was 63%. According to Kian et al. (2017), this phenomenon might be due to the hydronium ion penetration from HCl acid hydrolysis treatment into the amorphous cellulose region, followed by the hydrolytic fragmentation of glycosyl units that produced the highly ordered crystallites.

As reported by Kian et al. (2017), Fig. 18.16 and Table 18.11 show that NCC-III displayed the highest crystallinity degree of 79.5%, compared with NCC-I and NCC-II, with a crystallinity index of 76.2% and 77.7%, respectively. This phenomenon might be due to severe hydrolysis conditions, which occasioned the crystalline order rearrangement in (200) plane. Thus it can be summarized that the longer the hydrolysis time, the higher the crystallinity of the nanocellulose. This statement was supported by the work done by Ilyas et al. (2018, 2018a, 2018b), Ilyas, Sapuan, Atikah, Asyraf, et al. (2020), Ilyas, Sapuan, Atiqah, Ibrahim, et al. (2020), Ilyas, Sapuan, et al. (2020) on the isolation and characterization of NCC from sugar palm fibers (*Arenga pinnata*).

5 POTENTIAL APPLICATIONS OF ROSELLE NANOCELLULOSE

The unique macro-physical and outstanding thermal and mechanical properties of NCC make it an ideal material to be used in various fields of advanced applications (Fig. 18.17). Nanocellulose possesses good thermal stability and low or no chronic inflammatory response, which attracted massive interest in the use of nanocellulose as a novel functional material in applications such as nonwoven fabriclike products and paper (Basta & El-Saied, 2009), including flexible freestanding nanocellulose paper-based Si anodes for lithium-ion batteries (Zhang et al., 2015), bacterial cellulose paper (Buruaga-Ramiro et al., 2020), and ginger nanocellulose paper (Abral, Ariksa, et al., 2020). Nanocellulose is also used as a binder in advanced paper technology owing to its nanosized structure, a property that significantly improves the durability and strength of pulp when reinforced into paper (El-Saied et al., 2004). Besides, one of the main reasons it is being used in biomedicine is its excellent biocompatibility (Picheth et al., 2017). Nanocellulose possesses nanofibrillar and ultrafine structured material with an ideal combination of properties, such as high flexibility and tensile strength (Young modulus of 114 GPa) (Grande et al., 2009), as well as high crystallinity (84%–89%) (Gorgieva & Trček, 2019). In terms of biomedical applications, nanocellulose has been used as a never-dried microbial cellulose membrane because it has outstanding conformability to various body contours, maintains a moist environment, and significantly reduces pain (Czaja et al., 2007); as microbial cellulose dressing to be applied on a wounded hand (Czaja et al., 2006); as a nanocellulose mask (Czaja et al., 2007); and in 3D bioprinting of human chondrocytes with reinforcement of nanocellulose-alginate bioink (Markstedt et al., 2015).

Nanocellulose had also been utilized in the electronic applications, as shown in Fig. 18.17E, in the luminescence of an organic light-emitting diode deposited onto a flexible, low-CTE (coefficient of thermal expansion), and optically transparent cellulose nanocomposite (Okahisa et al., 2009).

Coatings are used extensively in pipelines, machines, vehicles, construction, packaging, furniture, and other military fields such as tanks, ships, artillery, and weaponry as the surface layer of items. Functional coatings may be applied to alter or modify the surface properties of the substrate, such as wettability, adhesion, wear

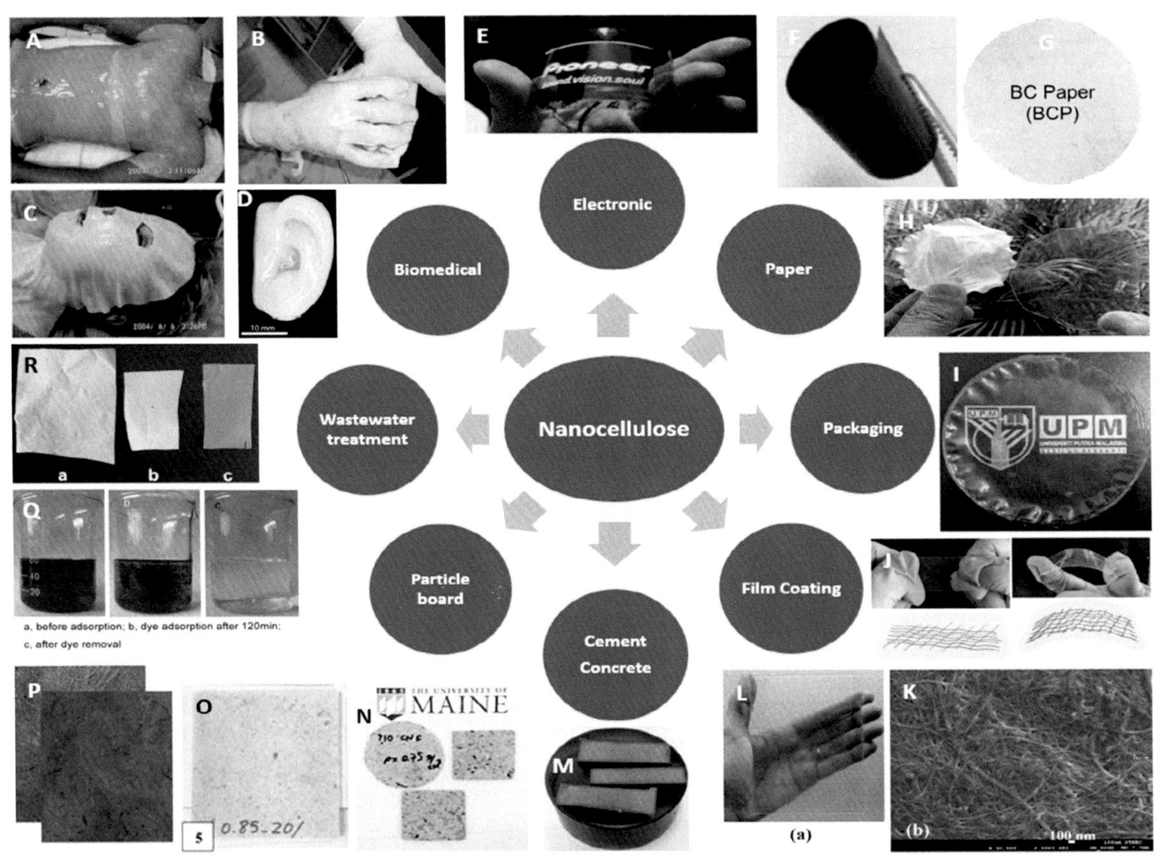

FIG. 18.17 Advanced applications of bacterial cellulose. **(A)** A never-dried microbial cellulose membrane showed remarkable conformability to the various body contours, maintained a moist environment, and significantly reduced pain (Czaja et al., 2007). **(B)** Microbial cellulose dressing applied on a wounded hand (Czaja et al., 2006). **(C)** Nanocellulose mask (Czaja et al., 2007). **(D)** 3D bioprinting of human chondrocytes with nanocellulose-alginate bioink (Markstedt et al., 2015). **(E)** Luminescence of an organic light-emitting diode deposited onto a flexible, low-CTE (coefficient of thermal expansion), and optically transparent cellulose nanocomposite (Okahisa et al., 2009). **(F)** Flexible freestanding nanocellulose paper-based Si anodes for lithium-ion batteries (Zhang et al., 2015). **(G)** Bacterial cellulose (BC) paper (Buruaga-Ramiro et al., 2020). **(H)** Ginger nanocellulose paper (Abral, Ariksa, et al., 2020). **(I)** Sugar palm nanocellulose for plastic packaging applications (Ilyas et al., 2018c). **(J)** Nanocellulose-reinforced polyurethane for waterborne wood coating (Pan et al., 2019). **(K)** Surface of nanocellulose film at 30,000× magnification by field emission scanning electron microscopy (Pacaphol & Aht-ong, 2017). **(L)** Nanocellulose film coated on glass (Pacaphol & Aht-ong, 2017). **(M)** Cellulose nanocrystal-reinforced cement (Purdue University, 2018). **(N)** Cellulose nanofiber-bound particleboard (Tayeb et al., 2018). **(O)** Particleboard from nanocellulose (Ezatollah Amini et al., 2017). **(P)** Adhesive application on particleboard from nanocellulose (Adedoja et al., 2020). **(Q)** Adsorptive removal of anionic dyes from aqueous solutions using microgel based on nanocellulose and polyvinylamine (Jin et al., 2015). **(R)** Electrospun biocomposite of nanocellulose and chitosan entrapped within a poly(hydroxyalkanoate) matrix for Congo red removal (Yong et al., 2019).

resistance, or corrosion resistance. They are able to protect materials and to withstand external factors such as climate, microorganisms, and harm to people; in addition, they create a decorative effect on materials, by giving them glossiness and beautiful color. Nanocellulose had been used as reinforcement in polyurethane for waterborne wood coating (Pan et al., 2019), and as a coating material for glass (Pacaphol & Aht-ong, 2017).

Another application of nanocellulose is nanobinder for cement concrete (Purdue University, 2018). The

Purdue University researchers studied whether the strength of the concrete could be improved by infusing it with microscopic nanocrystals from wood. The researchers have been working with cellulose nanocrystals, byproducts generated from paper, bioenergy, agriculture industry, and pulp industry to find the best mixture to strengthen concrete, the most common human-made material in the world. They found that the incorporation of nanocellulose has made the concrete stronger through a chemical reaction that increased the hydration of the cement particles.

The use of natural adhesives such as nanocellulose, starch, tannins, lignin, and soy protein to reduce the health threat related to the use of formaldehyde-based particleboard has contributed toward the reduction in the impending health challenges and the cost of building construction. Some of the examples of nanocellulose used with particleboard include the particleboard panels fabricated using cellulose nanofiber (CNF) that was produced through a refining process as a sole binder by Amini et al. (2017), preliminary samples made by mixing/pressing southern pine wood particles with a 3 wt% CNF suspension by Tayeb et al. (2018), and adhesive application on particleboard from nanocellulose by Adedoja et al. (2020).

Nanocellulose can be modified to enhance its chemical and physical properties. Nanocellulose is uniquely distinguished by its high crystallinity; excellent mechanical properties, including increased strength and stiffness; nanosized fiber; abundant hydroxyl group; high aspect ratio of surface-to-volume ratio; high surface area; large specific surface area; lightweight; biodegradability; and good chemical and thermal resistance. These outstanding properties make nanocellulose the materials with enormous potential to be used for wastewater treatment.

Some of the examples of nanocellulose application in wastewater treatment include adsorptive removal of anionic dyes from aqueous solutions using microgel based on nanocellulose and polyvinylamine by Jin et al. (2015). Another study conducted by Tong et al. (2019) showed that the electrospun biocomposite of nanocellulose and chitosan entrapped within a poly(-hydroxyalkanoate) matrix could be used for Congo red removal. Several studies had indicated the possible usage of nanocellulose as an agent in wastewater treatment (Kardam & Rohit, 2014; Karim et al., 2016; Korhonen et al., 2011; Paltakari et al., 2011; Suopajärvi et al., 2015; Wang et al., 2013).

Moreover, owing to its outstanding mechanical properties, nanocellulose has been utilized by reinforcing it with other polymers in various applications, including paper (Basta & El-Saied, 2009), treating tympanic membrane perforation (Biskin et al., 2016; Lou, 2016), the shielding film (Lang et al., 2014), food packaging films (Azeredo et al., 2017; Ilyas et al., 2018c), audio speaker diaphragm (Gellner et al., 1993), and so on. Nanocellulose has gained traction in many fields, thereby opening the novel horizon for scientific research, as well as for scientists and engineers dealing with nanotechnology and nanoscience.

6 CONCLUSIONS

This chapter focuses on the top-down techniques (i.e., acid hydrolysis, ultrasonification, HPH, supermasscolloider, steam explosion, TEMPO oxidation, microfluidizer, and a combination of chemomechanical processes) used to prepare nanocellulose from various sources of natural fibers including crops, agricultural residues, forest woods, and forest by-products. This chapter also aims to provide knowledge and data that were collected from the experiments conducted by several researchers in the nanocellulose extraction field. In 20 years, an increasing number of scientists in nanocellulose area were reported. Besides, this chapter summarized some selected works done in the field of nanocellulose extracted from different sources using different methods. It can be summarized that NCC and NFC can be extracted from various sources of (1) **hardwood and softwood** and (2) **plant fiber**. Among all sources, agro-industrial wastes are seen as high-potential materials and can be extensively utilized for producing NCC and NFC. This might be due to the higher NCC and NFC productivities and large-scale accessibility, as well as the abundance of resources, from agro-industrial wastes.

This chapter also presents the current status of roselle nanocellulose research for various applications. Roselle is a suitable fiber that can be utilized for the development of nanocellulose. To overcome the inherent defaults in the properties of raw roselle fibers due to their incompatibility in polymer composites, numerous modifications and extraction methods were employed, including acid hydrolysis treatment by using sulfuric acid under vigorous stirring. The new trend in improving the functional properties of roselle fiber is the use of chemical and mechanical treatments, which are a promising approach for the isolation of roselle fiber. Nanocellulose from roselle fiber are prospective nanoreinforcements or nanofillers to enhance the properties of polymer composites as a high-performance material.

ACKNOWLEDGMENTS

The authors would like to thank the Universiti Putra Malaysia for financial support through Geran Putra Berimpak (GBP), UPM/800-3/3/1/GBP/2019/9679800.

REFERENCES

Abitbol, T., Rivkin, A., Cao, Y., Nevo, Y., Abraham, E., Ben-Shalom, T., Lapidot, S., & Shoseyov, O. (2016). Nanocellulose, a tiny fiber with huge applications. *Current Opinion in Biotechnology, 39*(1), 76–88. https://doi.org/10.1016/j.copbio.2016.01.002

Abral, H., Ariksa, J., Mahardika, M., Handayani, D., Aminah, I., Sandrawati, N., Pratama, A. B., Fajri, N., Sapuan, S. M., & Ilyas, R. A. (2020). Transparent and antimicrobial cellulose film from ginger nanofiber. *Food Hydrocolloids, 98*(August 2019), 105266. https://doi.org/10.1016/j.foodhyd.2019.105266

Abral, H., Ariksa, J., Mahardika, M., Handayani, D., Aminah, I., Sandrawati, N., Sapuan, S. M., & Ilyas, R. A. (2019a). Highly transparent and antimicrobial PVA based bionanocomposites reinforced by ginger nanofiber. *Polymer Testing,* 106186. https://doi.org/10.1016/j.polymertesting.2019.106186

Abral, H., Atmajaya, A., Mahardika, M., Hafizulhaq, F., Kadriadi, Handayani, D., Sapuan, S. M., & Ilyas, R. A. (2020). Effect of ultrasonication duration of polyvinyl alcohol (PVA) gel on characterizations of PVA film. *Journal of Materials Research and Technology, 9*(2), 2477–2486. https://doi.org/10.1016/j.jmrt.2019.12.078

Abral, H., Basri, A., Muhammad, F., Fernando, Y., Hafizulhaq, F., Mahardika, M., Sugiarti, E., Sapuan, S. M., Ilyas, R. A., & Stephane, I. (2019). A simple method for improving the properties of the sago starch films prepared by using ultrasonication treatment. *Food Hydrocolloids, 93,* 276–283. https://doi.org/10.1016/j.foodhyd.2019.02.012

Abral, H., Dalimunthe, M. H., Hartono, J., Efendi, R. P., Asrofi, M., Sugiarti, E., Sapuan, S. M., Park, J. W., & Kim, H. J. (2018). Characterization of tapioca starch biopolymer composites reinforced with micro scale water hyacinth fibers. *Starch Staerke, 70*(7–8), 1–8. https://doi.org/10.1002/star.201700287

Adedoja, A., Junidah, O., & Rokiah, L. (2020). Adhesive application on particleboard from natural fibers: A review. *Polymer Composites,* 1–13. https://doi.org/10.1002/pc.25749

Adel, A. M., Abd El–Wahab, Z. H., Ibrahim, A. A., & Al–Shemy, M. T. (2010). Characterization of microcrystalline cellulose prepared from lignocellulosic materials. Part I. Acid catalyzed hydrolysis. *Bioresource Technology, 101*(12), 4446–4455.

Aisyah, H. A., Paridah, M. T., Sapuan, S. M., Khalina, A., Berkalp, O. B., Lee, S. H., Lee, C. H., Nurazzi, N. M., Ramli, N., Wahab, M. S., & Ilyas, R. A. (2019). Thermal properties of woven kenaf/carbon fibre-reinforced epoxy hybrid composite panels. *International Journal of Polymer Science, 2019,* 1–8. https://doi.org/10.1155/2019/5258621

Akil, H. M., Omar, M. F., Mazuki, A. A. M., Safiee, S., Ishak, Z. A. M., & Abu Bakar, A. (2011). Kenaf fiber reinforced composites: A review. *Materials and Design, 32*(8–9), 4107–4121. https://doi.org/10.1016/j.matdes.2011.04.008

Akonda, M. H., Shah, D. U., & Gong, R. H. (2020). Natural fibre thermoplastic tapes to enhance reinforcing effects in composite structures. *Composites Part A: Applied Science and Manufacturing, 131*(November 2019), 105822. https://doi.org/10.1016/j.compositesa.2020.105822

Alemdar, A., & Sain, M. (2008a). Biocomposites from wheat straw nanofibers: Morphology, thermal and mechanical properties. *Composites Science and Technology, 68*(2), 557–565. https://doi.org/10.1016/j.compscitech.2007.05.044

Alemdar, A., & Sain, M. (2008b). Isolation and characterization of nanofibers from agricultural residues—wheat straw and soy hulls. *Bioresource Technology, 99*(6), 1664–1671.

Alemdar, A., & Sain, M. (2008c). Isolation and characterization of nanofibers from agricultural residues - wheat straw and soy hulls. *Bioresource Technology, 99*(6), 1664–1671. https://doi.org/10.1016/j.biortech.2007.04.029

Amini, E., Tajvidi, M., Gardner, D. J., & Bousfield, D. W. (2017). Utilization of cellulose nanofibrils as a binder for particleboard manufacture Ezatollah. *BioResources, 12*(Elbert 1995), 4093–4110. https://doi.org/10.15376/biores.12.2.4093-4110

Arulmurugan, M., Prabu, K., Rajamurugan, G., & Selvakumar, A. S. (2019). Impact of $BaSO_4$ filler on woven Aloevera/Hemp hybrid composite: Dynamic mechanical analysis. *Materials Research Express, 6*(4), 045309. https://doi.org/10.1088/2053-1591/aafb88

Asrofi, M., Abral, H., Kasim, A., Pratoto, A., Mahardika, M., & Hafizulhaq, F. (2018). Characterization of the sonicated yam bean starch bionanocomposites reinforced by nanocellulose water hyacinth fiber (Whf): The effect of various fiber loading. *Journal of Engineering Science and Technology, 13*(9), 2700–2715.

Asrofi, M., Sapuan, S. M., Ilyas, R. A., & Ramesh, M. (2020). Characteristic of composite bioplastics from tapioca starch and sugarcane bagasse fiber: Effect of time duration of ultrasonication (Bath-Type). *Materials Today: Proceedings.* https://doi.org/10.1016/j.matpr.2020.07.254

Asrofi, M., Sujito, Syafri, E., Sapuan, S. M., & Ilyas, R. A. (2020). Improvement of biocomposite properties based tapioca starch and sugarcane bagasse cellulose nanofibers. *Key Engineering Materials, 849,* 96–101. https://doi.org/10.4028/www.scientific.net/KEM.849.96

Asyraf, M. R. M., Ishak, M. R., Sapuan, S. M., Yidris, N., & Ilyas, R. A. (2020). Woods and composites cantilever beam: A comprehensive review of experimental and numerical creep methodologies. *Journal of Materials Research and Technology, 9*(3), 6759–6776. https://doi.org/10.1016/j.jmrt.2020.01.013

Athijayamani, A., Thiruchitrambalam, M., Natarajan, U., & Pazhanivel, B. (2009). Effect of moisture absorption on the mechanical properties of randomly oriented natural fibers/polyester hybrid composite. *Materials Science and Engineering: A, 517,* 344–353. https://doi.org/10.1016/j.msea.2009.04.027

Athinarayanan, J., Alshatwi, A. A., & Subbarayan Periasamy, V. (2020). Biocompatibility analysis of Borassus flabellifer biomass-derived nanofibrillated cellulose. *Carbohydrate Polymers, 235*, 115961. https://doi.org/10.1016/j.carbpol.2020.115961

Atikah, M. S. N., Ilyas, R. A., Sapuan, S. M., Ishak, M. R., Zainudin, E. S., Ibrahim, R., Atiqah, A., Ansari, M. N. M., & Jumaidin, R. (2019). Degradation and physical properties of sugar palm starch/sugar palm nanofibrillated cellulose bionanocomposite. *Polimery, 64*(10), 27–36. https://doi.org/10.14314/polimery.2019.10.5

Atiqah, A., Jawaid, M., Sapuan, S. M., Ishak, M. R., Ansari, M. N. M., & Ilyas, R. A. (2019). Physical and thermal properties of treated sugar palm/glass fibre reinforced thermoplastic polyurethane hybrid composites. *Journal of Materials Research and Technology, 8*(5), 3726–3732. https://doi.org/10.1016/j.jmrt.2019.06.032

Ayu, R. S., Khalina, A., Harmaen, A. S., Zaman, K., Isma, T., Liu, Q., Ilyas, R. A., & Lee, C. H. (2020). Characterization study of empty fruit bunch (EFB) fibers reinforcement in poly(butylene) succinate (PBS)/Starch/Glycerol composite sheet. *Polymers, 12*(7), 1571. https://doi.org/10.3390/polym12071571

Azammi, A. M. N., Ilyas, R. A., Sapuan, S. M., Ibrahim, R., Atikah, M. S. N., Asrofi, M., & Atiqah, A. (2020). Characterization studies of biopolymeric matrix and cellulose fibres based composites related to functionalized fibre-matrix interface. In *Interfaces in particle and fibre reinforced composites* (1st ed., pp. 29–93). Elsevier. https://doi.org/10.1016/B978-0-08-102665-6.00003-0.

Azeredo, H. M. C., Mattoso, L. H. C., Wood, D., Williams, T. G., Avena-Bustillos, R. J., & McHugh, T. H. (2009). Nanocomposite edible films from mango puree reinforced with cellulose nanofibers. *Journal of Food Science, 74*(5). https://doi.org/10.1111/j.1750-3841.2009.01186.x

Azeredo, H. M. C., Rosa, M. F., & Mattoso, L. H. C. (2017). Nanocellulose in bio-based food packaging applications. *Industrial Crops and Products, 97*, 664–671. https://doi.org/10.1016/j.indcrop.2016.03.013

Azrina, Z. A. Z., Beg, M. D. H., Rosli, M. Y., Ramli, R., Junadi, N., & Alam, A. K. M. M. (2017). Spherical nanocrystalline cellulose (NCC) from oil palm empty fruit bunch pulp via ultrasound assisted hydrolysis. *Carbohydrate Polymers, 162*, 115–120.

Babaee, M., Jonoobi, M., Hamzeh, Y., & Ashori, A. (2015). Biodegradability and mechanical properties of reinforced starch nanocomposites using cellulose nanofibers. *Carbohydrate Polymers, 132*(3), 1–8. https://doi.org/10.1016/j.carbpol.2015.06.043

Bandera, D., Sapkota, J., Josset, S., Weder, C., Tingaut, P., Gao, X., Foster, E. J., & Zimmermann, T. (2014). Influence of mechanical treatments on the properties of cellulose nanofibers isolated from microcrystalline cellulose. *Reactive and Functional Polymers, 85*, 134–141.

Bano, S., & Negi, Y. S. (2017). Studies on cellulose nanocrystals isolated from groundnut shells. *Carbohydrate Polymers, 157*, 1041–1049. https://doi.org/10.1016/j.carbpol.2016.10.069

Barbash, V. A., Yaschenko, O. V., & Shniruk, O. M. (2017). Preparation and properties of nanocellulose from organosolv straw pulp. *Nanoscale Research Letters, 12*(1), 241. https://doi.org/10.1186/s11671-017-2001-4

Barbash, V. A., & Yashchenko, O. V. (2020). Preparation and application of nanocellulose from non-wood plants to improve the quality of paper and cardboard. *Applied Nanoscience, 10*(8), 2705–2716. https://doi.org/10.1007/s13204-019-01242-8

Basta, A. H., & El-Saied, H. (2009). Performance of improved bacterial cellulose application in the production of functional paper. *Journal of Applied Microbiology, 107*(6), 2098–2107. https://doi.org/10.1111/j.1365-2672.2009.04467.x

Bendahou, A., Habibi, Y., Kaddami, H., & Dufresne, A. (2009). Physico-chemical characterization of palm from phoenix dactylifera-L, preparation of cellulose whiskers and natural rubber-based nanocomposites. *Journal of Biobased Materials and Bioenergy, 3*(1), 81–90. https://doi.org/10.1166/jbmb.2009.1011

Benini, K. C. C. de C., Voorwald, H. J. C., Cioffi, M. O. H., Rezende, M. C., & Arantes, V. (2018). Preparation of nanocellulose from Imperata brasiliensis grass using Taguchi method. *Carbohydrate Polymers, 192*, 337–346. https://doi.org/10.1016/j.carbpol.2018.03.055

Biskin, S., Damar, M., Oktem, S. N., Sakalli, E., Erdem, D., & Pakir, O. (2016). A new graft material for myringoplasty: Bacterial cellulose. *European Archives of Oto-Rhino-Laryngology, 273*(11), 3561–3565. https://doi.org/10.1007/s00405-016-3959-8

Bondeson, D., Mathew, A., & Oksman, K. (2006). Optimization of the isolation of nanocrystals from microcrystalline cellulose by acid hydrolysis. *Cellulose, 13*(2), 171–180. https://doi.org/10.1007/s10570-006-9061-4

Braun, B., Dorgan, J. R., & Chandler, J. P. (2008). Cellulosic nanowhiskers. Theory and application of light scattering from polydisperse spheroids in the Rayleigh–Gans–Debye regime. *Biomacromolecules, 9*(4), 1255–1263. https://doi.org/10.1021/bm7013137

Brito, B. S. L., Pereira, F. V., Putaux, J.-L., & Jean, B. (2012). Preparation, morphology and structure of cellulose nanocrystals from bamboo fibers. *Cellulose, 19*(5), 1527–1536. https://doi.org/10.1007/s10570-012-9738-9

Buruaga-Ramiro, C., Valenzuela, S. V., Valls, C., Roncero, M. B., Pastor, F. I. J., Díaz, P., & Martínez, J. (2020). Bacterial cellulose matrices to develop enzymatically active paper. *Cellulose, 27*(6), 3413–3426. https://doi.org/10.1007/s10570-020-03025-9

Camarero Espinosa, S., Kuhnt, T., Foster, E. J., & Weder, C. (2013). Isolation of thermally stable cellulose nanocrystals by phosphoric acid hydrolysis. *Biomacromolecules, 14*(4), 1223–1230. https://doi.org/10.1021/bm400219u

Capron, I., Rojas, O. J., & Bordes, R. (2017). Behavior of nanocelluloses at interfaces. *Current Opinion in Colloid and Interface Science, 29*, 83–95. https://doi.org/10.1016/j.cocis.2017.04.001

Chan, C. H., Chia, C. H., Zakaria, S., Ahmad, I., & Dufresne, A. (2013). Production and characterisation of cellulose and nano- crystalline cellulose from kenaf core wood. *Bio-Resources, 8*(1), 785–794. https://doi.org/10.15376/biores.8.1.785-794

Chauhan, A., & Kaith, B. (2012). Versatile roselle graft-copolymers: XRD studies and their mechanical evaluation after use as reinforcement in composites. *Journal of the Chilean Chemical Society, 57*(3), 1262–1266. https://doi.org/10.4067/s0717-97072012000300014

Chen, D., Lawton, D., Thompson, M. R., & Liu, Q. (2012). Biocomposites reinforced with cellulose nanocrystals derived from potato peel waste. *Carbohydrate Polymers, 90*(1), 709–716. https://doi.org/10.1016/j.carbpol.2012.06.002

Cherian, B. M., Leão, A. L., de Souza, S. F., Thomas, S., Pothan, L. A., & Kottaisamy, M. (2010). Isolation of nanocellulose from pineapple leaf fibres by steam explosion. *Carbohydrate Polymers, 81*(3), 720–725. https://doi.org/10.1016/j.carbpol.2010.03.046

Chirayil, C. J., Joy, J., Mathew, L., Mozetic, M., Koetz, J., & Thomas, S. (2014). Isolation and characterization of cellulose nanofibrils from *Helicteres isora* plant. *Industrial Crops and Products, 59*, 27–34. https://doi.org/10.1016/j.indcrop.2014.04.020

Clemons, C. M. (2010). Natural fibers. In M. Xanthos (Ed.), *Functional Fillers for plastics* (pp. 211–223). Wiley-VCH Verlag GmbH & Co. KGaA. https://doi.org/10.1002/9783527629848.ch11.

Corrêa, A. C., de Morais Teixeira, E., Pessan, L. A., & Mattoso, L. H. C. (2010). Cellulose nanofibers from curaua fibers. *Cellulose, 17*(6), 1183–1192. https://doi.org/10.1007/s10570-010-9453-3

Criado, P., Hossain, F. M. J., Salmieri, S., & Lacroix, M. (2018). Nanocellulose in food packaging. *Composites Materials for Food Packaging*, 297–329. https://doi.org/10.1002/9781119160243.ch10

Cui, S., Zhang, S., Ge, S., Xiong, L., & Sun, Q. (2016). Green preparation and characterization of size-controlled nanocrystalline cellulose via ultrasonic-assisted enzymatic hydrolysis. *Industrial Crops and Products, 83*, 346–352.

Czaja, W., Krystynowicz, A., Bielecki, S., & Brown, J. R. M. (2006). Microbial cellulose—the natural power to heal wounds. *Biomaterials, 27*(2), 145–151. https://doi.org/10.1016/j.biomaterials.2005.07.035

Czaja, W. K., Young, D. J., Kawecki, M., & Brown, R. M. (2007). The future prospects of microbial cellulose in biomedical applications. *Biomacromolecules, 8*(1), 1–12. https://doi.org/10.1021/bm060620d

De Aguiar, J., Bondancia, T. J., Claro, P. I. C., Mattoso, L. H. C., Farinas, C. S., & Marconcini, J. M. (2020). Enzymatic deconstruction of sugarcane bagasse and straw to obtain cellulose nanomaterials. *ACS Sustainable Chemistry and Engineering*. https://doi.org/10.1021/acssuschemeng.9b06806

Debiagi, F., Faria-Tischer, P. C. S., & Mali, S. (2020). A green approach based on reactive extrusion to produce nanofibrillated cellulose from oat hull. *Waste and Biomass Valorization*. https://doi.org/10.1007/s12649-020-01025-1

El Achaby, M., El Miri, N., Aboulkas, A., Zahouily, M., Bilal, E., Barakat, A., & Solhy, A. (2017). Processing and properties of eco-friendly bio-nanocomposite films filled with cellulose nanocrystals from sugarcane bagasse. *International Journal of Biological Macromolecules, 96*, 340–352.

El-Saied, H., Basta, A. H., & Gobran, R. H. (2004). Research progress in friendly environmental technology for the production of cellulose products (bacterial cellulose and its application). *Polymer-Plastics Technology and Engineering, 43*(3), 797–820. https://doi.org/10.1081/PPT-120038065

Elanthikkal, S., Gopalakrishnapanicker, U., Varghese, S., & Guthrie, J. T. (2010). Cellulose microfibres produced from banana plant wastes: Isolation and characterization. *Carbohydrate Polymers, 80*(3), 852–859.

Errokh, A., Magnin, A., Putaux, J.-L., & Boufi, S. (2018). Morphology of the nanocellulose produced by periodate oxidation and reductive treatment of cellulose fibers. *Cellulose, 25*(7), 3899–3911. https://doi.org/10.1007/s10570-018-1871-7

Fahma, F., Lisdayana, N., Abidin, Z., Noviana, D., Sari, Y. W., Mukti, R. R., Yunus, M., Kusumaatmaja, A., & Kadja, G. T. M. (2020). Nanocellulose-based fibres derived from palm oil by-products and their in vitro biocompatibility analysis. *Journal of the Textile Institute, 111*(9), 1354–1363. https://doi.org/10.1080/00405000.2019.1694353

Faradilla, R. H. F., Lee, G., Arns, J. Y., Roberts, J., Martens, P., Stenzel, M. H., & Arcot, J. (2017). Characteristics of a free-standing film from banana pseudostem nanocellulose generated from TEMPO-mediated oxidation. *Carbohydrate Polymers, 174*, 1156–1163. https://doi.org/10.1016/j.carbpol.2017.07.025

Favier, V., Chanzy, H., & Cavaille, J. Y. (1995). Polymer nanocomposites reinforced by cellulose whiskers. *Macromolecules, 28*(18), 6365–6367. https://doi.org/10.1021/ma00122a053

Ferreira, F. V., Pinheiro, I. F., de Souza, S. F., Mei, L. H. I., & Lona, L. M. F. (2019). Polymer composites reinforced with natural fibers and nanocellulose in the automotive industry: A short review. *Journal of Composites Science, 3*(2), 51.

Flauzino Neto, W. P., Silvério, H. A., Dantas, N. O., & Pasquini, D. (2013). Extraction and characterization of cellulose nanocrystals from agro-industrial residue – soy hulls. *Industrial Crops and Products, 42*, 480–488. https://doi.org/10.1016/j.indcrop.2012.06.041

Fortunati, E., Puglia, D., Luzi, F., Santulli, C., Kenny, J. M., & Torre, L. (2013). Binary PVA bio-nanocomposites containing cellulose nanocrystals extracted from different natural sources: Part I. *Carbohydrate Polymers, 97*(2), 825–836. https://doi.org/10.1016/j.carbpol.2013.03.075

Gellner, P. E. L., Dang, A. E., & Jay, H. (1993). *United States patent (19)*.

Ghalehno, M. D., & Nazerian, M. (2013). The investigation on chemical and anatomical properties of Roselle (*Hibiscus sabdariffa*) stem. *The International Journal of Agriculture and Crop Sciences, 5*(15), 1622–1625.

Gorgieva, S., & Trček, J. (2019). Bacterial cellulose: Production, modification and perspectives in biomedical applications. *Nanomaterials, 9*(10), 1–20. https://doi.org/10.3390/nano9101352

Grande, C. J., Torres, F. G., Gomez, C. M., Troncoso, O. P., Canet-Ferrer, J., & Mart??nez-Pastor, J. (2009). Development of self-assembled bacterial cellulose-starch nanocomposites. *Materials Science and Engineering: C, 29*(4), 1098–1104. https://doi.org/10.1016/j.msec.2008.09.024

Grunert, M., & Winter, W. T. (2002). Nanocomposites of cellulose acetate butyrate reinforced with cellulose nanocrystals. *Journal of Polymers and the Environment, 10*(April), 27–30.

Haafiz, M. K. M., Hassan, A., Zakaria, Z., & Inuwa, I. M. (2014). Isolation and characterization of cellulose nanowhiskers from oil palm biomass microcrystalline cellulose. *Carbohydrate Polymers, 103*(1), 119–125. https://doi.org/10.1016/j.carbpol.2013.11.055

Habibi, Y., & Vignon, M. R. (2008). Optimization of cellouronic acid synthesis by TEMPO-mediated oxidation of cellulose III from sugar beet pulp. *Cellulose, 15*(1), 177–185. https://doi.org/10.1007/s10570-007-9179-z

Halimatul, M. J., Sapuan, S. M., Jawaid, M., Ishak, M. R., & Ilyas, R. A. (2019a). Effect of sago starch and plasticizer content on the properties of thermoplastic films: Mechanical testing and cyclic soaking-drying. *Polimery, 64*(6), 32–41. https://doi.org/10.14314/polimery.2019.6.5

Halimatul, M. J., Sapuan, S. M., Jawaid, M., Ishak, M. R., & Ilyas, R. A. (2019b). Water absorption and water solubility properties of sago starch biopolymer composite films filled with sugar palm particles. *Polimery, 64*(9), 27–35. https://doi.org/10.14314/polimery.2019.9.4

Hazrol, M. D., Sapuan, S. M., Ilyas, R. A., Othman, M. L., & Sherwani, S. F. K. (2020). Electrical properties of sugar palm nanocrystalline cellulose reinforced sugar palm starch nanocomposites. *Polimery, 65*(05), 363–370. https://doi.org/10.14314/polimery.2020.5.4

He, Y., Boluk, Y., Pan, J., Ahniyaz, A., Deltin, T., & Claesson, P. M. (2019). Corrosion protective properties of cellulose nanocrystals reinforced waterborne acrylate-based composite coating. *Corrosion Science*. https://doi.org/10.1016/j.corsci.2019.04.038

Herrera, M. A., Mathew, A. P., & Oksman, K. (2012). Comparison of cellulose nanowhiskers extracted from industrial bio-residue and commercial microcrystalline cellulose. *Materials Letters, 71*, 28–31. https://doi.org/10.1016/j.matlet.2011.12.011

Hietala, M., Mathew, A. P., & Oksman, K. (2013). Bionanocomposites of thermoplastic starch and cellulose nanofibers manufactured using twin-screw extrusion. *European Polymer Journal, 49*(4), 950–956. https://doi.org/10.1016/j.eurpolymj.2012.10.016

Ibrahim, M. I., Edhirej, A., Sapuan, S. M., Jawaid, M., Ismarrubie, N. Z., & Ilyas, R. A. (2020). Extraction and characterization of Malaysian cassava starch, peel, and bagasse, and selected properties of the composites. In R. Jumaidin, S. M. Sapuan, & H. Ismail (Eds.), *Biofiller-reinforced biodegradable polymer composites* (1st ed., pp. 267–283). CRC Press.

Ibrahim, M. I., Sapuan, S. M., Zainudin, E. S., Zuhri, M. Y., Edhirej, A., & Ilyas, R. A. (2020). Characterization of corn fiber-filled cornstarch biopolymer composites. In R. Jumaidin, S. M. Sapuan, & H. Ismail (Eds.), *Biofiller-reinforced biodegradable polymer composites* (1st ed., pp. [illegible]). CRC Press.

Ilyas, R. A., & Sapuan, S. M. (2020a). The preparation methods and processing of natural fibre bio-polymer composites. *Current Organic Synthesis, 16*(8), 1068–1070. https://doi.org/10.2174/157017941608200120105616

Ilyas, R. A., & Sapuan, S. M. (2020b). Biopolymers and biocomposites: Chemistry and technology. *Current Analytical Chemistry, 16*(5), 500–503. https://doi.org/10.2174/157341101605200603095311

Ilyas, R. A., Sapuan, S. M., Atikah, M. S. N., Asyraf, M. R. M., Rafiqah, S. A., Aisyah, H. A., Nurazzi, N. M., & Norrrahim, M. N. F. (2020). Effect of hydrolysis time on the morphological, physical, chemical, and thermal behavior of sugar palm nanocrystalline cellulose (*Arenga pinnata* (Wurmb.) Merr). *Textile Research Journal*, 1–16. https://doi.org/10.1177/0040517520932393

Ilyas, R. A., Sapuan, S. M., Atiqah, A., Ibrahim, R., Abral, H., Ishak, M. R., Zainudin, E. S., Nurazzi, N. M., Atikah, M. S. N., Ansari, M. N. M., Asyraf, M. R. M., Supian, A. B. M., & Ya, H. (2020). Sugar palm (*Arenga pinnata* [Wurmb.] Merr) starch films containing sugar palm nanofibrillated cellulose as reinforcement: Water barrier properties. *Polymer Composites, 41*(2), 459–467. https://doi.org/10.1002/pc.25379

Ilyas, R. A., Sapuan, S. M., Ibrahim, R., Abral, H., Ishak, M. R., Zainudin, E. S., Atikah, M. S. N., Mohd Nurazzi, N., Atiqah, A., Ansari, M. N. M., Syafri, E., Asrofi, M., Sari, N. H., & Jumaidin, R. (2019). Effect of sugar palm nanofibrillated cellulose concentrations on morphological, mechanical and physical properties of biodegradable films based on agro-waste sugar palm (*Arenga pinnata*(Wurmb.) Merr) starch. *Journal of Materials Research and Technology, 8*(5), 4819–4830. https://doi.org/10.1016/j.jmrt.2019.08.028

Ilyas, R. A., Sapuan, S. M., Ibrahim, R., Abral, H., Ishak, M. R., Zainudin, E. S., Atiqah, A., Atikah, M. S. N., Syafri, E., Asrofi, M., & Jumaidin, R. (2020). Thermal, biodegradability and water barrier properties of bio-nanocomposites based on plasticised sugar palm starch and nanofibrillated celluloses from sugar palm fibres. *Journal of Biobased Materials and Bioenergy, 14*(2), 234–248. https://doi.org/10.1166/jbmb.2020.1951

Ilyas, R. A., Sapuan, S. M., & Ishak, M. R. (2018). Isolation and characterization of nanocrystalline cellulose from sugar palm fibres (*Arenga pinnata*). *Carbohydrate Polymers, 181*, 1038–1051. https://doi.org/10.1016/j.carbpol.2017.11.045

Ilyas, R. A., Sapuan, S. M., Ishak, M. R., & Zainudin, E. S. (2017). Effect of delignification on the physical, thermal, chemical, and structural properties of sugar palm fibre. *BioResources, 12*(4), 8734–8754. https://doi.org/10.15376/biores.12.4.8734-8754

Ilyas, R. A., Sapuan, S. M., Ishak, M. R., & Zainudin, E. S. (2018a). Water transport properties of bio-nanocomposites reinforced by sugar palm (*Arenga pinnata*) nanofibrillated cellulose. *Journal of Advanced Research in Fluid Mechanics and Thermal Sciences Journal, 51*(2), 234–246.

Ilyas, R. A., Sapuan, S. M., Ishak, M. R., & Zainudin, E. S. (2018b). Sugar palm nanocrystalline cellulose reinforced sugar palm starch composite: Degradation and water-barrier properties. *IOP Conference Series: Materials Science and Engineering, 368*(1), 012006. https://doi.org/10.1088/1757-899X/368/1/012006

Ilyas, R. A., Sapuan, S. M., Ishak, M. R., & Zainudin, E. S. (2018c). Development and characterization of sugar palm nanocrystalline cellulose reinforced sugar palm starch bionanocomposites. *Carbohydrate Polymers, 202*, 186–202. https://doi.org/10.1016/j.carbpol.2018.09.002

Ilyas, R. A., Sapuan, S. M., Ishak, M. R., & Zainudin, E. S. (2019). Sugar palm nanofibrillated cellulose (*Arenga pinnata* (Wurmb.) Merr): Effect of cycles on their yield, physic-chemical, morphological and thermal behavior. *International Journal of Biological Macromolecules, 123*. https://doi.org/10.1016/j.ijbiomac.2018.11.124

Ilyas, R. A., Sapuan, S. M., Ishak, M. R., Zainudin, E. S., & Atikah, M. S. N. (2018e). Characterization of sugar palm nanocellulose and its potential for reinforcement with a starch-based composite. In *Sugar palm biofibers, biopolymers, and biocomposites* (pp. 189–220). CRC Press. https://doi.org/10.1201/9780429443923-10.

Ilyas, R. A., Sapuan, S. M., Ishak, M. R., Zainudin, E. S., & Atikah, M. S. N. (2018f). Nanocellulose reinforced starch polymer composites : A review of preparation, properties and application. In *Proceeding: 5th international conference on applied sciences and engineering (ICASEA, 2018)* (pp. 325–341).

Ilyas, R. A., Sapuan, S. M., Sanyang, M. L., Ishak, M. R., & Zainudin, E. S. (2018g). Nanocrystalline cellulose as reinforcement for polymeric matrix nanocomposites and its potential applications: A review. *Current Analytical Chemistry, 14*(3), 203–225. https://doi.org/10.2174/1573411013666171003155624

Imai, T., Putaux, J., & Sugiyama, J. (2003). Geometric phase analysis of lattice images from algal cellulose microfibrils. *Polymer, 44*(6), 1871–1879. https://doi.org/10.1016/S0032-3861(02)00861-3

Islam, F., Islam, N., Shahida, S., Karmaker, N., Koly, F. A., Mahmud, J., Keya, K. N., & Khan, R. A. (2019). Mechanical and interfacial characterization of jute fabrics reinforced unsaturated polyester resin composites. *Nano Hybrids and Composites, 25*, 22–31. https://doi.org/10.4028/www.scientific.net/NHC.25.22

Jasmani, L., & Adnan, S. (2017). Preparation and characterization of nanocrystalline cellulose from Acacia mangium and its reinforcement potential. *Carbohydrate Polymers, 161*, 166–171. https://doi.org/10.1016/j.carbpol.2016.12.061

Jiang, Y., Zhou, J., Zhang, Q., Zhao, G., Heng, L., Chen, D., & Liu, D. (2017). Preparation of cellulose nanocrystals from *Humulus japonicus* stem and the influence of high temperature pretreatment. *Carbohydrate Polymers, 164*, 284–293. https://doi.org/10.1016/j.carbpol.2017.02.021

Jin, L., Sun, Q., Xu, Q., & Xu, Y. (2015). Adsorptive removal of anionic dyes from aqueous solutions using microgel based on nanocellulose and polyvinylamine. *Bioresource Technology, 197*, 348–355. https://doi.org/10.1016/j.biortech.2015.08.093

Jonoobi, M., Harun, J., Shakeri, A., Misra, M., & Oksmand, K. (2009). Chemical composition, crystallinity, and thermal degradation of bleached and unbleached kenaf bast (*Hibiscus cannabinus*) pulp and nanofibers. *BioResources, 4*(2), 626–639. https://doi.org/10.15376/biores.4.2.626-639

Jonoobi, M., Khazaeian, A., Tahir, P. M., Azry, S. S., & Oksman, K. (2011). Characteristics of cellulose nanofibers isolated from rubberwood and empty fruit bunches of oil palm using chemo-mechanical process. *Cellulose, 18*(4), 1085–1095. https://doi.org/10.1007/s10570-011-9546-7

Julie Chandra, C. S., George, N., & Narayanankutty, S. K. (2016). Isolation and characterization of cellulose nanofibrils from arecanut husk fibre. *Carbohydrate Polymers, 142*, 158–166. https://doi.org/10.1016/j.carbpol.2016.01.015

Jumaidin, R., Ilyas, R. A., Saiful, M., Hussin, F., & Mastura, M. T. (2019). Water transport and physical properties of sugarcane bagasse fibre reinforced thermoplastic potato starch biocomposite. *Journal of Advanced Research in Fluid Mechanics and Thermal Sciences, 61*(2), 273–281.

Jumaidin, R., Khiruddin, M. A. A., Asyul Sutan Saidi, Z., Salit, M. S., & Ilyas, R. A. (2020). Effect of cogon grass fibre on the thermal, mechanical and biodegradation properties of thermoplastic cassava starch biocomposite. *International Journal of Biological Macromolecules, 146*, 746–755. https://doi.org/10.1016/j.ijbiomac.2019.11.011

Jumaidin, R., Saidi, Z. A. S., Ilyas, R. A., Ahmad, M. N., Wahid, M. K., Yaakob, M. Y., Maidin, N. A., Rahman, M. H. A., & Osman, M. H. (2019). Characteristics of cogon grass fibre reinforced thermoplastic cassava starch biocomposite: Water absorption and physical properties. *Journal of Advanced Research in Fluid Mechanics and Thermal Sciences, 62*(1), 43–52.

Kardam, A., & Rohit, K. (2014). Nanocellulose fibers for biosorption of cadmium, nickel, and lead ions from aqueous solution. *Clean Technologies and Environmental Policy*, 385–393. https://doi.org/10.1007/s10098-013-0634-2

Karim, Z., Claudpierre, S., Grahn, M., Oksman, K., & Mathew, A. P. (2016). Nanocellulose based functional membranes for water cleaning: Tailoring of mechanical properties, porosity and metal ion capture. *Journal of Membrane Science, 514*, 418–428. https://doi.org/10.1016/j.memsci.2016.05.018

Karimi, S., Tahir, P., Dufresne, A., Karimi, A., & Abdulkhani, A. (2014). A comparative study on characteristics of nanocellulose reinforced thermoplastic starch biofilms prepared with different techniques. *Nordic Pulp and Paper Research Journal, 29*(1), 41–45.

Kaushik, A., Singh, M., & Verma, G. (2010). Green nanocomposites based on thermoplastic starch and steam exploded cellulose nanofibrils from wheat straw. *Carbohydrate Polymers, 82*(2), 337–345. https://doi.org/10.1016/j.carbpol.2010.04.063

Khan, M. N., Rehman, N., Sharif, A., Ahmed, E., Farooqi, Z. H., & Din, M. I. (2020). Environmentally benign extraction of cellulose from dunchi fiber for nanocellulose fabrication. *International Journal of Biological Macromolecules, 153*, 72–78. https://doi.org/10.1016/j.ijbiomac.2020.02.333

Kian, L. K., Jawaid, M., Ariffin, H., & Alothman, O. Y. (2017). Isolation and characterization of microcrystalline cellulose from roselle fibers. *International Journal of Biological*

Macromolecules, 103, 931–940. https://doi.org/10.1016/j.ijbiomac.2017.05.135

Kian, L. K., Jawaid, M., Ariffin, H., & Karim, Z. (2018). Isolation and characterization of nanocrystalline cellulose from roselle-derived microcrystalline cellulose. *International Journal of Biological Macromolecules, 114*, 54–63. https://doi.org/10.1016/j.ijbiomac.2018.03.065

Kian, L. K., Jawaid, M., Ariffin, H., Karim, Z., & Sultan, M. T. H. (2019). Morphological, physico-chemical, and thermal properties of cellulose nanowhiskers from roselle fibers. *Cellulose, 26*(11), 6599–6613.

Korhonen, J. T., Kettunen, M., Ras, R. H. A., & Ikkala, O. (2011). Hydrophobic nanocellulose aerogels as floating, sustainable, reusable, and recyclable oil absorbents. *ACS Applied Materials & Interfaces*, 1813–1816.

Kumar, V., Pathak, P., & Bhardwaj, N. K. (2020). Facile chemo-refining approach for production of micro-nanofibrillated cellulose from bleached mixed hardwood pulp to improve paper quality. *Carbohydrate Polymers, 238*, 116186. https://doi.org/10.1016/j.carbpol.2020.116186

Ku, H., Wang, H., Pattarachaiyakoop, N., & Trada, M. (2011). A review on the tensile properties of natural fiber reinforced polymer composites. *Composites Part B: Engineering, 42*(4), 856–873. https://doi.org/10.1016/j.compositesb.2011.01.010

Lamaming, J., Hashim, R., Sulaiman, O., Leh, C. P., Sugimoto, T., & Nordin, N. A. (2015). Cellulose nanocrystals isolated from oil palm trunk. *Carbohydrate Polymers, 127*, 202–208. https://doi.org/10.1016/j.carbpol.2015.03.043

Langari, M. M., Nikzad, M., Ghoreyshi, A. A., & Mohammadi, M. (2019). Isolation of nanocellulose from broomcorn stalks and its application for nanocellulose/xanthan film preparation. *Chemistry, 4*(41), 11987–11994. https://doi.org/10.1002/slct.201902533

Lang, N., Merkel, E., Fuchs, F., Schumann, D., Klemm, D., Kramer, F., Mayer-Wagner, S., Schroeder, C., Freudenthal, F., Netz, H., Kozlik-Feldmann, R., & Sigler, M. (2014). Bacterial nanocellulose as a new patch material for closure of ventricular septal defects in a pig model. *European Journal of Cardio-Thoracic Surgery, 47*(6), 1013–1021. https://doi.org/10.1093/ejcts/ezu292

Latifah, J., Nurrul-Atika, M., Sharmiza, A., & Rushdan, I. (2020). Extraction of nanofibrillated cellulose from kelempayan (neolamarckia cadamba) and its use as strength additive in papermaking. *Journal of Tropical Forest Science, 32*(2), 170–178. https://doi.org/10.26525/jtfs32.2.170

Li, R., Fei, J., Cai, Y., Li, Y., Feng, J., & Yao, J. (2009). Cellulose whiskers extracted from mulberry: A novel biomass production. *Carbohydrate Polymers, 76*(1), 94–99. https://doi.org/10.1016/j.carbpol.2008.09.034

Li, M., Wang, L. J., Li, D., Cheng, Y. L., & Adhikari, B. (2014). Preparation and characterization of cellulose nanofibers from de-pectinated sugar beet pulp. *Carbohydrate Polymers, 102*(1), 136–143. https://doi.org/10.1016/j.carbpol.2013.11.021

Liu, D., Dong, Y., Bhattacharyya, D., & Sui, G. (2017). Novel sandwiched structures in starch/cellulose nanowhiskers (CNWs) composite films. *Composites Communications, 4*, 5–9.

Liu, C., Li, B., Du, H., Lv, D., Zhang, Y., Yu, G., Mu, X., & Peng, H. (2016). Properties of nanocellulose isolated from corncob residue using sulfuric acid, formic acid, oxidative and mechanical methods. *Carbohydrate Polymers, 151*, 716–724. https://doi.org/10.1016/j.carbpol.2016.06.025

Liu, D., Zhong, T., Chang, P. R., Li, K., & Wu, Q. (2010). Starch composites reinforced by bamboo cellulosic crystals. *Bioresource Technology, 101*(7), 2529–2536. https://doi.org/10.1016/j.biortech.2009.11.058

Llanos, J. H. R., & Tadini, C. C. (2017). Preparation and characterization of bio-nanocomposite films based on cassava starch or chitosan, reinforced with montmorillonite or bamboo nanofibers. *International Journal of Biological Macromolecules*. https://doi.org/10.1016/j.ijbiomac.2017.09.001

Lou, Z. C. (2016). A better design is needed for clinical studies of chronic tympanic membrane perforations using biological materials. *European Archives of Oto-Rhino-Laryngology, 273*(11), 4045–4046. https://doi.org/10.1007/s00405-016-4019-0

Lu, P., & Hsieh, Y. (2012). Preparation and characterization of cellulose nanocrystals from rice straw. *Carbohydrate Polymers, 87*(1), 564–573. https://doi.org/10.1016/j.carbpol.2011.08.022

Lu, Y., Weng, L., & Cao, X. (2006). Morphological, thermal and mechanical properties of ramie crystallites—reinforced plasticized starch biocomposites. *Carbohydrate Polymers, 63*(2), 198–204. https://doi.org/10.1016/j.carbpol.2005.08.027

Maisara, A. M. N., Ilyas, R. A., Sapuan, S. M., Huzaifah, M. R. M., Nurazzi, N. M., & Saifulazry, S. O. A. (2019). Effect of fibre length and sea water treatment on mechanical properties of sugar palm fibre reinforced unsaturated polyester composites. *International Journal of Recent Technology and Engineering, 8*(2S4), 510–514. https://doi.org/10.35940/ijrte.b1100.0782s419

Marinho, N. P., Cademartori, P. H. G. de, Nisgoski, S., Tanobe, V. O. de A., Klock, U., & Muñiz, G. I. B. de (2020). Feasibility of ramie fibers as raw material for the isolation of nanofibrillated cellulose. *Carbohydrate Polymers, 230*, 115579. https://doi.org/10.1016/j.carbpol.2019.115579

Markstedt, K., Mantas, A., Tournier, I., Martínez Ávila, H., Hägg, D., & Gatenholm, P. (2015). 3D bioprinting human chondrocytes with nanocellulose–alginate bioink for cartilage tissue engineering applications. *Biomacromolecules, 16*(5), 1489–1496. https://doi.org/10.1021/acs.biomac.5b00188

Mazani, N., Sapuan, S. M., Sanyang, M. L., Atiqah, A., & Ilyas, R. A. (2019). Design and fabrication of a shoe shelf from kenaf fiber reinforced unsaturated polyester composites. *Lignocellulose for Future Bioeconomy*, (2000), 315–332. https://doi.org/10.1016/B978-0-12-816354-2.00017-7

Megashah, L. N., Ariffin, H., Zakaria, M. R., & Hassan, M. A. (2018). Properties of cellulose extract from different types of oil palm biomass. *IOP Conference Series: Materials Science and Engineering, 368*. https://doi.org/10.1088/1757-899X/368/1/012049

Mohamad Haafiz, M. K., Eichhorn, S. J., Hassan, A., & Jawaid, M. (2013). Isolation and characterization of microcrystalline cellulose from oil palm biomass residue

Carbohydrate Polymers, 93(2), 628–634. https://doi.org/10.1016/j.carbpol.2013.01.035

Mohamed Zakriya, G., & Ramakrishnan, G. (2020). In *Natural fibre composites*. https://doi.org/10.1201/9780429326738

de Morais Teixeira, E., Corrêa, A. C., Manzoli, A., de Lima Leite, F., de Ribeiro Oliveira, C., & Mattoso, L. H. C. (2010). Cellulose nanofibers from white and naturally colored cotton fibers. *Cellulose, 17*(3), 595–606. https://doi.org/10.1007/s10570-010-9403-0

Morais, J. P. S., Rosa, M. D. F., De Souza Filho, M. de sá M., Nascimento, L. D., Do Nascimento, D. M., & Cassales, A. R. (2013). Extraction and characterization of nanocellulose structures from raw cotton linter. *Carbohydrate Polymers, 91*(1), 229–235. https://doi.org/10.1016/j.carbpol.2012.08.010

Morán, J. I., Alvarez, V. A., Cyras, V. P., & Vázquez, A. (2008). Extraction of cellulose and preparation of nanocellulose from sisal fibers. *Cellulose, 15*(1), 149–159. https://doi.org/10.1007/s10570-007-9145-9

Morandi, G., Heath, L., & Thielemans, W. (2009). Cellulose nano-crystals grafted with polystyrene chains through surface-initiated atom transfer radical polymerization (SI-ATRP). *Langmuir, 25*, 8280–8286.

Nadlene, R., Sapuan, S. M., Jawaid, M., Ishak, M. R., & Yusriah, L. (2016). A review on roselle fiber and its composites. *Journal of Natural Fibers, 13*(1), 10–41. https://doi.org/10.1080/15440478.2014.984052

Nadlene, R., Sapuan, S. M., Jawaid, M., Ishak, M. R., & Yusriah, L. (2018). The effects of chemical treatment on the structural and thermal, physical, and mechanical and morphological properties of roselle fiber-reinforced vinyl ester composites. *Polymer Composites, 39*(1), 274–287. https://doi.org/10.1002/pc.23927

Naduparambath, S., Jinitha, T. V., Shaniba, V., Sreejith, M. P., Balan, A. K., & Purushothaman, E. (2018). Isolation and characterisation of cellulose nanocrystals from sago seed shells. *Carbohydrate Polymers, 180*, 13–20. https://doi.org/10.1016/j.carbpol.2017.09.088

Nasri-Nasrabadi, B., Behzad, T., & Bagheri, R. (2014). Preparation and characterization of cellulose nanofiber reinforced thermoplastic starch composites. *Fibers and Polymers, 15*(2), 347–354. https://doi.org/10.1007/s12221-014-0347-0

Nazrin, A., Sapuan, S. M., Zuhri, M. Y. M., Ilyas, R. A., Syafiq, R., & Sherwani, S. F. K. (2020). Nanocellulose reinforced thermoplastic starch (TPS), polylactic acid (PLA), and polybutylene succinate (PBS) for food packaging applications. *Frontiers in Chemistry, 8*(213), 1–12. https://doi.org/10.3389/fchem.2020.00213

Ni, Y., Li, J., & Fan, L. (2020). Production of nanocellulose with different length from ginkgo seed shells and applications for oil in water Pickering emulsions. *International Journal of Biological Macromolecules*. https://doi.org/10.1016/j.ijbiomac.2020.01.263

Norizan, M. N., Abdan, K., Ilyas, R. A., & Biofibers, S. P. (2020). Effect of fiber orientation and fiber loading on the mechanical and thermal properties of sugar palm yarn fiber reinforced unsaturated polyester resin composites. *Polimery, 65*(2), 34–43. https://doi.org/10.14314/polimery.2020.2.5

Nurazzi, N. M., Khalina, A., Sapuan, S. M., & Ilyas, R. A. (2019). Mechanical properties of sugar palm yarn/woven glass fiber reinforced unsaturated polyester composites: Effect of fiber loadings and alkaline treatment. *Polimery, 64*(10), 12–22. https://doi.org/10.14314/polimery.2019.10.3

Nurazzi, N. M., Khalina, A., Sapuan, S. M., Ilyas, R. A., Rafiqah, S. A., & Hanafee, Z. M. (2019). Thermal properties of treated sugar palm yarn/glass fiber reinforced unsaturated polyester hybrid composites. *Journal of Materials Research and Technology*. https://doi.org/10.1016/j.jmrt.2019.11.086

Nurazzi, N. M., Khalina, A., Sapuan, S. M., Ilyas, R. A., Rafiqah, S. A., & Hanafee, Z. M. (2020). Thermal properties of treated sugar palm yarn/glass fiber reinforced unsaturated polyester hybrid composites. *Journal of Materials Research and Technology, 9*(2), 1606–1618. https://doi.org/10.1016/j.jmrt.2019.11.086

Okahisa, Y., Yoshida, A., Miyaguchi, S., & Yano, H. (2009). Optically transparent wood-cellulose nanocomposite as a base substrate for flexible organic light-emitting diode displays. *Composites Science and Technology, 69*(11–12), 1958–1961. https://doi.org/10.1016/j.compscitech.2009.04.017

Oksman, K., Etang, J. A., Mathew, A. P., & Jonoobi, M. (2010). Cellulose nanowhiskers separated from a bio-residue from wood bioethanol production. *Biomass and Bioenergy, 35*(1), 146–152. https://doi.org/10.1016/j.biombioe.2010.08.021

Oun, A. A., & Rhim, J.-W. (2016). Characterization of nanocelluloses isolated from Ushar (*Calotropis procera*) seed fiber: Effect of isolation method. *Materials Letters, 168*, 146–150. https://doi.org/10.1016/j.matlet.2016.01.052

Owolabi, A. F., Haafiz, M. K. M., Hossain, M. S., Hussin, M. H., & Fazita, M. R. N. (2017). Influence of alkaline hydrogen peroxide pre-hydrolysis on the isolation of microcrystalline cellulose from oil palm fronds. *International Journal of Biological Macromolecules, 95*, 1228–1234.

Pacaphol, K., & Aht-ong, D. (2017). Surface & coatings technology the influences of silanes on interfacial adhesion and surface properties of nanocellulose fi lm coating on glass and aluminum substrates. *Surface and Coatings Technology, 320*, 70–81. https://doi.org/10.1016/j.surfcoat.2017.01.111

Paltakari, J., Jin, H., Kettunen, M., Laiho, A., Pynn, H., Marmur, A., Ikkala, O., & Ras, R. H. A. (2011). Superhydrophobic and superoleophobic nanocellulose aerogel membranes as bioinspired cargo carriers on water and oil. *Langmuir, 27*(14), 1930–1934. https://doi.org/10.1021/la103877r

Pan, X., Dai, X., Dong, X., Liu, B., & Li, Y. (2019). Nanocellulose-reinforced polyurethane for waterborne wood coating. *Molecules, 24*(3151), 1–13. https://doi.org/10.3390/molecules24173151

Pandey, J. K., Kim, C., Chu, W., Lee, C. S., Jang, D.-Y., & Ahn, S. (2009). Evaluation of morphological architecture of cellulose chains in grass during conversion from macro to nano dimensions. *E-polymers, 9*(1), 1–15. https://doi.org/10.1515/epoly.2009.9.1.1221

Paralikar, S. A., Simonsen, J., & Lombardi, J. (2008). Poly(vinyl alcohol)/cellulose nanocrystal barrier membranes. *Journal of Membrane Science, 320*(1–2), 248–258. https://doi.org/10.1016/j.memsci.2008.04.009

Park, N.-M., Choi, S., Oh, J. E., & Hwang, D. Y. (2019). Facile extraction of cellulose nanocrystals. *Carbohydrate Polymers, 223*, 115114. https://doi.org/10.1016/j.carbpol.2019.115114

Pereda, M., Dufresne, A., Aranguren, M. I., & Marcovich, N. E. (2014). Polyelectrolyte films based on chitosan/olive oil and reinforced with cellulose nanocrystals. *Carbohydrate Polymers, 101*, 1018–1026. https://doi.org/10.1016/j.carbpol.2013.10.046

Pereira, P. H. F., Waldron, K. W., Wilson, D. R., Cunha, A. P., Brito, E. S. de, Rodrigues, T. H. S., Rosa, M. F., & Azeredo, H. M. C. (2017). Wheat straw hemicelluloses added with cellulose nanocrystals and citric acid. Effect on film physical properties. *Carbohydrate Polymers, 164*, 317–324. https://doi.org/10.1016/j.carbpol.2017.02.019

Picheth, G. F., Pirich, C. L., Sierakowski, M. R., Woehl, M. A., Sakakibara, C. N., de Souza, C. F., Martin, A. A., da Silva, R., & de Freitas, R. A. (2017a). Bacterial cellulose in biomedical applications: A review. *International Journal of Biological Macromolecules, 104*, 97–106. https://doi.org/10.1016/J.IJBIOMAC.2017.05.171

Plengnok, U., & Jarukumjorn, K. (2020). Preparation and characterization of nanocellulose from sugarcane bagasse. *Biointerface Research in Applied Chemistry*. https://doi.org/10.33263/BRIAC103.675678

Poiini, P., Subramanian, K. S., Janavi, G. J., & Subramanian, J. (2020). Synthesis of nano-film from nanofibrillated cellulose of banana pseudostem (Musa spp.) to extend the shelf life of tomato. *BioResources*. https://doi.org/10.15376/biores.15.2.2882-2905

Prathapan, R., Thapa, R., Garnier, G., & Tabor, R. F. (2016). Modulating the zeta potential of cellulose nanocrystals using salts and surfactants. *Colloids and Surfaces A: Physicochemical and Engineering Aspects, 509*, 11–18. https://doi.org/10.1016/j.colsurfa.2016.08.075

Purdue University. (2018). Researchers show microscopic wood nanocrystals make concrete stronger. *Physical Oceanography*. https://phys.org/news/2018-02-microscopic-woodnanocrystals-%0Aconcrete-stronger.html.

Purkait, B. S., Ray, D., Sengupta, S., Kar, T., Mohanty, A., & Misra, M. (2011). Isolation of cellulose nanoparticles from sesame husk. *Industrial & Engineering Chemistry Research, 50*(2), 871–876. https://doi.org/10.1021/ie101797d

Radzi, A. M., Sapuan, S. M., Jawaid, M., & Mansor, M. R. (2018a). Mechanical and thermal performances of roselle fiber-reinforced thermoplastic polyurethane composites. *Polymer - Plastics Technology and Engineering, 57*(7), 601–608. https://doi.org/10.1080/03602559.2017.1332206

Radzi, A. M., Sapuan, S. M., Jawaid, M., & Mansor, M. R. (2018b). Mechanical performance of roselle/sugar palm fiber hybrid reinforced polyurethane composites. *BioResources, 13*(3), 6238–6249.

Radzi, A. M., Sapuan, S. M., Jawaid, M., Noor, M. R. M. A. M., & Ilyas, R. A. (2019). A review on roselle fibre reinforced polymer composite. *Prosiding Seminar Enau Kebangsaan, 2019*, 101–103.

Rajinipriya, M., Nagalakshmaiah, M., Robert, M., & Elkoun, S. (2018). Importance of agricultural and industrial waste in the field of nanocellulose and recent industrial developments of wood based nanocellulose: A review. *ACS Sustainable Chemistry and Engineering, 6*(3), 2807–2828. https://doi.org/10.1021/acssuschemeng.7b03437

Razali, N., Salit, M. S., Jawaid, M., Ishak, M. R., & Lazim, Y. (2015). A study on chemical composition, physical, tensile, morphological, and thermal properties of roselle fibre: Effect of fibre maturity. *BioResources, 10*(1), 1803–1823. https://doi.org/10.15376/biores.10.1.1803-1824

Revol, J. F. (1982). On the cross-sectional shape of cellulose crystallites in Valonia ventricosa. *Carbohydrate Polymers, 2*(2), 123–134. https://doi.org/10.1016/0144-8617(82)90058-3

Rigg-Aguilar, P., Moya, R., Oporto-Velásquez, G. S., Vega-Baudrit, J., Starbird, R., Puente-Urbina, A., Méndez, D., Potosme, L. D., & Esquivel, M. (2020). Micro- and nanofibrillated cellulose (MNFC) from pineapple (*Ananas comosus*) stems and their application on polyvinyl acetate (PVAc) and urea-formaldehyde (UF) wood adhesives. *Journal of Nanomaterials, 2020*, 1–12. https://doi.org/10.1155/2020/1393160

Robles, E., Fernández-Rodríguez, J., Barbosa, A. M., Gordobil, O., Carreño, N. L. V., & Labidi, J. (2018). Production of cellulose nanoparticles from blue agave waste treated with environmentally friendly processes. *Carbohydrate Polymers, 183*, 294–302. https://doi.org/10.1016/j.carbpol.2018.01.015

Rosa, M. F. M., Medeiros, E. S., Malmonge, J. A. J., Gregorski, K. S., Wood, D. F., Mattoso, L. H. C., Glenn, G., Orts, W. J., & Imam, S. H. (2010). Cellulose nanowhiskers from coconut husk fibers: Effect of preparation conditions on their thermal and morphological behavior. *Carbohydrate Polymers, 81*(1), 83–92. https://doi.org/10.1016/j.carbpol.2010.01.059

Salajková, M., Berglund, L. A., & Zhou, Q. (2012). Hydrophobic cellulose nanocrystals modified with quaternary ammonium salts. *Journal of Materials Chemistry, 22*(37), 19798. https://doi.org/10.1039/c2jm34355j

Salminen, R., Reza, M., Pääkkönen, T., Peyre, J., & Kontturi, E. (2017). TEMPO-mediated oxidation of microcrystalline cellulose: Limiting factors for cellulose nanocrystal yield. *Cellulose, 24*(4), 1657–1667.

Sanyang, M. L., Ilyas, R. A., Sapuan, S. M., & Jumaidin, R. (2018). Sugar palm starch-based composites for packaging applications. In *Bionanocomposites for packaging applications* (pp. 125–147). Springer International Publishing. https://doi.org/10.1007/978-3-319-67319-6_7.

Sanyang, M. L., Sapuan, S. M., Jawaid, M., Ishak, M. R., & Sahari, J. (2016). Recent developments in sugar palm (*Arenga pinnata*) based biocomposites and their potential industrial applications: A review. *Renewable and Sustainable Energy Reviews, 54*, 533–549. https://doi.org/10.1016/j.rser.2015.10.037

Sapuan, S. M., Aulia, H. S., Ilyas, R. A., Atiqah, A., Dele-Afolabi, T. T., Nurazzi, M. N., Supian, A. B. M., &

Atikah, M. S. N. (2020). Mechanical properties of longitudinal basalt/woven-glass-fiber-reinforced unsaturated polyester-resin hybrid composites. *Polymers, 12*(10). https://doi.org/10.3390/polym12102211

Sari, N. H., Pruncu, C. I., Sapuan, S. M., Ilyas, R. A., Catur, A. D., Suteja, S., Sutaryono, Y. A., & Pullen, G. (2020). The effect of water immersion and fibre content on properties of corn husk fibres reinforced thermoset polyester composite. *Polymer Testing, 91*, 106751. https://doi.org/10.1016/j.polymertesting.2020.106751

Shaheen, T. I., & Emam, H. E. (2018). Sono-chemical synthesis of cellulose nanocrystals from wood sawdust using acid hydrolysis. *International Journal of Biological Macromolecules, 107*, 1599–1606. https://doi.org/10.1016/j.ijbiomac.2017.10.028

Sheltami, R. M., Abdullah, I., Ahmad, I., Dufresne, A., & Kargarzadeh, H. (2012). Extraction of cellulose nanocrystals from mengkuang leaves (*Pandanus tectorius*). *Carbohydrate Polymers, 88*(2), 772–779. https://doi.org/10.1016/j.carbpol.2012.01.062

Sonia, A., & Priya Dasan, K. (2013). Chemical, morphology and thermal evaluation of cellulose microfibers obtained from *Hibiscus sabdariffa*. *Carbohydrate Polymers, 92*(1), 668–674. https://doi.org/10.1016/j.carbpol.2012.09.015

Soni, B., Hassan, E. B., & Mahmoud, B. (2015). Chemical isolation and characterization of different cellulose nanofibers from cotton stalks. *Carbohydrate Polymers, 134*, 581–589. https://doi.org/10.1016/j.carbpol.2015.08.031

Stanislas, T. T., Tendo, J. F., Ojo, E. B., Ngasoh, O. F., Onwualu, P. A., Njeugna, E., & Junior, H. S. (2020). Production and characterization of pulp and nanofibrillated cellulose from selected tropical plants. *Journal of Natural Fibers*. https://doi.org/10.1080/15440478.2020.1787915

Sumaiyah, B.,W., Karsono, M. P.,N., & Saharman, G. (2014). Preparation and characterization of nanocrystalline cellulose from sugar palm bunch. *Interantional Journal of PharmTech, 6*(2), 814–820.

Suopajärvi, T., Liimatainen, H., Karjalainen, M., Upola, H., & Niinimäki, J. (2015). Lead adsorption with sulfonated wheat pulp nanocelluloses. *Journal of Water Process Engineering, 5*, 136–142. https://doi.org/10.1016/j.jwpe.2014.06.003

Suopajärvi, T., Ricci, P., Karvonen, V., Ottolina, G., & Liimatainen, H. (2020). Acidic and alkaline deep eutectic solvents in delignification and nanofibrillation of corn stalk, wheat straw, and rapeseed stem residues. *Industrial Crops and Products*. https://doi.org/10.1016/j.indcrop.2019.111956

Syafri, E., Kasim, A., Abral, H., & Asben, A. (2018). Cellulose nanofibers isolation and characterization from ramie using a chemical-ultrasonic treatment. *Journal of Natural Fibers*, 1–11. https://doi.org/10.1080/15440478.2018.1455073

Syafri, E., Kasim, A., Abral, H., & Asben, A. (2019). Cellulose nanofibers isolation and characterization from ramie using a chemical-ultrasonic treatment. *Journal of Natural Fibers, 16*(8), 1145–1155. https://doi.org/10.1080/15440478.2018.1455073

Syafri, E., Sudirman, M., Yulianti, E., Deswita, Asrofi, M., Abral, H., Sapuan, S. M., Ilyas, R. A., & Fudholi, A. (2019). Effect of sonication time on the thermal stability, moisture absorption, and biodegradation of water hyacinth (*Eichhornia crassipes*) nanocellulose-filled bengkuang (*Pachyrhizus erosus*) starch biocomposites. *Journal of Materials Research and Technology, 8*(6), 6223–6231. https://doi.org/10.1016/j.jmrt.2019.10.016

Tayeb, A. H., Amini, E., Ghasemi, S., & Tajvidi, M. (2018). Cellulose nanomaterials — binding properties and applications: A review. *Molecules, 23*(2684), 1–24. https://doi.org/10.3390/molecules23102684

Teixeira, E. D. M., Bondancia, T. J., Teodoro, K. B. R., Corrêa, A. C., Marconcini, J. M., & Mattoso, L. H. C. (2011). Sugarcane bagasse whiskers: Extraction and characterizations. *Industrial Crops and Products, 33*(1), 63–66. https://doi.org/10.1016/j.indcrop.2010.08.009

Teixeira, E. D. M., Lotti, C., Corrêa, A. C., Teodoro, K. B. R., Marconcini, J. M., & Mattoso, L. H. C. (2011). Thermoplastic corn starch reinforced with cotton cellulose nanofibers. *Journal of Applied Polymer Science, 120*(4), 2428–2433. https://doi.org/10.1002/app.33447

Teixeira, E. D. M., Pasquini, D., Curvelo, A. A. S. S., Corradini, E., Belgacem, M. N., & Dufresne, A. (2009). Cassava bagasse cellulose nanofibrils reinforced thermoplastic cassava starch. *Carbohydrate Polymers, 78*(3), 422–431. https://doi.org/10.1016/j.carbpol.2009.04.034

Thiruchitrambalam, M., Athijayamani, A., Sathiyamurthy, S., & Thaheer, A. S. A. (2010). A review on the natural fiber-reinforced polymer composites for the development of roselle fiber-reinforced polyester composite. *Journal of Natural Fibers, 7*(4), 307–323. https://doi.org/10.1080/15440478.2010.529299

Tibolla, H., Pelissari, F. M., & Menegalli, F. C. (2014). Cellulose nanofibers produced from banana peel by chemical and enzymatic treatment. *Lebensmittel-Wissenschaft und -Technologie- Food Science and Technology, 59*(2P2), 1311–1318. https://doi.org/10.1016/j.lwt.2014.04.011

Tonoli, G. H. D. H. D., Teixeira, E. M. M., Corrêa, A. C. C., Marconcini, J. M. M., Caixeta, L. A. A., Pereira-Da-Silva, M. A. A., & Mattoso, L. H. C. H. C. (2012). Cellulose micro/nanofibres from Eucalyptus kraft pulp: Preparation and properties. *Carbohydrate Polymers, 89*(1), 80–88. https://doi.org/10.1016/j.carbpol.2012.02.052

Velásquez-Cock, J., Castro, C., Gañán, P., Osorio, M., Putaux, J.-L., Serpa, A., & Zuluaga, R. (2016). Influence of the maturation time on the physico-chemical properties of nanocellulose and associated constituents isolated from pseudostems of banana plant c.v. valery. *Industrial Crops and Products, 83*, 551–560. https://doi.org/10.1016/j.indcrop.2015.12.070

Wahono, S., Irwan, A., Syafri, E., & Asrofi, M. (2018). Preparation and characterization of ramie cellulose nanofibers/$CaCO_3$ Unsaturated polyester resin composites. *ARPN Journal of Engineering and Applied Sciences, 13*(2), 746–751. https://doi.org/10.1039/c7nr02736b

Wang, R., Guan, S., Sato, A., Wang, X., Wang, Z., Yang, R., Hsiao, B. S., & Chu, B. (2013). Nano fibrous microfiltration membranes capable of removing bacteria , viruses and heavy metal ions. *Journal of Membrane Science, 446*, 376–382. https://doi.org/10.1016/j.memsci.2013.06.020

Wang, H., Pudukudy, M., Ni, Y., Zhi, Y., Zhang, H., Wang, Z., Jia, Q., & Shan, S. (2020). Synthesis of nanocrystalline cellulose via ammonium persulfate-assisted swelling followed by oxidation and their chiral self-assembly. *Cellulose, 27*(2), 657–676. https://doi.org/10.1007/s10570-019-02789-z

Wang, J., Xu, J., Zhu, S., Wu, Q., Li, J., Gao, Y., Wang, B., Li, J., Gao, W., Zeng, J., & Chen, K. (2021). Preparation of nanocellulose in high yield via chemi-mechanical synergy. *Carbohydrate Polymers, 251*, 117094. https://doi.org/10.1016/j.carbpol.2020.117094

Wardhono, E., Wahyudi, H., Agustina, S., Oudet, F., Pinem, M., Clausse, D., Saleh, K., & Guénin, E. (2018). Ultrasonic irradiation coupled with microwave treatment for eco-friendly process of isolating bacterial cellulose nanocrystals. *Nanomaterials, 8*(10), 859. https://doi.org/10.3390/nano8100859

Wulandari, W. T., Rochliadi, A., & Arcana, I. M. (2016). Nanocellulose prepared by acid hydrolysis of isolated cellulose from sugarcane bagasse. *IOP Conference Series: Materials Science and Engineering, 107*(1). https://doi.org/10.1088/1757-899X/107/1/012045

Xiang, L. Y., Mohammed, M. A. P., & Baharuddin, A. S. (2016). Characterisation of microcrystalline cellulose from oil palm fibres for food applications. *Carbohydrate Polymers, 148*, 11–20.

Xia, Z. P., Yu, J. Y., Cheng, L. D., Liu, L. F., & Wang, W. M. (2009). Study on the breaking strength of jute fibres using modified Weibull distribution. *Composites Part A: Applied Science and Manufacturing, 40*(1), 54–59. https://doi.org/10.1016/j.compositesa.2008.10.001

Yang, X., Han, F., Xu, C., Jiang, S., Huang, L., Liu, L., & Xia, Z. (2017). Effects of preparation methods on the morphology and properties of nanocellulose (NC) extracted from corn husk. *Industrial Crops and Products, 109*, 241–247. https://doi.org/10.1016/j.indcrop.2017.08.032

Yong, C., Abdul, N., Bond, Y., Talib, R. A., Hui, C., Abdan, K., Wei, E., & Chan, C. (2019). Electrospun biocomposite: Nanocellulose and chitosan entrapped within a poly (hydroxyalkanoate) matrix for Congo red. *Journal of Materials Research and Technology, 8*(6), 5091–5102. https://doi.org/10.1016/j.jmrt.2019.08.030

Yorseng, K., Rangappa, S. M., Pulikkalparambil, H., Siengchin, S., & Parameswaranpillai, J. (2020). Accelerated weathering studies of kenaf/sisal fiber fabric reinforced fully biobased hybrid bioepoxy composites for semi-structural applications: Morphology, thermo-mechanical, water absorption behavior and surface hydrophobicity. *Construction and Building Materials, 235*, 117464. https://doi.org/10.1016/j.conbuildmat.2019.117464

Yusriah, L., Sapuan, S. M., Zainudin, E. S., & Mariatti, M. (2014). Characterization of physical, mechanical, thermal and morphological properties of agro-waste betel nut (*Areca catechu*) husk fibre. *Journal of Cleaner Production, 72*, 174–180. https://doi.org/10.1016/j.jclepro.2014.02.025

Zhang, R., & Liu, Y. (2018). High energy oxidation and organosolv solubilization for high yield isolation of cellulose nanocrystals (CNC) from Eucalyptus hardwood. *Scientific Reports, 8*(1), 16505. https://doi.org/10.1038/s41598-018-34667-2

Zhang, X., Liu, D., Yang, L., Zhou, L., & You, T. (2015). Self-assembled three-dimensional graphene-based materials for dye adsorption and catalysis. *Journal of Materials Chemistry A, 3*(18), 10031–10037. https://doi.org/10.1039/c0xx00000x

Zhang, K., Su, Y., & Xiao, H. (2020). Preparation and characterization of nanofibrillated cellulose from waste sugarcane bagasse by mechanical force. *BioResources*. https://doi.org/10.15376/biores.8.3.6636-6647

Index

Note: Page numbers followed by "f" indicate figures and "t" indicate tables.